*World Survey of Climatology Volume 11*

# CLIMATES OF NORTH AMERICA

*World Survey of Climatology*

*Editor in Chief:*

H. E. LANDSBERG, College Park, Md. (U.S.A.)

*Editors:*

H. ARAKAWA, Tokyo (Japan)
R. A. BRYSON, Madison, Wisc. (U.S.A.)
H. FLOHN, Bonn (Germany)
J. GENTILLI, Nedlands, W.A. (Australia)
J. F. GRIFFITHS, College Station, Texas (U.S.A.)
F. K. HARE, Ottawa, Ont. (Canada)
P. E. LYDOLPH, Milwaukee, Wisc. (U.S.A.)
S. ORVIG, Montreal, Que. (Canada)
D. F. REX, Boulder, Colo. (U.S.A.)
W. SCHWERDTFEGER, Madison, Wisc. (U.S.A.)
K. TAKAHASHI, Tokyo (Japan)
H. THOMSEN, Charlottenlund (Denmark)
C. C. WALLÉN, Geneva (Switzerland)

*World Survey of Climatology Volume 11*

# Climates of North America

edited by

Reid A. BRYSON

*Institute for Environmental Studies*
*University of Wisconsin*
*Madison, Wisc. (U.S.A.)*

F. Kenneth HARE

*Research and Coordination Directorate*
*Environment Canada*
*Ottawa, Ont. (Canada)*

ELSEVIER SCIENTIFIC PUBLISHING COMPANY
Amsterdam-London-New York 1974

ELSEVIER SCIENTIFIC PUBLISHING COMPANY
335 Jan van Galenstraat
P.O. Box 211, Amsterdam, The Netherlands

AMERICAN ELSEVIER PUBLISHING COMPANY, INC.
52 Vanderbilt Avenue
New York, New York 10017

Library of Congress Card Number 78-477739
ISBN 0-444-41062-7
With 140 illustrations and 148 tables

Printed in The Netherlands

*World Survey of Climatology*

*Editor in Chief:* H. E. LANDSBERG

*List of Contributors to this Volume*

R. A. Bryson
Institute for Environmental Studies
University of Wisconsin
Madison, Wisc. (U.S.A.)

A. Court
San Fernando Valley State College
Northridge, Calif. (U.S.A.)

E. García
Universidad Nacional Autónoma
de México
Mexico City (Mexico)

F. K. Hare
Research and Coordination Directorate
Environment Canada
Ottawa, Ont. (Canada)

J. E. Hay
Department of Geography
University of Canterbury
Christchurch (New Zealand)

P. A. Mosiño Alemán
Universidad Nacional Autónoma
de México
Mexico City (Mexico)

# Contents

Contents

Chapter 3. THE CLIMATE OF THE CONTERMINOUS UNITED STATES
by ARNOLD COURT

*Contents*

*Chapter 4.* THE CLIMATE OF MEXICO
by PEDRO A. MOSIÑO ALEMÁN and ENRIQUETA GARCÍA

Contents

*Chapter 1*

# The Climates of North America

R. A. BRYSON AND F. KENNETH HARE

**General controls of North American climate**

Two features of the geography of North America dominate the character of its climatic pattern: the high, long, north–south Cordillera in the west, and the vast plains of low relief in middle and high latitudes to the east. The former is a most significant obstacle to both zonal westerlies and trade winds, while the plains of the north and east provide an unobstructed path of low thermal admittance for great meridional excursions of both Arctic and tropical air masses.

Unlike Europe with its deep eastward penetration of little altered maritime air, the interior of North America can be penetrated by the westerlies from the Pacific only over the high western mountains and plateaus, and only with significant air mass modification. Unlike Asia, North America has no towering Himalayan barrier to impede meridional air mass exchange. Unlike Africa and South America, most of its lands are at higher latitudes with seas between it and the Equator. Australia, unlike North America, lies mostly in the subtropics.

This unique distribution of land, sea, and mountains produces a distinctive climatic pattern, but while the climatic contrasts are not greater than may be found elsewhere, the weather variability which goes into the makeup of North American climates is maximized by the geography. Mild, sunny, dry subsident flow down the east face of the Rocky Mountains may replace moist, warm, cloudy tropical air one day, but give way to cold Arctic air the next.

The general climatic situation of North America may be visualized in terms of a simple model of the topography embedded in the general circulation of the atmosphere, and stretching from tropical seas at 15°N (50°N in the east) with an excess of net radiation, to polar lands and ice-covered sea at latitude 70°–80°N, with a small deficit of net radiation (Fig.1).

The major effect of the western Cordillera is mechanical, while the major effect of the eastern plains is thermal. The mountain barrier consists largely of a plateau 1,500–2,000 m high with mountain ranges superimposed that run to double that altitude. The low-level winds that impinge on this barrier are mostly deflected, unless they have sufficient kinetic energy or sufficient instability to cross the crest. Only a few significant passes allow the passage of the normal zonal flow. North of the maximum westerlies, higher static stability and weaker zonal flow result in almost total northward deflection of the low-level flow (Fig.2). Near the region of maximum westerly winds, the low-level flow usually divides, being deflected southward on the equatorward side. Between 30° and 50°N, a

*1*

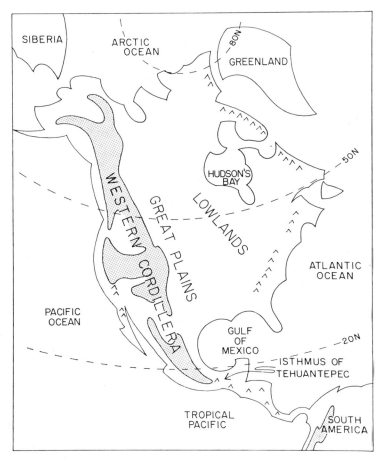

Fig.1. North America and its major topographic regions.

combination of strong westerlies, low stability, and a more broken mountain barrier allows the maximum eastward penetration of the westerlies from the Pacific (MITCHELL, 1970).

In Mexico, the stable southward-deflected westerlies usually do not cross the Sierra Madre Occidental, while the trades reaching the east coast are more unstable, stronger perpendicular to the coast, and may top the lower Sierra Madre Oriental to enter the central plateau. Usually even the low-level trade winds are deflected northward before reaching the Pacific, except in the south where the unstable air crosses the lower mountains and turns equatorward.

There are three main routes for low-level westerlies to cross the Cordillera: (*1*) where the westerlies reaching the coast are on the average strongest, near latitudes 45°–50°N; (*2*) through the Columbia River–Snake River–Wyoming Basin gap; and (*3*) through the lower region along the Mexican–United States border (Fig.3). This westerly flow may be visualized as a jet penetrating into eastern North America in midlatitudes, entraining air from the Arctic and tropics over the plains, and eventually mixing with those air masses in vortices along the edges of the jet (Fig.4).

Most of the air which reaches the interior of North America from the Pacific is not from the lower portion of the air columns which reached the coast. Several times as much air flows away from the east face of the Cordillera at 1,700 m as enters the western edge at

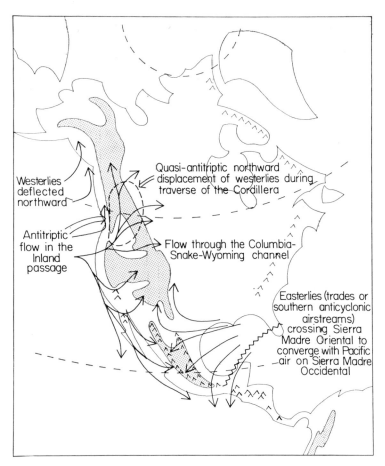

Fig.2. Schematic representation of various flow patterns affecting the Cordilleran climate, based on October resultant surface winds.

that height (BORCHERT, 1950). The excess is air which enters the continent at an altitude above the level of the western plateau, then descends the eastern slope of the mountains with warming due to adiabatic compression and with decreasing relative humidity. This dry, mild character of the Pacific airstream when it reaches the interior contrasts with the cool, moist character it had when it reached the coast. That low-level air which does cross the mountains is similarly modified by the classical foehn process.

East of the mountains air sweeps southward from the frozen Arctic and northward from the tropical seas. No significant mountain barriers impede the flow. No underlying sea of high thermal admittance softens the chill of the Arctic air. It may cross the entire length of North America to the isthmus of Tehuantepec in a few days time, arriving there still much colder than the normal tropical air.

Tropical air may penetrate far northward, too, but very rarely beyond southern Canada at the ground. The gradient of net radiation from tropics to Arctic and the limited transfer of heat from air to soil maintains sufficient slope to the isentropic surfaces that by the time tropical air has reached the northern United States it has usually left the surface and can only continue northward aloft, rising over Arctic and Pacific air.

On the other hand, the same prevailing slope of the isentropic surfaces and higher density of the Arctic air means that it stays near the surface throughout its southward journey.

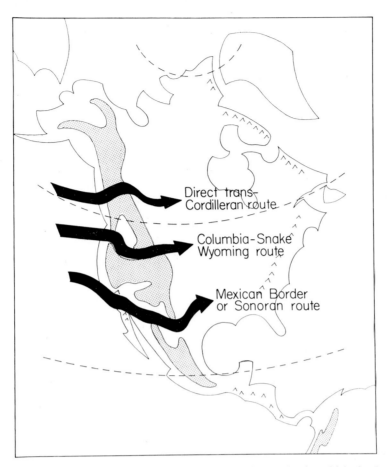

Fig.3. Schematic representation of the three major modes by which the Pacific airstream crosses the Cordillera at low levels.

On the average, Arctic air, cold and dry, reaches south to mid-Canada in summer and the northern United States in winter. Tropical air, warm and moist, is normally restricted to Mexico and the extreme southern United States in winter but reaches the northern United States in summer. Between these air masses, the modified westerlies of Pacific air, mild and dry, penetrate eastward. On the average, the region of Pacific air dominance forms a wedge pointing eastward across southern Canada in summer and the north-central United States in winter.

One complication of this simple picture is found in the southeastern United States in winter. There, between the westerlies to the north and the trade winds to the south, at the normal latitude of the subtropical anticyclone belt, there is normally an anticyclonic eddy. Sometimes this is the remains of an outbreak of Arctic air. Sometimes it is composed of Pacific air but rarely in winter is it of tropical air. When originally of Arctic or Pacific air, the air of this anticyclone will be called "southern anticyclonic" air in the remainder of this introductory chapter.

The overall climatic pattern of the North American continent is dependent on its size—lying athwart both westerlies and trades, and reaching from the source region of Arctic air deep into the realm of tropical airstreams. It is also dependent on the particular gross topography—a high barrier of mountains and plateaus on the west and lowlands to the east.

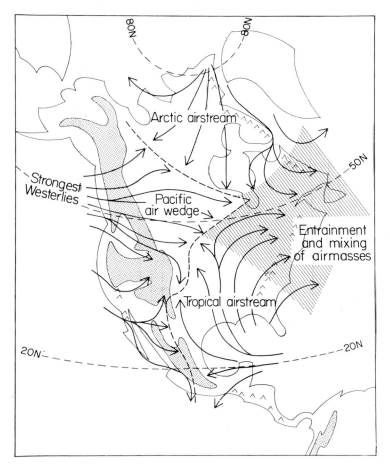

Fig.4. Schematic representation of the surface flow across North America, illustrating the wedge of Pacific air east of the Cordillera and the downstream entrainment and mixing of airstreams. Based on July resultant surface winds.

to the east. The details of the resulting airflows, and eddies, will be elaborated in the following sections.

## The climatic pattern of North America

It is clear from an examination of the monthly resultant flow (Fig.5–16) that the airstream pattern over North America consists of streams from three distinct sources: the Pacific, the Atlantic and the Arctic. The relative dominance of these airstreams, their latitudes, and their characteristics change with the seasons, but the three are identifiable throughout the year. In fall and winter a fourth can be defined: the eddy of modified Pacific or Arctic air in the southeastern United States which we have called the "southern anticyclonic" regime.

### West coast climates

The Pacific airstream may be further subdivided into that northern portion which entered the Pacific from Asia and is most prominent in winter, and that southern portion

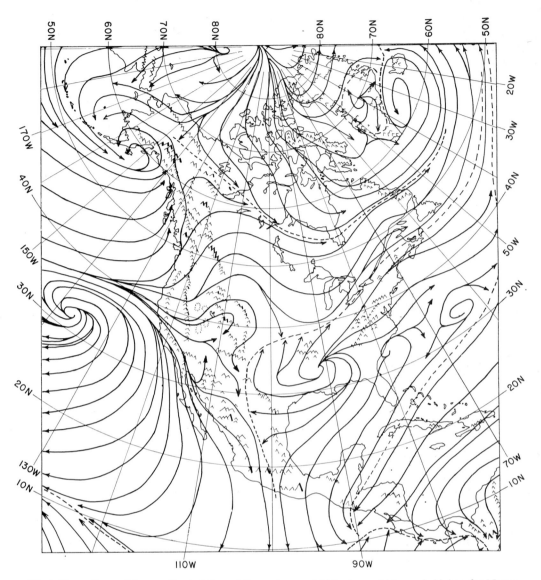

Fig.5. Streamlines of the resultant airflow at the surface or 1,000 mbar, whichever is higher, for November. Based on data from NAVAIR, 1966. Large departures from winds observed at individual stations are to be expected in mountainous areas.

which emerged from an oceanic anticyclone over the Pacific. This southern portion is most prominent in summer and dominates western Mexico during most of the year.

The northern Pacific westerlies, colder originally and moving over a warmer, but still rather cold, sea, arrive on the west coast cool, with a near moist-adiabatic lapse rate, and with high humidity to a considerable depth.

The southern portion of the westerlies subsides from a subtropical anticyclone, and then either moves rather directly eastward to the coast (by a more direct trajectory than the cooler westerlies to the north) or flows southeastward along the Mexican west coast. This airstream, stable and with a shallow moist layer, subsides still more along its southward track. The subsidence, despite surface warming as the airstream moves southward, keeps the airstream stable and the moist layer shallow until convergence at

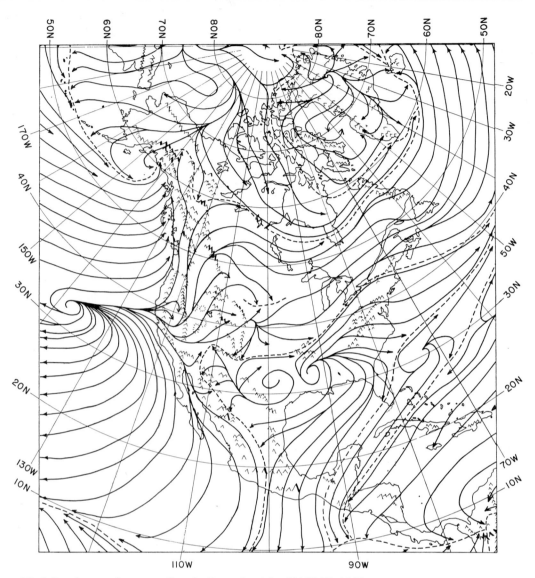

Fig.6. Resultant surface streamlines for December (after NAVAIR, 1966).

the Intertropical Convergence or with the mountainous west coast counteracts the stability and deep vertical mixing occurs.

Between the two portions of the Pacific airstream there is, in the mean, a baroclinic zone which intersects the west coast at latitude 50°–55°N in summer and 40°–45°N in winter, judging by the jet stream location, but farther north as indicated by surface airstreams.

The stability of the northward-deflected cooler portion of the westerlies is decreased along the coastal zone by the convergence due to the rougher surface to the right of the wind while the southward moving branch, stable as it reaches the coast, is made more stable by divergence resulting from the rougher surface to the left of the wind and the upwelling of cold water (BRYSON and KUHN, 1961). Thus the northern west coast of North America (northern British Columbia and southern Alaska) is cool and wet the whole year, while extreme southern California and the west coast of northern Mexico are dry throughout the year.

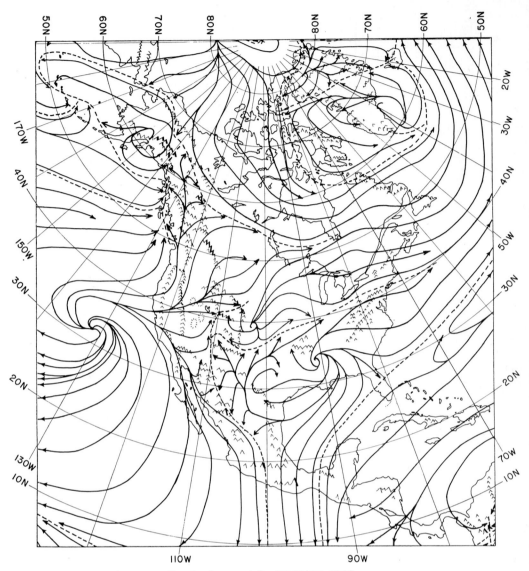

Fig.7. Resultant surface streamlines for January (after NAVAIR, 1966).

In southern Mexico the west coast lies close to the mountains, on which showers form and drift coastward embedded in the upper trades. It is also the site of convergence between the southeastward flow from the Pacific anticyclone and the southwestward flow from the Atlantic trade winds. Here the coastal region receives 800 to 1,000 mm of rainfall per year more or less.

In between the wet northern coast and the dry coast of northwestern Mexico, the regime changes with the seasons as the air mass stability and wind directions change. Northern California is wet only in winter when the westerlies are farthest south and the Washington and southern British Columbia coasts are dry only in summer with the most northerly position of the cool westerly airstream.

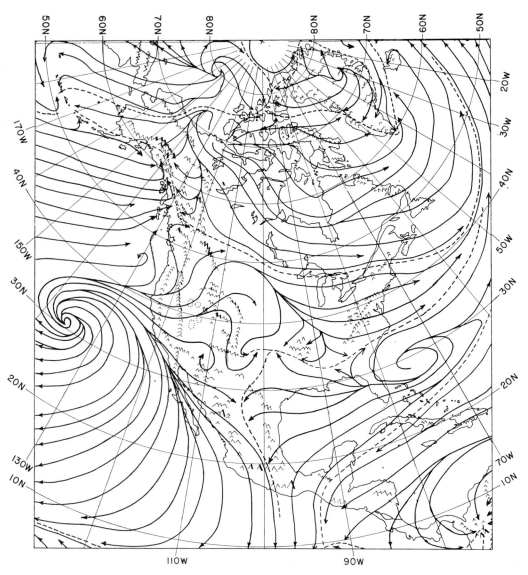

Fig.8. Resultant surface streamlines for February (after NAVAIR, 1966).

**Intermontane climates**

During winter, when the westerlies are far south, there is a strong flow of air through the low section of the Cordillera along the northern border of Mexico—the Sonoran route. However, a low inversion usually caps the moist surface layer along the California coast and only the dry air above the subsidence inversion crosses the low mountains of southern California. The dry air that does pass through the Sonoran route dominates the climate of west Texas during part of the winter. The winter precipitation of northern Mexico and the southwestern United States is almost entirely associated with the few cyclonic disturbances that pass that far south.

Of course, the Sonoran gap in the Cordillera is open at both ends and occasional deep outbreaks of Arctic air passing south into Texas and northeastern Mexico will spill through it towards the west.

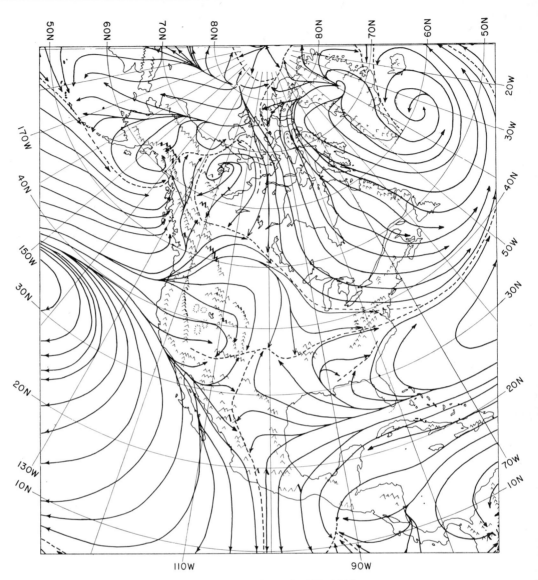

Fig.9. Resultant surface streamlines for March (after NAVAIR, 1966).

The double mountain chain structure of the Cordillera is a most significant climatic factor. The Coast Ranges–Sierra Nevada–Sierra Madre Occidental chain rims the western edge of the Cordilleran plateau, and the Rockies–Sierra Madre Oriental chain rims it on the east. The upland region between the two is broken into a series of irregular basins by transverse montane areas. From south to north these basins are: the Central Plateau of northern Mexico; the Great Basin of Utah, Nevada and eastern California; the Columbia Plateau of southwesterr Idaho, eastern Washington and eastern Oregon; and the Central Plateaus of British Columbia.

All of these basins may fill with cold air from the north and east in winter, but a long-lasting lake of cold air is most likely in the Great Basin. Here the cold air filling the basin may dominate the circulation of November through January and the milder westerlies pass over the region with little circulation effect at the surface. Cold air

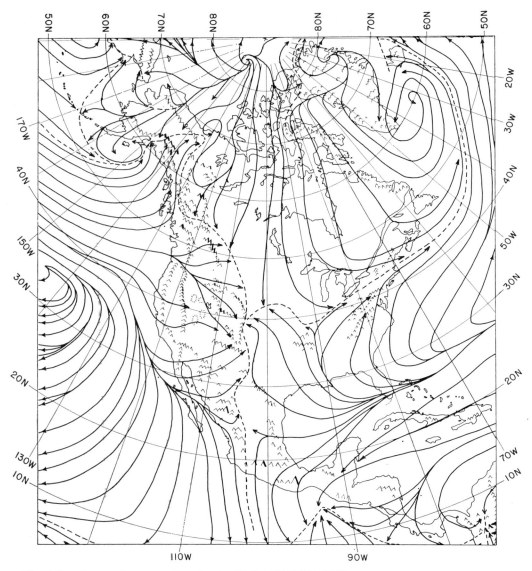

Fig.10. Resultant surface streamlines for April (after NAVAIR, 1966).

filling is least common in the Central Plateau of Mexico which is too high and too far south for this to be a common occurrence.

The more northerly basins of Oregon–Washington–Idaho and British Columbia are more often invaded by cold air from the east, but rapidly flushed by the westerlies flowing over coastal mountains. Here in the winter westerlies, the topographic roughness is such that the airflow becomes quasi-antitriptic with a strong northward and north-westward component through the basins and passes as it flows towards the subpolar low (Fig.2).

In high summer the northern basins are filled with dry air which originated in the Pacific anticyclone, while the Central Plateau of Mexico and the southern Great Basin are invaded by tropical air from the Gulf of Mexico. In this unstable airstream, with a deep moist layer, thunderstorms develop over the upland to provide a summer rainfall period.

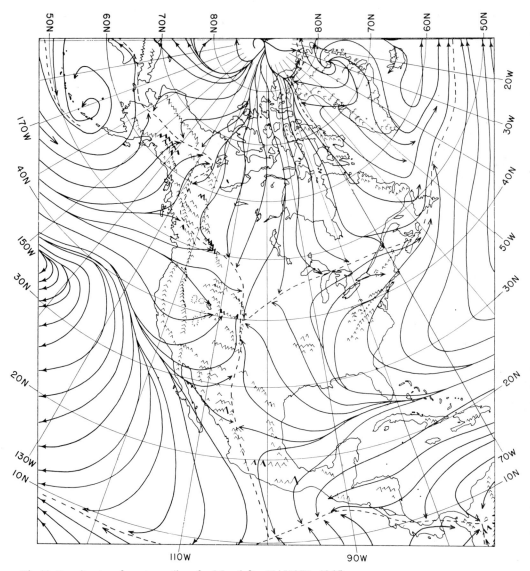

Fig.11. Resultant surface streamlines for May (after NAVAIR, 1966).

In the southwestern United States and northwestern Mexico the region of these summer rains overlaps the region of winter cyclonic disturbances in the westerlies to give two rainy seasons per year.

In the northern intermontane basins winter precipitation is associated with cyclonic disturbances, but this precipitation decreases when the westerlies and subtropical anticyclones shift northward in mid-summer and the southwestern rains begin. All of the basins of the intermontane region, nearly surrounded by mountains, tend to be dry, but winter cyclonic storms produce more uniform precipitation than do thundershowers which concentrate on the mountains. Cold season precipitation is also more effective in maintaining soil moisture. These two factors make the northern and southern basins more unlike in biota than the absolute rainfall amounts would suggest.

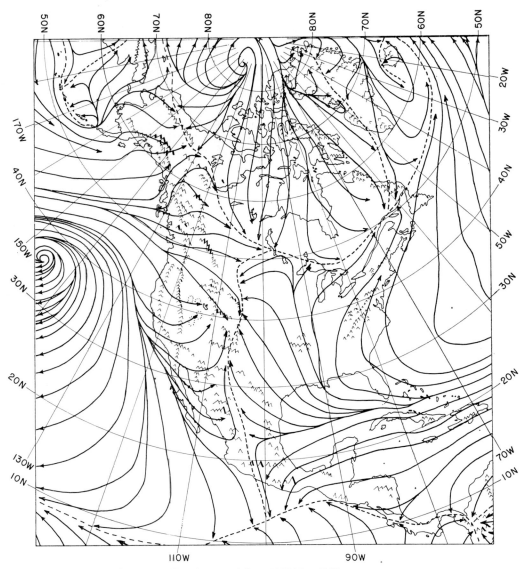

Fig.12. Resultant surface streamlines for June (after NAVAIR, 1966).

### The climates of eastern North America

East of the Cordillera, over the plains, lowlands and low mountains of eastern North America, local topography is less important and the climate is dominated by the great sweeps of major airstreams–cold, dry air from the frozen Arctic Ocean, moist warm air from the southern tropical seas, or the intrusion of mild, dry westerlies between the two.

During the winter, strong net outward radiation from the snow in the region of the polar night rapidly converts invading air masses into Arctic air which in a few days to a week may build dynamically and thermally into a cold anticyclone from which polar outbreaks periodically erupt. These may penetrate to the isthmus of Tehuantepec, underrunning all other airstreams, but the high cooling rate in the source region rapidly replenishes the reservoir. Individual streams of Arctic air usually are either "entrained" into the wester-

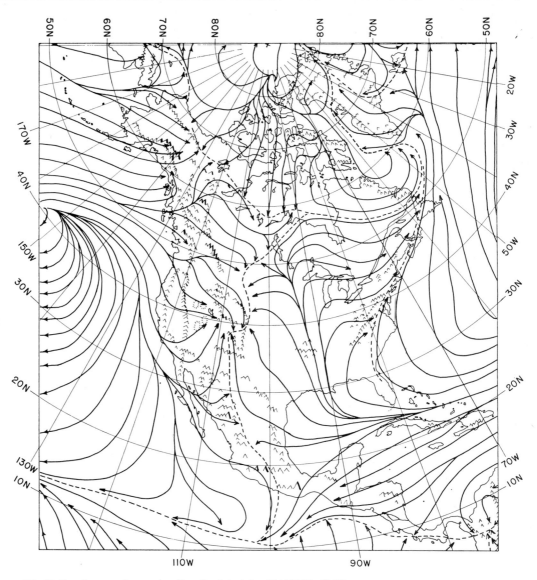

Fig.13. Resultant surface streamlines for July (after NAVAIR, 1966).

lies to mix with air from the Pacific or tropical Atlantic over eastern North America or the North Atlantic, or settle into anticyclonic eddies over the southern United States. Here in this sunnier snow-free region they are transformed into the mild dry air mass of the southern anticyclonic region. However, cyclones along the frontal boundary between this air mass and the warmer air over the tropical seas to the south produce excursions of tropical air northward over the cooler southern anticyclonic air to give abundant winter rains in a broad band reaching from Texas towards New York.

The winter westerlies enter eastern North America primarily across Wyoming, Montana and Alberta in a series of surges led by fronts that are usually rather weak. However, the cyclonic storms that may develop in the troughs associated with these Pacific incursions are often the most dramatic events of the winter in the eastern half of the continent. They may draw tropical air aloft into southern Canada producing the

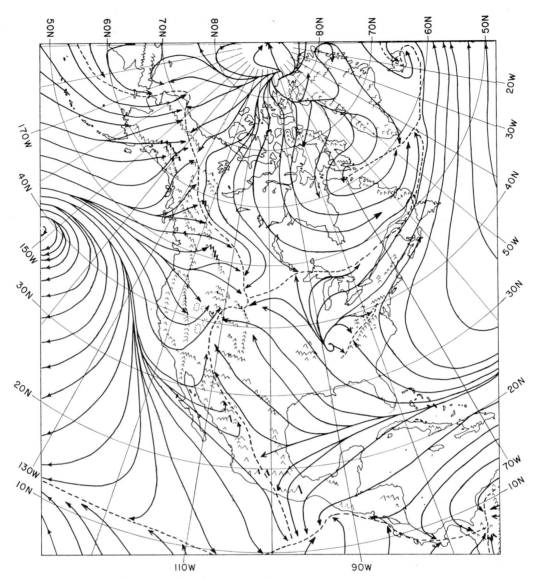

Fig.14. Resultant surface streamlines for August (after NAVAIR, 1966).

heaviest snows of the northeastern third of the continent, and following the storm the polar outbreaks may sweep southward with high winds drifting the new snow in blizzards that may leave drifts on the plains two or three meters deep with horizontal dimensions of many kilometers.

The Pacific air itself, modified by its passage over the western mountains, may be quite mild for winter if it has traversed snow-free plains, or quite cold in its northern portion where the shorter days, lower sun angle and snow-covered surface produce more rapid modification. In the mean this air dominates a wedge between two mean confluences: one between Pacific air and the southern anticyclonic air—modified Arctic or Pacific originally—and the other at the mean southern edge of Arctic air (Fig.17). These confluences merge and become more indistinct east of the Great Lakes and the Pacific air is found more often only aloft.

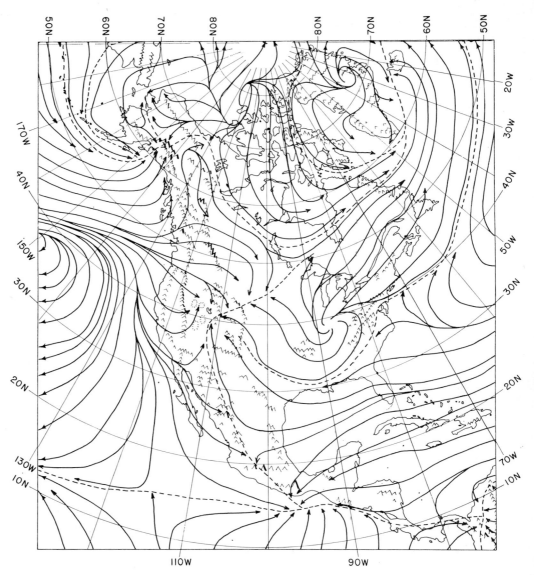

Fig.15. Resultant surface streamlines for September (after NAVAIR, 1966).

In late spring and summer far eastward incursions of Pacific air are less common, Arctic outbreaks do not extend as far south and the changing inclination of the sun reduces the polar area of net radiative cooling. Decreasing meridional temperature gradients reduce the strength of the westerlies and the eastern half of the continent is dominated by more meridional flow—tropical air northward from the Gulf and Arctic air flowing almost straight southward to meet it in the mean. The convergence of these airstreams is normally about 35°–40°N in early spring and triggers violent thunderstorms in the very unstable tropical air. In summer the convergence is weaker and usually between 50° and 60°N.

In autumn, before the polar outbreaks become very strong, the westerlies strengthen and dominate southern Canada from coast to coast. Stagnant anticyclonic eddies, of tropical or Pacific air, also return to the southeastern United States. This would make a

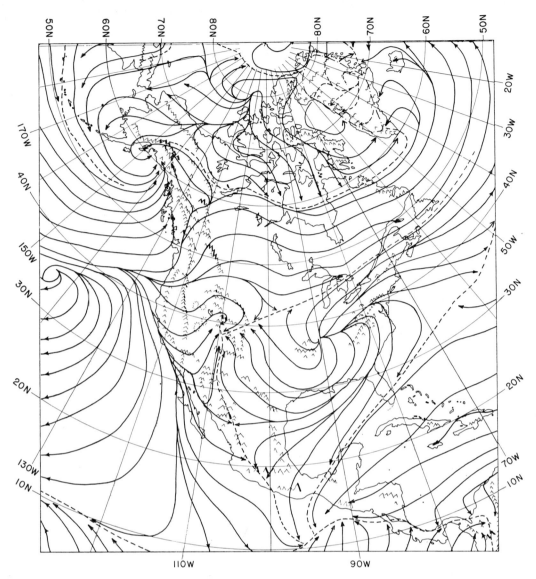

Fig.16. Resultant surface streamlines for October (after NAVAIR, 1966).

dry season there were it not for the hurricanes that disrupt the tranquility to produce an autumn precipitation maximum on the Gulf and southern Atlantic coasts of the United States and northeastern Mexico.

Over eastern North America the heaviest precipitation, throughout the year, falls where tropical air is lifted by convergence:

(*1*) Over the southern anticyclonic air as storms move along the Gulf coast in winter.

(*2*) Over the Atlantic coast as these storms turn northward along the Gulf stream.

(*3*) In the frontal convergences and storms of spring and summer in the mid-continent.

(*4*) In the deep storms that move from Alberta and Colorado—Texas towards the St. Lawrence River in winter.

(*5*) Year-round where all of these systems seem to congregate in the maritime provinces of Canada and New England.

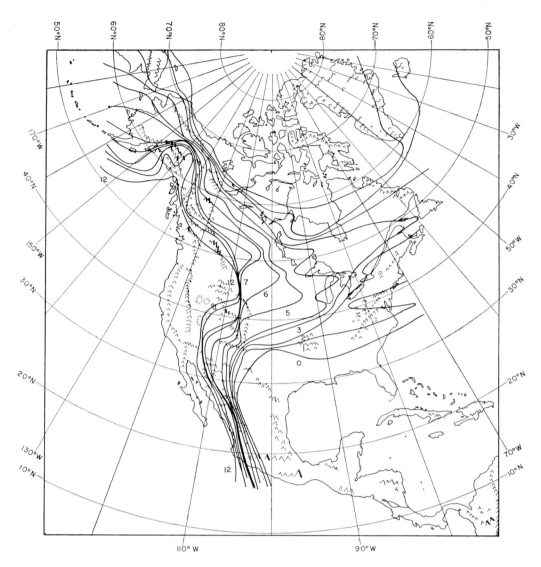

Fig.17. Duration of dominance of the Pacific airstream in months.

## Genetic climatic regions

The atmospheric circulation has distinctive monthly patterns which interact with surface thermal and topographic characteristics to give each place on earth a particular annual sequence of dominant airstreams. These airstreams, with their physical properties derived from their previous history, plus the dynamics of the locale and its radiation regime, largely determine the climatic character of the place. This defines an annual sequence of assemblages of meteorological parameters which characterize regions delineated by the preferred location of the airstream boundaries (Fig.18). We shall call these distinctive climatic assemblages *climata*, and the regions they occupy *genetic climatic regions*.

The climatic regions defined by the annual sequence of airstream dominance correspond

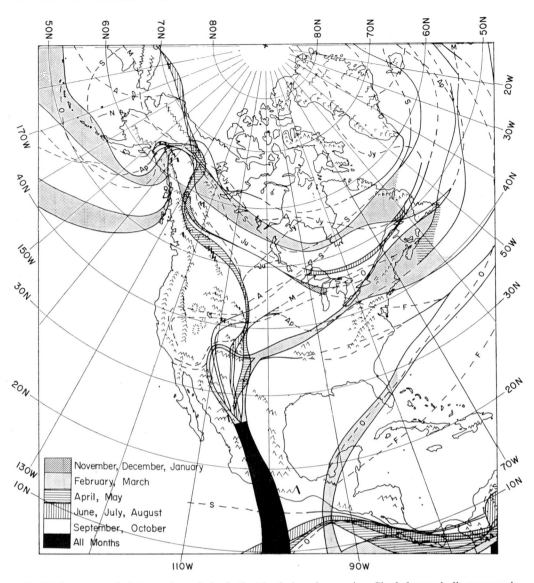

Fig.18. Locations of airstream boundaries in the North American region. Shaded areas indicate zones in which the boundaries lie very close to the same position during a climatic season or more. Based on Fig.5–16. The dashed boundaries shown in the western United States are from MITCHELL, 1969.

quite well with many of the major biotic regions (Fig.19)[1]. This is hardly surprising since the *biota* developed under the influence of corresponding *climata* (BRYSON, 1966).

The region occupied in the mean by the Arctic airstream for 10–12 months is the region

---

[1] The monthly streamline charts of Fig.5–16 are derived from NAVAIR (1966) based on data for five or more years, 1960–1964 or 1953–1964. The pattern of mean airstream boundaries in Fig.18 is derived from these data. The airstream boundaries shown on Fig.19 are from an earlier set of data, largely the 1930–1950 period, and the climatic data used in this chapter are from the standard 1931–1960 period. Certain discrepancies will be apparent to the reader. The NAVAIR data have the best areal coverage, the earlier data of Fig.19 are more detailed for the United States and Canada, have less smoothing and interpolation, and fit the standard climatic data period better. The differences between the two patterns appear to be consistent with the recent intensification of the Aleutian Low and weakening of the Icelandic Low, in winter, identified by KUTZBACH (1970).

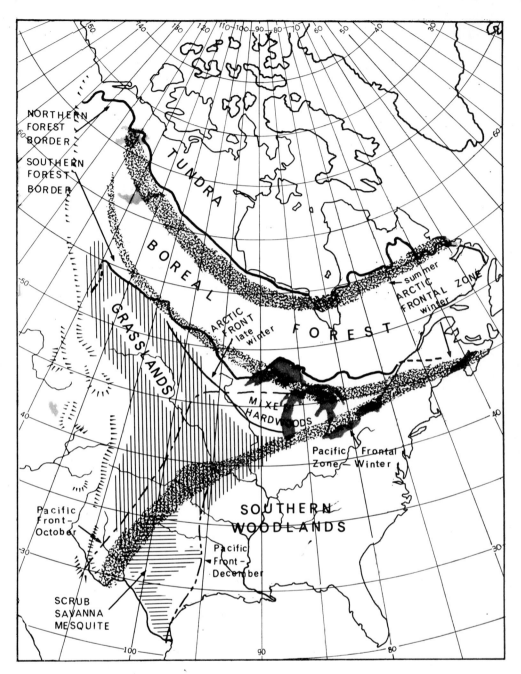

Fig.19. Correspondence of climata and biota in eastern North America north of Mexico, from BRYSON et al., 1970. Stippled zones indicate selected seasonal airstream boundaries.

characterized by *tundra*, while that occupied by the Arctic airstream less than that but more than about 3½ months is largely *boreal forest* (Fig.20). In the plains areas occupied by air of Pacific origin in winter, there are grasslands, *steppe* where the duration of dry Pacific air is greater than half the year and *prairie* where the dominance of this dry air is cut to as little as four months by a spring incursion of moist tropical air (Fig.17). Other similarly defined regions are listed in Table I.

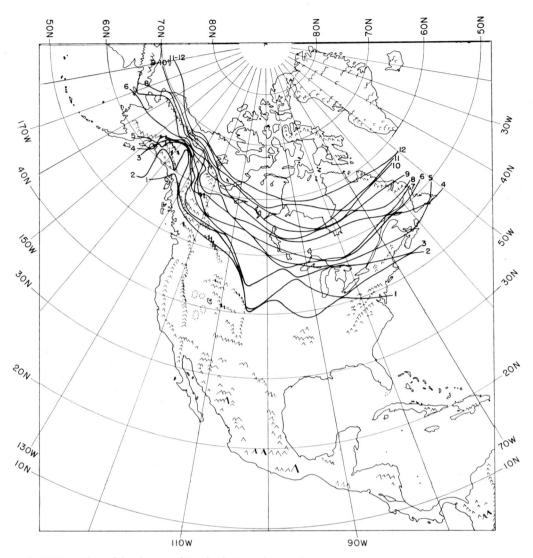

Fig.20. Duration of dominance of Arctic airstream in months.

Of course, the match of these genetically defined climatic regions, and their *climata*, with the corresponding *biota* is far from perfect; the climata are macroclimatic in origin, but microclimatic, topographic, and edaphic factors modify their boundaries somewhat. It is beyond the scope of this introductory chapter to detail even the many mesoscale departures from the macroclimatic pattern.

In the montane areas of the Cordillera, airstream regions may also be defined, but they are strongly dependent upon the topography, as previously suggested, and the airstreams involved are largely varieties of Pacific air modified on their various routes through and over the mountains. Details are not yet completely elucidated but some broad features are given in MITCHELL (1969).

In general the high elevations of the western Cordillera have rather maritime climates, with smaller temperature range than the lower elevations and with delayed seasons. Except for eastern and southern Mexico the interior mountains north of latitude 53°N, and the southern Rockies–Colorado plateau region, immersion of the high elevations in

TABLE I

EASTERN NORTH AMERICAN CLIMATES[1]

| | Early winter (Nov.–Jan.) | Late winter (Feb.–Mar.) | Spring (Apr.–May) | Early summer (June) | Late summer (July–Aug.) | Autumn (Sept.–Oct.) | Biotic region |
|---|---|---|---|---|---|---|---|
| I | A | A | A | A | A | A | tundra |
| II | A | A | A | P or T | P or T | P or SA | Boreal forest |
| III | P | A | A | T | T | P or SA | Great Lakes forest |
| IV | P | P | P or A | P | P | P | northern steppe |
| V | P | P | T | T | T | SA | prairie |
| VIa | P (Sonoran) | P (Sonoran) | T | T | T | P or SA | southern steppe |
| VIb | P (Sonoran) | SA | T | T | T | SA | mesquite |
| VII | SA | SA | T | T | T | SA (without hurricane rains) | southeastern deciduous |
| VII | SA | SA | T | T | T | SA (with hurricane rains) | southern pine |
| VIII | SA | T | T | T | T | T | pine–cypress |
| IX | T | T | T | T | T | T | southern Florida tropical |

[1] A = Arctic airstream, including Canadian varieties; P = Pacific airstream, dry from passage of the Cordillera; SA = southern anticyclonic airstream; T = moist tropical airstream.

fresh Pacific air is almost continuous. North of latitude 52° or 53°N the Arctic airstream is deep enough in late winter to top the mountains. In the southern Rockies–Colorado plateau region deep, moist tropical air pushes northwestward over the mountains in summer—bringing thundershowers to New Mexico and Colorado in May and June and to eastern Arizona and Utah in July and August. This region may also be in the southern anticyclonic airstream in autumn and early winter.

The mountains of eastern and southern Mexico are immersed in either tropical air with deep moisture (February–September) or in southern anticyclonic air which has crossed part or all of the Gulf of Mexico and has a shallower moist layer (October–January). The former season is wetter and the latter, drier. The mountains of western Mexico have climates related to the convergence of the Pacific airstream and the Atlantic trades or the southern anticyclonic airstream. During the October through January period the convergence is farther east and the southern anticyclonic airstream has a shallower moist layer. During this season the northern portion of the Sierra Madre Occidental is mostly in stable Pacific air, only the most southerly of cyclonic disturbances in the westerlies bringing winter rains. In summer, with the convergence farther west and a deep moist tradestream involved there is abundant precipitation in those mountains.

The mountains of southern Mexico lie close to this convergence throughout the year. At

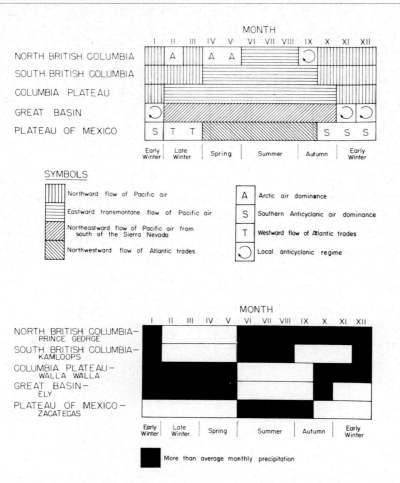

Fig.21. Annual march of circulation types and precipitation seasons for the intermontane basins of the Cordillera.

this latitude both Pacific and southern anticyclonic airstreams are somewhat less stable and rains may occur throughout the year, but the greatest intensity by far is in the summer season when the air on the western side of the convergence is least stable, the deep Atlantic trades reach the mountains, the intertropical convergence is near, and there are many tropical disturbances.

The airstream seasons of the intermontane plateaus are summarized in Fig.21. There are distinct circulation seasons, but they do not, of course, exactly coincide with either the astronomical seasons or precipitation seasons. The northernmost intermontane plateau in British Columbia is conveniently divided into a northern portion (north of about 53°N) which is occupied during part of the late winter and spring by Arctic air. This is, not surprisingly, a season of less than average monthly precipitation amounts. The southern part of the British Columbia intermontane plateau has no month with Arctic air dominance, but is also drier than average during that same season and in autumn as well. Both areas are characterized by coniferous trees, but there is also some open grassland in the southern area.

The Columbia Plateau, as described previously, is dominated by the Pacific airstream throughout the year, lying in the path of the strongest westerlies. In the cool season when the westerlies are far south, cyclonic storms bring widespread moderate precipitation but in summer and early autumn the storms are too far north and the region is dry. The plateau is largely grassland with some shrubs.

The Great Basin has a similar regimen but differs in that the enclosure of the basin is more complete, stagnant pools of cold air develop in early winter, and the dominant flow of Pacific air the remainder of the year is from around the south end of the Sierra Nevada and is warmer and drier. The basin is south of the storms longer, and the dry season includes November and December. The vegetation, at comparable elevation, is more xeric than on the Columbia Plateau.

The airstream regimen of the Central Plateau of Mexico is quite different. Southern anticyclonic air dominates most of the plateau from October through January and the region is dry. Despite the invasion of trade-wind air from the Atlantic the rest of the year, the rains do not start until the Atlantic subtropical anticyclone shifts northward in June and the southern portion of the trades with its deeper moist layer reaches the plateau. The vegetation is xerophytic scrub.

A glance at the circulation charts (Fig.5–16) shows that at the latitude of Alaska, where the Pacific air is cold and much of it is of recent Arctic origin, a shorter duration of Arctic air is sufficient to suppress arboreal plant forms. Here more than about 6 months of fresh Arctic air, which is usually cloudy here in summer, results in tundra. The remainder of Alaska is a coniferous forest as are those other Pacific coastal regions with convergent cool Pacific air throughout most of the year.

Of course, the relationships between the airstream regions and standard climatic variables are not simple, and as will be noted in the following chapters, these regions are not homogeneous. However, there are characteristics of the airstreams that are consistent: Arctic air is cold in January and cold for the season in July. It has, therefore, low absolute humidity and rather shallow cloud. Tropical air is moist and warm, all seasons, and deep convection is common, etc. The sequence of air types throughout the year, with their distinctive assemblages of characteristics, and the frequency of airstream alternations largely sets the atmospheric flavor, or climate, of each region.

## Mean circulation patterns and their seasonal variation

The foregoing account of the broad features of the North American climate was given in descriptive terms. We shall now reinforce it by presenting analyses of the mean flow patterns. The treatment will be based on climatological analyses of the wind-fields by LAHEY et al. (1960), NAVAIR (1966) and CRUTCHER and HALLIGAN (1967). These analyses are based on periods differing from (and hence not wholly consistent with) the streamline charts of surface winds given in Fig.9–20. Moreover, we shall have to adopt the meteorological seasons (Dec.–Feb., Mar.–May, June–Aug., Sept.–Nov.) rather than the natural seasons we have defined, since statistics are not prepared for the latter.

Time averaged or *mean* circulation provides valuable information about the fundamental nature of the atmospheric motion. But it also hides much of the complexity of day-to-day changes caused by the movement of synoptic disturbances. These will be treated in a later section.

**Winter (December–February)**

The mean flow at 300 mbar (about 9–10 km) for January is illustrated in Fig.22. The flow in the other months is similar. The streamlines on this chart (and also on Fig.23–25) were drawn so that they were everywhere parallel to the vector mean (resultant) winds. The streamlines were selected so as to illustrate the general configuration of the flow, and are not at strictly equal intervals of the stream-function; i.e., their spacing is not a direct measure of speed. The time-average value of the latter is given in meters per second by the plotted isotachs.

Although the flow is generally zonal, certain standing waves appear on the chart. There is a ridge over Alaska in the northern westerlies, and a trough in the southern westerlies, west of California. Both are persistent features. A very strong and equally persistent trough affects the current north of about 35°N, the trough-axis being about 85°W. One

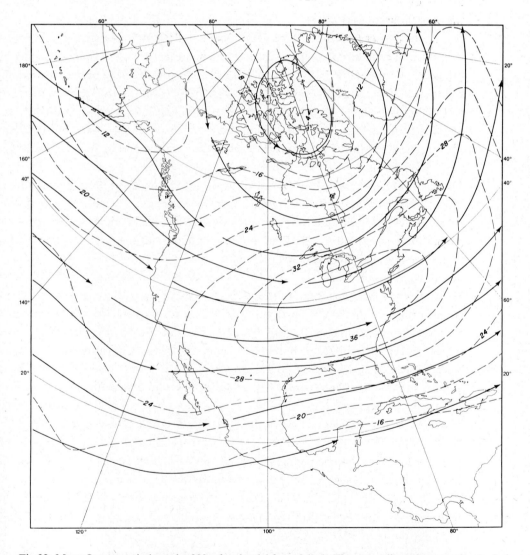

Fig.22. Mean January winds at the 300-mbar level (about 9 km). The streamlines show resultant winds, and are chosen to illustrate characteristic patterns. Speeds (in m/sec) are shown by the dashed isotachs.

of the cyclonic centers of the westerly vortex lies over the Canadian Arctic Archipelago, at the northern end of this eastern trough.

Flow is diffluent over the eastern Pacific, strongly confluent from 120°W to mid-continent, and diffluent thence eastwards across the North Atlantic. A strong speed maximum (over 36 m/sec) lies over the eastern United States. There are three major axes of wind-maximum. One axis runs from the California–Mexico border via St. Louis and Washington to the Atlantic south of Newfoundland. It can thence be traced to north of Scotland. This is the primary westerly maximum, or midwesterly *jet*. The second axis crosses the Pacific and enters North America in about 40°N. Over and south of the Great Lakes region it merges with the first maximum. The third or Arctic maximum is much weaker, and can be detected from near Wrangell Island via lakes Great Bear, Great Slave and Athabasca to Lake Winnipeg.

Each of these wind maxima lies above a zone of strong mid-tropospheric baroclinity, and the fourfold thermal divisions of the troposphere thereby implied is a characteristic feature of the winter and spring months. Surface fronts are usually present below each of the baroclinic zones (Fig.18).

In each of the seasons the surface wind regime underlying these strong upper westerlies is weaker but far more complex, because of the influences of the physiography already discussed. For winter (Fig.6–8), as for the other seasons, we have prepared streamlines of resultant flow based on data published in NAVAIR (1966), which refer to the surface or to 1,000 mbar, whichever is higher. Characteristic lines of diffluence and confluence are identified; in many cases, as can readily be seen, these are also lines of divergence and convergence. The lines of confluence–convergence correspond in many cases to the mean position of fronts.

The winds of the Arctic regime form a vast cyclonic vortex centered near south Greenland. West of 85°W the streamlines originate over the Arctic Ocean, but turn east towards the Atlantic across Labrador–Ungava. An apparently independent system provides the circulation over inland Alaska. The southern boundary of the Arctic regime corresponds to the *Arctic front* of the daily weather map.

The Pacific regime covers the west coast from Alaska to Mexico, with a deep eastward penetration to the Canadian prairies. The analysis suggests that the southerly flow over inland valleys of the Pacific northwest and British Columbia comes from far south. The westerlies approaching the British Columbian coast must rise across the Cordilleran ridges. In the immediate offshore belt, inside the Queen Charlotte and Vancouver islands, canalized southeasterly flow is persistent in the lowest layers. The Pacific regime borders another mid-continental belt of westerlies whose streamlines emerge from the United States high plains, and which covers the Great Lakes–St. Lawrence–maritime provinces sector. Clearly the air of this belt is derived by subsidence from overrunning Pacific westerlies.

The southern anticyclonic regime covers all southern parts of the U.S.A. east of 110°W, and also encompasses the prevailing northeasters of central and northern Mexico. Along its northern flank, from Nevada to Lake Erie and the Gulf of Maine, the winds of this system are in contact with the mid-continental westerlies along another of the major surface frontal boundaries of winter.

The remaining current is the trade wind regime, which in the mean covers southern Florida, the West Indies, Yucatan and other southern areas of Mexico and central

America. The boundary between this regime and the southern anticyclonic regime is often a front on daily charts. The trades circulation of winter is rather shallow, being overlain by the southern part of the westerly stream.

### Spring (March–May)

This is a season of rapid change in circulation. In many years, March is a full winter month in pattern. During the month, however, the westerly speed maximum at 300 mbar begins to drift southwestward and the quasi-permanent southern trough off the California coast attains peak amplitude. The eastern trough and the Alaskan ridge both drift eastwards. The chart for April (Fig.23) shows how these long-wave adjustments simplify and reorganize the flow patterns over North America.

The most noteworthy result is the intensification and expansion of the Arctic regime at low levels (Fig.10). Arctic air makes its deepest penetrations of eastern and central North America in late March, April and early May, because of weakening westerlies and

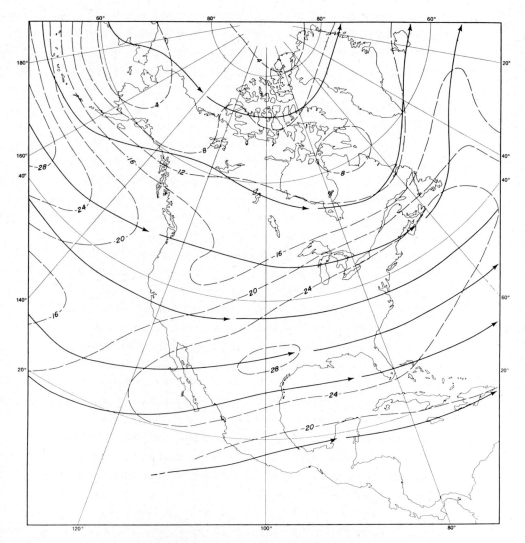

Fig.23. Mean April winds at the 300-mbar level (uniform with Fig.22).

persistent high pressure over the Arctic Ocean, Hudson's Bay and Keewatin. Most of Alaska, all Canada east of the Alberta foothills, and the United States midwest north of 40°N, are dominated by the Arctic regime, though the streamlines east of 90°W trace back to a divergence axis over north Greenland and Baffin Bay. At 300 mbar (Fig.23) there is still an identifiable Arctic westerly maximum from Wrangell Island to northern Saskatchewan.

At the same time the true tropical regime replaces the southern anticyclonic regime of winter. The southerlies that flood into the Mississippi Valley and Atlantic states in spring are of trade wind origin, and import much humidity and warmth. At low altitudes the warm, moist advance reaches the Arctic front across the midwest and southern New England. This is far north of the main westerly jet (Fig.23), especially west of the Mississippi, implying extreme instability that expresses itself in the tornado maxima of spring. The westerly jet itself constitutes a single system the entire distance from the California coast across the United States and Atlantic into northern Europe and Siberia.

The Pacific regime commands the whole coast from the Kenai Peninsula to western Mexico, but is less vigorous than in winter. At low levels it makes its least deep penetration of the continental interior. The highly divergent and stable southern wing of the westerlies meets the tropical trade-wind regime along a north–south axis in about 103°W, leaving much of Mexico within trade-wind air. The flow over the Caribbean and Gulf of Mexico is, however, highly divergent, and hence fairly stable aloft.

**Summer (June–August)**

Summer begins with a further major reorganization of circulation, whose date varies from year to year (like that of the Asian monsoons, which the process resembles). By mid-June, however, the basic processes have usually become apparent: a northward shift of the subtropical high in the upper troposphere, so that deep easterlies cover the Gulf of Mexico; renewed deep invasion by Pacific airstreams; and a considerable weakening and retreat of the Arctic regime.

The 300-mbar chart for July (Fig.24) illustrates these effects. A single mid-latitude westerly maximum in 45°–50°N dominates the flow. Tropical easterlies control all areas south of about 25°N. The Arctic maximum is still present, and extends from Wrangell Island to Churchill. The Alaskan ridge no longer deserves the title, lying in 100°–105°W, and the eastern trough lies near or off the Atlantic coast. Circulation is everywhere much weaker than in winter, kinetic energy per unit volume being only ca. one-third as high. Summer marks the maximum northward extent of the tropical regime, its July boundary running at the surface (Fig.13) from Texas to southern James Bay, and thence via Cape Breton Island to the Atlantic. The axis of this great influx is east of the Mississippi at low levels, but the greatest flux of water vapor is west of the Mississippi Valley (RASMUSSON, 1968). A well-marked surface axis of convergence off the Atlantic coast is a frequent breeder of rainy disturbances (see below). In Mexico, the westerlies of Pacific origin push east again towards the Sierra Madre Oriental. The trades are now at their deepest, least stable and most humid.

The Pacific regime at low levels is dominated by the subtropical high pressure center in about 43°N 1,500 km off the Oregon coast. Maximum inland penetration at low levels is across the Canadian prairies, as in winter.

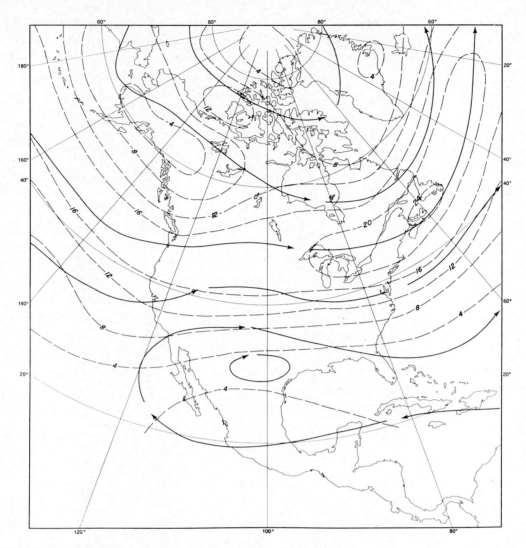

Fig.24. Mean July winds at the 300-mbar level (uniform with Fig.22).

The Arctic regime at low levels consists of a divergent flow from a center near Ellesmere Island. The surface Arctic front within the mean circulation extends from the Beaufort Sea via the great Mackenzie lakes to northern Ontario, James Bay and thence to Cape Breton Island. The cool, cloudy summer of the eastern Arctic and sub-Arctic reflects this dominance by Arctic airstreams flowing off thawing ice surfaces.

**Autumn (September–November)**

Autumn is marked by the rapid collapse sometime in mid- or late September of the deep tropical penetration of summer, and its replacement by the southern anticyclonic regime that persists through the winter. This is accompanied by a rapid strengthening and expansion of the westerlies, which are more purely zonal at this time of year than at any other.

The 300-mbar chart for October (Fig.25) illustrates the chief features. The main westerly maximum is at its farthest north over the Pacific and western North America, but from

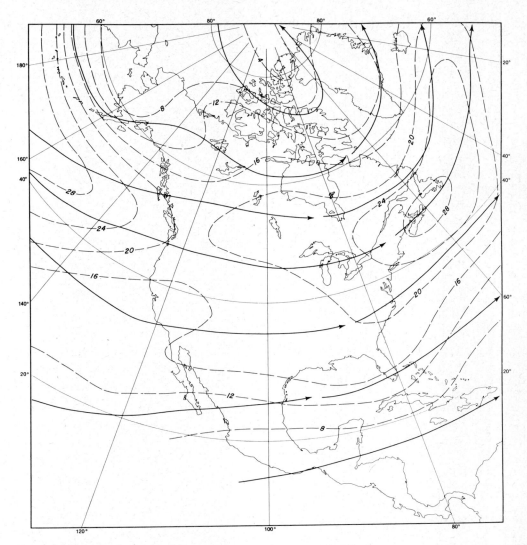

Fig.25. Mean October winds at the 300-mbar level (uniform with Fig.22).

the Great Lakes eastwards lies near its summer locus. A substantial gain in kinetic energy has taken place. Westerlies once again cover all tropical areas down to about 10°N.

The surface flow (Fig.16) reflects the dominance of zonal flow. The Pacific regime is at its annual peak, the westerly airstream of Pacific origin covering all southern Canada. The northeastern United States is controlled by the southern anticyclonic regime, with strongly divergent flow. The southern boundary of this regime, where the continental air meets the trade winds (both flowing from northeast), lies from Yucatan to southern Florida and Bermuda.

The northeasterly circulation over the Gulf of Mexico, Caribbean and Mexico is clearly composite at low levels. All of it, however, has passed over very warm ocean surfaces, and the current is slow to cool and stabilize. The layer of easterly flow becomes rapidly shallower, however, and by the end of the season the trades are once again capped by the southern westerlies in the upper troposphere even in southern Mexico.

**Disturbances**

Some mention was made in the second section of the role of disturbances in determining the sequence of weather events typical of the several climata. We shall now present formal average distributions for the chief types of synoptic scale disturbances, viz.: (*1*) cyclones and anticyclones associated with the transient waves of the westerlies; (*2*) tropical storms and hurricanes. Brief mention will also be made of other types of tropical disturbances. (*1*) Cyclones and anticyclones of the westerlies are the synoptic-scale perturbations familiar on the daily weather map. The closed high and low pressure centers of the sea-level charts correspond to moving waves in the general westerly current, or in parts of it. The eastward travel of these synoptic-scale disturbances causes large day-to-day weather changes, and also brings about large-scale meridional transports of heat, moisture and momentum. The disturbances are, in fact, the effective driving mechanisms of the westerlies.

The zone affected by the disturbances varies with the season, since they are associated with the fronts and jet-streams in the heart of the westerly current. The area of highly disturbed circulation is large, and is loosely associated with the main westerly maximum. The latitude of greatest disturbance over all North America is as follows: in winter 45°N; in spring 43°N; in summer 54°N; and in autumn 48°N. In winter and spring the variance of the flow is high across the whole continent, whereas in summer, and to some extent autumn, the interior has much less disturbed flow; maximum disturbance activity is then over the oceans. Over Mexico and the tropical zone generally the disturbance level is much lower, reaching a maximum in mid-winter.

In Fig.26–29 we show for January and July the frequency with which cyclone and anti-cyclone centers occur over North America (after KLEIN, 1957). On each chart we have also indicated regions of high frequency of cyclogenesis or anticyclogenesis. The continent stands out as a region of high frequency in all of these distributions; there is more synoptic activity than in most other parts of the hemisphere. We have added generalized tracks to the charts, but stress that the moving systems choose a wide variety of directions and speed of motion. Moreover, stationary blocking highs, cold lows and "heat" lows are common in some areas.

The winter is the season of maximum intensity (Fig.26, 27). Pacific cyclones approach the North American coast, and tend to stagnate over the Gulf of Alaska or off Vancouver Island. Separate centers often develop in the coastal belt. These cyclones give the coast a violently stormy winter, though the storminess does not go far inland. Strong cyclogenesis also occurs inland, chiefly in the lee of the Rockies, notably over Alberta, Colorado, Texas and along the Gulf coast. Yet another belt of strong cyclogenesis occurs along the Atlantic coast, with secondary effects west of the Appalachians. The general direction of motion of cyclones is parallel to the 300-mbar flow.

Because of their intensity, winter cyclones are great weather bringers. Those of the Pacific deluge the coast and mountainous littoral belt with heavy rains and snows. Those of the interior produce snows and blizzards in the north, and thunderstorms in the south. Atlantic coast cyclones, and also interior systems moving northeast out of the Mississippi Valley produce the great snow, freezing rain and rainstorms of the eastern winter. Only in the dry western interior and the high Arctic are the cyclones largely ineffective as producers of precipitation.

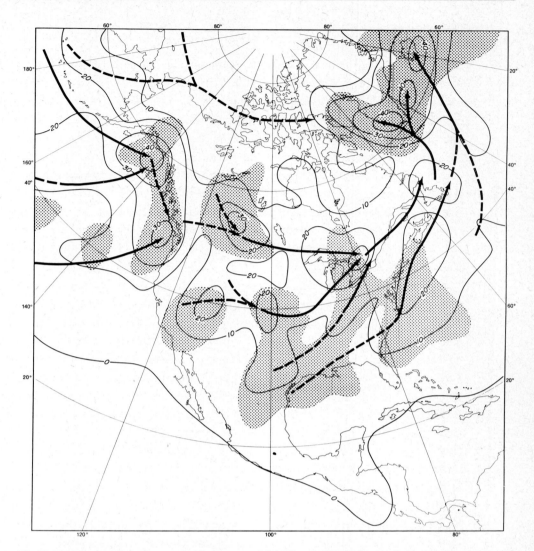

Fig.26. Transient cyclone frequencies for January (after KLEIN, 1957). The isopleths show the number of cases in a chosen 20-year period (1909–14, 1924–37) when cyclones were centered in five degree squares (adjusted to the size of such squares at 45°N). Since no cyclone was counted in a given square more than once, stationary systems like heat lows were filtered out. Hurricanes were also excluded. The shaded area had more than 4 cases per five degree square of cyclogenesis during the same 20 years. Characteristic tracks of the cyclone centers are also shown. Dashed tracks are of minor importance.

Winter anticyclones are of two main classes. Persistent highs tend to dominate two areas—the Mackenzie–Yukon–Beaufort Sea area of the north, and the Great Basin states of the U.S. Migratory cold highs move southeast out of the Mackenzie–Yukon area, bringing cold waves to the midwest and east. A secondary area of cold high formation is over west-central Quebec. Eastward moving warm-cored highs also form over the southern states, especially over Texas and the southern Appalachians. Over populous eastern North America the moving high is as much a normal part of the winter scene as the cyclone. The cold-cored systems bring the cold waves, the warm systems a good deal of fog and smog. In all cases clear skies are the norm.

In spring (not illustrated) the stream of Pacific cyclones weakens and shifts further north. Cyclogenesis inland becomes concentrated in western districts, especially over Alberta and the Mackenzie Valley, the Great Basin and the central High Plains. By far the most important of the resulting cyclones are the numerous systems that move east and north-east across the middle west and Great Lakes regions to the St.Lawrence Valley. Involving both Arctic and tropical airstreams, these vigorous disturbances bring late season snows to the prairies and central Canada, and the heavy thunderstorms and tornadoes near their cold fronts in central U.S. areas. Cyclonic activity also continues high all along the Atlantic seaboard.

Spring is the chief season of true Arctic anticyclones, which tend to dominate the Arctic Basin, Keewatin and Hudson's Bay. Keewatin is in fact a strongly anticyclogenetic region. Most spring highs, however, form as eastward moving systems below ridges in the upper westerlies, especially in the Pacific northwest and the High Plains. These highs move out into the subtropical high-pressure belt of the Atlantic.

The main areas of cyclonic activity shift rapidly north in late June, and by July (Fig.28, 29) lie at their furthest north. Pacific storms now affect primarily Alaska and northern British Columbia; the Oregon and California coasts are almost immune. Cyclogenesis is

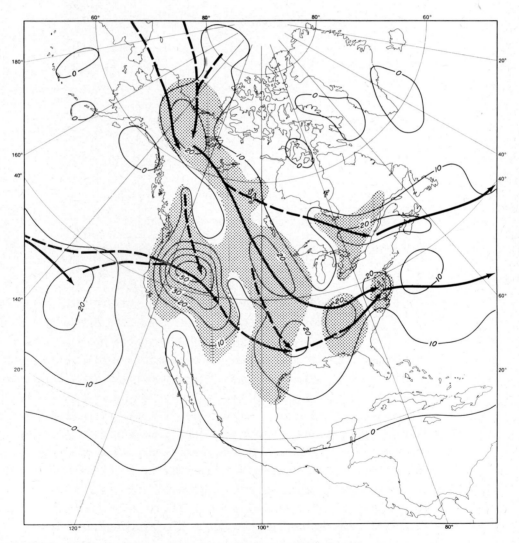

Fig.27. Transient anticyclone frequencies for January, compiled like Fig.26.

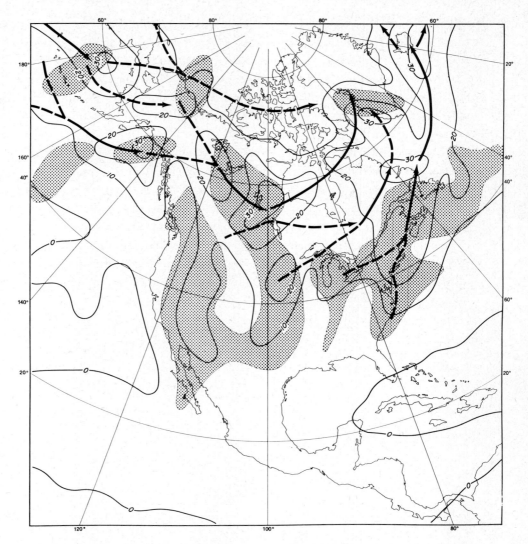

Fig.28. Transient cyclone frequencies for July, compiled like Fig.26.

common in western districts inland from Alberta to Colorado and New Mexico. "Heat" lows, i.e., formless, shallow cyclones over the hot basins of the arid southwest, are present on most days. Atlantic coastal cyclogenesis is still frequent, but much less intense than at other seasons. The typical summer cyclones move northeast out of the western breeding grounds across central Canada, Hudson's Bay and Labrador–Ungava, often involving the Arctic front. Heavy thunderstorm rains occur in the cold front belt, rain showers further north. The Atlantic and Gulf coast states are affected by increasing numbers of tropical storms or hurricanes as the season progresses (see below).

Summer highs are chiefly eastward-moving warm systems in Pacific airstreams developing primarily over the Plains states. The southeastern states are normally covered by the western flank of the stationary subtropical Bermuda High at this season.

The autumn season continues these tendencies. The main cyclonic motion is eastward or northeastward across central Canada from western cyclogenetic areas. There is, however, a marked recrudescence of Pacific storms on the coasts of Washington and British Co-

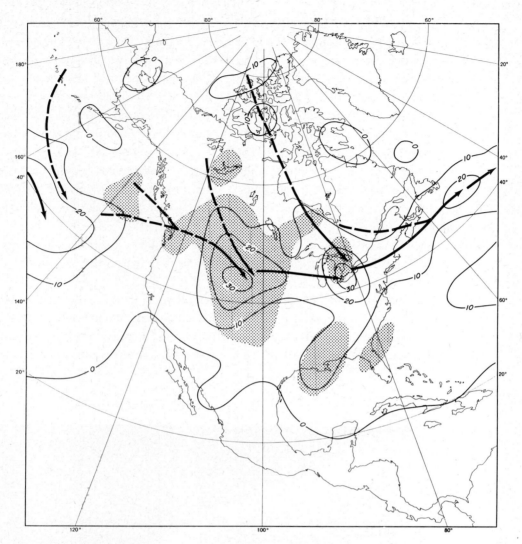

Fig.29. Transient anticyclone frequencies for July, compiled like Fig.26.

lumbia. Atlantic coast cyclogenesis also becomes more frequent and intense. This is the season, moreover, of peak tropical storm intrusion into the westerly belt (see below).

As in summer, the anticyclones of autumn are primarily eastward moving systems in the 30°–50°N latitude belt. Cold anticyclogenesis in the far northwest becomes increasingly common late in the season, but the stationary highs of the Mackenzie–Yukon belt in winter do not normally appear before late November.

In all seasons the structure of North American weather systems tends to be more complex than that of the classical frontal cyclone or baroclinic wave. There is typically more than one front across the continent: we have already stressed the frequent occurrence of Arctic, midwesterly and subtropical frontal zones and jets. When waves in the westerlies move inland and across North America they may disturb one, two or all of these zones, and very complex disturbances ensue. A familiar event, for example, is the development of a new cyclone on the Atlantic coast near Cape Hatteras as an old cyclone moves northeast from the Mississippi–Ohio valleys. Such details cannot be elaborated further here, but are touched upon in the regional chapters.

(2) Tropical storms and hurricanes are the second major group of disturbances affecting North and Central America, chiefly between June and November. These are initially small, warm-cored cyclones containing dense cloud from which heavy rain falls, with inward spiralling bands of cumulus-scale showers and thunderstorms along their periphery. Forming over warm ocean surfaces (>28°C) in the tropical easterlies, they normally move slowly with that current, growing larger with time. Although a mature hurricane remains quite small (~100 km radius) as a region of violent winds (>50 m/sec in gusts), it is ordinarily surrounded by quite a large area of enhanced convective shower activity (Fig.30). Rainfall amounts during a single hurricane passage may exceed 50 cm, and 10–15 cm is commonplace. Many tropical storms of this nature fail to develop the strong winds of a true hurricane (>35 m/sec), but still yield enormous rainfalls.

On the average (WINSTON, 1959) 5–10 tropical storms and hurricanes affect the North Atlantic and Caribbean areas per annum, but in 1933 there were 21, and in 1914 only one. (See also Table II.)

Tropical storms may develop anywhere over the Gulf of Mexico, the Caribbean, the eastern tropical Pacific and the Atlantic near the Bahamas, but at the peak of the season (August–October) the preferred areas of origin are the trade wind belt east of the Antilles (including many from near the Canary Islands), and the southwestern parts of the

Fig.30. Apollo-7 view of hurricane Gladys west of Florida on 17 October, 1968, showing organized convection bands. Photo courtesy of NASA. For details see SOULES and NAGLER, 1969.

TABLE II

MONTHLY DISTRIBUTION OF TROPICAL STORMS AND HURRICANES

|  | Jan.–Apr.[1] | May | June | July | Aug. | Sept. | Oct. | Nov. | Dec. |
|---|---|---|---|---|---|---|---|---|---|
| Mean number per annum | * | 0.1 | 0.5 | 0.5 | 1.8 | 2.7 | 1.9 | 0.4 | 0.1 |
| % attaining hurricane strength | – | 22 | 42 | 53 | 74 | 64 | 46 | 42 | – |

[1] One each in 70 years in February and March; none in January or April.

Gulf of Mexico and Caribbean. Satellite observation is now making early detection and tracking of these storms a far easier and more efficient process.

As the storms move beneath the axis of the subtropical highs in the upper troposphere they tend to change their direction of motion from northwestward to northeastward, often after a period of vacillation. Aberrant tracks are, however, common. In 1966 hurricane Faith was first detected east of the Canaries on August 21. Subsequently it followed a normal northwestward course, recurving August 31 before beginning a rapid northeastward movement across the North Atlantic, reaching the Faeroe Islands on September 5 still with hurricane winds, and dying over the Arctic Ocean north of the New Siberian Islands on September 15. In 1971 hurricane Ginger, after moving east far beyond Bermuda, abruptly reversed direction due to the development of a blocking high south of Greenland; she struck the mid-Atlantic coast of the United States 72 h later.

The risk of heavy rain, damaging winds and flooding by the sea is high along the entire coast of Central and North America from southern Mexico to Newfoundland. Even the Pacific coast of Mexico is occasionally struck by systems from the Pacific. The highest risk is along the Gulf coast of the U.S., Florida (whose low, smooth topography allows the ready passage of the storms), the mid- and south Atlantic states and the Bahamas. Hurricanes and tropical storms nearly always die out quickly when they move inland, but many of them interact with the westerly current's baroclinity to start fresh developments that may become intense. Thus Camille in 1969 passed inland from the Gulf of Mexico. Later it moved across Virginia to become one of the greatest rainstorms in American history. In 1954 Hazel initiated fresh cyclonic development on a stationary front across Lake Ontario to give Ontario its worst recorded storm, with a large loss of life.

Hurricanes and tropical storms, though infrequent, produce such great rainfalls that the late summer and fall are the wettest seasons in some areas exposed to their passages.

Other tropical disturbances include cold-cored lows, easterly waves, shear-lines and various ill-defined mesoscale shower areas. We shall not treat these here, because as Mosiño and García show (p.357) they do not appear as coherent elements in the Mexican climate, and do not figure largely in that of the United States. The large body of satellite evidence now available has in any case made necessary a re-evaluation of tropical disturbance models.

## An historical perspective on the climates of North America

This book is about the climates of North America, and as such presents discussion and data based on the standard period 1931–1960 as much as possible. But these climates are not the only climates North America has known, for it has had and will continue to have a varied climatic history. Actually, very nearly the same *range* of climates has prevailed over the last twenty millennia or so, but the distribution of the climates has changed markedly. A brief account of the general sequence of climatic episodes during the past 12,000 years will put the 1931–1960 pattern of climates into perspective.

This time span may be divided into two major portions: that more than 10,800 or so years ago, the end of the Wisconsinan ice age, and the later or Holocene Period. This date is chosen because it, or a date not far from it, marks the dramatic change in climate that initiated the essentially destructive retreat of the vast ice sheet that covered most of North America north of 45°N. This ice sheet was so large and so deep that its waning took about 6,000 years—from a mass larger and about as deep as the Antarctic ice sheet to a few ice cap remnants such as those still found on Baffin and Ellesmere Islands (BRYSON et al., 1969). Although world climates changed from glacial to post-glacial about 10,800 years ago, for nearly half of the post-glacial period there was a considerable area of glacial ice in Canada.

Weather records span about 1 % of this Holocene time period, so that reconstruction of past climates must be done from fossil evidence—biological, chemical and geomorphic. There is a large literature on past climates and paleoclimatic methods (Brooks, 1950; Mitchell and Kiss, 1964), but only in recent years have quantitative methods developed that give us an insight into *how much* variation of climate there has been, *how little* climatic change is necessary to be environmentally significant, and *how fast and when* climates have changed.

## Climatic subdivisions of the Holocene

Apparently, the Pleistocene ice-age climate ended suddenly, for nearly every fast-responding indicator of climate that has been studied in North America shows an abrupt change at very nearly the same time (OGDEN, 1967; BRYSON et al., 1970). The expression "nearly the same time" needs some elaboration, however, for dates in this time range of 10,000 to 12,000 years ago are based on radiocarbon dating. It now appears that there are systematic interlaboratory differences which make the choice of a date for the end of the Pleistocene dependent upon which laboratory did the dating (WENDLAND and DONLEY, 1971).

This introduces an uncertainty of about a thousand years, but the dates appear to cluster about 10,800 radiocarbon years ago. However, on time series of climatic indicators where the relative accuracy of dating is high, the change was compressed into a couple of centuries or less. Here the uncertainty is due to an unknown lag time of the indicators, such as biotic assemblages. Assuming that the half-response time of a forest community to a sudden climatic change is about a century, one may calculate from pollen diagrams that the change from a glacial to a non-glacial climatic pattern occurred in much less than a century. If glacial climates are in part a response to atmospheric turbidity caused by volcanic activity (LAMB, 1970), then this would hardly be surprising, for volcanic dust

TABLE III

DATES OF SIGNIFICANT HOLOCENE ENVIRONMENTAL CHANGE IN RADIOCARBON YEARS BEFORE THE PRESENT
(After BRYSON et al., 1970)

| European terminology | Major discontinuities | Minor discontinuities |
|---|---|---|
| Sub-Atlantic | 820 B.P. | |
| | 1,680 B.P. | |
| | 2,710 B.P. | |
| Sub-Boreal | | 3,660 B.P. |
| | | 4,430 B.P. |
| | 5,180 B.P. | |
| Atlantic | | 5,970 B.P. |
| | | 6,740 B.P. |
| | | 7,060 B.P. |
| | 7,900 B.P. | |
| Boreal | | 8,490 B.P. |
| | 9,160 B.P. | |
| Pre-Boreal | (?)  9,970 B.P. | |
| | (?)10,800 B.P. | |

in Antarctica and Greenland terminated quite abruptly at the end of the Pleistocene (HAMILTON, 1972). There is no meteorological or geological reason to believe that the Holocene could not end just as abruptly, and indeed the ice-core data suggest that the Wisconsin segment of the Pleistocene started climatically as rapidly as it terminated.

While no Holocene climatic changes as dramatic as that at its start have been identified, the last ten millennia of North American (or world) climatic history have not been uniform. Analyzing the distribution of dates of environmental events suggesting climatic change, Wendland found that these dates clustered about a series of times that are similar for North America and Europe, and indeed for the world (Table III). Of course, just as present-day anomalies are not uniform, or even of the same sign, one would expect that while times of climatic change might be synchronous in our single interconnected atmosphere, the nature of the change itself would vary from place to place.

While the climatic episodes of the Holocene can be distinguished in North America and Europe by the use of environmental indicators, the changes in the usual meteorological variables appear to be rather small. To illustrate the general magnitude of the changes, a brief description of the general character of two climatic episodes follows.

**Late Wisconsin (latest Pleistocene) climate**

North America in Late Pleistocene time was dominated by a great continental ice sheet, about the size of Antarctica, that reached from coast to coast and reached heights of probably 3,100–3,300 m. It was centered in Canada but extended southward into the northern tier of the United States, and a little farther in the Great Lakes and New

England areas. This barrier against the southward sweep of air from the Arctic would, one would reason, result in milder winters than today in central North America. This would be especially so since north winds would suffer nearly 300 mbar compression and adiabatic warming in descending from the ice cap. Even if the temperature at the ice crest were as low as that at the South Pole today, north winds in the Dakotas in January would have been no colder than today, and such low crest temperatures at latitude 50°N or so are unlikely to have prevailed (MORAN, 1972). The fossil evidence appears to bear this out.

At the same time, oxygen isotope measurements in the American tropics suggest temperatures only slightly cooler (perhaps 2°C) than today.

The fossil evidence, both floral and faunal, suggests considerably cooler weather (3°–6°C) in central North America during summer. One example is given in Fig.31, which gives the mean July temperature in Dakota County, Minnesota, as calculated by multivariate analysis from fossil pollen assemblages (WEBB and BRYSON, 1972). Other evidence found in numerous references, such as MEHRINGER (1967), supports the idea of milder winters and cooler summers resulting in an overlap of heat-limited and cold-limited species.

MORAN (1972) found, on carefully summarizing the evidence, that the mid-summer circulation pattern of late-glacial time should have been very similar to the present March circulation pattern. The July precipitation pattern then appears to have been similar to the present March pattern also. The temperatures were not as low as present March temperatures, even though colder than present July temperatures. He also found

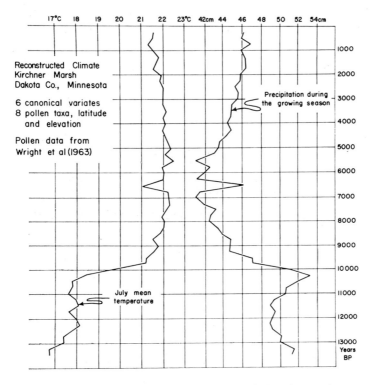

Fig.31. July mean temperature and precipitation during the growing season in Dakota Co., Minnesota (Kirchner Marsh) reconstructed from a pollen profile, using the canonical correlation technique of WEBB and BRYSON (1972). No correction has been made for lag of pollen rain change behind climatic change.

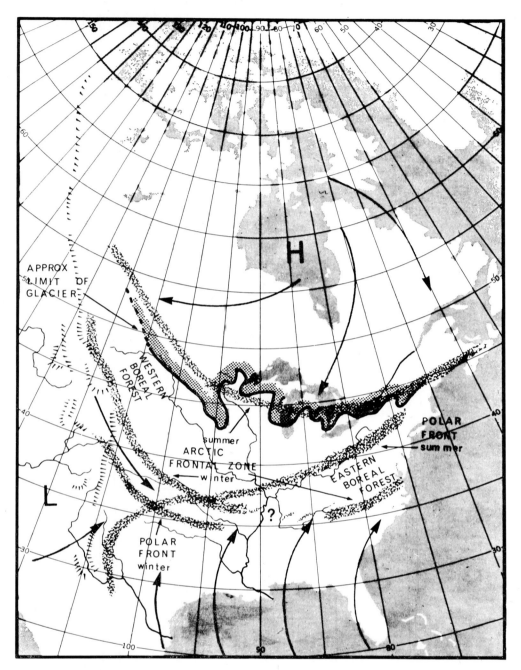

Fig.32. Dominant mean airstream boundaries in late-glacial time, tentatively reconstructed from a variety of fossil data. From BRYSON and WENDLAND, 1967.

that the extant literature indicated a compression of the climatic regions of eastern North America into a narrower band between the ice front and the Gulf of Mexico. From this type of data BRYSON and WENDLAND (1967) reconstructed the late glacial mean "frontal" positions shown in Fig.32.

It is especially significant that the Late Pleistocene monthly climatic means appear to have been not much different than extreme individual months at the present time. It apparently takes only a changed frequency and combination of present-day weather patterns to produce an ice-age climate.

**The historic climate of North America**

Fig.31, which was introduced to show the Late Pleistocene summer temperature depression, also shows that after the rapid change at the beginning of the Holocene there was rather little change in mean mid-summer temperatures. However, there were important changes in other climatic parameters such as rainfall during the growing season. Looking

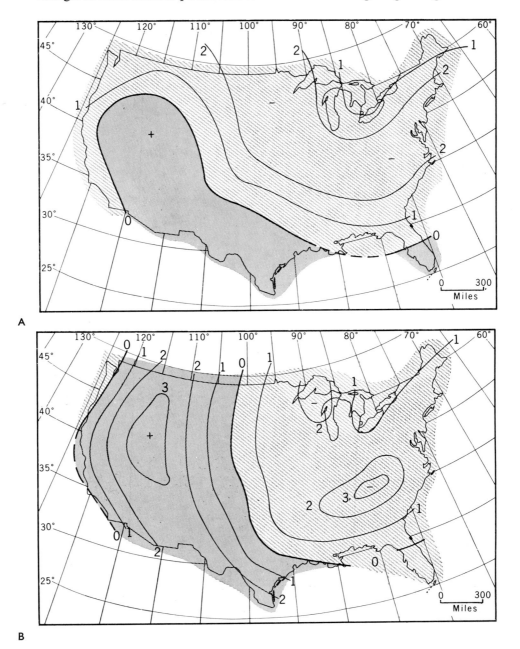

A

B

at the most recent two millennia, however, one sees that they are more like the Pleistocene than the intervening 8,000 years. Evidence on glacier advances suggests that of the past two millennia, the period between about 1600 and 1900 has been the most "glacial". The hemispheric warming of the first half of the 20th century appears to have been a temporary respite, according to LAMB (1966).

If Lamb is correct, and later evidence suggests that he is (MITCHELL, 1970), then an examination of the climatic character of the mid-19th century might provide a better estimate of the climate of coming decades than the 1930–1960 normals. This has been done in part by WAHL and LAWSON (1970).

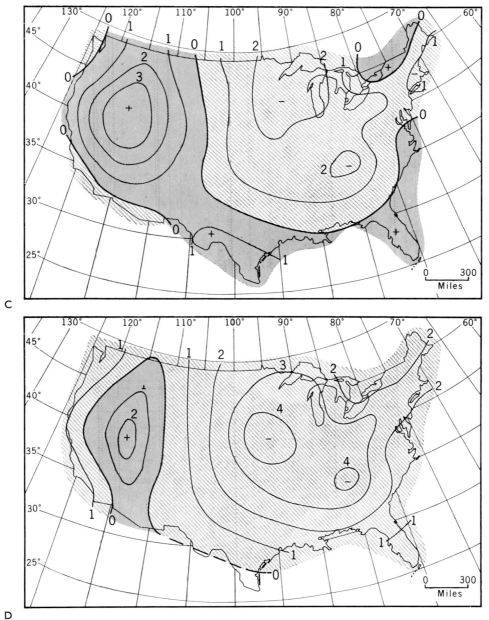

Fig.33. Temperature deviations (in °F) of the 1850's and 1860's from the 1931–1960 climatic normals, after WAHL and LAWSON (1970). A. Winter (January through March); B. Spring (April through June); C. Summer (July–August); D. Early fall (September–October).

43

Wahl and Lawson found that the 1850–1869 temperature anomaly patterns over the United States in most seasons were similar, warmer in the intermontane region and Texas and cooler on the west coast and the eastern half of the country than the 1931–1960 normals (Fig.33). Only in September and October (early autumn) was Texas cooler like the eastern regions, and Arizona–New Mexico in November and December. Seasonal differences ranged up to about 2°C warmer in the west in summer and 2°C cooler in the middle of the country in early autumn. These are significant departures for seasonal means, and must have meant a shorter growing season in the "corn belt" of the United States.

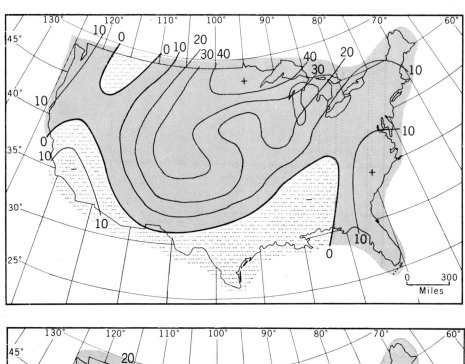

A

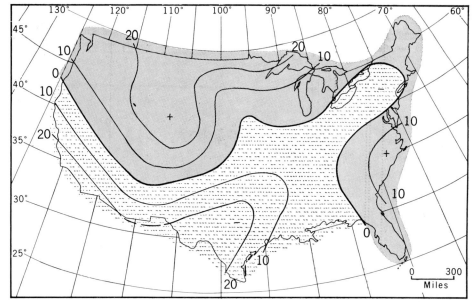

B

The precipitation differences between the means for the 1850–1869 period and the 1931–1960 normals are even more striking (Fig.34). The annual means were up to 20% higher in the Cordilleran region, 40% higher in the northern Great Lakes area in winter, 30% higher in the southern Rockies in summer, but 20% lower in California in fall, winter and spring. If the present climatic change indeed brings a return to mid-19th century precipitation patterns, this drier "wet season" would be most unfortunate for southern California which was quite dry even in the 1931–1960 normal period.

When one considers that the climatic changes illustrated by these differences over a century are part of a pattern of global changes, and that the changes were perhaps a third

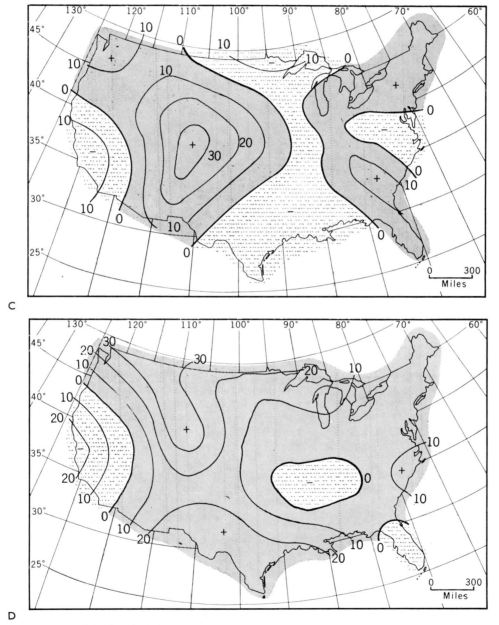

Fig.34. Precipitation deviations (%) of the 1850's and 1860's from the 1931–1960 climatic normals, after WAHL and LAWSON (1970). Seasons as in Fig.33. A. Winter; B. Spring; C. Summer; D. Early fall.

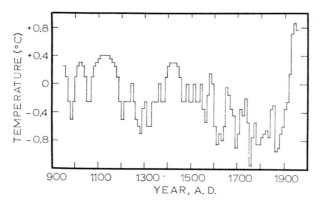

Fig.35. Running 20-year means of mean annual temperature in Iceland, derived primarily from sea ice data, after BERGTHORSSON, 1962.

of those that would mark the difference between the present and a glacial climate, one is struck by the sensitivity of the climatic system.

In order to put these changes in a longer perspective, one might consider that these changes are part of a global pattern of change which was associated with a mean annual temperature change for the Northern Hemisphere of about half a degree Celsius (MITCHELL, 1970). If one compares this hemispheric mean annual temperature variation with that for Iceland reconstructed by BERGTHORSSON (1961), one finds that the two variations were rather parallel for the last century, but with the amplitude of the Icelandic variation several times as great (Fig.35). Since Bergthorsson's reconstructed series of Icelandic mean annual temperatures goes back a thousand years, one may put the climatic variation of this century in the perspective of a millennium. A brief glance at Fig.35 suggests that the "normal" period of 1931–1960 was the most abnormal thirty-year period in the last thousand years! There is clearly much research ahead for the climatologist.

## References

BERGTHORSSON, P., 1962. Preliminary notes on past climate of Iceland. *Informal Notes, Conference on the Climate of the 11th and 16th Centuries*. Aspen, Colo., June 16–24, 1962.

BLODGET, L., 1857. *Climatology of the United States and of the Temperate Latitudes of the North American Continent*. Lippincott, Philadelphia, N.Y., 536 pp.

BORCHERT, J. R., 1950. Climate of the central North American grassland. *Ann., Assoc. Am. Geograph.*, 40: 1–39.

BROOKS, C. E. P., 1950. Selective annotated bibliography on climatic changes. *Meteorol. Abstr. Bibliogr.*, 1: 446–475.

BRYSON, R. A., 1966. Air masses, streamlines and the Boreal forest. *Geograph. Bull. (Can.)*, 8: 228–269.

BRYSON, R. A. and KUHN, P. M., 1961. Stress-differential induced divergence with application to littoral precipitation. *Erdkunde*, 15: 287–294.

BRYSON, R. A. and WENDLAND, W. M., 1967. Tentative climatic patterns for some late-glacial and post-glacial episodes in central North America. In: W. J. MAYER-OAKES (Editor), *Life, Land and Water*. Univ. of Manitoba Press, Winnipeg, xvi + 414 pp.

BRYSON, R. A., WENDLAND, W. M., IVES, J. D. and ANDREWS, J. T., 1969. Radiocarbon isochrones and the disintegration of the Laurentide ice sheet. *Arctic Alpine Res.*, 1(1): 1–13.

BRYSON, R. A., BAERREIS, D. A. and WENDLAND, W. M., 1970. The character of late-glacial and post-glacial climatic changes. In: W. DORT JR. and J. K. JONES JR. (Editors), *Pleistocene and Recent Environments of the Central Great Plains*. University Press of Kansas, Lawrence, Kans., pp.53–74.

CRUTCHER, H. L. and HALLIGAN, D. K., 1967. Upper wind statistics of the northern Western Hemisphere. *ESSA Tech. Rep.*, EDS-1, 20 pp. + 100 charts.

HAMILTON, W. L., 1972. Atmospheric turbidity and surface temperature on the polar ice sheets. *Nature*, 235: 320–322.

KLEIN, W. H., 1957. Principal tracks and mean frequencies of cyclones and anticyclones in the Northern Hemisphere. *U.S. Dept. Commerce, Weather Bur., Res. Pap.*, 40: 60 pp.

KUTZBACH, J. E., 1970. Large-scale features of monthly mean Northern Hemisphere anomaly maps of sea level pressure. *Monthly Weather Rev.*, 98: 708–716.

LAHEY, J. F., BRYSON, R. A., CORZINE, H. A. and HUTCHINS, C. W., 1960. *Atlas of 300 mb Wind Characteristics for the Northern Hemisphere*. University of Wisconsin Press, Madison, Wisc.

LAMB, H. H., 1966. Climate in the 1960's. *Geograph. J.*, 132: 183–212.

LAMB, H. H., 1970. Volcanic dust in the atmosphere; with a chronology and assessment of its meteorological significance. *Phil. Trans. Roy. Soc. London, Ser. A*, 266: 425–533.

MEHRINGER JR., P. J., 1967. The environment of extinction of the Late Pleistocene megafauna in the arid southwestern United States. In: P. S. MARTIN and H. E. WRIGHT JR. (Editors), *Pleistocene Extinctions*. Yale University Press, New Haven, Conn., pp.321–336.

MITCHELL, V. L., 1969. *The Regionalization of Climate in Montane Areas*. Ph.D. thesis, University of Wisconsin, Madison, Wisc., 147 pp.

MITCHELL JR., J. M., 1970. A preliminary evaluation of atmospheric pollution as a cause of the global temperature fluctuation of the past century. In: S. F. SINGER (Editor), *Global Effects of Environmental Pollution*. Springer-Verlag, New York, N.Y., pp.139–155.

MITCHELL JR., J. M. and KISS, E., 1964. Annotated bibliography on climatic changes. *Meteorol. Geoastrophys. Abstr.*, 20: 2236–2284; 2434–2478.

MORAN, J. M., 1972. *An Analysis of Periglacial Climatic Indicators of Late Glacial Time in North America*. Ph.D. thesis, University of Wisconsin, Madison, Wisc., 160 pp.

NAVAIR, 1966. Components of the 1000-mb winds (or surface winds) of the Northern Hemisphere. *NAVAIR Document 50-1C-51, Wash.*, 74 charts.

OGDEN, J. G., III, 1967. Radiocarbon and pollen evidence for a sudden change in climate in the Great Lakes region approximately 10,000 years ago. In: E. J. CUSHING and H. E. WRIGHT JR. (Editors), *Quaternary Paleoecology*. Yale Univ. Press, New Haven, Conn., pp.117–127.

RASMUSSON, E. M., 1968. Atmospheric water vapor transport and the water balance of North America, II. Large-scale water balance investigations. *Monthly Weather Rev.*, 96: 720–734.

SOULES, S. D. and NAGLER, K. M., 1969. Two tropical storms viewed by Apollo 7. *Bull. Am. Meteorol. Soc.*, 50: 58–65.

WAHL, E. W. and LAWSON, T. L., 1970. The climate of the mid-19th century United States compared to the current normals. *Monthly Weather Rev.*, 98: 259–265.

WEBB, T., III and BRYSON, R. A., 1972. Late- and post-glacial climatic change in the northern midwest, U.S.A.: quantitative estimates from fossil pollen spectra by multivariate statistical analysis. *Quaternary Res.*, 2: 70–115.

WENDLAND, W. M. and DONLEY, D. L., 1971. Radiocarbon–calendar age relationship. *Earth Planetary Sci. Letters*, 11: 135–139.

WINSTON, J. S. (Editor), 1959. Hurricane forecasting. *U.S. Dept. Commerce, Weather Bur., Forecasting Guide*, 3: 108 pp.

WRIGHT JR., H. E., WINTER, T. C. and PATTEN, H. L., 1963. Two pollen diagrams from southeastern Minnesota: problems in the late- and post-glacial vegetational history. *Geol. Soc. Am. Bull.*, 74: 1371–1396.

*Chapter 2*

# The Climate of Canada and Alaska[1]

F. KENNETH HARE AND JOHN E. HAY

**Preface**

We present in these pages a description of the physical climate of northern North America. We have confined ourself, for the most part, to the larger-scale distributions, and have not attempted much local detail. We have also avoided a treatment of ecological questions, because these seem to us to be too uncertain for a broad-ranging review. Our central object has been to analyze, map and then account for the chief physical distributions of the atmospheric or surface-exchange parameters.

Canada and Alaska are huge countries. Counting the enclosed seas and lakes we have treated an area of over 13 million square kilometers, of which over 11 million are claimed as U.S. or Canadian territories. These figures represent 2.5% of the earth's entire surface and 7.3% of the earth's land surface, respectively. Faced with this immensity we have not attempted regional or local treatments. Suitable references on these scales are included in the list of references on pp. 188–192. The limitations of our large-scale approach are very obvious in both text and charts.

We have used, for the most part, the standard climatological elements and parameters used in the other volumes of the survey, and prescribed by the World Meteorological Organization. Our view of world climate is considerably colored by this standardization of observational technique and statistical practice. Originally these were designed to minimize local influences, and to serve the needs of the synopticians. If we had begun our study of world climate with an ecological objective, we should probably have adopted a quite different scheme of standardization. This bias is unavoidable, and does little harm if one keeps it in mind.

In one major area, however, we have gone well beyond standard climatology and have presented the results of extensive research. This is the moisture and energy balance of the surface, which we have treated starting with John Hay's detailed study of the climatology of radiation, both solar and terrestrial. In this we have been trying to follow the magnificent lead given by Mikhail Budyko and the U.S.S.R. school of climatologists. We have also been much influenced by the teaching on this continent of Warren Thornthwaite, who advocated such an approach nearly 30 years ago; of Heinz Lettau, who has recently shown us how to escape from too much empiricism; and of Reid Bryson, who has regained for broad-scale climatology the reputation it lost after the golden age of Von Hann and Köppen.

---

[1] With a contribution by Wolfgang Baier (pp.178–187).

We have presented an analysis of various climatic hazards, such as freezing rain and blowing snow. These are quite properly regarded as elements of the physical climate. They are, however, of trivial importance except in the context of man's economic activities, which has not been our usual criterion of choice. Our only excuse is that they could be treated very readily on the large scale by the methods that we were using, and it seemed a pity not to include them. But there are many other aspects of applied climatology that we have ignored. In particular we have been forced to omit the rapidly growing and important field of urban climatology. To treat this and related questions properly would call for another volume.

We have received help from so many persons that it is embarrassing to be able to mention only a few. Morley Thomas and Gordon McKay have been endlessly generous in providing climatological data for Canada, and have put at our disposal many research results obtained by the Climatological Division of the Atmospheric Environment Service. Bruce Findlay and David Phillips, also of that Service, have helped by providing data and reading our text. In Alaska we acknowledge the ready help of the Regional Climatologist, Harold Searby, and of Harry Hulsing, District Chief of the Geological Survey (Water Resources Division). Wolfgang Baier of the Canada Department of Agriculture (Agrometeorology Section) has helped us greatly in the treatment of seasonal moisture regimes. Finally, Harold Crutcher at the Environmental Data Service at Asheville has provided much of the remaining background material.

We make the point in closing that this account, like all the others of the World Survey, depends on the members of the climatological services of the two countries, from local observers to senior officers. Without their work, nothing of what we have done could have been started.


## Introduction

The immense size of Canada and Alaska staggers the imagination, and complicates the task of writing the area's climatology. St. Johns, Newfoundland, is as close to Moscow as it is to the mouth of the Yukon River. Venezuela and Ellesmere Island are equidistant from Lake Erie. Alaska and Quebec Province are each larger than France, Spain, Portugal and Germany (east and west) combined. We are concerned with the latitude belt 45°–80°N between the longitudes 55° and 170°W. The 11 million km² of land contained within that belt contain only a little over 20 million persons. So we shall be dealing not merely with vastness, but with largely empty vastness.

The southern limit of the area happens to lie close to the mean annual core of the westerly belt. Canada and Alaska may thus be expected to have a climate typical of the cyclonic flank of the westerlies, or (in the north) of the core of the westerly vortex itself. They do indeed have such a climate, in a guise sharply influenced by continentality, for the highest mountain barriers of the area lie along the Pacific rim. The usual type of continental climate is one where the temperature cycle dictated by the radiative heat balance is not tempered by free air mass exchanges with the oceanic areas. We should hesitate to say that this was Canada's case, or even Alaska's, because Pacific air actually penetrates quite deeply. Yet the mountain barrier still gives the climate its unique flavor, by imposing long-wave disturbances and other dynamical constraints on the free flow of the wester-

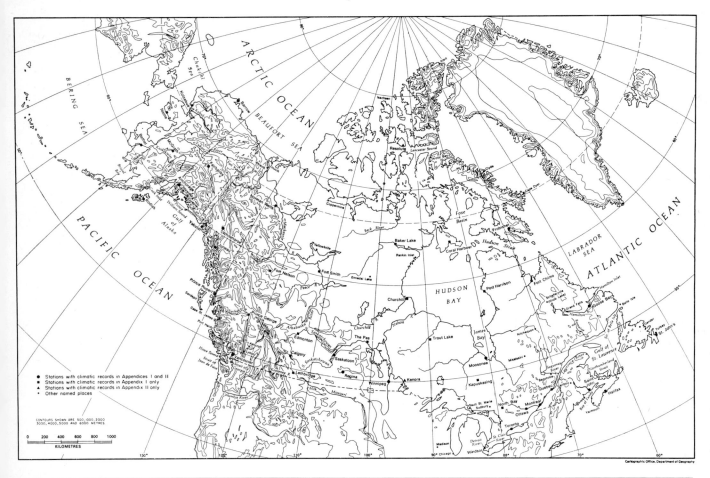

Fig.1. Place-name charts, Canada and Alaska.

lies. It is in this special sense that the climate is continental.

In all except the extreme southwestern parts of Canada, the area's climates are harsh and prone to violent contrasts. Their closest parallels are in northern and eastern Siberia, from the Korean border to the Arctic Ocean. Harsh and austere though they may be, these microthermal and Arctic climates have allowed a rich and varied human economy to develop. The northern forests, for example, happen to be well adapted to use in the pulp and paper industry, and the winter snowfall helps not only in the movement of logs to mill, but in the generation of power. The Canadian prairies, in spite of their cold and prolonged winter, have a combination of high radiation income and moderate summer rainfall ideal for the cultivation of high protein wheats. What they lack in comfort, the climates of northern North America thus make good in their challenge to the ingenuity of man.

**Geographical controls**

The climates of the area depend primarily on the radiative exchanges between surface, sun and atmosphere, and on the circulation of the troposphere. The precise manner in which these controls work, however, is very much influenced by certain aspects of the physical geography, notably the relief, vegetation and surface wetness, and on the surface properties of the surrounding seas. For a comprehensive review of these questions, we refer to works by HARE (1950a; 1968) PÉWÉ (1965), BIRD (1967), WARKENTIN (1968), SATER (1969) and DUNBAR (1951). Here we shall confine ourselves to the briefest sketch of the geographical controls.

The chief relief features of concern to the climatologist are the western Cordillera, the interior plains, the Canadian Shield, the Arctic Archipelago, the St. Lawrence lowlands and Atlantic Canada.

Alaska, Yukon, parts of Mackenzie district (the river itself flowing within the Cordilleran province for part of its journey to the Arctic) and British Columbia lie within the western Cordillera of North America. This massive western bulwark is a major control of the hemispheric circulation, blocking low-level penetrations by moist, cloudy airstreams, and imposing long-wave perturbations on the tropospheric flow over the entire continent. Pacific air as such is not excluded by the barrier, but the air that does penetrate the interior tends to come from well above sea-level, and hence has a rather lower vapor content.

In detail the western Cordillera are made up of three primary units. The coastal ranges extend from the Aleutians through the Kenai, Alaska, Chugach, Wrangell and St. Elias ranges to the Coast Range of British Columbia, and the offshore islands. This coastal barrier is high and massive. East and north of it are high, rolling plateaus drained by such great rivers as the Yukon, Peace, Fraser and Columbia. The easternmost part of the Cordillera is a further mountain barrier including the Brooks and Richardson ranges, and the Mackenzie, Rocky, Purcell, Selkirk and Monashee Mountains. The Cordilleran barrier is thus dual, and its duality can be detected in both the dynamic and the physical climatology of the continent.

The plains to the east extend from the Arctic coast east of the Mackenzie Valley to the prairies along the U.S. border. The southern part of the plains is tilted eastwards, falling from about 1,300 m in western Alberta to about 250 m in Manitoba.

The Canadian Shield, making up the vast core of Canada, with Hudson's Bay in its sunken center, is a rigid shield of ancient metamorphic rock (chiefly granite) whose surface is a monotonous plain much roughened in local relief by recent glaciation. Its surface is strewn with a chaos of lakes, bogs and thin deposits of glacial sand and till. The outer rims tend to be uplifted and rugged, especially north of Lake Superior, and along the Atlantic coast and the Gulf of St. Lawrence. These rims have some effect on precipitation distribution, but do not much affect the pattern of circulation.

Southernmost Ontario and the Montreal–Ottawa–Quebec lowland areas are parts of the subdued central lowland of North America. The Atlantic provinces and the eastern Townships-Gaspésie region of Quebec form a series of plateaus and basins of considerable relief. Again the climatic influence is mainly on precipitation distribution.

Off the mainland of Canada is the unique Arctic Archipelago, where the drowning of ancient river systems has produced a tangle of islands separated by narrow, ice-choked channels of the sea connecting Baffin Bay with the Arctic Ocean. The largest eastern islands—Ellesmere, Devon and Baffin—have extensive glacierized mountain areas, and beyond Baffin Bay the huge mass of Greenland contributes a barrier effect. But this area does not lie across the main course of the westerlies, and it may be that the abundance of high land and ice affects the radiation climatology more than the flow of the atmosphere.

The major vegetation belts of Canada and Alaska are, of course, themselves functions of the ambient climate, but the effect is to some extent mutual. The formations widely represented include the tundra, the treeless response to the Arctic climate; the Boreal forest, the belt of spruce–fir–larch–pine forest extending from interior western Alaska to Labrador and Newfoundland, continued southward in the western Cordillera (at considerable altitudes) by the very similar Subalpine forest; the Pacific rainforest, dominated by giant conifers such as the Sitka spruce, the western red cedar and the Douglas fir; the dry forest and semi-arid scrub of the British Columbia dry belt; the grassy prairies of southern Alberta, Saskatchewan and Manitoba; and the Great Lakes–St. Lawrence–Acadian forest, dominated by maple, beech, basswood, oak, white pine and hemlock (with red spruce in the east) from the Lakes to Nova Scotia.

Permanently frozen ground underlies much of northern Canada and Alaska. Continuous permafrost extends north of a line from the Seward Peninsula and the southern flanks of the Brooks range to the eastern tips of Great Bear and Great Slave lakes, and then to Churchill, Manitoba. East of Hudson's Bay the Ungava Peninsula has continuous permafrost. A belt of patchy permafrost extends 500–1,200 km south of this line. Well-developed forests grow in this patchy zone, and it is possible that permafrost may actually be created by the shading of the floor caused by thick spruce forest (VIERECK, 1970).

The marine influence in the area's climate depends on the surface characteristics of the sea concerned. The Pacific is clearly of paramount importance, because Pacific airstreams are dominant in the continental climate for much of the year. The Gulf of Alaska and the nearby open ocean, from which these airstreams come, are ice-free throughout the year, and there is a considerable flux inland of heat and moisture derived from the ocean surface during the cooler seasons. Actually the North Pacific is not notably warm for its latitude, and the effect is less strong than the equivalent Atlantic effect in northwest Europe. In summer especially onshore airstreams are relatively cool, though they do carry much moisture into Alaska and Yukon at this time of year.

The Bering Sea is much colder, and has an extensive winter ice-cover. Its bleak, fog-shrouded surface in summer keeps the Aleutians and west coastal Alaska cold, moist and cheerless—an influence that prevents the well-developed Boreal forest of the interior valleys reaching the west coast, whose surface is a soggy tundra. Hudson's Bay has a similar influence inland. It is frozen almost completely from mid or late December until late June, and during the brief summer melting pack-ice keeps surface temperatures near freezing until late August. The coastal regions are hence badly chilled, and the Arctic tree-line is pushed south to the northern shores of James Bay near 55°N. This enormous cold water body has a marked effect on the summer climate of all eastern Canada. In autumn its cooling but unfrozen surface contributes to the high cloudiness and frequent snowfall of nearby areas (an effect also seen in the Bering Sea).

The Atlantic plays a smaller role in the area's climate, partly because onshore flow is common only in spring and summer, and partly because there is a broad belt of south-ward-drifting Arctic water for over 250 km offshore, the Canadian and Labrador currents. Heavy pack and some bergs (from Greenland) are carried south in this current as far as northern and eastern Newfoundland (reached in January). Only between July and November does the pack thin out, and the water remains cold and berg-laden throughout this period. Onshore Atlantic airstreams hence reach eastern Canada over a cold surface, and are laden with fog and heavy stratus in summer. On the other hand in winter prolonged spells of easterly wind partially break up the pack-ice, and permit unusually high temperatures and humidities to penetrate inland.

Finally, the Arctic Ocean, to which Alaska and the outer Arctic islands of Canada present coastlines, is covered by permanent pack-ice about 3 m thick. The pack partially thaws in summer, when net radiation is positive, and its surface is then wet and hummocky. At the same time a narrow channel of open water extends along the Alaska coast to well east of the Mackenzie delta. For the rest of the year the pack-ice is frequently broken by various stresses, but the numerous leads so created rapidly refreeze. The channels of the Arctic Archipelago are choked partly by heavy pack drifting out from the Arctic Ocean, and partly by locally formed winter ice. The drifting "ice-islands" of the Arctic Ocean, which have played a major role as weather stations for the past twenty years, are fragments of thick shelf-ice carried away from the shore of Ellesmere Island.

## Airstreams and synoptic-scale disturbances

### Nature of airstreams and disturbances

The seasonal variations of the mean flow described in Chapter 1 establish the broad background of the dynamic climate. They also account for some of the larger aspects of precipitation and temperature distribution. But they fail to explain the rapid day-to-day weather changes so characteristic of northern North America. These are caused by various types of disturbances, to whose behavior we now turn.

The major weather-forming disturbances are the traveling cyclones and anticyclones of the daily weather chart. These surface systems correspond to moving waves—the so-called cyclone waves—in the westerly current, of circumpolar wave-numbers 5 to 11, with wavelength of about 70° to 30° longitude. A trough in the westerlies moves east-

TABLE I

CANADIAN OPERATIONAL AIRMASS SCHEME

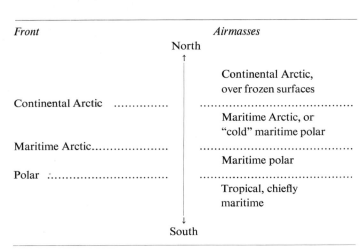

| Front | | Airmasses |
|---|---|---|
| | North ↑ | |
| | | Continental Arctic, over frozen surfaces |
| Continental Arctic ............... | ........................................... | |
| | | Maritime Arctic, or "cold" maritime polar |
| Maritime Arctic.................... | ........................................... | |
| | | Maritime polar |
| Polar ................................ | ........................................... | |
| | | Tropical, chiefly maritime |
| | South ↓ | |

wards just behind each cyclone center, and a ridge follows closely each moving anti-cyclone. The cyclones form along the more baroclinic parts of the westerly current, and hence involve the jet-streams in the moving waves, with associated warm and cold fronts at lower levels. Strong uplift of the warmer air, chiefly near these fronts, produces most of the dense cloud and precipitation typical of the systems.

Northern North America is affected in all seasons by a constant procession of these synoptic-scale disturbances. Coastal Alaska and British Columbia receive great numbers of Pacific cyclones, usually near the peak of their development, and hence have a very disturbed climate. Canada east of about 90°W receives another stream of vigorous and growing cyclones from the southwest, and along the Atlantic seaboard these may be very intense, since they tend to travel up along the United States coast, deepening over the open water. In between these two great cyclonic realms is a large belt in which the disturbances are still very numerous, but are fast-moving and much less intense. Western Canada east of the Rockies and interior Alaska also see many traveling anticyclones, moving chiefly out of the northwest.

Canada and Alaska are affected at various times of year by all four major airstreams; i.e., Arctic, cool Pacific, mild Pacific and tropical, the latter in Canada only. The airstreams do not move about the maps as solid masses, as do airmasses of the traditional kind. They possess shear along the vertical, and any given column of air is tilted and then in effect ruptured by this shear, so that it is quite misleading to think of them as being homogeneous masses drawn from source regions (MCINTYRE, 1950, 1955). Canadian operational analysts systematically divide the broad westerly flow by means of three conventionalised fronts (Table I) (PENNER, 1955; ANDERSON et al., 1955).

This scheme was devised primarily to allow frontal contour charts to be drawn that would objectively demonstrate the three-dimensional structure of weather systems affecting North America. It succeeds in this purpose very well (PENNER, 1955; LONGLEY, 1959) and has given a new insight into these structures. In our account of the mean flow we have already identified the principal airstreams and fronts. They relate to those shown in Table I as follows:

| *Canadian operational usage* | *Usage in this volume* |
|---|---|
| Continental Arctic airmass | Arctic airstreams |
| Continental Arctic front | Arctic front |
| Maritime Arctic airmass | Cool Pacific airstreams |
| Maritime Arctic front | Midwesterly front |
| Maritime polar airmass | Mild Pacific airstreams |
| Polar front | Pacific front |
| Maritime tropical airmass | Tropical or southern anticyclonic airstreams |

The splitting of Pacific airstreams into two classes at most seasons is necessary to explain the existence of the midwesterly jet, which enters North America near Vancouver Island (see pp.5–8).

Finally, airstreams from the northern North Atlantic may penetrate Canada from the east, occasionally reaching the west coast of Hudson's Bay. Associated with anomalies in the mean flow, these Atlantic streams have to be added to the four main streams above.

Many of the traveling disturbances described below involve several if not all the above fronts and airstreams, and hence appear far more complex on analyzed charts than do European systems.

*The Pacific coast*

The entire Pacific coast of North America, from the Mexican border to the Aleutians, is exposed to a stream of cyclones that often attain great intensity. Some develop over the western Pacific, often in the intensely baroclinic zone south of Japan. These tend to be large, occluded vortices, often very intense, and are commonest in the Aleutian–Gulf of Alaska area. Others develop in mid-Pacific, and may strike the British Columbian coast at the peak of their development, with severe outer coastal gales. Still others form not far off the coast, and are in the early stages of deepening as they arrive: many of those entering North America south of the United States–Canadian border are of this kind. Stationary cold lows may also lie off or near the coast for periods of days. Those occurring south of 45°N are mainly of the cut-off variety, with a westerly flow entering British Columbia or Alaska well to the north.

In autumn and winter the frequency of such disturbances is very high. Fig.2 shows the annual course of the meridional temperature and pressure gradients in mid-troposphere (5.0–5.5 km, 500-mbar surface), as indicated by height and temperature differences between Port Hardy and Aklavik-Inuvik, which are 1,950 km apart, roughly along the 130°W meridian. For comparison's sake comparable differences are given for the Caribou, Me.–Frobisher, Baffin, meridian in eastern Canada. The height differences are roughly proportional to the mean westerly zonal wind passing inland from the Pacific and off-shore to the Atlantic, respectively. The temperature differences are a rough measure of the baroclinity, and hence of the increase of zonal wind with height. This level in the troposphere is a good indicator of the "steering" of the synoptic systems.

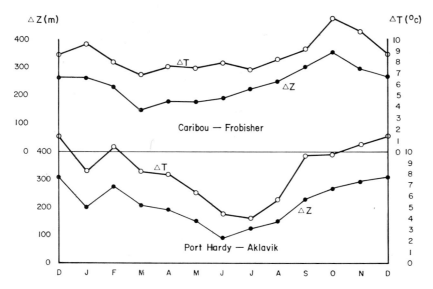

Fig.2. Differences of mean temperature ($\Delta T$) and mean height ($\Delta Z$) at 500 mbar across westerlies. The differences are taken along meridians over distances of 1,950 km, the chosen radiosonde stations being shown in the diagram. The $\Delta Z$ curve is an approximate index of the mean strength of the westerlies, and the $\Delta T$ curve governs the rate of increase with height at this level, in each case averaged over the whole distance (data from TITUS, 1967, vol.2).

Clearly there is a sharp increase in baroclinity and mean zonal flow in September, which is a month of rapidly increasing storminess, cloud and rainfall along the whole Pacific coast. The baroclinity reaches a broad cooler season plateau thereafter, although the highest value of the crude index employed is in December, which is also the month of strongest zonal flow, and also of heaviest rainfall in the southern third of the coast. Further north, maximum rainfall tends to come earlier in the autumn. Throughout the season fresh Pacific storms affect the coast at about 48-h intervals, and prolonged interruptions in the cycle are rare.

By late November, however, the Arctic front often lies along the Coast Range and the coastal mountains of Alaska, Arctic air occupying the valley systems and covering the chief plateaus, often with high sea-level pressures and appreciable ridging up to mid-troposphere west of the Pacific coast. At such times the Pacific storms tend to stagnate in the Gulf of Alaska, or to affect the United States coast. Heavy snow may occur in southern British Columbia from such systems.

In spring—essentially from mid-February onwards—zonal circulation and baroclinity both decline rapidly, as does the frequency with which Pacific cyclones approach the coast. By June the zonal flow has reached its annual minimum, kinetic energy over western Canada being only about a tenth of the mid-winter value. Rainfall decreases, as does cloudiness, although from the Queen Charlotte Islands round to the Aleutians coastal Alaska gets substantial rainfall even in June.

Summer is the season of least cyclonic activity along the coast. At this season the sub-tropical high pressure cell lies far north, and the locus of maximum variability of mid-tropospheric wind (CRUTCHER and HALLIGAN, 1967), is in about 55°N across the Aleutian Chain, 10° north of the winter locus. The Pacific cyclones, moreover, come from well north of their winter breeding grounds. Hence the summer is a season of slacker zonal flow, with less frequent and vigorous cyclonic storms than in winter; and the main activity lies over the Gulf of Alaska, the Aleutians and inland Alaska. Southern British Columbia lies south of the main disturbance zone in most summers.

The above account omits to stress the occasional large-scale breakdown of the normal patterns. At all seasons, but especially in spring and summer, strong warm-cored ridges or blocking anticyclones may develop over the Gulf of Alaska, the Pacific west of British Columbia, or the northern interior. These blocking ridges or highs screen the region from Pacific storms, and if persistent can create striking precipitation and temperature anomalies. These aberrations from the normal pattern appear on world charts as strongly amplified long westerly waves. Pacific blocking is less common than Atlantic blocking, but is still a potent climatic factor. It is also common for persistent warm-cored ridges or highs to lie over the interior plateaus of the northwestern United States. If this occurs, the flow of mild Pacific air onshore in British Columbia is very persistent. Periods of drought, like that of the summer of 1967, also usually trace back to this cause. Extreme winter cold, in turn, requires that the blocking ridge lie well out over the Pacific, with a high-level trough over British Columbia (BAUDAT and WRIGHT, 1969).

*The Cordilleran interior*

As each successive Pacific storm strikes the coast, the trough in the mid-troposphere westerlies behind the surface fronts usually continues eastwards across the Cordillera, passing over the valleys and plateaus of the Yukon and inland British Columbia. It is often difficult to trace surface fronts across this physiographically complex region, but the system's passage shows itself in the observed weather. In the deep valley systems only high and medium cloud may be observed, many of the frontal passages producing no rain or snow. In winter especially, however, dense cloud masses and snow occur over the higher mountain systems, whose icefields and cirque glaciers are mainly fed from this kind of moving disturbance. In summer thundershowers commonly break out along the trough line and behind.

The broad interior valleys of Alaska south of the Brooks Range lie well to the north of such passages, except in summer, but troughs corresponding to the surface occlusion sometimes move into the interior from the south, even when bitterly cold air occupies the lower levels. Light snow and thin altostratus or altocumulus cloud decks ensue. In summer the direction of tropospheric flow favors more ready access for these cyclones, whose centers may even move into the interior from the west. The brief summer rains of Alaska arise partly in this way.

The cooler season climate of all parts of the Cordillera, especially from northern British Columbia northward, is greatly influenced by the high frequency of anticyclogenesis. In part this arises from the effects of prolonged radiative cooling of Arctic air trapped in the valleys (WEXLER, 1937; BODURTHA, 1952). But stable anticyclones over the Cordillera require that there be, in addition, strong amplification of the familiar longwave ridge, usually well off the coast; warm mid-tropospheric advection over the Bering Sea and western Alaska; and a steady rise and cooling of the tropopause due to the consequent uplift (HARE and ORVIG, 1958). In this sense, the most intense and stable north Cordilleran highs represent special modifications of the blocking process referred to above. Pacific air may even, in such circumstances, sweep directly round Alaska, reinforcing the

Arctic front over the Beaufort Sea, and allowing mid-winter temperatures near 0°C on the Arctic shores of Alaska and Mackenzie district.

*The Mackenzie–Athabasca valleys, Keewatin and the prairies*

East of the Cordillera lies the broad sweep of physiographically subdued country falling eastwards to the great depression occupied by Hudson's Bay. The upper troughs and intervening ridges move onwards across the huge expanse of land, lake and inland sea. Since the westerlies over this tract normally have a component from north, and are moderately or strongly confluent, the synoptic-scale disturbances typically move rapidly eastwards, often with a drift towards the south. Since vertical motion is not strong, the entire belt has a low precipitation frequency and amount; yet paradoxically it has a high frequency of cyclone passages and frontal activity, and precipitable water is quite high in summer.

The eastern foot of the Cordillera is a strongly cyclogenetic belt. As each synoptic-scale trough arrives from the west, an organized low center forms somewhere in the lee of the mountains. These *Alberta Lows*, as they are familiarly called in Canada, are important throughout the year. They travel rapidly across from the foot of the Cordillera, usually towards the east or southeast, and are the main sources of precipitation west of the 90th meridian.

The exact details of the cyclogenesis are very complex, depending on the location of jet axes, on the amplitude, phase velocity and location of the triggering trough, and on the accidents of the physiography. A study by MOKOSCH (1961) of the 10-year period ending 1960 showed that the frequency of low centers near the mountains was high in the entire belt between 40°N and 65°N, with strong and persistent maxima near 50°N and 59°N. The former was most prominent between October and March, the latter between March and October (in both cases inclusive). Between July and September another maximum occurred south of the international boundary, between 45°N and 50°N. At all seasons there was a perceptible minimum centered on 56°N.

It is not yet possible convincingly to relate these inequalities to relief, or to the baroclinity of the westerlies, though both are almost certainly involved. The preferred latitudes for cyclone centers seem to bear no simple relation to the relief. The maximum near 59°N is opposite the Liard gap, known (BRYSON, 1966) to be a preferred route for Pacific air penetration. The summer maximum in 45°–50°N occurs opposite the relatively open Montana Rockies. On the other hand the 50°N maximum—the most important—lies opposite the massive Alberta Rockies–Selkirk–Purcell complex of high mountains. The relationship to baroclinity, or jet maxima, looks more convincing; the summer maximum at 59°N lies fairly close to the average position of the Arctic front, and the combined 45°–50°N maxima are close to the midwesterly jet axis. A distinct confluence is visible in or near these latitudes in the surface windfield over the prairies for much of the year. But these relationships have not yet been established by a long enough analysis for one to be more specific.

In any case, the Alberta Lows, as they move eastwards, involve the Arctic front in their circulation as well as the midwesterly jet, if present. Strong airstream contrasts are typical, with cool (in summer) or bitterly cold (in winter) Arctic air moving from the east north of the center, and with Pacific air south of the center traveling east or southeast across the plains. The cloud systems are normally not very thick, and precipitation is generally light. Heavy rains or snow do occur, however, at times, especially in Alberta where the easterlies encounter the Rocky Mountain barrier.

As the Pacific air descends from the Cordilleran barrier it is warmed dynamically. In some of the deeper valley systems of Alberta local concentrations of the current become true foehn wind systems, called locally the chinook. Temperatures rise well above freezing, even in winter, melting the thin snow-cover and giving the impression of dryness because of low relative humidity.

The broad warm sectors of Pacific air typical of the Alberta Lows move eastwards across the plains, giving the rather cloudy mild spells of winter (with surface temperatures of −5°C to 10°C) and sunny, warm (25°C to 30°C) and often windy weather in summer. Precipitation falls in the usual fashion through the wedge of cooler air north of the fronts involved, so that to the prairie farmers it seems to come from the east.

The area between the Cordillera, Hudson's Bay and the Great Lakes is also subject to frequent anticyclone passages throughout the year. These highs occur north of the Arctic front in the outbreaks of Arctic air behind many of the Alberta Lows. They are hence in part cold anticyclones, diminishing in intensity with height. Almost always, however, a corresponding ridge occurs in the overlying westerlies (the ridge crest lying behind the surface center), with a high cold tropopause. These anticyclones are, in fact, just as much part of the wave perturbations of the westerlies as the lows with which they alternate. Some develop over northwest Canada, but others can be traced back to the Beaufort Sea or even Siberia (BERRY et al., 1954). Their typical path is southeastwards across the great northwestern lakes—Great Bear, Great Slave, Athabasca—or Keewatin into northern Ontario or the Great Lakes region, and then (often weakening) to the Atlantic or the southeastern United States. A few cross Hudson's Bay and northern Quebec. They bring clear skies, brilliant visibilities and (in winter) eastern Canada's spells of "zero weather".

*Inland and northern Alaska*

The grain of the country turns westwards in Alaska, whose chief mountain ranges—the Brooks Ranges in about 68°N, and the Alaskan Ranges in 60°–63°N—are separated by the deep, broad river basins and plateaus drained by the Yukon, Kuskokwim, Kobuk, Tanana and Porcupine rivers. West of the 150th meridian the grain even swings towards the southwest. The state might therefore be thought highly accessible to airstreams and disturbances moving inland from the Bering Sea, Chukchi Sea and the northeastern tip of Siberia.

In practice such access is common only in the warmer seasons. Mention has already been made of the occasional movement inland of Pacific storms, which produce the chief summer rains. The state also lies, however, on or near the Arctic front for much of the year, and in the summer season this sustains a stream of rather weak cyclones moving from Siberia or the Arctic Ocean. Fig.3 (in part after HARE, 1968) shows the axis of maximum summer frontal frequency, which we identify with the mean Arctic front. REED and KUNKEL (1960) and KREBS and BARRY (1970) show that central and northern Alaska have a distinctly cyclonic regime at this season, causing frequent periods of overcast weather and light rain derived from the moist Pacific air in the warm sectors. Such cyclones also

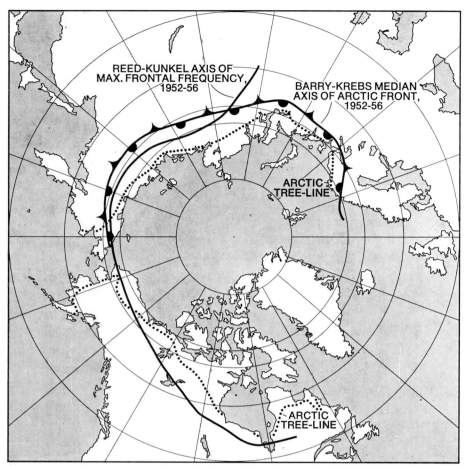

Fig.3. Summer position of Arctic front, in relation to Arctic tree-line (after REED and KUNKEL, 1960; HARE, 1968; KREBS and BARRY, 1970).

move eastwards across the Yukon, the Arctic islands of Canada and nearby Keewatin. From these two sources central Alaska gets a substantial summer rainfall, with a maximum usually in August. Significant precipitation continues well into September or early October. The light summer precipitation of the Yukon, Mackenzie, Keewatin and the Canadian Arctic Archipelago also comes primarily from such cyclones.

*Eastern Canada*

The annual rainfall map (Fig.15, p.88) shows that the southeastern third of Canada—essentially Ontario east of 90°W, Quebec and the Atlantic provinces—have much more abundant rainfall than the regions immediately to the west, and the run of the isohyets suggests correctly that this is connected with the flow into the region of moisture from the Gulf of Mexico, the Caribbean and the subtropical Atlantic. It is often in fact, from airstreams originating over these seas, involved in vigorous cyclonic storms, that the heavier precipitation over eastern Canada falls—sometimes, however, at one remove, in that the water may well have been precipitated and re-evaporated over United States territories before reaching Canada.

61

At times of strong zonal flow throughout the year, eastern Canada is affected by alternations of Pacific and Arctic airstreams, with the fast-moving Alberta Lows involving the Arctic front sweeping across the area. Especially in winter the passage of such systems often alternates with moving highs associated with southward pushes of Arctic air (though with strong zonal flow such pushes do not sweep very far south). Such periods bring to eastern Canada alternations of cold and warmer weather, with cloud and precipitation on the fronts of the moving cyclones. In winter the precipitation is light snow; in summer it comes in the form of showers or thunderstorms from complex and often rather broken cloud systems. It is unusual for heavy precipitation to develop, although occasionally this happens.

When, however, the zonal flow is less vigorous, and especially when the standing long-wave trough over eastern Canada shifts west to lie along or west of the Mississippi Valley, conditions are favorable for cyclonic development in the United States midwest involving in the lower troposphere Pacific air and moist tropical air over the southeast. In summer this latter air is often derived from the tropical Atlantic or Gulf of Mexico, having been swept into the interior on the western flank of the Bermuda High. In autumn and winter the warmer air is more often returning Pacific or Arctic air that has picked up heat and moisture during a prolonged anticyclonic sweep into low latitudes. We have already called these southern anticyclonic airstreams. In spring the airstream contrasts are particularly strong because at this time the cooler stream is often Arctic, and the warmer Atlantic tropical; hence the proneness of spring to violent thunderstorms and tornadoes.

Systems of this sort move into eastern Canada at intervals throughout the year, and are major bringers of precipitation and high winds. In late autumn, winter and early spring their northeastward courses usually lie across the Great Lakes or even further east, and the centers typically travel across southern Quebec and the Gulf of St. Lawrence. They bring heavy snow to regions affected by their warm fronts and in a wide arc north and northwest of their centers. Freezing rain is common near the front itself, since the warm sector air is often well above 0°C at ground level. The mid-winter thaws so typical of the St. Lawrence Valley and maritime provinces are mostly of this origin. As the storms pass out to the Atlantic they often become very intense, bringing strong winds, blowing snow and low temperatures in their wake.

In late spring and summer, however, the more northerly course taken by the centers allows longer periods of very warm and humid weather in southeastern Canada, with maximum temperatures of 30°C or more in the warm sector. Heavy and often violent thunderstorms are characteristic of the cold front belt, often preceded by a squall-line some 100 to 250 km ahead; much of the summer precipitation of the area is of this type. The precipitation along the warm front and north of the center falls from turbulent layer clouds containing local cumulonimbus cells. Intermittent rain, often heavy, with occasional thunderstorms, is typical. Tornadoes occur infrequently, chiefly in spring, primarily in the cold front belt or ahead of it.

An important variant of this class of storm is the Atlantic coast cyclone, which travels up the coast to affect the Atlantic provinces. These storms are often intense, and bring strong winds or gales. Rainfall (or snow and freezing rain in winter) is often very heavy along the entire seaboard, and may also produce moderate or heavy falls along the St. Lawrence Valley, some of whose worst blizzards come from this type of storm. From time to time tropical hurricanes follow a similar path, chiefly between late August and the end of October. Very heavy rain and strong winds result. All classes of Atlantic cyclone, tropical and frontal alike, occasionally turn inland over New York or New England. Thus hurricanes Carol and Hazel in 1954 crossed Montreal and Toronto, respectively, producing intense rainfall, considerable damage and loss of life.

Eastern Canada is also subject to occasional major breakdowns of the above patterns. During periods of Atlantic blocking—when the normal sweep of the westerlies out to sea is impeded by persistent ridges of high pressure in the central and western Atlantic—it is common for easterly airstreams from the Atlantic to spread westward across the area, within the circulation of stationary cold lows over the Gulf of St. Lawrence or further inland. In winter this brings cloudy, very mild weather far inland, thawing temperatures occasionally reaching Baffin Island, all Quebec and all the Atlantic provinces. In the warmer seasons such weather is anomalously cool and showery. Once established the pattern is very persistent, and may in extreme cases dominate the flow for an entire month, as in February, 1947, when positive temperature anomalies of over 10°C were widespread.

A further anomaly that has striking effects is the high incidence in spring—from late February until late May—of strong, stationary anticyclones over Hudson's Bay or vicinity (JOHNSON, 1948). This tendency is persistent enough to dominate the mean charts. Once established, these highs maintain a flow of cold, clear and extremely dry Arctic air across eastern Canada, often from a northeasterly point.

*The Canadian Arctic*

Northern Baffin, the Arctic Archipelago and the enclosed seas usually lie north of all the features described in the foregoing pages, and are affected by the homogeneous Arctic core of the westerly vortex (HARE, 1968). In summer moving cyclones of the Arctic frontal belt often traverse the area, producing the light precipitation typical of the season. At other seasons, however, fast moving wave cyclones, and the zonal currents required to generate them, are usually absent, and the region tends to lie with the circulation of persistent cold lows centered over Baffin Bay, Foxe Basin or even further south. Little weather occurs in such systems, although thin altostratus or cirrostratus sheets are common. Surface temperatures are very low, and in the cold center of the cyclones temperatures may occasionally be below −40°C at all levels from the surface to the upper stratosphere.

**Local wind systems**

Since this study is concerned with the broadest scale of climate we have not attempted to treat in detail the local wind systems of Canada and Alaska, which are of great complexity. A few words about the relations of locally observed winds to the synoptic-scale disturbances just treated will, however, be offered.

In general, surface wind is related to the wind at the top of the planetary boundary layer according to the well-known Ekman spiral relationship. Over uniform, flat inland sites the wind at anemometer-level (10 m) tends to be backed 30°–60° with respect to the wind at 1 km, and to have a speed between one-third and two-thirds as great. The passage of moving cyclones and anticyclones hence produces a highly variable wind regime at ground-level, the wind-rose being rotated cyclonically by about the above angle, and speeds reduced as indicated.

In some areas, however, the surface wind-roses show a high frequency of winds that cannot be related in this way to the movement of synoptic systems. These anomalous surface winds are usually shallow, but may be deep enough to reflect quite deep perturbations of the windflow. We give the name *local winds* to such systems. They are especially common in mountain areas and along coastlines, but can occur in many other sites.

We can classify those observed in Canada and Alaska as follows:

(*1*) Large-scale mountain barrier effects, divisible into upstream and downstream cases. These effects are due primarily to the breakdown of the normal geostrophic relationship with pressure near barriers, and to the imposed vertical motion.

(*2*) Canalized wind systems due to motion through deep valleys along which pressure gradients are imposed by the movement of synoptic systems, where the valley walls prevent the establishment of geostrophic balance.

(*3*) Diurnal wind-systems, usually reversing, such as land-and-sea breezes, and the complex of katabatic-mountain and anabatic-valley winds within mountain areas.

(*4*) Arctic coastal winds.

The first class affords spectacular examples. West of the Coast Range of British Columbia and the Alaskan mountains continuing it northwestward is the well-known cyclonic eddy of the Pacific coast, superimposed on most cases where the general flow is onshore. The effect dominates the surface streamline field at most seasons, and must be regarded as a deep, sub-synoptic eddy imposed by the barrier action of the mountains. From Vancouver to the southeastern coast of Alaska nearly all observing stations on the mainland coast have an overwhelming preponderance of easterly or southeasterly winds. At Vancouver itself these winds account for 64% of all winds in winter, the air being derived from the lower mainland (the Fraser lowlands) to the east. Elsewhere the southeasterlies are usually cloud-laden Pacific air, and may attain severe gale-force as Pacific storms approach the coast.

East of the Rockies, the descent of Pacific airmasses on to the foothills and the prairies leads to the well-known chinooks, which are excellent examples of foehn winds. Recent studies by LONGLEY (1967), BRINKMANN and ASHWELL (1968) and BRINKMANN (1969) have considerably clarified our understanding of these winds. The trans-mountain flow is not itself a local effect, and belongs to the synoptic scale. Brinkmann shows that it occurs under at least two contrasted situations. In the first, true trans-mountain flow occurs as Pacific air traverses British Columbia and the Rockies. The descent of the east face of the Rockies is highly complex, favoring certain valleys, and neglecting areas backed by high mountains; it takes the form of a turbulent leewave, leading in winter to the sudden replacement of much colder, very dry Arctic air by warm, moister (but with lower relative humidity) air from the Pacific. Strong warm advection melts the snow, and gusty surface winds cause some discomfort. The second class, analogous to the anticyclonic foehns of the Alpine forelands, occurs when high pressure dominates the Cordillera, and a strong pressure gradient exists east of the Rockies. Brinkmann finds that the two types produce very similar weather at Calgary, but the source of the warmth is now entirely subsidence of air with high potential temperatures over the foothills and nearby plains.

The second class, where a strong regional pressure-gradient has a component down a conspicuous valley system, consists of strong downgradient winds along the valleys, under antitriptic (i.e., frictional) balance rather than geostrophic. When, for example, high pressure in Arctic air covers the interior valleys and plateaus of British Columbia, pressure remains much lower over the nearby sea, and strong pressure gradients exist across the coast range. These create strong, often violent outblowing winds in the fjords like Howe Sound, whose Squamish wind is of this origin. Strong up-fjord winds sometimes occur in these same valleys in the warmer season when ridges build in from the Pacific.

Even quite shallow valleys demonstrate this canalization. For much of the year, for example, the surface wind at Montreal is from the northeast—i.e., up the St. Lawrence Valley—whenever there is an upvalley component of pressure-gradient (POWE, 1968); and this effect is visible also in such trenches as the St. John River of New Brunswick and the Saguenay. Spectacular examples of the canalized type of local wind are known from southern Alaska, in the Matanuska and Knik River valleys at the head of Cook Inlet. Both Matanuska and Knik winds develop as strong, gusty winds down their respective valleys when there is a low over the Gulf of Alaska (Matanuska wind) or the eastern Aleutians (both Matanuska and Knik winds) and much higher pressure over interior Alaska and the Yukon.

The coastal gales of various sites in the Arctic, chiefly on the hillier islands, are notorious. Many of them are of the canalized type just described, reinforced by katabatic effects, but the species is not enough understood for us to treat it in detail.

Mountain and valley wind systems of the diurnally reversing kind are undoubtedly well-developed in the mountain valleys and slopes of both Canada and Alaska. There is, however, little knowledge of details, and we have again omitted any treatment.

## The thermal regime

All parts of Canada and Alaska lie within what we can properly call the Arctic and microthermal zones of the Northern Hemisphere, short, comparatively warm summers alternating with long and very cold winters (except where oceanic influences allow heat advection to offset the radiatively-induced cold). There are, however, very striking regional differences of thermal regime, and to these we turn in this chapter.

### Temperatures in the free air

As a background to a detailed analysis of surface temperature distribution we shall look briefly at temperatures in the free air, that is to say from about 1 km up.

Fig.4 gives, for January and July, mean temperatures at the 850-mbar and 500-mbar aerological levels. In January temperatures at 850 mbar (1.3–1.5 km) are higher than at ground-level over most parts of the country, the 850-mbar surface being close to the top of the normal winter inversion. Coldest conditions (below $-20°C$) occur over the Arctic islands, Hudson's Bay, Keewatin and Ungava. The meridian of greatest cold is about 80°W. The temperature gradient at 500 mbar (5.0–5.5 km) is less intense, but the cold trough-line is in about the same longitude. At both levels Alaska is warmer than the same latitude of Canada. In both Alaska and Canada this characteristic winter

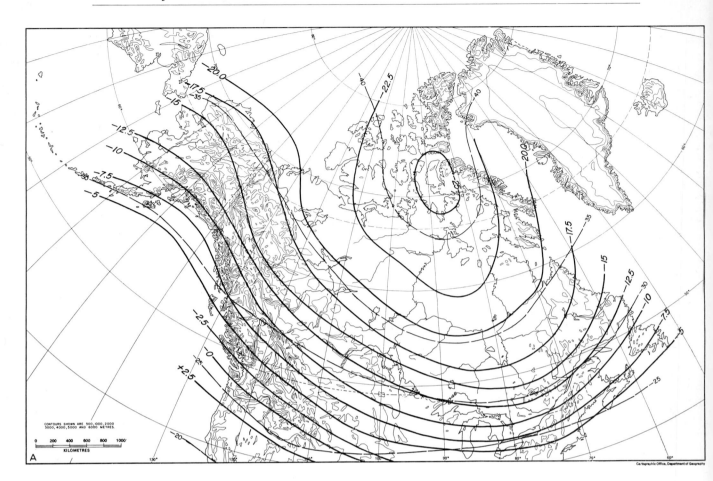

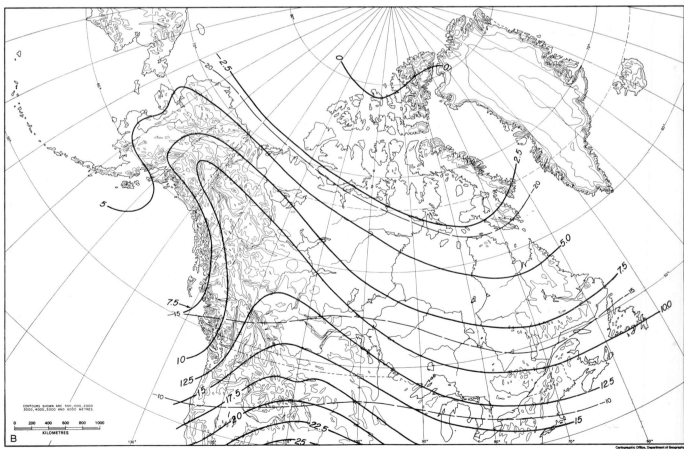

Fig.4. Mean temperature at 850 and 500 mbar, 1951–60. A. January; B. for July. Heavier isotherms are for 850 mbar, which is close to the ground in the western Cordillera. Note the strong warm belt over the Cordillera in July at 850 mbar (mainly after TITUS, 1967).

structure is subject to large anomalies lasting some days or even weeks due to the blocking effect discussed on pp.58–59.

In July the pattern is only crudely similar. At 850-mbar (1.4–1.6 km) the cold eastern trough remains a dominant feature, but lies about 15° of longitude further east, in 65°W. In western Canada and Alaska a remarkable warm belt covers the western Cordillera, temperatures being about 6°C higher than in the eastern trough. This western warmth is sharply cut off on the Pacific coast, the Gulf of Alaska remaining cool. The warmth is presumably the effect of strong heating over the mountain and plateau surfaces, many of which are at or above this level. This view is supported by the fact that the warm belt forms in April and disappears in September. At 700 mbar (about 3 km) and 500 mbar the warmest area lies well east of the mountains, highest temperatures occurring near the Saskatchewan–Manitoba border.

In July the ground surface is substantially warmer than the 850-mbar surface almost everywhere. The lower troposphere has normal lapse rates that are often unstable during midday hours. An exception must be made of the Arctic coastlines and channels, where a shallow inversion is very persistent due to the cold sea surface.

**Air temperature at screen level (1.5 m)**

*The mean fields*

Fig.5 gives the fields of mean air screen temperature for the four standard months, for the period 1931–1960. We shall deal with the various scales of variability later. Needless to say the complexity of relief in mountainous areas makes the maps highly approximate in those areas.

*Winter*

In January lowest mean temperatures (below −30°C) occur over the Queen Elizabeth Island, northwest Baffin, the northern Arctic islands and Keewatin, the coldest meridian being about 100°W. In interior Alaska temperatures are rather warmer with values below −25°C only in isolated upland valleys. South of 55°N there is a distinctly different regime. Coldest conditions are now in about 80°W (apart from a small warm cell over the residual open water in southeastern Hudson's Bay, which vanishes in February). This eastward shift of phase is due: (*1*) to the eastward chilling of Pacific air as it traverses southern Canada; and (*2*) to the high frequency of Arctic outbreaks over and east of Hudson's Bay. Along 50°N, in fact, mean temperatures in southern Alberta are 10°C *higher* than over northern Ontario, although the latter is about 750 m nearer sea-level. Densely populated southern Quebec and Ontario are, for their latitudes, the coldest parts of North America and among the coldest parts of the earth.

The barrier effect of the Cordilleran ranges nearest the Pacific is very clear. Temperatures are above 0°C in the southernmost Aleutians, and in the outermost Alaskan Panhandle. They are above −8°C along the entire coast, except for the deep inlets (e.g., Anchorage, −11°C). In British Columbia above-freezing temperatures extend to the mountain-girt marine waterways. Temperatures are near +3°C at coastal sites all the way from the Queen Charlotte Islands to the U.S. border. These warm coastal temperatures reflect the

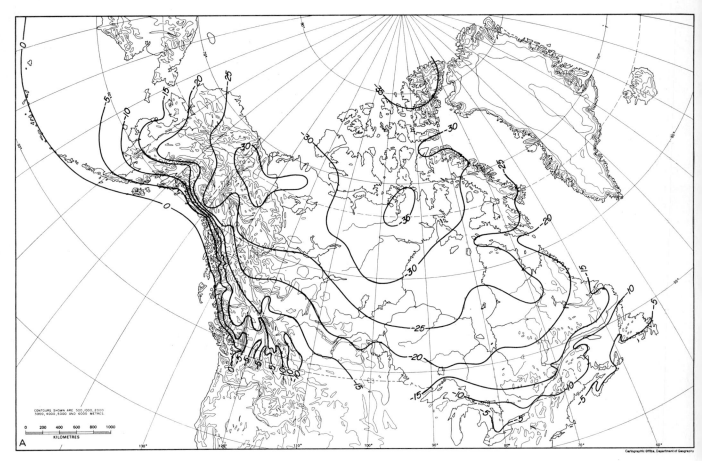

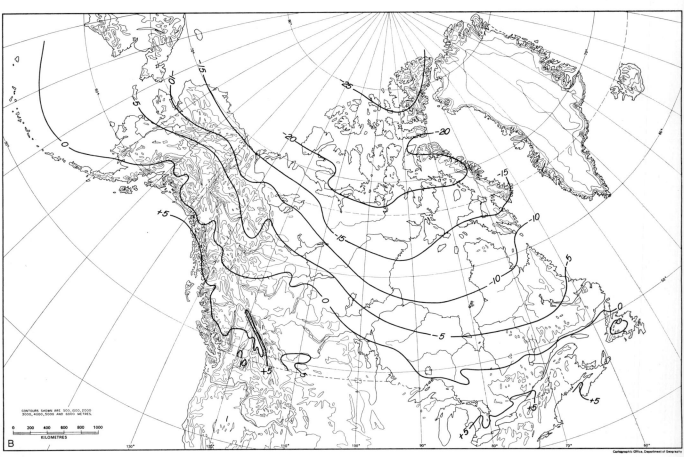

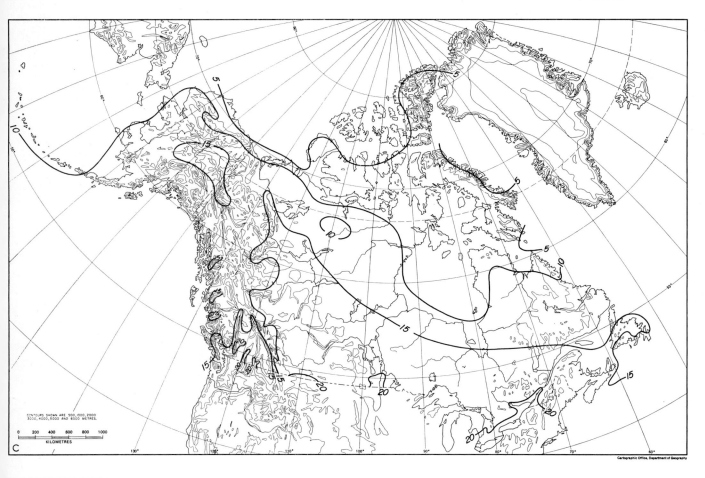

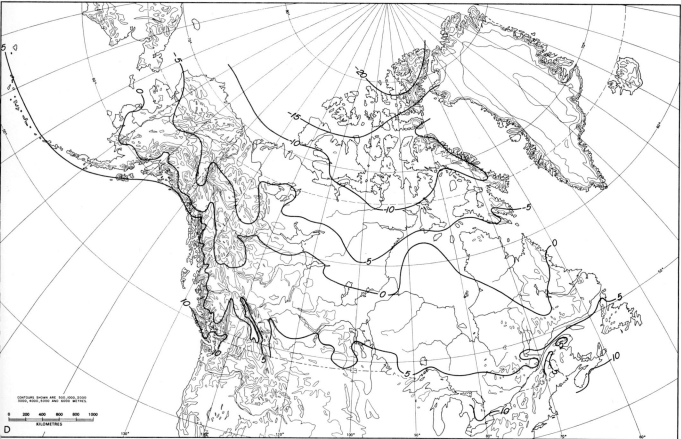

Fig. 5. Mean air temperature at screen-level (ca. 1.5 m), 1931–60. Isotherms show mean daily air temperature for: A. January; B. April; C. July; and D. October (after CANADA DEPARTMENT OF TRANSPORT, 1967; U.S. WEATHER BUREAU, 1968).

dominance of onshore Pacific airstreams (though Pacific waters are not strikingly warm for their latitudes, and coastal British Columbia is cooler than the corresponding region of Europe). The coastal warmth disappears on the occasions (rare in some winters, common and prolonged in others) when Arctic airstreams from the interior spill over the mountains and through the passes. Even then dynamical warming of the descending air from the interior, and the release of heat stored in the marine channels, keeps coastal temperatures higher than immediately east of the mountains.

There is no real Atlantic equivalent, since the prevailing flow is offshore. Occasional invasions of Atlantic air bring much higher temperatures, however, and the Gulf of St. Lawrence, Northumberland Strait and the Bay of Fundy also raise the temperature of eastward-moving continental airstreams. Hence Nova Scotia and the south and southeast coasts of Newfoundland have milder January mean temperatures (above −5°C, with Sable Island at 0°C) than inland points in the same latitude. The relative mildness of the Canadian east coast extends all the way north to Lancaster Sound, which with Hudson Strait form small gulfs of relative warmth in the cold Arctic, due to partially open water (HARE and MONTGOMERY, 1949).

The remaining anomaly is the warmth of the east coast of Hudson's Bay from fall into mid-winter. This warmth is related to the delayed freeze-up of the bay which ensures that the prevailing northwesterly Arctic airstreams are strongly heated before they arrive on the Quebec shore. The residual effect noted above for January marks the end of this anomaly. From mid-January on the snow-covered winter ice of Hudson's Bay effectively prevents heating of cold air moving across. The freeze-up of the bay was not realised until the end of World War II; an open Hudson's Bay was one of Canada's most persistent myths.

*Spring*

Although March is conventionally a spring month, its pattern of mean temperature is similar to that of winter. A slow general rise of temperature after mid-February shifts the isotherms north, but March mean temperatures are still below −30°C over the Arctic islands, and in eastern Canada they rise to 0°C only in southernmost Ontario and Nova Scotia. The rise is slightly more impressive in Alberta and southern British Columbia, where the gardens on the coast begin to bloom. Most of Alaska and Canada remains bitterly cold, however, and one faces the annual paradox of a northern country—that the sun is strong and the days are longer, yet the thaw does not come.

April behaves differently (see Fig.5B). The warming of the western prairies and the Cordillera continues rapidly whereas the heavily snow-covered east warms more slowly. April mean temperatures are as high in the Peace River country as in Prince Edward Island. Fairbanks, Alaska, is 15°C warmer than Keewatin and Baffin Island in the same latitude. This western warming creates a zone of strong thermal gradient from the Lower Mackenzie to James Bay, which intensifies in May. In the latter month mean temperatures are near 10°C through much of southern Ontario, the prairies, the Peace and Liard valleys, and southern British Columbia. It is slightly cooler in the interior valleys of Alaska. The 0°C isotherm runs from the Brooks Range of Alaska to the Lower Mackenzie, just north of Churchill, the Belcher Islands, and the Labrador coast near 58°N—a 12° spread of latitude between warm Alaska and cold Labrador–Ungava. These east–west differences show up well on the threshold-date maps for the 0°C mean isotherm (Fig.11).

*Summer*

Summer's pattern is established in June when the thaw finally reaches the Arctic coast plain of Alaska and traverses the Arctic islands. The south attains temperatures near or above the 15°C level in Canada. In Alaska the pattern of warm interior and cool coasts is firmly established.

The most striking feature of Fig.5C for July is the strong temperature gradient from the Brooks Range to Churchill and James Bay—essentially the alignment of the Arctic tree-line and the forest-tundra to the south. This gradient is established here in May, when the snowmelt line, and hence the jump to high values of net radiation, reaches the zone. For the rest of the summer the gradient of mean temperature remains fixed in position, and the Arctic front (see pp.54–55) usually lies along it. Thus an alignment which comes into being because of the northward march of the thaw becomes anchored by the deep-seated structure of the troposphere.

South of the Arctic front, much of Alaska and Canada has a warm summer. Mean July temperatures are highest (>20°C) in southernmost Ontario and the Montreal region, in the Red River Valley of Manitoba, the Milk and South Saskatchewan River valleys, and in the interior dry-belt of British Columbia. Temperatures are above 15°C in almost all closely inhabited parts of Canada. In Alaska it is only slightly cooler in the Tanana, Yukon and Matanuska valleys. Coastal areas are everywhere, on Atlantic and Pacific littorals alike, a few degrees cooler because of stratiform cloud decks and cool onshore airstreams. There are also cooler temperatures on the higher ground.

North of the Arctic front—the Arctic coast plain of Alaska and all Canada north of the forest-tundra—the summer is much cooler, July mean temperatures being everywhere below 12°C, and on the coldest shorelines below 5°C. Conditions away from ice-fringed coasts are a few degrees warmer, but the difference is not large. These low mean temperatures occur in spite of net radiation incomes of over 200 ly day$^{-1}$ in high summer; they reflect the continued dominance of Arctic airstreams and the coldness of the seas and permafrost layers. The coastal chill extends southward along the Bering coast of Alaska, and includes the Aleutians themselves. It also includes the shores of Hudson's Bay, where the 10°C isotherm for July lies in 55°N on the 80°W meridian.

The summer regime thus reflects two over-riding controls—the coldness of the encircling seas, and (in the Arctic) of the permafrost layers that resist melting in summer; and the persistence near the forest-tundra sub-zone of the Arctic front. The advection of higher temperatures by warm airstreams becomes progressively less likely as one travels poleward from this front's usual position.

*Autumn*

Autumn—called *the fall* in Canada from the beauty of the deciduous leaves at the end of the growing season—has well-defined characteristics. It begins with the cooling of the high Arctic at the end of August and beginning of September, when the Arctic front still lies across northern Hudson's Bay. It ends when that front is forced southward, usually in late November, to lie from the Alaskan and Wrangell ranges to the Rockies and eastwards to the Great Lakes. Between these times Pacific air makes its most pronounced and continuous appearance across Canada. October is the most zonal month in circulation.

This zonality is reflected in the October mean isotherms (Fig.5D) and net radiation (Fig.28). At no time of year is there a more east–west arrangement of the climatic isolines. In the high Arctic very rapid cooling takes temperatures down from +5°C in mid-August to −15° or −20°C during October, and to below −25°C in November. The 0°C isotherm (Fig.9), whose passage marks the beginning of the persistent freeze-up, reaches the southern Alaskan mountains in the first half of October, and in Canada the 0°C isotherm runs across the country a little south of 60°N, with some deeper southward movement over northern British Columbia and Labrador–Ungava, and with some retardation over the open waters of Hudson's Bay. There are less local anomalies at this season, however, than in the other seasons. The onset of fall is, in fact, rapid and uniform. Not until late November, when anticyclogenesis over the far northwest becomes common, do the characteristic non-zonal features of winter appear.

*Variability of temperature*

Temperature varies widely about these mean values throughout the year. This variation is in part cyclic, in part irregular, and occurs on a wide variety of time-scales. Cyclic variations are due to the diurnal and annual changes in solar radiation intensity and duration, whereas the irregular variations are the results of advection, chiefly horizontal (airstream changes), but occasionally along the vertical because of heat transports induced by turbulence. There are also interannual differences of more obscure origin, and in the far background is the slow secular variation of temperature due to world-wide climatic changes (treated separately elsewhere in the Survey).

*Diurnal variation and daily range*

The usual parameter of daily changes is the mean daily range of temperature, the difference between mean daily maxima and minima. As so defined the parameter includes both the true diurnal rhythm and the non-cyclic advective effects. Values of the mean daily range can be obtained for all climatological stations in Appendix I.

These effects are illustrated graphically in Fig.6 which presents for four representative stations the annual courses of (*1*) mean daily range and (*2*) range of the 24-hourly mean temperatures (CUDBIRD, 1964). The former gives the total variation typical of an arbitrary 24-h period, the second the effect of the true diurnal solar heating wave. If we assume (not quite correctly) that advective effects are independent of the time of day the difference between the curves should represent the advective component of the daily range, whereas the lower curve gives the true early morning–early afternoon temperature range (these being the hours of minimum–maximum temperatures). A much more detailed analysis by CATCHPOLE (1969) of the Winnipeg diurnal effects may also be consulted. The true diurnal effect is a maximum in summer, when the range of mean hourly temperature is between 8° and 12°C throughout inland Alaska and Canada, except in the coastal areas and the Arctic, where values are 5°C or even less. Because of airstream changes one must add 2°–4°C to these values in all areas to get the mean daily range. In winter the diurnal variation drops to low values, and virtually to zero in the area of the polar night. Advective effects are stronger, however, so that the net variability remains high. In all areas which have strong surface inversions increases in wind produce rises;

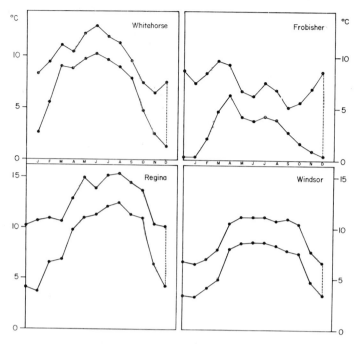

Fig.6. Diurnal variation of temperature at selected stations. Upper curves show mean daily range, which includes cyclic and advective changes; lower curves show range of mean hourly temperatures (data from CUDBIRD, 1964).

and a dropping of wind encourages strong radiative cooling. Hence there are large changes within most 24-h periods, even in the Arctic area.

The advective effect also shows itself in a very high level of day-to-day temperature change. In the St. Lawrence Valley, for example, a day of thawing temperatures may be followed by two days with maxima below −20°C, following a cold front passage. The intense airstream contrasts, and the rapidity of cyclonic and frontal passages, are responsible. These changes are seen most dramatically along the foothills of the Rockies, where temperatures below −30°C in Arctic air may be followed by values of +10°C in Pacific (chinook) air descending.

In many areas with abundant snow-cover, there are secondary minima in the mean daily range at times of widespread melt and thaw (i.e., chiefly in spring and fall). The high cloudiness of autumn further reduces variability.

*Interannual variability*

One season may be very unlike its counterpart the next year in Canada and Alaska. Mean annual temperature, a very stable parameter, has a standard deviation of 1°–1.5°C in most areas, maximum values being along the mean alignment of the Arctic front, i.e., roughly Fairbanks–Yellowknife–James Bay–Newfoundland. Much more striking is the seasonal change in interannual variability.

Fig.7 gives the four mid-season month fields of the standard deviation of mean monthly temperature. In January there are clearly two areas of strong variability—from Alaska south to the prairies, and along the Atlantic coast. Both areas also occur in December and February. The first suggests that the Arctic front varies in its position from year to year:

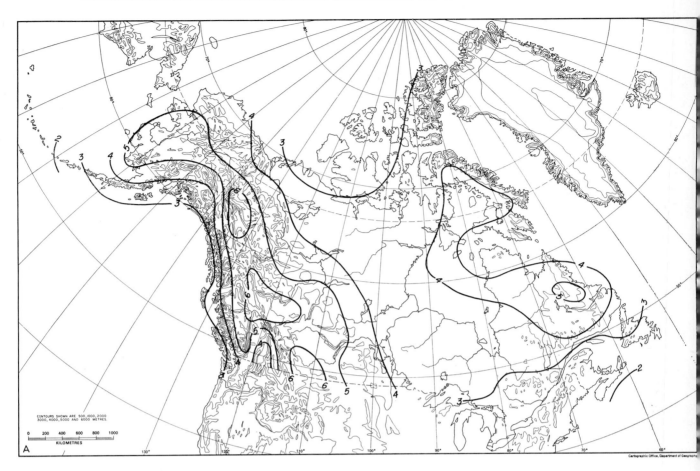

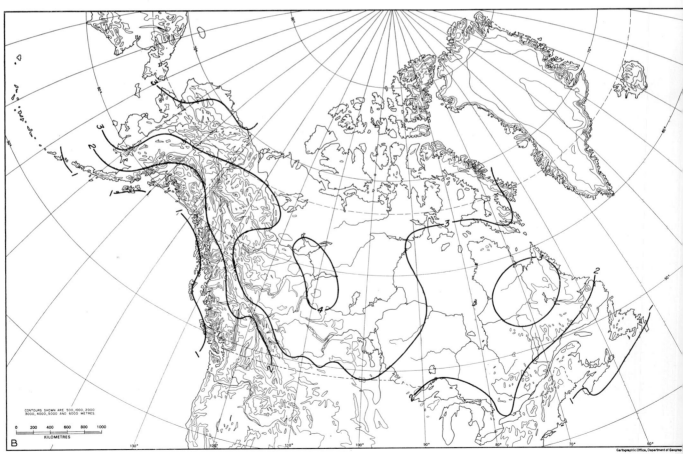

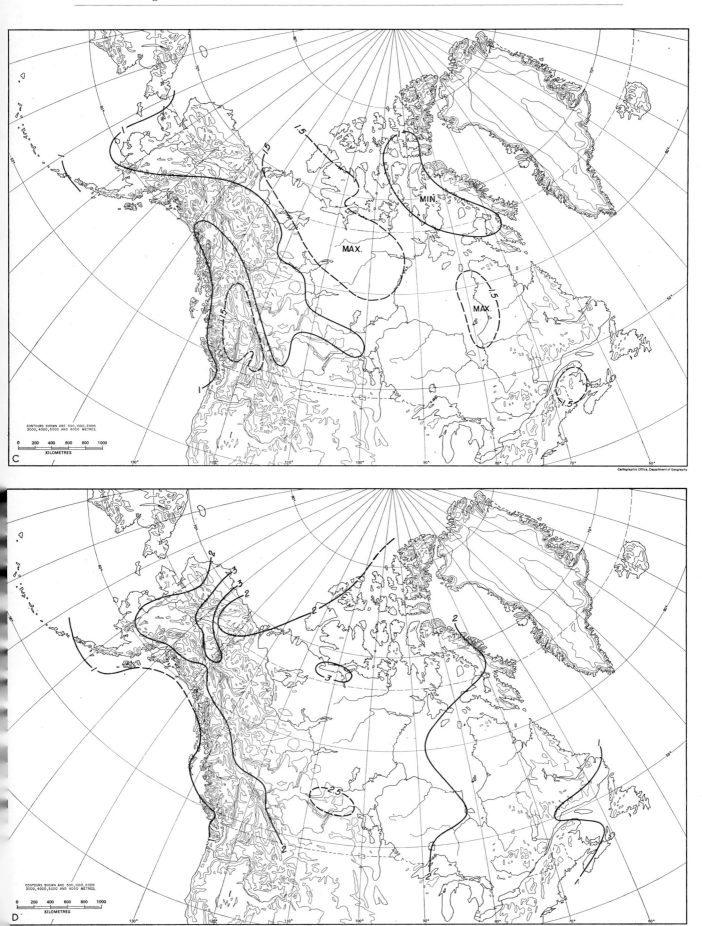

Fig.7. Standard deviation of mean monthly air temperature (°C). Mainly based on period 1931–60, but adjustment of Alaskan records to this period is not exact. (After KENDALL and ANDERSON, 1966; THOM, 1968.) A. January; B. April; C. July; D. October.

if it lies well east, Cordilleran and foothill stations have mild spells; if it is forced west, intense cold occurs. The high values along the Atlantic coast must reflect the frequent recurrence of prolonged blocking which in some winters allows Atlantic maritime air, mild and damp, to cross Baffin and Labrador–Ungava for long periods, giving large positive anomalies of mean temperatures. An area of lower variability from Banks Island across Keewatin to northern Ontario separates the two most variable regions.

This pattern changes abruptly in March, when the pattern resembles April's (Fig.7), as does May's. In June, July and August the standard deviation is much lower (1°–2°C). Maximum values lie over or just north of the mean locus of the Arctic front, at least west of Hudson's Bay. The same is broadly true of September and October, when, however, the rapid airstream changes diffuse the pattern and increase the variability.

In summary, the chief reasons for the considerable interannual differences typical of the northern part of the continent seem to be the proneness of the Arctic front to persistent displacement, and the frequent occurrence in eastern Canada of invasions of warm air from the Atlantic, chiefly in Arctic and sub-Arctic latitudes.

*Extreme values*

Appendix I gives tabulated values of extremes and little further discussion is needed. Extreme cold in winter may occur in almost all parts of the two countries, open-water coasts excepted. Values below −40°C have been recorded nearly everywhere inland, except in southernmost Ontario and Quebec, and −50°C has been recorded in many parts of the Canadian Shield and the prairies. Values below −60°C are sometimes recorded in the northwest, most often in high valleys of Alaska and the Yukon. In summer, most inland stations have recorded 35°C or above, and values of 40°C are occasionally reported in the southern prairies and the Cordilleran valleys of British Columbia.

*Accumulated winter temperatures*

For the purpose of building design, and in the estimation of fuel requirements, some integral of the year's lower temperatures is needed. North American practice is to use accumulated temperatures (degrees Fahrenheit) below 65°F (18.3°C) for this purpose. In Fig.8 we have used published figures for this parameter for 1931–1960, but have chosen isolines at 1,800 degree-days Fahrenheit intervals, with equivalent Celsius values at successive 1,000's.

Fuel costs are usually assumed to have a simple linear relationship to accumulated temperatures. For comparative purposes it may be noted that Chicago's annual mean of 6,115 degree-days F is equivalent to roughly 3,400 degree-days C. Values three times this figure apply to the belt from the Brooks Range through Keewatin to southern Baffin.

**Advance of the seasons**

Throughout Alaska and Canada a very strong seasonality is evident in the climate. The succession consists of the frozen and snowcovered winter, bitter except on the Pacific shore; the spring thaw, marked by the clearing of the snow, then of lake, river and sea-ice; the warm summer; and the rapid freeze-up of autumn. The dates and durations of

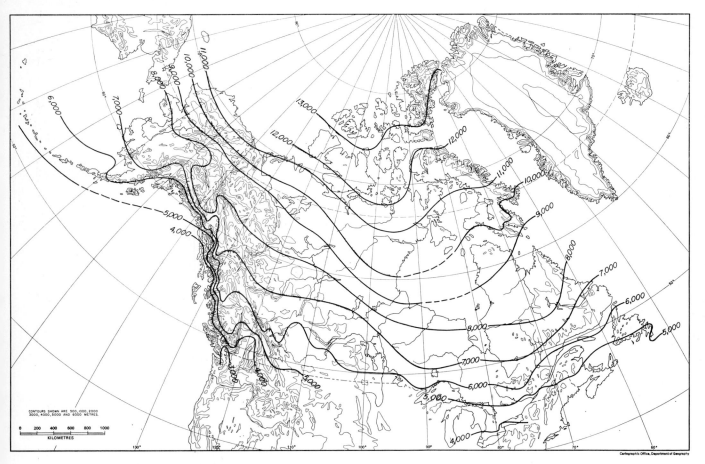

Fig.8. Mean annual heating degree-days (deg-days °C). Derived from base of 65°F (18.3°C). (After CANADA DEPARTMENT OF TRANSPORT, 1964; U.S. WEATHER BUREAU, 1968.)

these seasons vary from place to place, but they are everywhere distinct. Each imposes its special mark on the landscape and on economic and social affairs. Canada and Alaska closely resemble Scandinavia and the U.S.S.R. in these climatic effects and the social response that is made to them. The northern countries have a close kinship to one another.

Obviously the key to this seasonality is the annual freeze–thaw cycle. This cycle depends on the energy balance of the various surfaces, and on the storage of heat in lakes, sea, rivers and soil. Much attention has been given to the physical climatology of these processes in recent years, and large-scale research is now in progress into the freeze–thaw phase changes over glaciers, freshwater bodies and the sea-ice. The subject is very complex, however, and cannot yet be treated comprehensively. The following account is largely descriptive and empirical. An attempt will be made to relate the freeze–thaw cycle to the closely related cycle of mean air temperature and the date when it passes the 0°C threshold value.

*The autumn freeze-up*

Fig.9 gives the dates on which mean daily air temperature falls to 0°C, the freezing point for fresh water. This date effectively marks the start of the persistent air frost of winter.

77

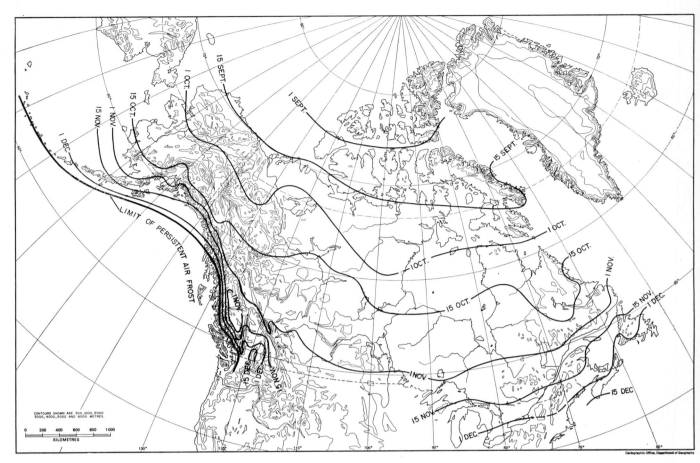

Fig.9. Mean date of fall of mean daily air temperature to 0°C, 1931–60.

Mean daily maximum temperature usually falls to 0°C about two weeks later, after which temperature is ordinarily below freezing continuously. In individual years departures from these dates may be two to four weeks on either side. Moreover, there is rarely a simple plunge to sub-zero temperature, mild spells interrupting the cooling.

The freeze-up of rivers and lakes is closely related to these atmospheric threshold dates. MCFADDEN (1965) showed that shallow lakes in central and northern Canada freeze when 3-day running mean air temperature falls to 0°C, whereas the large deep lakes of the same region (extending from Baker Lake to Madison, Wisconsin) freeze only when 40-day running mean air temperature falls to 0°C. The much longer lag for the large lakes reflects their greater heat storage. WILLIAMS (1968) demonstrated for McKay Lake, near Ottawa, Ontario, that the cooling proceeds in two stages. In stage 1 the lake becomes isothermal at 4°C, cooling according to the relation (due to BILELLO, 1964):

$$dT_W = K(T_W - T_A) \cdot dt \tag{1}$$

where $T_W$ and $T_A$ are water and air temperatures respectively, and $K$ is a constant peculiar to the locality expressing the depth of convective mixing and the heat transfer coefficient for the water–air interface. In stage 2 the surface water cools rapidly to 0°C, with ice forming quickly thereafter. Accumulated freezing degree-days (below 0°C) for air temperature correlate highly with the time taken to form a coherent ice layer thick enough to resist thawing during the short warmer spells. WILLIAMS (1968) found that 40–55 degree-days C are accumulated between the end of state 1 (lake isothermal at 4°C) and the formation of a stable ice layer of thickness 10–15 cm, which he found to be proof against subsequent melting in mild spells. Values of a freezing index based on this type of accumulated temperature have been published in a valuable summary by THOMPSON (1963).

Those findings are broadly consistent with the results of ALLEN (1964) who published break-up and freeze-up dates for all available Canadian stations. He distinguished between *freeze-up* when ice first forms, and *freeze-over,* when the ice-cover becomes complete. He found that the period between these dates, the freeze-over period, averages 18.7 days for rivers, though the period is highly variable from stream to stream and from year to year. For lakes the freeze-over period averages 12.6 days, and is less variable than for rivers. In general, lakes freeze-over 30 to 60 days earlier than rivers, where currents delay the process. On the average, lake freeze-over occurs 10 to 20 days after the passage of the 0°C threshold mean air temperature. The larger, deep lakes, however, take much longer; thus Great Bear Lake and Lake Athabasca freeze-over on the average near November 20, and Great Slave Lake not until after December 1.

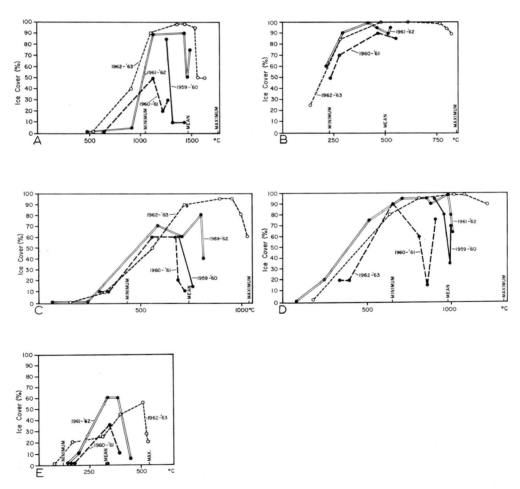

Fig.10. Growth and decay of Great Lakes ice. Ordinate is ice-cover (%), abscissa is accumulated freezing degree-days (°C) from start of season. (After RICHARDS, 1964.) A. Lake Superior; B. Lake Erie; C. Lake Huron; D. Georgian Bay; E. Lake Ontario.

The Great Lakes are very special cases because of their size, their variation in depth and their low latitudes, which allows only a short season of negative net radiation. Regular reconnaissance over the lakes began in 1960, and RICHARDS (1964) has analyzed the available evidence. He treated the problem as being one of balance between heat storage and winter heat loss, expressing both in the familiar form of accumulated temperature below 0°C for the heat loss and, above 0°C for heat gain. His regression equations explain from 63 to 90% of the variance of percentage ice-cover at maximum extent. Fig.10 is based on his results, showing the growth of ice cover on each of the lakes. Striking differences occur:

(*1*) *Lake Erie* freezes earliest and most completely, being very shallow; 170 to 220 freezing degree-days C give a 50% ice cover, above 400 degree-days (reached in late January as a rule) an almost complete cover. *Georgian Bay* has a similar regime, and is rarely less than 80% covered even in mild winters.

(*2*) Only bay ice forms in *Lake Superior*, which is very deep, until 550 to 775 degree-days C have accumulated, and a virtually complete cover (not common) requires 1,100 degree-days C. The lake frequently partially clears even while degree-days are still accumulating, because of wind and wave action. *Lake Huron*, though shallower, has a similar regime.

(*3*) *Lake Ontario*, which is also deep, rarely has more than a 50% cover, and is usually fairly well open.

Because of the freeze-over period in Canada and Alaska of roughly 13 to 19 days (ALLEN, 1964), the southward march of the freeze-up creates a wide geographical zone of part-open, part-covered water bodies, which remain anchored to 0°C while the land areas between may cool to values 10°C or more below freezing. Strong, shallow convection over this zone creates a heavy stratocumulus pall, with considerable low-level turbulence and snowflurry activity. Cloudiness reaches its annual maximum at nearly all stations in the month immediately following the threshold date for 0°C (HARE, 1950b; RAGOTZKIE and McFADDEN, 1965). In the area west of Hudson's Bay McFADDEN (1965) found the zone between the freeze-lines for shallow and deep lakes to be typically 250 to 300 km wide, but varied from 83 to 626 km on specific dates during two years of detailed aerial reconnaissance.

*The spring break-up*

The date on which mean air temperature again rises to the threshold of 0°C is shown in Fig.11. The persistent thaw enters the Okanagan late in February, and Nova Scotia and southwestern Ontario in mid-March. Ten days later it is into the Montreal–Ottawa area, the Rocky Mountain Trench, southern Alberta and south coastal Alaska. Thereafter a remarkable asymmetry appears between east and west. In the prairies, the Athabasca–Mackenzie Basin and the interior plateaus of British Columbia the thaw travels rapidly northward. By April 25 it has passed Fairbanks, Alaska, and by May 10 has reached the eastern shores of Great Slave and Great Bear lakes. In the east progress is much slower, and by May 10 thawing temperatures have only reached central Labrador–Ungava and the Hudson's Bay coast of Ontario.

There is then a prolonged retardation near the Arctic tree-line. The Arctic coast plain of Alaska attains 0°C about June 10, about a week after the isotherm has crossed the

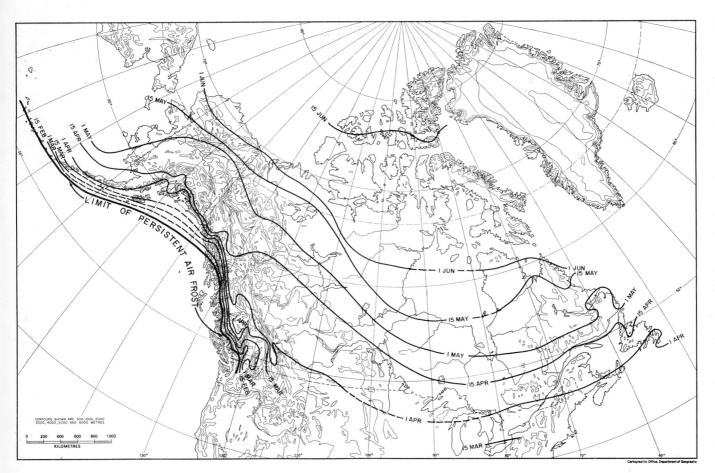

Fig.11. Mean date of rise of mean daily air temperature to 0°C, 1931–60. Note the early warming of western districts, especially inland Alaska.

Mackenzie delta. It takes about 25 days for the thaw to advance 200 km northeast of the eastern heads of Great Slave and Great Bear lakes. McFADDEN (1965, fig.16, and p.44) noted the slowness with which ice break-up occurred in this zone; reconnaissance in 1964 showed that only 30 km separated large lakes with zero ice cover from those still half covered. Finally the advance is resumed, and by June 15 all but the high glacierized plateaus of the Arctic islands have been reached. June is thus the month of thaw in all parts of the true Arctic.

We have already noted that the winter snow-cover disappears quite rapidly after the rise of mean air temperature to 0°C. In the case of lake ice, however, the break-up takes much longer, and there is a period each spring in which the snow-free land surfaces are 10°–15°C warmer by day than the still-frozen lakes (PETERSON, 1965), an effect that persists well after final thawing, because of cold lake temperatures.

River ice tends to clear fairly rapidly 10 to 20 days after mean air temperature rises to 0°C (ALLEN, 1964, fig.3), complete clearance taking an average of ten days. Except in interior southern British Columbia, where there is no lag, lakes generally take 10 to 30 days longer than rivers to clear of ice. North of the Arctic tree-line lakes normally retain some ice well into July (as does Great Bear Lake). The actual date of clearance of both rivers and lakes is highly variable from year to year. If early snow melt produces high run-off, it generally accelerates river break-up. Severe wind-storms (and hence wind-drag and wave action) may similarly advance the date on lakes.

McFadden (1965) related the break-up in shallow lakes to the rise of 3-day running mean air temperature to 5°C, and for deep lakes to the rise of 40-day temperature to the same threshold, though the fit was not good because of the complication just discussed. Williams (1965) considered the energy balance of McKay Lake, near Ottawa, over the decade 1952–1961, arriving at the following average values (in percentage of total melting-energy).

Contribution due to:

(*1*) Solar radiation (before reflection) +82%

(*2*) Net terrestrial radiation and atmospheric heat flux +29%

(*3*) Evaporation −11%

The surprisingly high value of evaporation reflects the fact that dewpoints in Arctic air, dominant at this season, are well below 0°C. From this analysis he obtained a simple heat exchange formula that fitted the observed data from a latitudinally wide range of Canadian lakes:

$$\Sigma Q = 0.5 \, \Sigma R_s + 18 \, \Sigma T \, \text{(1y)} \tag{2}$$

in which $\Sigma Q$ = accumulated heat for melting; $\Sigma R_s$ = accumulated solar radiation (before reflection); $\Sigma T$ = accumulated air degree-days above 1.67°C.

The Great Lakes again fail to conform to these simple laws (Richards, 1964). In general the Great Lakes ice clears earlier and more rapidly than the local air temperature regime would suggest. In detail:

(*1*) Lake Erie clears rapidly (55 to 85 degree-days C above 0°C) after the rise of air temperature to 0°C about 10–15 March, but may show a marked loss of ice-cover even before this date.

(*2*) Lake Superior may lose as much as 50% of its winter ice while freezing temperatures still persist, chiefly because of storm-induced up-welling of warm deep water. Dispersal of the remaining ice is slower than in Lake Erie, with mean air temperature rising to 0°C about April 5. Georgian Bay and Lake Huron behave similarly, about 5–10 days earlier on the average. In all three cases the occurrence of storm winds is the key to the removal of ice.

(*3*) Lake Ontario's winter ice is too sparse and too erratic in distribution for even approximate analysis along these lines. In general it clears early.

**Frost-free period**

The period normally free of sub-freezing temperatures is illustrated in Fig.12. Detailed analyses of the consequences of the short frost-free period in Canada have been published by Brown and Baier (1970), Baier and Ouellet (1970), Chapman (1970) and Findlay (1970). The frost-free period is, of course, substantially shorter than the period when mean daily air temperature is above 0°C.

Over the entire Arctic area, and on all high plateaus and ranges inland in Alaska and southern Canada, the frost-free period is below 50 days, and through the Boreal forest region is usually 60–110 days. In the small areas of agricultural land in Alaska it is about 90–120 days.

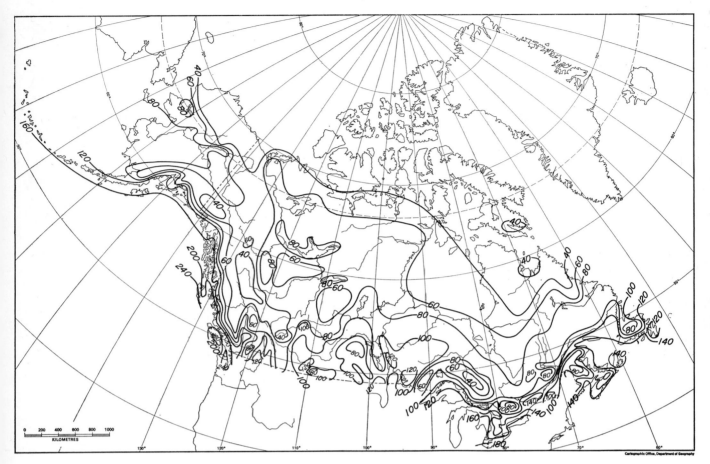

Fig.12. Mean length in days of frost-free period. Not rigorously standardized as to period, and heavily simplified in Alaska and much of the western Cordillera (after CANADA, DEPARTMENT OF TRANSPORT, 1956; U.S. WEATHER BUREAU, 1968).

The cool, cloudy Pacific coast has a much longer period, being as high as 240 days in parts of the Queen Charlotte Islands and in the Saanich Peninsula of Vancouver Island. Across the prairies values generally lie in the range 80–120 days, considerable differences arising from local site, altitude and exposure. Slightly higher values—usually in the range 120–160 days—occur in the agricultural lands in the Montreal plain and in southernmost Ontario, as well as in the maritime provinces.

**The moisture regime**

The moisture regime depends on sources for water vapor: mechanisms for inducing the condensation of clouds, and to make them precipitate; and a delicate interplay between that precipitation and the energy supply (which induces evapotranspiration), run-off and the consumptive use of stored water by the biosphere. Several aspects of this complex history—notably cloudiness, rain-forming disturbances and the net radiation income— are discussed elsewhere. In this section we shall cover moisture sources (which influence, but do not control, precipitation distribution), and the rainfall, snowfall and snowcover distributions.

The emptiness of so much of Canada and Alaska, and the consequent thinness of the observational network,make this subject particularly hard to treat in any detail. There are many difficulties, of which the worst is the unsolved problem of snowfall measurement, which is particularly acute in the windy tundra and forest–tundra areas. A detailed treatment of this problem is given in the section on the energy-moisture system.

**Moisture supply**

The precipitable water in the entire column of air over Canada and Alaska has recently been computed by HAY (1970) from radiosonde data for the period 1957–1964. We reproduce here the mean charts for January and July (Fig.13). In January mean values exceed 1 cm only in southwest coastal British Columbia. They exceed 0.6 cm over and southwest of the outer ranges of the Pacific littoral, and again in Nova Scotia and Newfoundland. Over the whole area between these moist flanks, Canada and Alaska are in the mean very dry, values falling below 0.2 cm in the central Canadian Arctic. The region of heavy winter precipitation (chiefly snowfall) in the St. Lawrence drainage basin has no larger average supply of precipitable water than the prairies. Obviously there is no simple relationship between mean precipitation and the mean supply of water available for it.

The same effect is apparent in July (Fig.13B). Mean values now exceed 2.0 cm along the Pacific littoral, and over a huge area from the lower Mackenzie Valley to Hamilton Inlet. Minima occur over the western Cordillera and, inevitably, over the cool Arctic. But in all parts of the Arctic which have little precipitation, values are at least equal to those of winter in excessively wet British Columbia. The most striking feature of the chart, typical of the entire summer season, is the near uniformity of precipitable water over the Mackenzie Valley, the central and eastern prairies and the main populated areas of eastern Canada. The chilling effect of the Great Lakes produces a small minimum. Also very striking is the strong vapor gradient along the axis Barrow–Churchill–Mistassini–Goose Bay, which corresponds to the mean position of the Arctic front (see p.55). Clearly the run of the isolines is quite unlike that of the summer isohyets indicating that, as in winter, there is no simple relationship between the mean supply of vapor and delivery to the ground as rain.

Nor is there a simple link with vapor transport. The water vapor is distributed through a considerable depth of the troposphere, whose wind systems continually move it about. The integrated transport has been computed by BENTON and ESTOQUE (1954) and RASMUSSON (1966, 1968) for all North America, and by BARRY (1966, 1967) and BARRY and FOGARASI (1968) for specific parts of Arctic and sub-Arctic Canada. These sources allow us to relate the transports to the moisture regions defined later.

Over Alaska the effective transport is very light, and comes from the southeast, from the Gulf of Alaska. There is little if any zonal transport. Much the same is true of Arctic Canada, though here the source may be the northwest Atlantic. Over all remaining parts of Canada throughout the year the resultant transport traces back to a zonal inflow from the Pacific, chiefly below 5 km. In summer the inflow is greatest across northern British Columbia and the Yukon, but in winter, when the transport is stronger, maximum inflow occurs between 50° and 55°N. Fig.14, derived from RASMUSSON (1966), shows the mean inflow across the Pacific coast during two recent years. These are probably close to

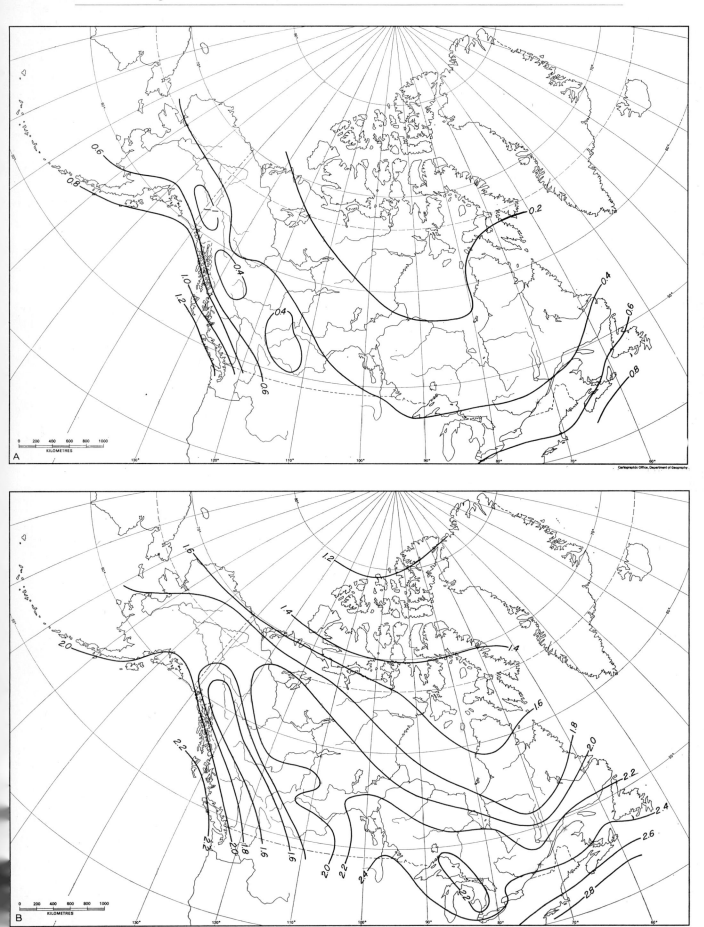

Fig.13. Mean precipitable water (cm), 1957–64. A. January; B. July. (After HAY, 1970.)

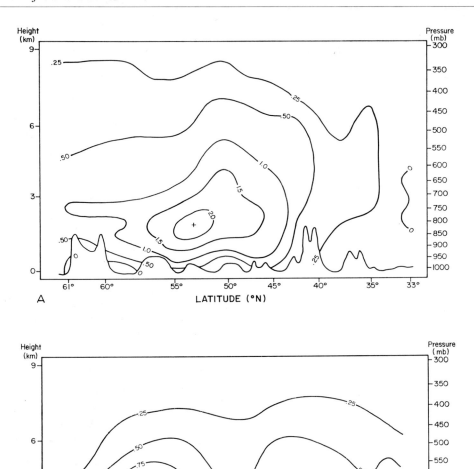

Fig.14. Mean rate of inflow of water vapor, Pacific coast of North America, 1961–62. Units are g(cm mbar sec)$^{-1}$. A. for January, B. for July, showing shift in latitude of maximum inflow. (After RASMUSSON, 1968.)

normal conditions. The eastward transport decreases slightly over the prairie provinces, but increases again in the east due to confluence and mixing with the stream of tropical moisture entering the United States west of the Mississippi and east of New Mexico. Again no simple relation with precipitation can be adduced; the dry climate of Brownsville, Texas, for example, occurs directly below the principal route of entry for tropical water vapor.

Over much of Canada and Alaska, precipitation distribution thus reflects the influence of generating disturbances more than it does the availability of water vapor, or the flux of that vapor. It is true that heavy rain or snow normally occur at times when the local value of precipitable water is well above the seasonal average. But the key to precipitation incidence in the entire region is the travel of convection systems on all scales from the local thunderstorm to the large cyclone, capable of producing the required strong uplift and convergence, on such a scale that a single day's rainfall may exceed five or even ten times the local mean value of precipitable water. Even orographic uplift, so obvious in coastal regions of Alaska and British Columbia, functions mainly to strengthen existing frontal precipitation belts, and to exaggerate a tendency towards shower-type precipitation due to instability. The mountains do not pluck much precipitation out of undisturbed air-streams.

## Moisture regions

### General pattern

Canada and Alaska contain parts of all four major moisture regions of North America. These regions depend primarily on the distribution of atmospheric disturbances, already discussed. A region is considered wet if it has a surplus of precipitation over water need for evaporation, transpiration and other purposes; surface humidities need not be high, nor need there be a large supply of precipitable water. A region is dry if it has a deficit of precipitation for such purposes. This does not necessarily imply low humidities. Fig.15 shows the basic annual patterns.

The dry regions are those of the American west and of the Arctic. The dryness of the Arctic arises from the weak uplift (on all scales) typical of disturbances, and in winter from a lack of precipitable water. The other area of dry climate, that of the American west and high plains, enters Canada only in the interior valleys of southern British Columbia, the southern parts of Alberta and Saskatchewan, and southwestern Manitoba. The core of this region is the desert landscape of Arizona, New Mexico, inland California and Nevada. The only approach to truly arid conditions in Canada—in the deepest Cordilleran valleys of British Columbia—is due to orographic intensification of the dry climate.

The wet regions are also two in number, the Cordilleran and the eastern. The Cordilleran region includes southern Alaska, much of British Columbia (exclusive, as just said, of the interior valleys), and the Rocky Mountain axis. Precipitation is generally abundant because of strong disturbances acting on moist Pacific airstreams, aided by orographic uplift.

The second region of high precipitation forms part of a vast area made up of the entire United States east of and including the Mississippi Valley, and all Ontario, Quebec and the Atlantic provinces of Canada. In this area an important source of moisture is the Gulf of Mexico and the sub-tropical Atlantic, although the admixture of moist Pacific air is strong in the north, especially in Canada. The region has a highly disturbed circulation, thus allowing strong uplift of humid air on a variety of scales–chiefly thunderstorms in summer, chiefly large-scale cyclonic motion in winter.

In these two wet regions the large moisture surpluses permit heavy forest growth, a major hydroelectric power industry, and a dairy-based agriculture. Paradoxically Canada's most celebrated farm region, the grain and beef producing prairies, lies on the margin of the southwestern dry region.

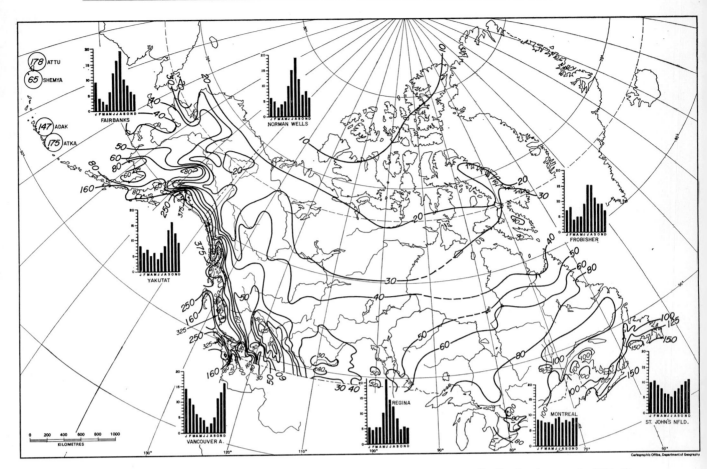

Fig.15. Mean annual measured precipitation (cm), with percentage distribution by month, 1931–60. Heavily oversimplified in western Cordillera and Alaska, where irregular isohyetal interval has been used. (After CANADA DEPARTMENT OF TRANSPORT, 1967; U.S. WEATHER BUREAU, 1968.)

*The Cordilleran wet region*

This region represents a very complex picture because of the orographic factor. This effect has been intensively studied by WALKER (1961), who applied mountain wave theory, combined with data from all available stations, to produce a more accurate picture of precipitation distribution over British Columbia. We reproduce here his annual precipitation map (Fig.16), and have made much use of his other results. As Fig.16 suggests, the mountainous relief creates, within the Cordilleran wet region, three quasi-parallel belts of high precipitation, the intervening ground being far drier.

The outermost belt consists of the summits and western flanks of the mountains of the Queen Charlotte and Vancouver islands. Annual totals exceed 200 cm, and precipitation is recorded on well over 200 days per annum at all stations. There is, however, a spell of drier weather in the late spring and summer; individual monthly totals then fall below 8 cm at low-lying stations. Rain becomes heavier again in September, and rises to a maximum in the fall and early winter, when monthly totals generally exceed 25 cm, and may locally exceed 40 cm. The rain comes in prolonged spells of moderate intensity from

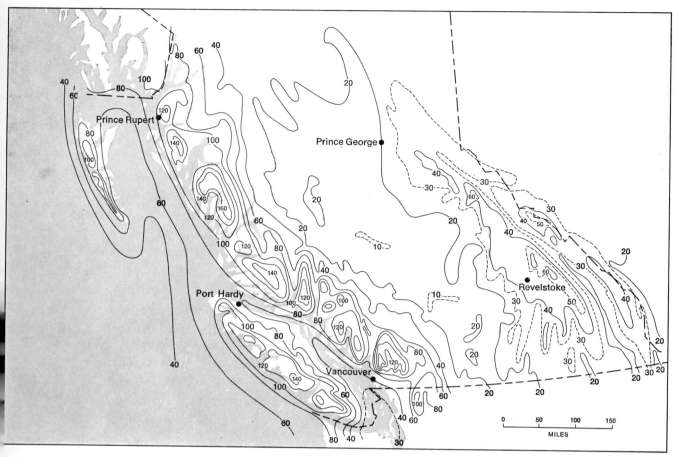

Fig.16. Estimated mean annual precipitation over British Columbia, for long term (inches); based on combination of observed values with estimates derived from mountain wave theory, and adjusted to long term. (After WALKER, 1961.)

westerly and southwesterly flow associated with the passage of Pacific disturbances and the instability of the airstreams behind each system. At high levels the precipitation falls as snow, and extremely high snowfalls occur on the upper slopes. Excessive falls of rain or snow occur when the set of the onshore flow in relation to relief magnifies locally the convergence typical of the Pacific disturbances. Henderson Lake (49°08′N 125°08′W, 4 m above sea level), in Vancouver Island, recorded 42 cm of rain on December 30, 1926, and Ucluelet Brynnor Mines (49°03′N 125°26′W, 90 m above sea level) received 49 cm on October 6, 1967, the heaviest recorded 24-h fall in Canada. This exposed mountainous coast is thus liable to rainfalls over a day as intense as those associated with tropical hurricanes, and all structures have to be designed to allow for excessive run-off. The luxuriant rain forest vegetation is scarcely ever dry at floor-level.

Much of the same is true of the second belt of high precipitation, that of the Coast Range of British Columbia, the Alaskan Panhandle, the St. Elias Range and the Chugach, Wrangell and Kenai ranges of Alaska. Annual precipitation exceeds 100 cm over the whole belt, and on the southern and western slopes generally exceeds 250 cm. Highest totals (over 300 cm) probably occur immediately north of Vancouver, opposite Queen Charlotte Sound, and in the central Chugach Range. Because of the much greater extent of high mountain surfaces, this belt gets far more snowfall than the Vancouver

Island–Queen Charlotte Island belt, and is spectacularly glacierized. The seasonal rhythm is similar: precipitation frequency and amount is lowest in summer (when, however, shower and thundershower activity is common), and rises to a maximum in fall (in the north) or early winter (in the south).

The heavy falls of this wet region extend southwestwards from the Chugach and Kenai ranges to include Kodiak Island, the Alaskan Peninsula, the Aleutians and the Alaskan mainland as far northwest as the summits of the Kuskokwim Mountains. In this area precipitation comes entirely from the same deep Pacific cyclones that affect the rest of the region, but here the exposure is to the southeast, and heaviest falls occur on that flank of the various ranges. Thus the Bristol Bay shore of the Alaskan Peninsula has no more than 60 cm fall, whereas the southeast coast has generally 200 cm or more.

The third belt of high precipitation consists of the high ground of the Purcell, Monashee, Selkirk and Cariboo Mountains, and the summit and west face of the Rockies, separated by a narrow belt of much lower falls along the Rocky Mountain Trench. Much of the precipitation falls as heavy snow, with a cold season maximum (SABBAGH and BRYSON, 1962), but considerable rainfall also occurs from rain showers and thunderstorms in the warmer months. Precipitation frequency is significantly lower than on the outer coast, however.

In all three of the wet regions total precipitation depends on aspect and relief; the orographic factor is very effective. Annual totals below 40 cm occur in some of the deep, enclosed valleys of southern Alaska—most notably the Matanuska Valley, and in the Copper River Valley. Skagway, at the head of Lynn Canal, has only 68 cm, in spite of the great icefields on the surrounding mountains. Within the Purcell–Monashee–Selkirk–Cariboo–Rocky belt the Rocky Mountain Trench has totals below 50 cm through much of its length, although the floor is marshy and often flooded by run-off of snowmelt from the mountains on either flank.

By far the most important of these drier areas surrounds the shores of the straits of Georgia and Juan de Fuca. This summer-dry climate (KERR, 1951) is Canada's only approach to a Mediterranean regime, and is associated with an unusual flora including the Garry oak (*Quercus garryana*) and Madrona (*Arbutus menziesii*) adapted to the high solar radiation income and low rainfall of summer. Annual totals fall well below 100 cm in the Gulf islands, on the southeast shore of Vancouver Island and a few localities on the lower mainland. The area of drier climate on the east shore is, however, severely limited; thus mean annual rainfall increases from a little over 100 cm at Vancouver Airport (in the Fraser delta) to 150 cm downtown and to over 250 cm on hills north of the city, within 20 km distance (WRIGHT and TRENHOLM, 1969).

*The western dry region*

The western dry region includes the interior of British Columbia (*1*) between the coast and Cascade ranges on the west and Monashee range on the east; and (*2*) south of the Cariboo plateaus. East of the Rockies it includes the true prairie and park belts of Alberta, Saskatchewan and southwestern Manitoba. The driest part of the latter has long been called the *Palliser triangle*, its lack of water having been commented on by Captain Palliser at the time of his celebrated survey in 1857–60.

The dryness of the British Columbian interior is directly a rainshadow effect. In winter the region gets only light and infrequent snowfall, in contrast to the heavy falls on the ranges on both flanks. In the warmer months showers and thunderstorms are frequent, but heavy falls and widespread general rain are rare. The region shares with the interior of Washington State to the south an extensive area of dry Ponderosa pine forest, steppe and even sagebrush landscapes. Summer is a baking season, and the surface is dusty and parched in spite of the shower activity. In parts of the floor of the Fraser and Thompson valleys, and again in the Okanagan and nearby valleys, mean annual precipitation drops below 25 cm, falling primarily in the warmer months. August is the wettest month (SABBAGH and BRYSON, 1962) of the annual cycle. The actual area of very dry conditions is, however, quite small, since precipitation increases rapidly with height. Only in the Thompson Trench near Kamloops and Ashcroft, and in the Okanagan region, is there any great expanse of xerophytic vegetation. The contrast on the northern trans-Canada highway is astonishing. One goes from high rain forest at Yale, at the southern entrance of the Fraser Canyon, to sagebrush and Ponderosa pine at Lytton, at the Thompson confluence, in less than 90 km.

The dryness of the prairies and park-belts is less marked, but covers a very much greater area. The paucity of precipitation reflects the weakness of disturbances, and their associated uplift. Eastward moving airstreams, moreover, fall steadily from the Rocky foothills to the low ground of Manitoba, a descent of well over 1 km in a trajectory of 1,000 km, or, at 10 m sec$^{-1}$ horizontal flow, a rate of descent of order 1 cm sec$^{-1}$. This is enough to reduce cyclonic precipitation appreciably, except when the surface airstream runs upslope, as it does during periods of considerable precipitation. There is ample moisture in the atmosphere, but precipitating mechanisms are not effective.

The region has a pronounced summer maximum, June being the wettest month in the south, July (locally August) along the southern Boreal forest margin. This summer rain regime meets the winter maximum of the high mountains to the west along the main Rocky crest (SABBAGH and BRYSON, 1962) in another of Canada's sharp climatic divides. Annual totals are everywhere below 50 cm, and locally (in southeast Alberta and southwest Saskatchewan) below 35 cm. The rainiest month of summer averages 8–10 cm. Winter snowfall contributes on the average less than 35% of the total water yield. All these figures are notoriously unreliable, the prairies being subject to a very high inter-annual variation, and even to droughts persisting for two or more successive years, interspersed with summer rains—or unseasonable snow— that comes close to ruining the grain crops.

### The eastern wet region

This region includes eastern Manitoba, all of Ontario, Quebec, the Atlantic provinces (including Labrador) and the high plateaus of southeast Baffin. It is the northward continuation of the wet belt of the United States east of about the 95th meridian. Throughout this area precipitation-making disturbances are vigorous and effective, and the precipitable water from Pacific, Atlantic and Gulf of Mexico sources is sufficient to give a reliable, seasonally well-distributed and usually abundant precipitation. Variations due to relief are much less pronounced than in the west, even in the mountainous areas, because moving airstreams can carry high vapor contents from a wide range of azimuths.

Along the northwestern margin of the region annual totals are near 50 cm, and in general they rise steadily southeastwards to near 150 cm in the wettest parts of Nova Scotia and southeast Newfoundland. Similar figures are also attained on the high ground of Gaspé, the eastern townships of Quebec and the Laurentide escarpment from Quebec City to the Romaine River. Interannual variability is low.

The northern two-thirds of the region has a pronounced warm season maximum, with August the wettest month of the annual cycle in most areas. Over Manitoba and northern Ontario, a considerable fraction of the total comes from summer thundershowers, whereas over northern Quebec, south Baffin and northern Labrador the bulk falls from frontal rain belts. Winter snowfall is not heavy, though high windspeeds produce heavy drifting and hence uneven pitching. Special effects occur along certain coasts. Hudson's Bay has a very retarded freeze-up (HARE and MONTGOMERY, 1949; BURBIDGE, 1951; HARE, 1951), and in autumn the prevailing northwesterly winds deposit heavy snowfalls on the exposed Quebec coast until the ice forms in December or later. Other Arctic coasts show this tendency to a lesser extent.

The remaining third of the region—essentially from the Great Lakes to Newfoundland and southern Labrador–Ungava, including southern Ontario, southern Quebec and the Atlantic provinces—has a seasonally well-distributed precipitation combining heavy and frequent winter snow with equally heavy summer rains, largely of cyclonic origin, though mainly showery and thundery in character. In no other area of the world is there so heavy a winter precipitation at so low a mean temperature and precipitable water.

Special effects are again visible in winter. Northwestern peninsular Ontario, for example, has a heavy snowfall due to instability showers formed over partly frozen Georgian Bay and Lake Huron; similar snowbelts occur south of the border in New York State because of lakes Erie and Ontario. They are also visible on the northwest facing coasts of Cape Breton Island, Pictou and Antigonish counties, Nova Scotia and Newfoundland. Orographic effects are otherwise subdued in this well-watered region.

In addition to the general southeastward increase towards Nova Scotia and the Avalon Peninsula of Newfoundland there are significant variations along the St. Lawrence Valley. Through much of populous southern Ontario annual precipitation is near 75 cm. Values of 100 cm and over occur between Montreal and Quebec, and again in many parts of the north shore below Quebec.

### The dry region of the Arctic and northern Alaska

A vast area of low and variable precipitation extends from the Pacific coast of Alaska north of Bethel to the coasts of Baffin Bay, Davis Strait and the Labrador Sea off Ellesmere, Devon and Baffin Islands. Throughout this area annual falls are light, though not as low as measured values indicate (HARE and HAY, 1971), and are largely derived from cyclones of the Arctic frontal group, or from slow-moving old occluded cyclones from the south.

In Alaska measured annual falls decline from about 40 cm on the Bering Sea coast to about 30 cm along the Yukon boundary. Heaviest falls occur on the southeast flanks of the hill ranges, with corresponding drought in enclosed basins on northwest facing slopes. Fort Yukon, with 17 cm, is the driest sub-Arctic site in North America, and the Arctic coast plain south of Barrow has less than 15 cm, in spite of its waterlogged surface. Winter snowfall is light in all areas, and the main rainfall comes in late summer and early fall, when monthly totals of 10 cm occur in some areas. August is usually the wettest month.

The Canadian Arctic has a similar regime, total measured falls being very light—below 30 cm over most areas, with greater falls largely confined to the high plateaus facing Baffin Bay, from which occasional heavy falls can be derived (BARRY and FOGARASI, 1968). Most of the precipitation falls in light summer rains, but there is a coastal maximum of snowfall in September and October when instability showers fall over the still unfrozen seas. Some stations in the deep fjords of Ellesmere, Devon, Baffin and Axel Heiberg Islands have mean annual falls of less than 15 cm.

It is in this northern region, where the season of sub-freezing temperatures is longest, that the uncertainties in the measurement of precipitation are most serious. The above figures may well be an underestimate of true precipitation, especially as regards snow. Qualitatively the account given is correct. Quantitatively it probably exaggerates the real droughtiness of the northern region.

## Snowfall and snowcover

Mention has been made throughout the previous account of winter snowfall. Canada and Alaska include two of the world's greatest snowbelts, and except for a tiny coastal strip in southern British Columbia all regions have a mid- and late-winter period of snowcover. We present a map of mean annual snowfall (after THOMAS, 1964; MANNING, 1968; U.S. WEATHER BUREAU, 1968) and maps of the duration and effective dates of the snowcover season (POTTER, 1965a; Fig.17–18). A rough measure of the water equivalent is that 10 cm of snow yield on melting 1 cm of water, i.e., a relative density of 0.1, but this figure is in practice very variable (POTTER, 1965b). The snowcover remaining at the end of winter may have relative densities as high as 0.40–0.48 (McKAY and THOMPSON, 1968; Fig.19).

The entire length of the Pacific littoral has heavy and prolonged snowfall along the ranges paralleling the coast. The Aleutian Islands have more rain than snow, but from the Alaskan Peninsula round to the Washington border falls exceed 250 cm everywhere except near sea-level, and are over 500 cm in most summit areas. Even greater falls occur on some slopes. Thompson Pass, in the Chugach Range behind Valdez, Alaska, has a mean fall estimated at 1,500 cm; Kildala Pass, between Kemano and Kitimat, in the coast range of British Columbia, averaged 1,940 cm over a short period, and recorded over 2,200 cm in a single winter (1956–57); in February, 1954, over 500 cm fell in a single month. Allison Pass, on the southern trans-Canada highway's crossing of the Cascades, has an average of 934 cm. With falls of this magnitude it is inevitable that intense glacierisation should occur. The plateau and valley glaciers are most extensive in Alaska around Prince William Sound, behind the coast between Cordova and Juneau, and on the summit plateaus of the St. Elias Range (MARCUS, 1964). But active cirque glaciers are visible from the campus of the University of British Columbia in Vancouver, and occur further south in the Pacific Northwest states.

Substantial glacierisation also occurs on the summits of the second part of the great western snowbelt, the Monashee–Selkirk–Purcell–Cariboo–Rocky complex. Glacier, in the Selkirks, has a mean annual fall of 930 cm, and most areas exceed 250 cm. Avalanches are especially common and destructive in this area, which is traversed by Canadian

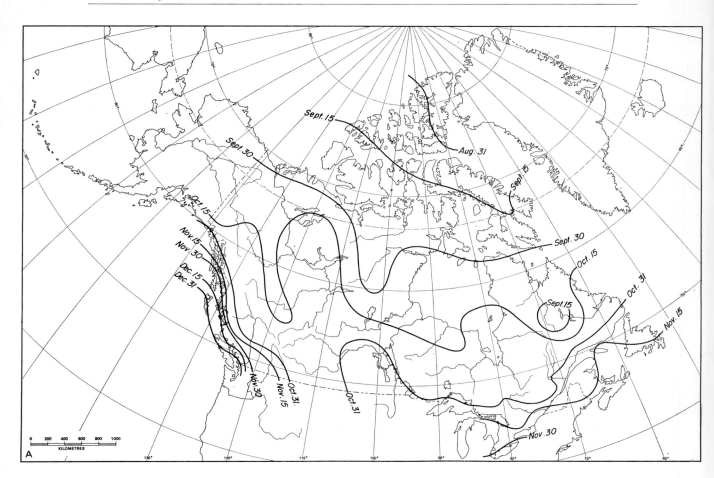

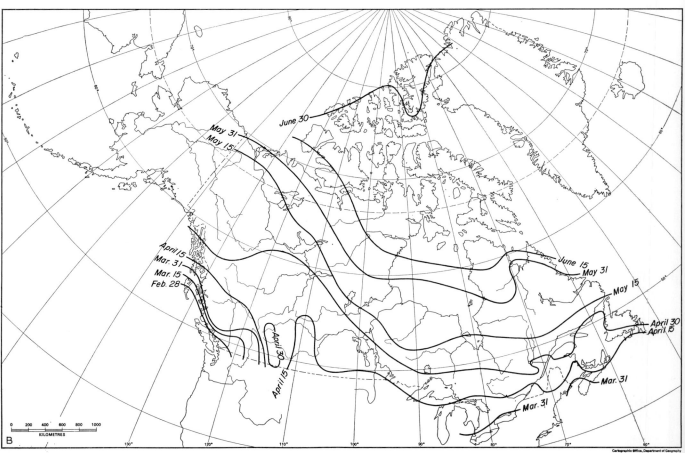

Fig.17. Median first (A) and last (B) dates of winter snowcover greater than 2.5 cm depth for Canada (after POTTER, 1965a).

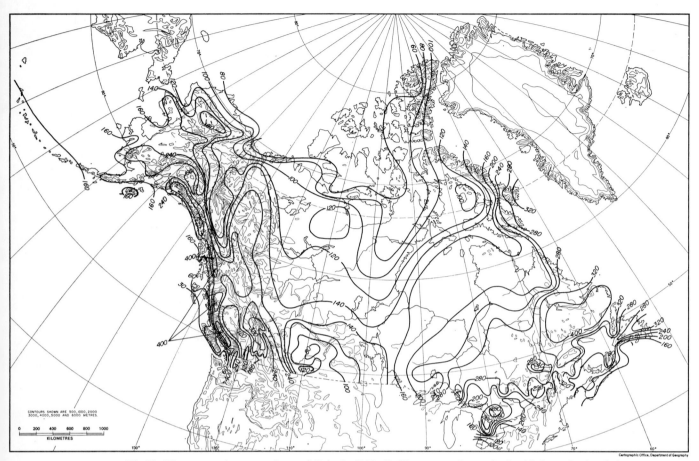

Fig.18. Mean annual measured snowfall (cm), 1931–60. Very approximate (and with irregular isopleth interval) in Alaska and the western Cordillera (after CANADA, DEPARTMENT OF TRANSPORT, 1967; U.S. WEATHER BUREAU, 1968).

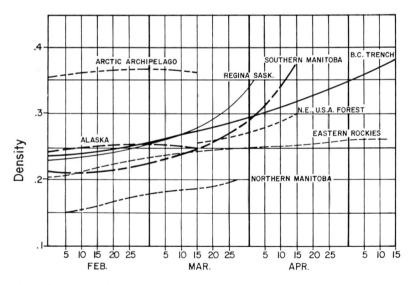

Fig.19. Seasonal variation in average snowcover density (after McKAY and THOMPSON, 1968).

*95*

National and Canadian Pacific mainlines as well as both routes of the trans-Canada highway.

The interior plateaus and valleys of British Columbia, the Yukon and Alaska have much lighter falls, generally between 140 and 180 cm, but locally much less, as in the British Columbian dry belt, and east of the St. Elias Range.

A vast area of light snowfall extends from the prairies to the Arctic coast. Falls are below 150 cm in the prairies, and over the entire Arctic and sub-Arctic northwest of a line roughly from central Baffin through Southampton Island, Ennadai Lake and thence south to near Winnipeg. On the Alaskan coastal plain and in the Queen Elizabeth Islands average falls are well below 100 cm. The windswept surface concentrates the light snow into drifts, and much of the surface blows free, as it does in the prairies where winter soil-drifting may cover snowbanks with black dust. Serious errors of measurement of snowfall are hence very likely.

Eastern and southeastern Canada form the second major snowbelt, joined by nearby parts of Michigan, Wisconsin, New York and the northern New England states. In many areas annual snowfall exceeds 250 cm, and in the snowiest belt—southern Labrador–Ungava—exceeds 400 cm over a wide area. The snow in these areas comes from deep cyclones moving east or northeast across the area, and is generally firm as well as deep, being mostly low temperature snow that is heavily wind-drifted. Freezing rain is also common near the St. Lawrence Valley and together with occasional thaws creates numerous crusts within the snow-pack. Median depths attain their maximum late in winter (POTTER, 1965a), when they exceed 100 cm from the upper Ottawa Valley to the Labrador coast between Hamilton Inlet and the Straits of Belle Isle.

Certain areas of lower snowfall within this belt are worthy of mention. The part of southernmost Ontario from the Detroit River near Windsor, the Erie shore counties, the Niagara Peninsula and the Lake Ontario shore to the Toronto area gets much of its winter precipitation as rain or freezing rain, and total snowfall averages below 120 cm in many areas. The same is true of the extreme southern and eastern coasts of Nova Scotia and Newfoundland. These are the only parts of eastern Canada where a winter snowcover is not reliably continuous. In mild winters it may occur only for brief, intermittent periods in southernmost peninsular Ontario and Nova Scotia.

Off-setting these, the belt also includes certain *great snowbelts*, as they are often called. Mention has been made of those of the Georgian Bay and Huron coasts of Ontario, and that occurring on the east shore of Hudson's Bay, where instability over warm water surfaces is the triggering mechanism. A similar belt occurs on the west coast of Newfoundland. Heavy falls extend up the east coast of Labrador, and on to the southern peninsulas of Baffin, where the snow falls from unstable airstreams coming off the Labrador Sea. A mean fall of 400 cm occurs at Cape Dyer, on the east Baffin shore near the Arctic Circle.

Snowcover becomes continuous, on the average, about 5 to 10 days after the fall of mean daily temperature below 0°C, though the date varies greatly from year to year, and comes earliest in regions of heavy and frequent snow. Fig.17 shows the probable dates. Snow-melt begins at or shortly before the date of return of mean air temperature to 0°C, and is generally completed within two to four weeks, depending on fresh snowfall, cloudiness and weather changes. Local run-off follows immediately, and usually very harshly, because of the frozen soil. The main impact on the large streams is delayed by a week to a month, depending on the size of the basin. In the western mountain belt, however, the high elevation of much of the snow delays the melt, and the great rivers of the Pacific coast are in spate in early summer.

The snowcover of Canada and Alaska has high economic importance. Some of this is negative. Overland transport (and airport operation) is greatly complicated by the frequent snow, whose clearance is costly; the city of Montreal, truly *Notre Dame des Neiges*, has sometimes faced clearance costs of over $10 million in a single winter, and the loss of production due to travel and transportation delays is many times greater. On the other hand watershed management has put the spring snow-melt to good use. Hydro-electric development of the St. Lawrence relies on even flows due to the Great Lakes storage, but the development of smaller streams from Alaska to Newfoundland depends on the use of barrages to spread the spring freshet over the year. The most spectacular of them is the Hamilton River barrage above Grand Falls, where an installation of over four million kilowatts capacity is under construction. In a real sense the deep snowcover lying on the surface of the northern continent, much of it on uninhabited drainage basins, is a major natural resource, readily convertible to power, and (unlike soil moisture storage) insulated against serious evaporation losses by low temperatures and radiation incomes. The lack of competing uses for so much of the surface on which the snow accumulates makes large-scale exploitation of this resource efficient and cheap.

**Thunderstorm incidence and severe weather**

Thunderstorms arise from two situations, general uplift and surface heating of unstable airmasses, which are often combined. Both situations are common in all southern and western parts of Canada, and in inland Alaska, but the supply of moisture and heat in the lower troposphere is rich enough only in the warmer months (May–October) to make thunderstorm activity common. North of the Arctic tree-line—which coincides with the summer locus of the Arctic front—such conditions rarely occur. Winter thunderstorms are observed almost exclusively in the vicinity of the Great Lakes, St. Lawrence Valley and the Atlantic coast south of the Avalon Peninsula; they occur in association with powerful cyclones, and are very infrequent even in this region.

Fig.20 shows the annual incidence of thunderstorm days, and also the incidence for July, generally the peak month (after KENDALL and PETRIE, 1962); the two are significant-ly different. Annual frequencies are moderately high—20 days per annum or more—across Canada from the eastern Cariboo and the Okanagan to the Eastern Townships of Quebec. Highest values occur in extreme southwestern Ontario, where a more detailed analysis by HAACK (1964) suggests that the highest values in Canada (over 30 days) may lie along the high ground immediately west of the Niagara cuesta. Significantly lower values occur in the lee of the Great Lakes, especially over the shield between lakes Nipigon and Nipissing, and along the north shore of Lake Ontario, probably because of the stabilizing effect of the cold surface water. In these eastern areas thunderstorms often occur in large (mesoscale) systems drifting slowly in sluggish, moist air, and oc-casional excessive falls occur in these indeterminate situations.

Thunderstorm frequency drops off rapidly northward, and in most tundra areas lies at one day or less. The outer coast of British Columbia is also largely free of such storms, though they can often be watched forming over the mountains behind the coast.

The midsummer distribution, typified by July, shows that at this season thunderstorm

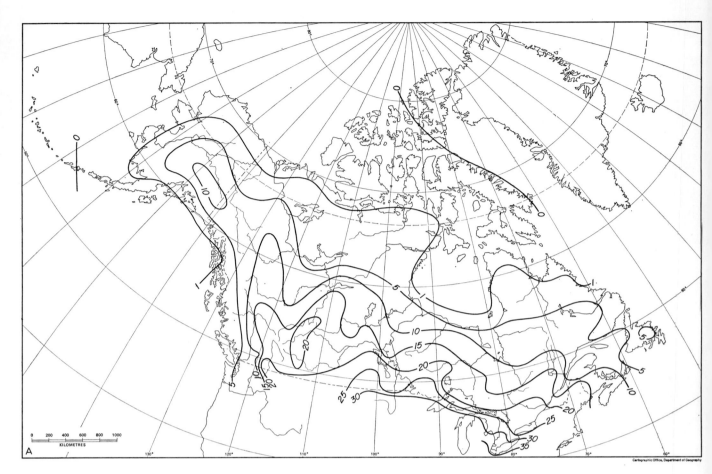

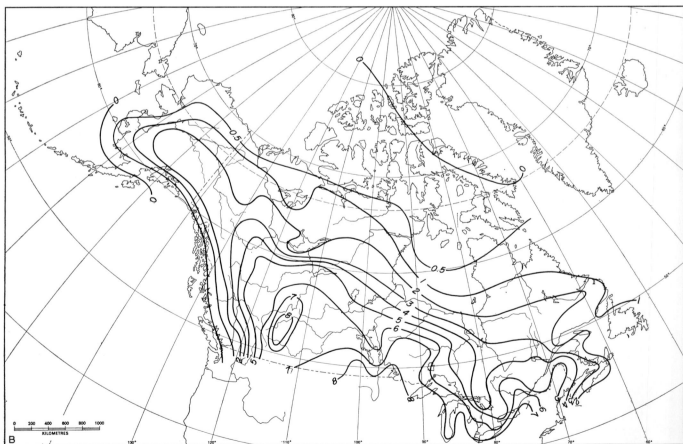

Fig.20. Mean number of days of thunderstorms per annum (A) and in July (B). 1941–60 in Canada, 1931–60 in Alaska (after KENDALL and PETRIE, 1962; MS data, Alaska).

activity is a maximum in northwestern Ontario and the prairies, although southwestern Ontario is nearly comparable. The thunderstorm season in the west is shorter, and a higher proportion of the storms is associated with generally clear weather and strong insolation.

Alaska presents an interesting picture. There is a coastal belt in which thunder is rarely or never heard (some Aleutian stations have never reported it). Maximum frequencies (probably over 10 days per annum) are in and along the flanks of the Wrangell and Alaskan ranges, but thunderstorms are not uncommon in the interior valleys and plateaus south of the Brooks Range. The season is virtually confined to the period late May to early September.

Many thunderstorms go unobserved, especially at night, except at hourly synoptic stations. The real frequencies are almost certainly higher than Fig.20 indicates.

**Rainfall intensity**

Intense rainfall is a threat to sewers, culverts and bridges, as well as a discomfort; and if it falls on melting snow, as it often does in the coastal mountains, it can contribute to avalanche disasters. In Canada and Alaska such heavy falls are of two chief kinds: (*1*) unusually prolonged and intense rainfall from Pacific cyclones, usually where lower tropospheric flow is from the southwest; and (*2*) thunderstorm rain in other areas. We have already seen that falls of over 40 cm in a day have been recorded from type (*1*). In eastern Canada and central Alaska falls of this magnitude have not been observed, but 8–12 cm per day is not uncommon in thunderstorm areas, and falls such as this can cause serious local damage and inconvenience.

The risk has been systematically estimated by MURRAY (1964) and BRUCE (1968). To illustrate the geographical variation of risk, Fig.21 presents the 24-h rainfall likely to be experienced once in ten years, and the ratio over all cases of the amount likely to fall in 6 h to the 24-h total. The latter is a measure of the liability to short falls of great intensity. Comparable data are not available from Alaska, but it may be expected that in the Kenai, Chugach, Alaskan and Wrangell ranges conditions will resemble those in the outer ranges of British Columbia, and those of the interior and north will likewise resemble their Canadian neighbors.

It is clear that the risk of heavy falls in no way follows the distribution of rainfall. It is high in the west coastal mountains, chiefly due to the occasional orographic intensification of slow-moving cyclonic rain areas, not exclusively frontal. It is low over the interior plateaus and ranges of British Columbia and the Yukon; and it is even lower across the entire extent of Arctic Canada. But it is very uniform—at about 8 cm—in the populous southern belt of the country from the continental divide to Newfoundland. A local center of rather higher risk—up to 11 cm—occurs in southwestern Ontario, chiefly in the Niagara cuesta belt northwest of Toronto (where disastrous flooding from over 20 cm of rain followed the arrival of hurricane Hazel in October, 1954). And it is also higher in the southeast coastal regions of New Brunswick, Nova Scotia and Newfoundland, because of exposure to hurricanes and very rainy Atlantic coast systems.

In nearly all populous southern parts of Canada east of the Alberta/Saskatchewan boundary the ratio of 6 h to 24-h falls is high—usually 0.7 or 0.8—showing that the chief risk is for short period shower type falls. Coastal British Columbia, Nova Scotia and

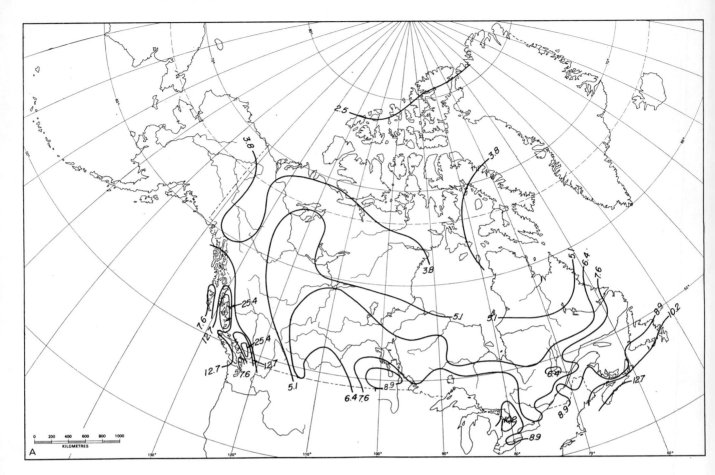

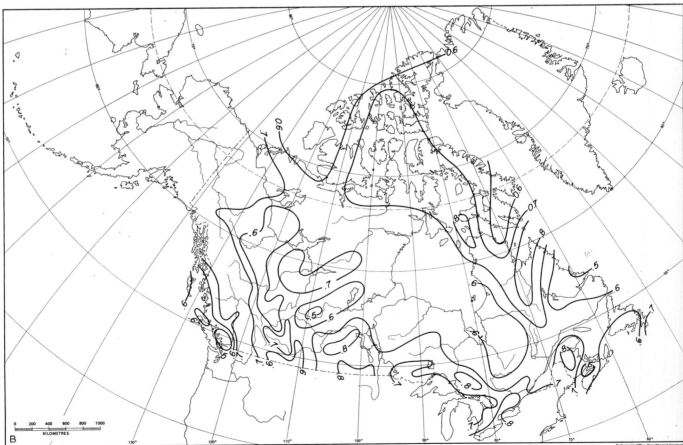

Fig.21. 24-h rainfall (cm) with 10-year return period (A) and ratio of corresponding 6-h to the 24-h fall (B), Canada (after MURRAY, 1964; BRUCE, 1968).

Newfoundland have ratios of 0.6 or below, indicating a greater proneness to prolonged falls.

## Precipitation variability

Large interannual variations in total precipitation are common in all parts of Canada and Alaska. We present Fig.22 as an indicator of this variability. The parameters are the root-mean-square deviation of annual totals about the long-term value, not in this case standardized to a common period because of data problems; and the ratio of this deviation to the annual mean (the coefficient of variation).

The chart speaks for itself. The largest values of the deviation are clearly associated with high mean values, but proportionately the variability is highest in the dry interior belts—a result consistent with world-wide tendencies. Experience suggests, however, that the time-spectrum of variability contains energy well beyond the interannual-scale. Several dry years or wet years in a row are not uncommon. In regions where rainfall is marginal for crop-growth (as on the prairies) or water supply, these little-understood longer variations are of very critical importance.

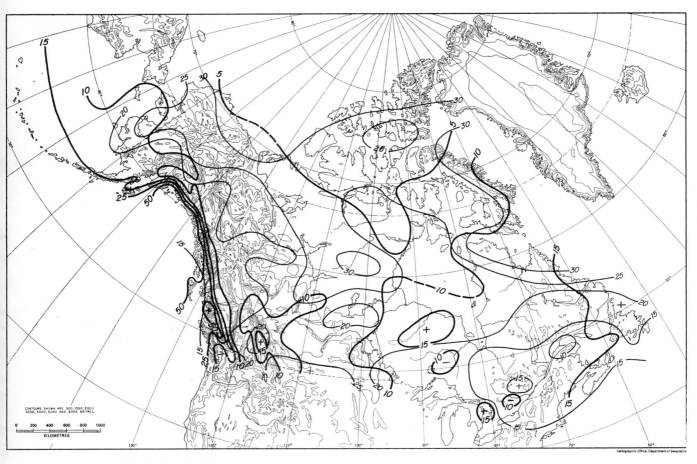

Fig.22. Standard deviation of annual precipitation (cm). Heavy isohyets give the standard deviation in cm, not standardized as to period, but including only stations with at least 20 years of record. The light curves give the coefficient of variation, $100\,\sigma/P$.

**Energy and moisture exchanges**

### General

We shall now try to relate the thermal and moisture regimes to the mass and energy exchange processes near the ground—the radiative fluxes, emitted and absorbed; the turbulent fluxes of enthalpy and water vapor; the run-off; and the sub-surface flux of heat.

That we can attempt this at all is to the credit of the Canadian Atmospheric Environment Service, whose network of radiation stations is excellent, and whose *Radiation Bulletin* makes the results readily available; to the recent extension of stream gauging by various bodies in both Canada and Alaska, and to the publication of the results, primarily by the U.S. Geological Survey (WILSON and ISERI, 1969) and the CANADIAN NATIONAL COMMITTEE FOR THE INTERNATIONAL HYDROLOGICAL DECADE (1969), but also by SANDERSON and PHILLIPS (1967), in a pioneer review of Canada's water surplus; and to the Agrometeorological Service of Canada's Department of Agriculture, who have systematized moisture budgeting techniques for farmed areas.

We have avoided the use of traditional techniques of approximation, such as Thornthwaite's scheme, because in northern climates the annual moisture cycle invalidates the assumptions on which they are based. Instead we have based ourselves on the simple but fundamental conservation laws for mass and energy. Many difficulties still occur, the worst being the doubtful accuracy of precipitation and run-off measurement in cold regions; in places this has stopped us from proceeding.

We begin by a brief examination of the water and heat balance equations, and of the parameters derived from them.

### Parameters of the energy and moisture balances

Disregarding changes in stored water and energy, we may write, per unit time:

$$P = N + E \tag{3}$$

$$R = H + LE \tag{4}$$

where $P$ = precipitation (precip-cm = g cm$^{-2}$); $N$ = run-off (precip-cm); $E$ = evapotranspiration (precip-cm); $H$ = convective heat-flux (ly); $R$ = net radiation (radiation balance) (ly); and $L$ = latent heat of vaporization ($\sim$590 cal. g$^{-1}$).

The use of these balance equations is valid for mean annual values, since the small omitted terms almost vanish over a single year (e.g., the soil heat flux, which reverses anually; and the photosynthesis–respiration balance. For shorter periods, especially to show seasonal variations, one must, of course, include these other processes.

Following BUDYKO (1958) and LETTAU (1969) we define the dimensionless parameters:

$$B = H/LE \text{ (the Bowen ratio)} \tag{5}$$

$$D = R/LP \text{ (the dryness ratio)} \tag{6}$$

$$C = N/P \text{ (the run-off ratio)} \tag{7}$$

Again following Lettau we combine these in the general balance equation for annual mean values:

$$(1 + B)(1 - C) = D \tag{8}$$

We note that for such annual means:

$$E = RL^{-1}(1 + B) = (1 - C)P \tag{9}$$

For much of our region $P$ and $N$ are measured values, though the density of observation is sparse. Where there are good enough observations we can compute $C$. We have computed $R$ for all main climatological stations, and thus we can get $D$. Since $E = P - N$ for annual mean values we can readily estimate $H$ and $B$.

The key energy input is, of course, $I$, the solar radiation, or more precisely the proportion absorbed at the surface, $I(1 - \alpha)$, where $\alpha$ is the albedo. The net radiation is then:

$$R = \underbrace{I(1 - \alpha)}_{\text{(net solar)}} + \underbrace{R{\downarrow} - \varepsilon\sigma T_s^4}_{\text{(net long-wave)}} \tag{10}$$

where $R{\downarrow}$ = atmospheric long-wave radiation received at surface; $\varepsilon$ = emissivity of surface (generally $> 0.9$); $\sigma$ = Stefan-Boltzmann constant ($1.355 \cdot 10^{-10}\,\text{cal} \cdot \text{cm}^{-2}\text{K}^{-4}\text{sec}^{-1}$) and $T_s$ = surface temperature ($^\circ$K).

We have also computed (Fig.27A,B, p.112) the Angström ratio:

$$A = (R{\downarrow} - \varepsilon\sigma T_s^4)(\varepsilon\sigma T_s^4)^{-1} \tag{11}$$

This leads to Lettau's (1969) form of the surface energy balance equation:

$$I(1 - \alpha) = A\varepsilon\sigma T_s^4 + H + S + LE \tag{12}$$

where $S$ = soil heat flux. Eq.12 is valid for all time intervals, provided that photosynthesis and respiration are disregarded, horizontal divergence is considered to contribute to $H$ and $LE$, and $L$ is taken to include the latent heat of fusion as well as of vaporization if thawing, melting or sublimation are involved. Most evaporation takes place at above-freezing temperatures, but the latent heat of fusion is important to the surface energy balance in the freeze-up and thaw seasons.

The non-linearity of the various flux equations imposes certain constraints on the use of time-averaged values. In particular it should be noted that the Bowen ratio of the annual means of $H$ and $LE$ is *not* the same as the annual means of monthly or daily values of $B$. The latter are, of course, rarely available, a notable exception being the work of Davies and Buttimor (1969) and associates at Simcoe, Ontario. The only other regional treatment on the present scale, and in the present context, is that of Bryson and Kuhn (1962), which gives Bowen ratios for parts of northern Canada.

**Solar radiation** *(I)*

Observations of solar radiation are available from 5 Alaskan and over 50 Canadian stations, mostly for rather short periods. Most of the records contain breaks, and variations in instrumentation add to the inhomogeneity. Mateer (1955) published a pioneer analysis of the insolation climate of Canada, and maps of the whole continent have recently been prepared by the U.S. Environmental Data Service (U.S. Weather Bureau, 1968)

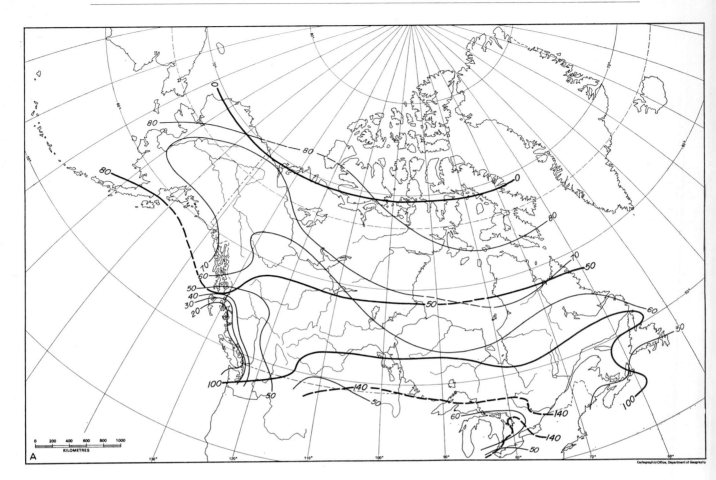

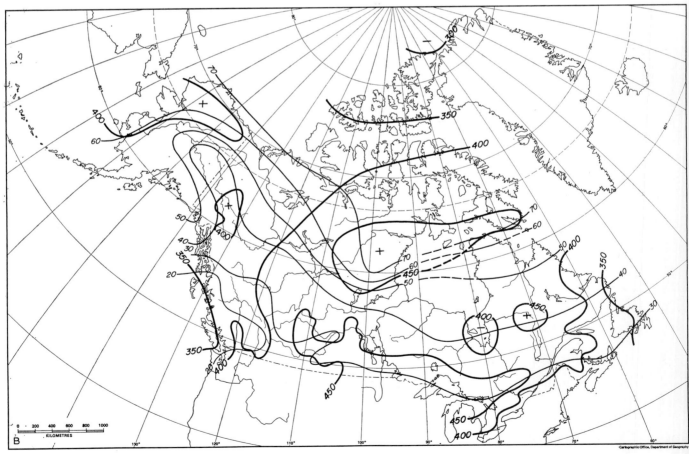

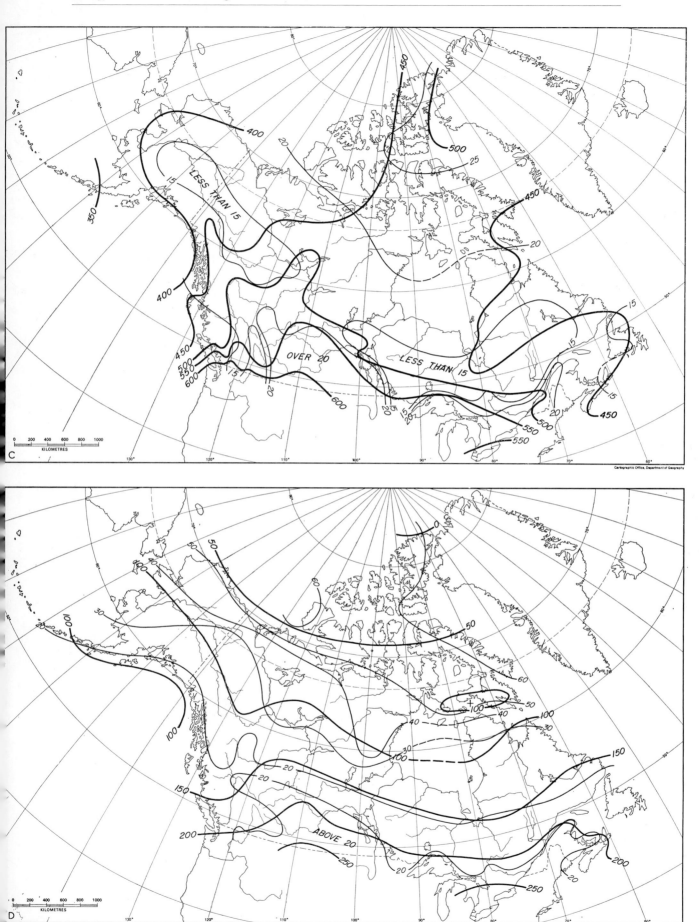

Fig.23. Mean global solar radiation received at surface, with generalized albedoes, 1957–64, mid-season months: A. January; B. April, C. July; D. October.
Heavy curves give radiation in ly day⁻¹. Lighter curves are percentage surface albedo (heavily generalized).
(Data from HAY, 1970.)

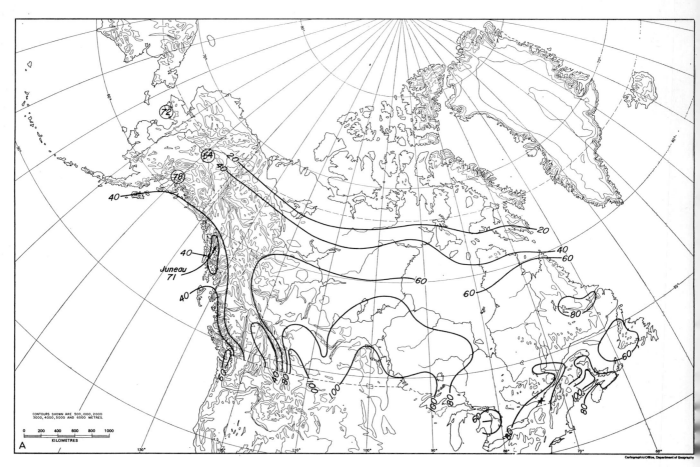

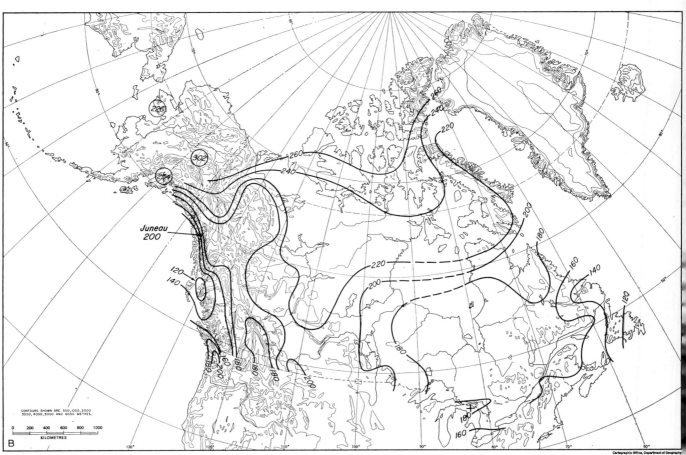

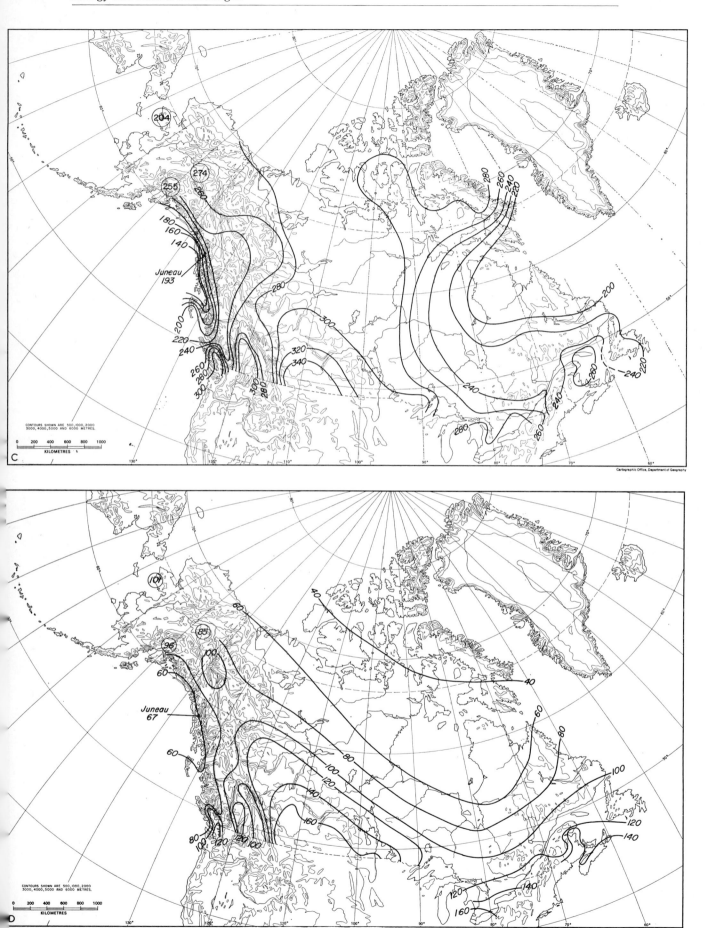

Fig.24. Mean hours of bright sunshine, 1931–60, mdi-season months: A. January; B. April; C. July; D. October.
Isopleths not drawn over Alaska because of paucity of stations. Values for Nome, Fairbanks, Anchorage, and Juneau are ringed. (After CANADA DEPARTMENT OF TRANSPORT, 1967; U.S. WEATHER BUREAU, 1968).

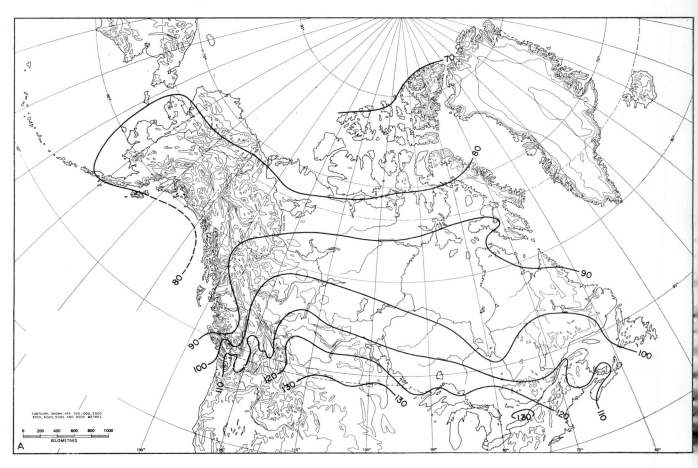

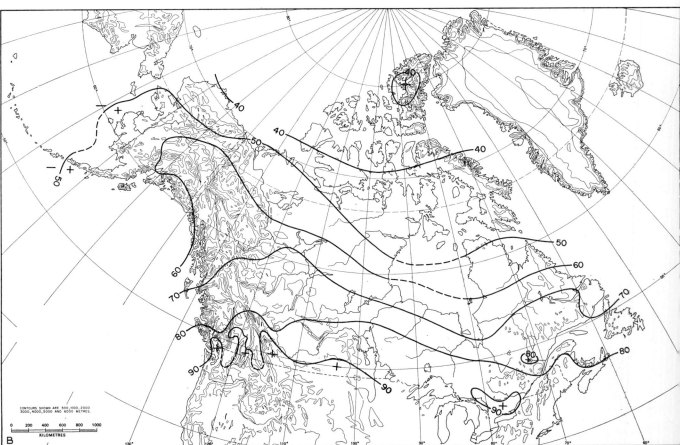

Fig.25. Annual mean global solar radiation (kly) (A) with absorbed values (B), 1957–64. (Data from HAY, 1970.)

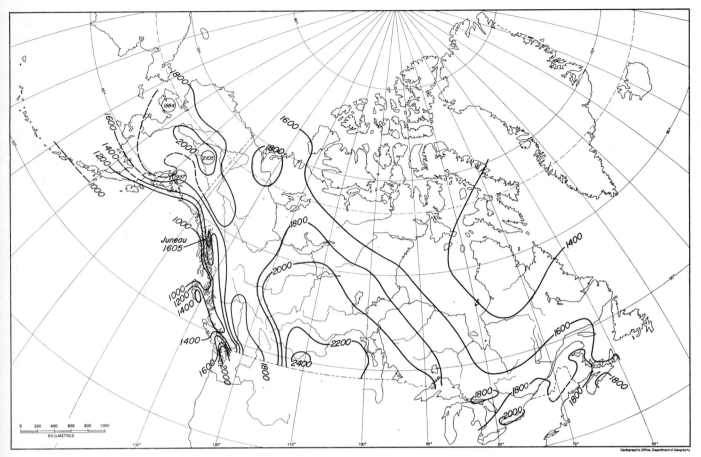

Fig.26. Annual mean hours of bright sunshine, 1931–60. Isopleths very approximate over Alaska (after CANADA DEPARTMENT OF TRANSPORT, 1967; U.S. WEATHER BUREAU, 1968).

and the Canadian Atmospheric Environment Service (TITUS and TRUHLAR, 1969). There are also published computations (from the same sources) of bright sunshine duration and mean cloudiness. HAY (1970) has re-examined the evidence from Canada and Alaska, and we have used his computed fields in this study, because we believe these to be regionally the most representative, and because of their higher spatial sampling density.

We present in Fig.23–24, for the four mid-season months, estimates of the mean fields of solar radiation at the surface ($I$); the seasonally appropriate value of the surface albedo ($\alpha$), from which the absorbed solar radiation $I(1 - \alpha)$ can be derived; and the duration of bright sunshine, as measured at climatological stations. Mean cloud amounts are given in Appendix I, but have not been mapped. For the annual totals (Fig.25–26) we have confined ourselves to solar radiation and sunshine duration.

We emphasize that the absorbed solar radiation $I(1 - \alpha)$, and to some extent the other relevant fluxes, display strong local variations on the topoclimatological scales. Nothing of this can be inferred from Fig.23–26. Excellent studies of the topoclimatological scale of variation have been published by ROUSE and WILSON (1969) and ROUSE (1970). Little is known of the vertical variation of the radiative fluxes with respect to a horizontal surface. Hence our isolines are highly tentative near mountains.

*Seasonal variation of solar radiation*

In January, when there is darkness north of about 70°N, there is also high cloudiness along the Atlantic (>7/10) and Pacific (>8/10) coasts. Elsewhere it is generally above 6/10 except in a broad central zone about the 95°W meridian, where it is below 5/10 in many areas. Because of snowcover, albedo is above 0.5 everywhere except in the mild coastal regions of southern British Columbia, where values below 0.2 permit substantial absorption. There are also areas of albedo below 0.5 in Nova Scotia and southwestern Ontario, due to interruptions of snowcover.

The distribution of solar radiation in January is dominated by the latitudinal gradients of daylength and zenith angle. In the least cloudy areas of central Canada values fall from 140 ly day$^{-1}$ along the U.S. border to zero in 70°N. On the Pacific and Atlantic littorals, however, cloudiness pulls values down below 100 ly day$^{-1}$. Sunshine duration accords with this picture, which is broadly representative of the whole winter.

April (Fig.23) shows the remarkably different patterns of spring, when the rapid increase of day length in the north, coupled with continued low mid-continental cloudiness, largely destroys the latitudinal gradient typical of winter. Over the high Arctic and the oceanic littorals values of solar radiation are still below 400 ly day$^{-1}$, but over a large area of Canada the level is 400–460 ly day$^{-1}$. This tendency is accentuated in May when a strong maximum occurs over Keewatin, with high values to Ellesmere Island and to Lake Winnipeg. Values are also high over inland Alaska. The effectiveness of these high inland values is diminished by the slow northward retreat of the snowline and the consequent high albedoes. In April albedo still exceeds 0.4 north of a line from Newfoundland through northern Ontario to northern Alberta and the Yukon Valley.

July (Fig.23) is marked by the broadening of the mid-continental cloudiness minimum, values falling below 5/10 in the prairies, the southern interior of British Columbia and parts of southern Ontario. Interior Alaska becomes cloudier, while both oceanic coasts remain fairly cloudy. Daylength exceeds 15 h everywhere, and from about 70°N to the Arctic shore daylight is continuous, though with high zenith angles. Albedo is broadly 20–25% over tundra, prairie and cultivated surfaces, and near 15% in the main coniferous forests, with no strong gradients.

Solar radiation exceeds 600 ly day$^{-1}$ from southern Vancouver Island to southern Saskatchewan, and is almost as high in southern Ontario. Values fall below 500 ly day$^{-1}$ northeast of a line from Great Slave Lake to northeastern Quebec, and northwest of a line from Great Slave to northern Vancouver Island, but remain well about 400 ly day$^{-1}$ to the Arctic coasts. The high cloudiness of Alaska brings values down to below 400 ly day$^{-1}$ all along the Bering Sea coast and the Aleutian chain, as well as along the southeast coast. Cloudiness also causes the low values over eastern Labrador–Ungava. The secondary minimum over the Lake Erie counties of Ontario must depend mainly on smoke attenuation (FLOWERS et al., 1969).

The autumn brings the snowline rapidly south, and hence the high albedoes typical of snowcover. In addition cloudiness rises to its annual maximum in most areas, exceeding 8/10 in October (Fig.23) in a broad belt near the 0°C isotherm. Values are somewhat lower from Yukon to the prairies and in extreme southeastern Canada. Rapidly shortening days, and increasing zenith angles, also introduce strong meridional gradients of clear-sky radiation. Average values for normal cloudiness range from 250 ly day$^{-1}$ in the southernmost prairies, and southern Ontario to near zero values over the Arctic islands north of Lancaster Sound.

Annual radiation and sunshine totals (Fig.25–26) generally accord with the above account. Alaska's low mid- and late-summer income is offset by the brilliance of spring and early summer, so that her annual income (except on the outermost Pacific coast and in the Aleutians) is comparable with that of central Canada in the same latitudes. Southern Ontario, however, has values significantly lower (primarily as regards radiation rather than the sunshine duration) than the cloudiness would suggest. Almost certainly this is a consequence of smoke attenuation.

Excellent agreement exists between Fig.23 and 25 and the independently prepared series of Titus and Truhlar (1969) for Canada, and with observed values at long-period radiation stations. The U.S. Weather Bureau does not attempt to draw isopleths over Alaska, but the published means also fit the present charts.

**Long-wave radiation**

For reasons of space we present only mid-winter and mid-summer values of the upward and downward long-wave fluxes. The sum of these fluxes, the *net long-wave flux*, is almost always negative, except for brief periods, and is always negative as a climatological entity. We also give the Angström ratio (eq.11 above) showing the net long-wave as a fraction of the upward long-wave flux at the surface. In estimating upward long-wave in the absence of reliable estimates of surface emissivity, we have put $\varepsilon = 1$, which may have over-estimated the upward flux up to a maximum of 10%, probably offset in *net* computations by compensating errors in the estimated downward flux and its absorption (with long-wave albedo assumed zero).

The January chart (Fig.27A) for the upward flux thus depends effectively on an estimate of the distribution of surface temperature. Values exceed 600 ly day$^{-1}$ along both unfrozen oceanic shores, and fall below 400 ly day$^{-1}$ only over the central Arctic Archipelago. The downward atmospheric radiation naturally shows a more emphatic pattern, from about 550 near the Pacific and Atlantic shores to below 300 in the cold, dry Arctic core region. Net long-wave cooling thus lies in the range 50–100 ly day$^{-1}$, corresponding to Angström ratios of near 0.10 along cloudy warm coasts and in much of Alaska to above 0.20 in the Arctic islands.

In July (Fig.27B) the principal contrasts are across the mean locus of the Arctic front. As regards upward flux, values exceed 800 ly day$^{-1}$ everywhere south of the axis Brooks Range–Great Bear Lake–James Bay–Belle Isle, and are above 850 ly day$^{-1}$ in most parts of closely settled Canada. Values over 900 ly day$^{-1}$ occur in the hot interior of British Columbia and southernmost Ontario. North of the axis overland emission is everywhere in the range 700–800 ly day$^{-1}$. Atmospheric radiation to the surface is in the 550–600 ly day$^{-1}$ range well north of the Arctic front, and generally above 650 ly day$^{-1}$ elsewhere (over 700 ly day$^{-1}$ in the British Columbia dry belt, the Great Lakes–St. Lawrence region and the maritime provinces).

Angström ratios are generally near 0.20 but are lower on the outer British Columbia coast and in Alaska.

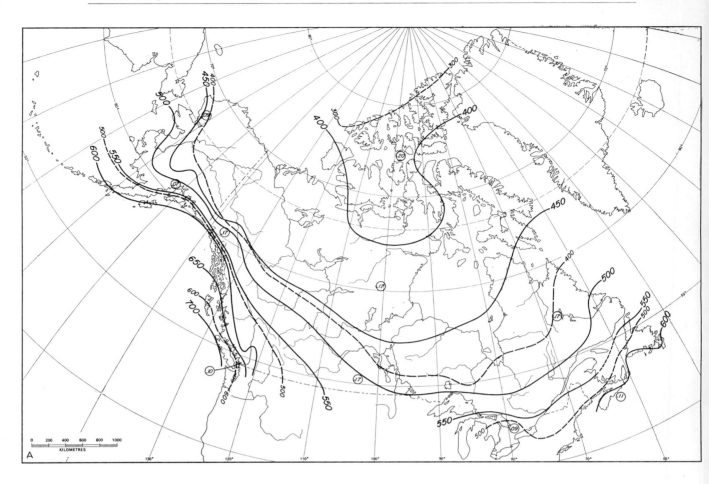

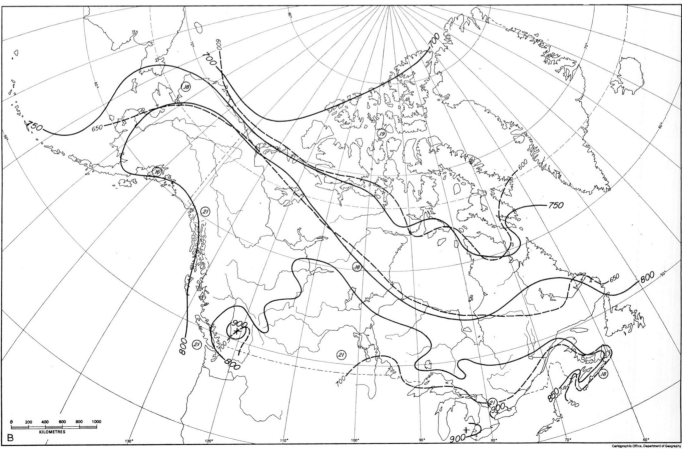

Fig.27. Mean *l*ong-wave radiation. Solid isopleths are long-wave radiation (ly day⁻¹) emitted by surface, dashed isopleths long-wave radiation from atmosphere absorbed by surface. Ringed figures are Angström ratios (net long-wave as a fraction of the emitted long-wave at surface). A. January, 1957–64; B. July 1957–64. (Data from HAY, 1970.)

## Net radiation

The net radiation, $R$, sometimes called the radiation balance, is one of the fundamental parameters of climate. Net radiometer measurements are made routinely at a few stations in Canada and Alaska, and mean values have been computed for them. There are, however, uncertainties about both the representativeness of the observing sites (especially as regards albedo) and the field performance of the net radiometers. HAY (1970) has accordingly calculated the net radiation for most Alaskan and Canadian first-order climatological stations by summing the component fluxes on the right-hand side of eq.10. In performing the calculations he used what we believe are regionally representative albedo values, which tend to be a little higher than observing stations values.

We have used Hay's analyses in Fig.28–29, since they offer the only available consistent estimates for our region. These charts present estimates for the mid-season months and for the year, in terms both of energy (ly or kly) and the equivalent evaporating power (in cm). Although some uncertainty remains as regards the actual net radiation income of the surface, chiefly because of the difficulty of albedo determination, we have found that the use of Hay's values makes possible a synthesis of the water and energy balances within Lettau's climatonomic framework.

The mid-winter chart, January (Fig.28), shows negative values everywhere, a condition normally true of the period late November–mid-February in all areas. In the Arctic, values are below $-70$ ly day$^{-1}$, but they are near zero in southern Ontario and the Montreal district. There is a notable change to less extreme values west of the mean Arctic front locus. Since mean daily temperature is near its annual minimum at this time, these negative values must be compensated by sensible heat advection.

In February and March net radiation in southern districts turns positive, and this ultimately leads to snowmelt, a fall of albedo, and a further consequent increase in net radiation. By April (Fig.28) negative net radiation is now confined to the Arctic islands and northern Keewatin. A strong gradient exists over much of Canada, and as the snowmelt moves north in April and May this gradient intensifies. In May it is extremely strong, lying along the line from the Brooks Range to Churchill and Hudson's Bay.

Highest values of net radiation occur in June at some stations, but for consistency's sake we have used July values for Fig.28. Distribution is very uniform. Values below 200 ly day$^{-1}$ are confined to southwestern Alaska and the plateau of the Yukon. Across southern and sub-Arctic Canada values are close to 250 ly day$^{-1}$. Highest values lie near the U.S. border in British Columbia, Alberta and Saskatchewan, on west coastal Vancouver Island, and in the Sault Ste. Marie–Sudbury–Val d'Or area. Significantly lower values occur in southernmost Ontario, in the middle St. Lawrence Valley and in the St. John River Valley of New Brunswick. A rapid decrease in net radiation occurs in August. (Fig.29 shows month-to-month variation for selected stations.)

The rapid downturn after July accelerates in September, negative values being established by mid-September down to about 70°N (a week later in Alaska and the Beaufort Sea coast). By October (Fig.28) negative values are down to about 55°N, and by November have reached all areas, a zonal pattern being maintained throughout.

Net radiation is only slightly lagged with respect to the solar forcing function $I(1 - \alpha)$. This arises mainly because the short-wave component is the largest part of $R$ in most months and in most areas. July is usually the month of strongest radiative heating,

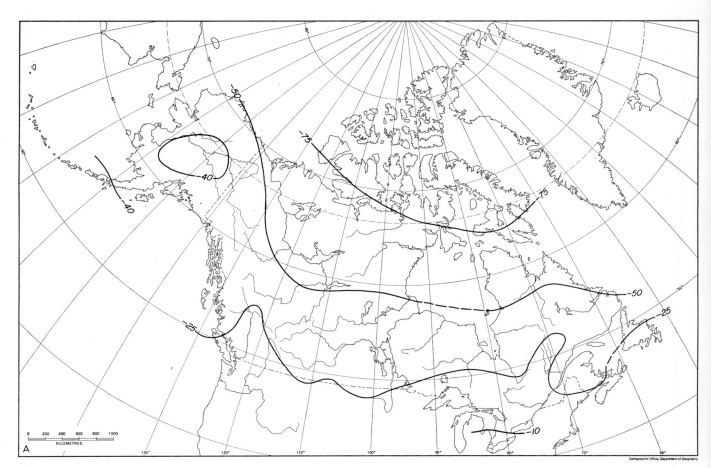

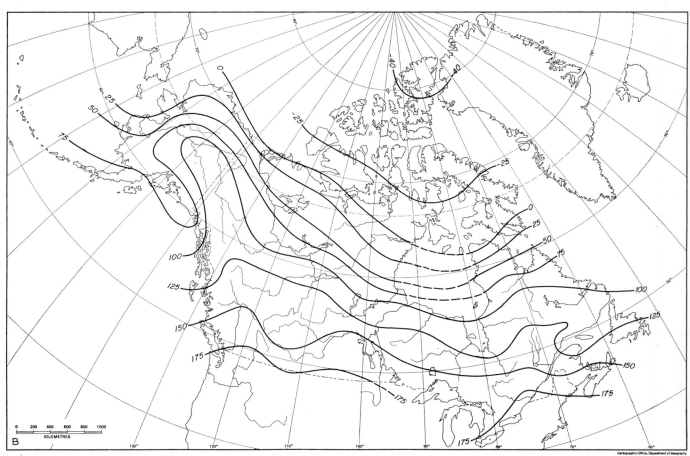

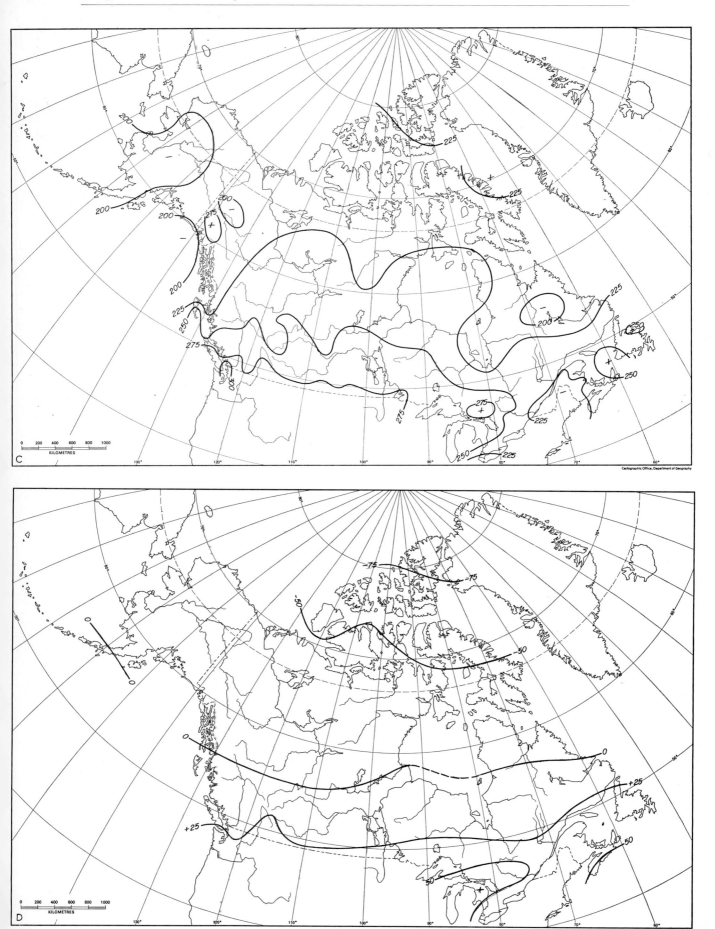

Fig.28. Net radiation for mid-season months, 1957–65. Isopleths give net radiation received at surface, positive or negative, in ly day⁻¹. A. January; B. April; C. July; D. October. (Data from HAY, 1970.)

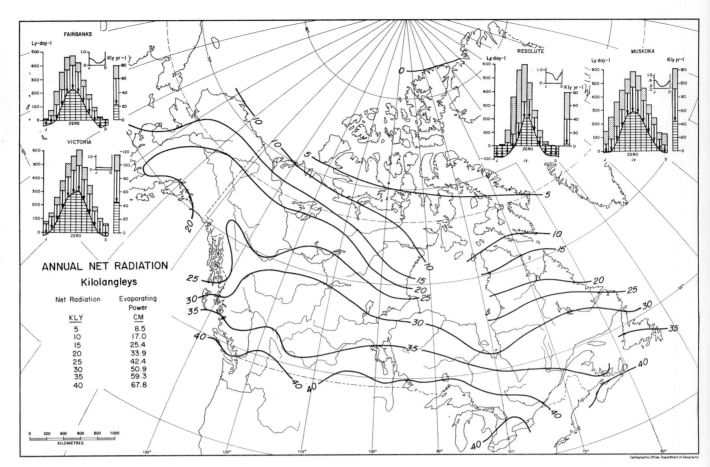

Fig.29. Mean annual net radiation, 1957–64. Isopleths are in kly year⁻¹, with equivalent evaporating power in cm in table. Inset diagrams for selected stations give (*i*) global solar radiation (entire column above zero); (*ii*) reflected solar (stippled); (*iii*) net long-wave cooling, downward arrow; and (*iv*) net radiation, ruled, month-by-month. Smaller inset shows equivalent albedo variation. (Data from HAY, 1970.)

December the time of strongest cooling. The lag in air temperature behind this cycle reflects the effect of thermal storage in the soil, water-bodies and the surrounding oceans. Probably 15% of all Canada is either lake, marsh or inlet, in which water storage effects accentuate the seasonal lag, and the presence of Hudson's Bay and the channels of the Arctic Archipelago add to this effect. Alaska has less surface water, and much less lag.

We have tabulated measured or observed values of the various component fluxes and the net radiation for principal stations in Appendix II.

**Evaporation and run-off**

We shall be concerned here with evaporation off open water, off soil and off vegetation. In the latter case the process is properly called *evapotranspiration*, since it involves water transpired from green tissues and evaporated off the soil beneath plants. The evaporation of rain or snow clinging to plants—interception—will be treated as a part of evapotranspiration, since it cannot be separately treated on this scale, and since its energy requirements are virtually the same.

There have been several previous attempts to estimate evaporation over Canada and Alaska. In Canada SANDERSON (1949, 1950) applied the Thornthwaite method to the entire country, and made experimental measurements of potential evapotranspiration (using watered lysimeters) at Norman Wells, Windsor and Toronto. With Phillips she recently published a survey of Canada's moisture surplus (SANDERSON and PHILLIPS, 1967), again using estimated potential evapotranspiration from Thornthwaite's empirical formula. A similar analysis for Alaska has been prepared by PATRIC and BLACK (1968).

There are serious difficulties in both defining and estimating an evaporative potential for cold, snowy countries, even where the net radiation is known. At low temperatures the Bowen ratio tends to be high even over wet surfaces, but its value can be predicted theoretically only for equilibrium evaporation (i.e., when saturation exists at both evaporating surface and screen level). The spring snow-melt complicates the annual energy cycle by consuming up to 3 kly of the net radiation. The soil heat flux is also large in early summer. On all counts we are therefore reluctant to use empirical formulae for the potential evapotranspiration. Even where reliable estimates of the potential can be made, the derivation of actual evapotranspiration depends on budgeting the potential against measured precipitation, which in northern areas is unreliable. In what follows we have hence relied mainly on direct observation, or simple conservation principles.

*Open-water evaporation*

Class-A pan measurements of open water evaporation are carried out at 39 stations in Canada and 8 in Alaska. FERGUSON et al. (1970) used the Canadian observations to estimate evaporation off small lakes in which stored heat plays only a negligible role in supplying energy, which must hence come from radiation or atmospheric advection. In well-watered areas such small lakes might therefore be expected to evaporate at a rate almost equivalent to the net radiation over the water, which will in general be slightly higher than over more reflective land surfaces. In drier areas of strong solar heating, such as the prairies, or the dry belts of British Columbia, Yukon and Alaska, small lakes should act as oases (as should Class-A pans), and should hence receive considerable advected energy. In such areas evaporation should exceed the net radiation, especially if the latter is measured or computed for dry-land or regionally representative albedoes. Fig.30 shows the chart for annual mean evaporation values. A rational relationship with the annual net radiation chart (Fig.29) is apparent. Evaporation does indeed exceed the regional values of net radiation over the prairies and in southern Ontario. In a wide area of the north, however, and especially around Hudson's Bay, evaporation falls below 70% of the net radiation. This arises chiefly from the spring snow and ice cover on the lakes, which reduces their net radiation income well below that of the land in spring and early summer, and from the high Bowen ratios typical of low temperatures.

*Evapotranspiration off land surfaces*

Because of the difficulties mentioned above we present no chart of actual evapotranspiration, and no analysis of month-by-month water budgets for the country as a whole. This is a serious gap in our analysis. Doubts as to the representativeness of precipitation data

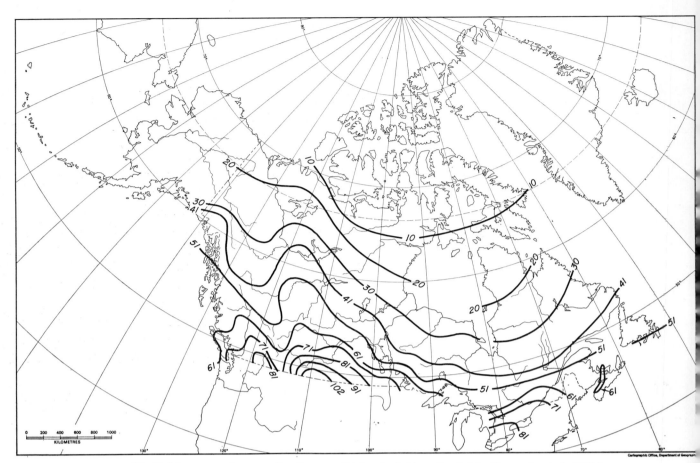

Fig.30. Mean annual evaporation from small lakes (cm). (As estimated by FERGUSON et al., 1970.)

are less justified, however, in the densely settled areas of southern Canada. Much of the land in these areas is used for arable farming or controlled pasture, in which a high degree of management of soil water is normal. There is considerable experimental knowledge in these areas of the water cycle over farmland and woodland (e.g., LAYCOCK, 1960; FRASER, 1965; DAVIES and McCAUGHEY, 1968; BAIER, 1969a; PELTON and KOR- VEN, 1969). From these and other sources, the Agrometeorological Service of the Canada Department of Agriculture has established techniques of water budgeting that can be applied directly to both estimates of evapotranspiration over cropped land and to crop development and yield (BAIER and ROBERTSON, 1965; BAIER, 1969b).

In Appendix III (by Wolfgang Baier) we present analyses of this kind for representative sites across southern Canada, using wheat as the reference crop.

### Run-off estimates and rainfall

Run-off is measured over about a third of the surface of Alaska and Canada (CANADA NATIONAL COMMITTEE FOR THE INTERNATIONAL HYDROLOGICAL DECADE, 1969; WILSON and ISERI, 1969). We have had access to extensive data tabulations prepared by D. W. Phillips, personal communication, 1969). Stream-gauging in these latitudes presents considerable difficulties. It is difficult to measure winter flow below the ice, and to cope with the spring flood, accompanied as it is by the break-up of the river ice.

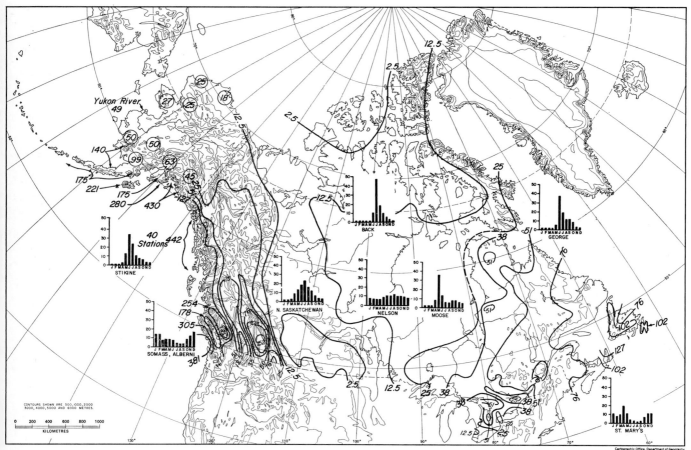

Fig.31. Mean annual run-off (cm). The isopleth interval is irregular in the west. Ringed figures over Alaska represent local basin estimates, since isopleths could not be attempted. Histograms show percentage distribution of run-off by month for representative streams. (After CANADIAN NATIONAL COMMITTEE FOR THE INTERNATIONAL HYDROLOGICAL DECADE, 1969; HULSING, 1969.)

In Fig.31 we present the best current estimate of run-off, accompanied by seasonal regime histograms for some representative streams. We have amended the chart for Canada along the Alaskan border, where we used the WILSON and ISERI (1969) estimates for the Panhandle region, as well as unpublished data provided by the U.S. Geological Survey (D. W. Hulsing, personal communication, 1969). In Alaska itself we have been unable to complete a run-off chart, and have followed Wilson and Iseri in providing the data by drainage slope. Two difficulties at once present themselves:

(*1*) Over Keewatin (except for its western margin) and the Canadian Arctic islands there are no measured data, and the chart has hence been prepared by its original authors by *assuming* a run-off ratio and probable values of evapotranspiration. Hence in these areas we could not estimate evapotranspiration from eq.3, even if we could rely on measured precipitation.

(*2*) In mountainous British Columbia, Yukon and Alaska, where run-off data are available, the streams concerned invariably rise in mountainous areas from which there is little or no information about precipitation. Hence the run-off figures are always greater than that typical of the extensive plateau and valley regions where the rainfall stations are located. Between these areas, however, essentially from the Mackenzie–Yukon areas to the prairie provinces and across eastern Canada, a more meaningful

TABLE II

ANNUAL MEAN RUN-OFF FROM CANADA AND ALASKA BY DRAINAGE SLOPE

| Drainage slope | Area<br>$10^3$ km$^2$ | Run-off<br>$10^3$ m$^3$ sec$^{-1}$ | Precip.<br>equivalent (cm) |
| --- | --- | --- | --- |
| *Canada:* | | | |
| Pacific (including Columbia, Yukon) | 1,086 | 21.3 | 67 |
| Arctic | 3,583 | 16.5 | 15 |
| Hudson's Bay | 4,054 | 29.4 | 23 |
| Atlantic | 2,045 | 33.5 | 52 |
| Gulf of Mexico (via Missouri) | 28 | 0.9 | 1 |
| Total | 10,796* | 100.7 | 29 |
| *Alaska:* | | | |
| Gulf of Alaska | 469 | 28.1 | 189 |
| Bering Sea (Kvichak Bay to Noatak River) | | | |
|     Yukon River | 849 | 7.3 | 27 |
|     Others | 356 | 6.2 | 55 |
| Arctic | 218 | 1.2 | 17 |
| Total | 1,890 | 42.8 | 29 |

* Includes 780,000 km$^2$ of United States territory draining into Canada.

Sources: CANADIAN NATIONAL COMMITTEE FOR INTERNATIONAL HYDROLOGICAL DECADE, 1969; U.S. Geological Survey (WILSON and ISERI, 1969; D. W. Hulsing, personal communication, 1969).

comparison between rainfall and run-off, and hence evaporation, can be attempted. The run-off accepted for Canada and Alaska is given in Table II. The aggregate annual total, $4,540 \cdot 10^9$ m$^3$, is only a little less than the $4,710 \cdot 10^9$ m$^3$ reported for the external drainage of the Soviet Union, whose territories are nearly twice as extensive (BOCHKOF et al., 1970).

HARE and HAY (1971) have recently shown that the fields of mean annual precipitation, net radiation and run-off, given here as Fig.15, 29 and 31, respectively, are incompatible. The fields of evapotranspiration and convective enthalpy flux derived from them by means of eq.3 and 4 are physically improbable, as are the Bowen, dryness, and run-off ratios. The authors suggest that the difficulty probably arises from under-measurement of precipitation, especially snowfall. Their findings reinforce the view expressed above that moisture budgeting is not yet possible, especially in the north.

We shall not reproduce their arguments, but in Table III we show a comparison of estimates of precipitation and evaporation made by standard climatological methods at Knob Lake, Quebec, and by experimental basin techniques in a small nearby basin that was carefully instrumented by the same observers. FINDLAY (1966, 1969) has analyzed the results, and in Table IV we give his basin values compared with those based on standard climatology. The basin values of run-off, dryness and Bowen ratios are far more readily comprehensible than the climatological values. The analysis found that snowfall was being under-measured by standard climatological techniques by about 25%, and rain by 15%. Findlay's values for the basin doubled the estimated evapotranspiration and

TABLE III

COMPONENTS OF THE MEASURED WATER BUDGET, KNOB LAKE BASIN, BASED ON NINE YEARS 1956–64, WITH ENERGY EQUIVALENTS

(Basin area 35 km²; after FINDLAY, 1966; HAY, 1970; D. W. Phillips, personal communication, 1969)

| | Precip. (cm) (water equiv.) | Energy values $R$, $LP$ or $LE$ ($10^3$ ly yr$^{-1}$)* | |
|---|---|---|---|
| *Measured rainfall;* basin | 48 | 28.2 | |
| Knob Lake, climatological station | 41 | 24.2 | |
| Difference | +7 | +4.0 | |
| *Measured snowfall;* basin (water equiv.) | 43 | 25.4 | (28.8)* |
| Knob Lake, climatological station | 32 | 18.9 | (21.4)* |
| Difference | +11 | +6.5 | |
| *Total precipitation;* basin | 90 | 53.3 | (57.0)* |
| Knob Lake, climatological station | 73 | 43.1 | (45.6)* |
| Difference | +17 | +10.2 | |
| Run-off, basin (measured) | 57 | 33.6 | |
| Run-off, Ashuanipi River, (Menihek Rapids) | 64 | 37.7 | |
| Evapotranspiration, basin (measured) | 33 | 19.4 | |
| Net radiation (present computed value) | | 28.0 | |

\* Bracketed figures include snow-melt energy for the reported snowfalls.

halved the convective enthalpy flux. Faced with evidence of errors of this magnitude one can hardly attempt water balance calculations under northern, high snowfall conditions. Considerable difficulty also arises in the mountains and plateau-basins of Alaska, the Yukon and British Columbia. The great bulk of the run-off in this huge area comes from large moisture surpluses on the mountains, snowfields and glaciers. Very little is contributed by the interior basins, yet measured precipitation records from these areas are almost exclusively on the dry valley bottoms. On the west flanks of the mountains nearly all the rainfall stations are close to sea-level, very often in sheltered bays away from the wettest exposed slopes.

Observed run-off in the Pacific coastal mountains (see Fig.31), principally off the western

TABLE IV

VALUES OF CLIMATOLOGICAL RATIOS FOR KNOB LAKE

| | Using basin values | | Using standard climatology | |
|---|---|---|---|---|
| C = run-off ratio (for Knob Lake) | 0.63 | | 0.78 | |
| D = dryness ratio | 0.53 | (0.49)* | 0.65 | (0.61)* |
| B = Bowen ratio | 0.44 | (0.27)* | 1.98 | (1.61)* |

\* Bracketed figures allow for energy use in snow-melt.

slopes, generally exceeds 200 cm from the Fraser Valley to the Kenai Peninsula. Highest run-offs from individual basins are generally in the 350–450 cm range (e.g., Sarita River at Bamfield, Vancouver Island, 404 cm; Capilano Creek, Vancouver, 377 cm; average of 40 gauging stations in Alaskan Panhandle, 442 cm; Power Creek, Cordova, 430 cm). At probable run-off ratios in the range 0.85–0.90 (tending to increase northward), these values imply annual precipitation in the 500–550 cm range on the mountain slopes near the level of maximum precipitation. In fact there are observed mean precipitations as high as 500 cm near sea-level in several areas.

In the inner ranges, the Mt. McKinley Massif and the Alaskan Ranges are markedly less productive of run-off than the Pacific flanks of the Kenai, Chugach and Wrangell ranges, because of effective rainshadow. On the other hand, the Purcell–Monashee–Selkirk–Cariboo complex in British Columbia yields run-off generally over 100 cm, and in the central part of the complex over 150 cm. The Columbia, Kootenay, Thompson and Fraser rivers get most of their flow (east of the Coast and Cascade ranges) from this inland maximum.

The interior dry-belt of British Columbia, from the Okanagan to the Nechako country, generally yields less than 20 cm, and in the true semi-arid country almost nil.

In Alaska and the Yukon, the Matanuska and Copper River valleys are completely dominated by run-off from the surrounding mountains, but the major streams of the interior—the Yukon, Kuskokwim and their tributaries—have huge non-mountainous areas in their catchments. The Yukon's flow corresponds to 25–30 cm run-off for enormous distances, yet measured precipitation along its course is not much greater than this, and at Fort Yukon and vicinity falls to 17 cm. Since net radiation in the interior valleys is equivalent to over 30 cm evaporating power, most of the available soil water in summer should go into evapotranspiration. Even here, unless precipitation is being seriously undermeasured, the streamflow must hence be derived primarily from the hilly areas in which there are few precipitation stations.

### Soil temperatures and heat flux

Soil temperatures are now measured on a routine basis at about 50 stations in Canada and Alaska, and for a few of these stations it is possible to portray with some confidence the annual course of temperature at various depths.

It is not yet possible to portray climatologically the distribution of the soil heat flux, though the rapidly accumulating data, and the satisfactory state of theory, will make such a portrayal possible in the near future. We can get some insight into the probable magnitudes involved from the work of Gold and his associates at the Division of Building Research, National Research Council, Ottawa. GOLD and BOYD (1965) derived for 1960–1961 (a two-year span) simple harmonic expressions for the various component heat fluxes of the surface energy balance at Ottawa (the atmospheric convective heat flux $H$ being taken as the residual):

$$I(1 - \alpha) = 252 - 211 \cos (wt + 13°)$$
$$R = 125 - 157 \cos (wt + 3°)$$
$$LE = 120 + 105 \cos (wt - 2°)$$
$$H = -5 + 35 \cos (wt + 7°)$$
$$S = 18 \cos (wt + 22°)$$

and where the units are ly day$^{-1}$, $I$, $\alpha$, $R$, $LE$, $H$ and $S$ have the meanings assigned in eq.12, $w$ is 360°, and $t$ is time in years from the beginning of January.

If these equations are integrated over the period April 1–September 30 they yield the net values shown in Table V. At this site (grass, with a prolonged winter snowcover) the soil heat flux (which is, of course, a function of the soil temperature gradient) was downwards from late March until early September, with maximum flux of heat into the soil about three weeks ahead of the maximum net radiation. The effect was therefore to divert energy from evapotranspiration early in the summer season, and to release it again in the fall winter. The net diversion was, however, less than 5% of the net radiation, which justifies the frequent neglect of the soil heat flux in the computation of the water balance.

Recent work by LESLIE (1971), using derived values of soil diffusivities and probable values of conductive capacity, has shown that the Gold-Boyd relation for $S$ is very typical of southern Quebec and Ontario as regards phase. Amplitude tends to be a little larger in southern Ontario than in Quebec, possibly because of deeper snowcover in Quebec.

Values of soil temperatures have been published for Canada by ASHTON (1969), and for Palmer and Fairbanks, Alaska, by the U.S. WEATHER BUREAU (1968).

TABLE V

VALUES OF THE VARIOUS ENERGY FLUXES (POSITIVE FOR FLUXES TOWARDS SURFACE), OTTAWA, APRIL 1ST–SEPT. 30TH, 1960–61

| Flux | Value (17) (gcal.cm$^{-2}$) | Equivalent evaporation (cm) | Fraction of net radiation (%) |
|---|---|---|---|
| Net radiation ($R$) | 41,000 | 69 | 100 |
| Evapotranspiration ($LE$) | −34,100 | −57 | 83 |
| Convective heat flux ($H$) | − 5,100 | − 9 | 13 |
| Soil heat flux ($S$) | − 2,000 | − 3 | 4.8 |

Source: GOLD and BOYD (1965).

## Some climatic hazards

The climates of Canada and Alaska contain much to challenge the stamina, patience and imagination of the people. Only in tiny areas of southern British Columbia—the Okanagan Valley and the southeast coast of Vancouver Island—are there climates genial and risk-free enough to invite settlement by retired people. Over the bulk of both countries climate is something to be fought against, withstood, tolerated, and occasionally submitted to. Though it often permits high economic yield, as in the rain forest of the Pacific coast, in the sunny prairies, and on the snow-covered watersheds of eastern Canada, it also creates high cost economically and in terms of human stress.

Much of this will have become apparent in the foregoing sections. Snowfall and snowcover, for example, are simultaneously hazards and resources (for power development).

The extreme variability of temperature, on a variety of time-scales, is largely a hazard. Yet it is so much a commonplace that technology and living habits have to a great extent overcome it. It is an accepted and provided-for cost, written into each individual's and each institution's budget.

In the present section we have grouped together, largely at random, certain remaining hazards that are not nearly so easily overcome, and that cause extensive disruption to travel, discomfort to persons and damage or destruction to property and production. Among these are obstructions to vision, including fog and blowing snow, both of which severely affect travel. Freezing precipitation can also affect travel, and cause heavy damage besides. Tornadoes and hailstorms are a further hazard. We have not treated wind-chill, since winter life in both countries is now largely spent indoors, or in protective clothing, and since in any case the measurement of convective heat loss from the body is not realistic for everyday purposes.

The applied climatology of such hazards is in its infancy, and in Canada has scarcely yet been born. Their impact on decisionmaking, on technological innovation, and on costs, is only now being studied. So also is the entire question of environmental perception. Subsequent editions of this study may well concern themselves with such human responses. Here we can only present a description of the hazards and their distribution, with a few remarks on causes.

**Obstructions to vision**

*Fog*

Fog is uncommon in most parts of Canada and Alaska, occurring on less than 25 days per annum. In such cases it is hardly a major climatic hazard, especially since it occurs chiefly during the few hours before and after dawn. It tends to form chiefly in low-lying areas, especially near (but not in) smoky towns, at times when moist air lies over a surface being strongly cooled radiatively. This happens mostly in late summer and autumn, and least often in late winter and early spring.

Fog is defined internationally as obstruction to vision by water-droplets or ice-crystal concentrations reducing visibility below 1 km, and we have used this definition. Where we have used "days of fog" statistics, they refer to the number of days when visibility at some hour or hours fell below the 1-km limit. Statistics of this sort are included in Appendix I.

Most fog is composed of water-droplets, which are often supercooled, depositing rime on exposed surfaces as long as temperatures remain below 0°C. Occasionally, however, ice-crystal fog occurs, almost exclusively at temperatures below −30°C, and chiefly near settlements where man-made humidity is abundant. In the following account we have confined ourselves to those areas where fog frequencies are high enough to create real problems for traffic or navigation, and to the ice-fog problem. We have used statistics published by MANNING (1965), COURT and GERSTON (1966), the Meteorological Service of Canada (CANADA, DEPARTMENT OF TRANSPORT, 1967b—data series H.D.S.), and the U.S. Weather Bureau.

*Areas of high fog frequency*

The Pacific coast of British Columbia and Alaska has a high frequency of sea-fog, which occurs when stable, moist air moves across the cool waters of the offshore belt. This happens at intervals throughout the year, but the effect is strongest in summer. The fog drifts ashore at many exposed localities, and tends to be densest and most frequent near the hilly shoreline itself, where orographic lifting adds to the effect of surface chilling by the sea. The zone of high frequency extends from the Aleutians to Vancouver Island. It does not, however, extend far inland. The fog may penetrate certain deep inlets like the Strait of Juan de Fuca, but quite small islands and promontories afford protection. Thus Cape St. James, at the southern tip of the Queen Charlotte Islands, has 78 days, whereas Sandspit (on the island's east coast) has only 25. Most of the harbors along the inland passage, from Juneau to Cape St. James, and again along the Straits of Georgia, are also largely protected, fog occurring on less than 30 days in most localities.

A small but important area of high fog frequency of different origin encircles the city of Vancouver and the nearby Fraser delta. In Fig.32 there are plotted the percentage frequencies of fog (observed hourly) for four of Canada's major airfields. Vancouver is the worst affected, fog occurring on 17% of all observations. In mid-winter the frequency rises to over one-third of all observations, and air traffic is often dislocated for many consecutive hours, and occasionally for several days. Navigation in the nearby waterways also becomes difficult. The fog arises from the combination of low average wind-speeds, abundant surface water and considerable local smoke sources (EMSLIE and SATTERTHWAITE, 1966; HARRY and WRIGHT, 1967).

The Arctic coast is similarly affected by widespread sea-fog that drifts ashore during the warmer months, particularly in the early summer. All parts of the coast, from the Bering Sea to the Labrador, are considerably affected. The fog is usually shallow, and breaks up

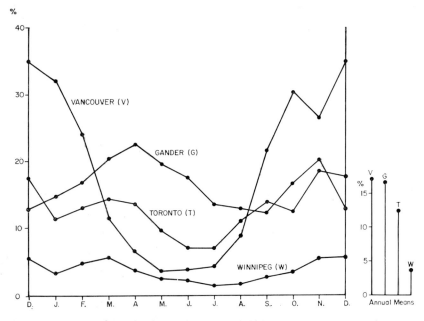

Fig.32. Percentage of hourly observations with visibility 1 km or less due to fog, selected Canadian airfields. (From CANADA DEPARTMENT OF TRANSPORT, METEOROLOGICAL BRANCH, 1967b.)

into fractostratus cloud as it moves inland. It often hangs over inlets and fjords, especially when broken sea-ice is present, while the nearby hills are clear. Some Arctic settlements, such as Eureka, Arctic Bay, Clyde and Frobisher, are protected by local site factors, and suffer little. Others, notably Alert and Barrow, are badly affected. The worst conditions seem to occur in and along the shores of eastern Hudson Strait, the northern Labrador and the southeastern headlands of Baffin.

"Sea-smoke", fog forming over open leads when cold air crosses them, is a commonplace feature of the Arctic winter, and occasionally affects narrow coastal zones.

The Atlantic coast is the third and worst of the areas of high sea-fog frequency. The Strait of Belle Isle, the Avalon Peninsula of Newfoundland, the southern and eastern coasts of Nova Scotia, and the islands and north shore of the Gulf of St. Lawrence are the areas of highest frequency. The source of the fog is the very cold sea-surface, sometimes with partial ice-cover, over which moist air from points between northwest and east moves with great frequency. Belle Isle itself has 143 days of fog; Cape Race and Argentia, in the Avalon Peninsula, over 150; Yarmouth, N.S., 119. These fogs may occur at all seasons, but have highest frequencies in early summer (Gulf of St. Lawrence) or midsummer (most other areas). They are frequently dense and call for the constant use of navigational aids. On land the airports at Halifax, Torbay and Gander are also seriously affected (see Fig.32).

The Great Lakes area and adjacent Ontario and Quebec are the remaining regions of high fog frequency. The area east of Rainy Lake, south of about 49°N and west of about 75°W, generally has over 30 days of fog per annum. Across northern Ontario and southwestern Quebec many of these fogs occur in summer, and are early morning events associated with nocturnal cooling of the very humid airstreams typical of the area. But in densely settled southern Ontario and the St. Lawrence lowlands to Montreal and Quebec, fog is more common in fall and winter. The frequency is not high enough to create serious problems, but interruptions to flying and highway hazards are occasionally felt. In the major cities, the downtown areas usually have considerably lower frequencies. Lake Ontario, however, is prone to fog in spring and summer, and this often affects the north shore.

*Ice fog*

In most parts of Canada and Alaska the air normally remains unsaturated with respect to ice throughout the winter in all except the most intense cases of radiative cooling. Visibility remains generally good, though a thin ice haze, that occasionally sparkles to the naked eye, is a commonplace event when visibility exceeds 20 km. Near settlements, however, where there is abundant release of water from cars, heating systems and aircraft exhausts, thin but persistent fog occurs at temperatures below −30°C, and it has been shown (THUMAN and ROBINSON, 1954; ROBINSON et al., 1957) that such fog consists primarily of ice particles formed by the freezing of supercooled droplets condensed from the various artificial sources. Above −30°C the effect is uncommon, probably because of the absence of freezing nuclei in sufficient concentrations.

Periods of ice-fog occur in the coldest spells of winter over the cities of the Canadian prairies, the northern interior plateaus of British Columbia, and less frequently in the larger towns of the mining belts of Ontario and Quebec. But it is in the far northwest

TABLE VI

SUMMARY OF ICE-FOG EVENT CHARACTERISTICS, FAIRBANKS, ALASKA

| Phenomenon | Normal event | Severe event |
|---|---|---|
| Unbroken duration | 1 to 8 days | about 14 days |
| Surface temperature | −38 to −48°C | about −50°C |
| Depth of cooling layer | 1 km | about 3 km |
| Flow at 700 mbar | Bering Sea ridge | high in N.E. Siberia |
| Surface pressure | migrating Siberian high | expanded Siberian high |

Source: BOWLING et al. (1968)

that ice-fog forms a significant climatic element. The settlements of the Mackenzie and Yukon valleys have a high mid-winter frequency. Because ice-fogs form under very calm conditions, and depend on water from the various exhaust systems, they are usually accompanied by high concentrations of pollutants, and thereby have a reputation for unpleasantness that has nothing to do with the ice-fogs themselves.

In Fairbanks, Alaska, ice-fog periods are frequent, and may become more so as the settlement, and especially the use of its airfields, grows. The fogs have been studied for many years, and their characteristics and origin have been summarized by BOWLING et al. (1968). They develop in association with strong anticyclogenesis of the sort described on pp.56–57 of this study, in which the advection of cold Siberian air clears away the cloud-cover, and allows strong surface cooling. Table VI summarizes their findings.

There is much current anxiety that increased activity in these areas will worsen the ice-fog incidence. The use of large ships in winter in the channels of the Arctic Archipelago, and in harbors in northwest Canada and Alaska, clearly poses the threat of still greater water-vapor pollution of frigid Arctic air.

*Blowing snow*

Blowing snow consists of snow particles raised by the wind so as to diminish visibility at eye-level, and in Canada (where it is a serious problem) a reduction of visibility to 6 miles or below is considered statistically to be a blowing snow event. Drifting snow is the term applied when snow moves along its own snow surface below eye-level, without affecting the visibility. When fresh snow falls with strong winds blizzard conditions exist, and it is often impossible to distinguish between blowing and precipitating snow.

Although blowing snow may occur throughout Canada and Alaska, it is characteristically a feature of the windy eastern Canadian Arctic, and of open areas of southern Canada east of the Rockies. In Alaska it occurs chiefly along windy parts of the Bering Sea coast, and on the Arctic coast plain, but nowhere attains the proportions of a major hazard. It does not occur over open water.

Fig.33 shows rather schematically the percentage frequency of blowing snow over Canada in January, in most areas the month of most frequent occurrence. There is a striking zone of high frequency extending from the central Queen Elizabeth Islands southwards to Keewatin and the Hudson's Bay coast of Manitoba and Ontario. South-

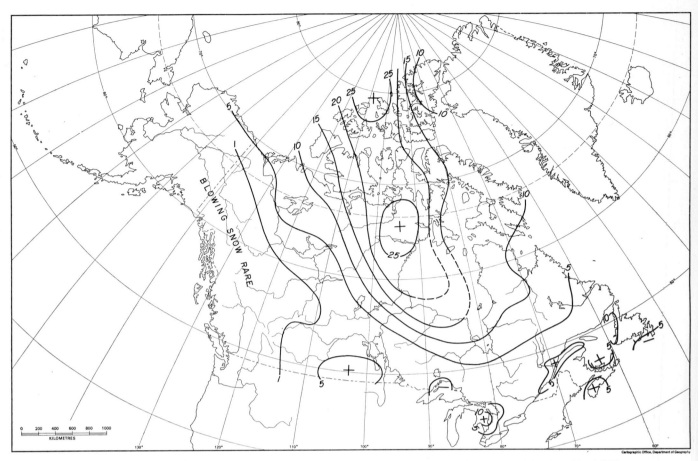

Fig.33. Percentage frequency of hourly observations in January (3 hourly in Arctic islands) with visibility 10 km or less because of blowing snow. (From CANADA DEPARTMENT OF TRANSPORT, METEOROLOGICAL BRANCH, 1967b.)

west of this zone frequency drops off rapidly, especially in the forest tundra zone. The plateaus and valleys of the western Cordillera (including most of Alaska in practice) have very low frequencies (below 0.5% in most areas), as does the Boreal forest belt of Alberta and Saskatchewan. Substantially higher frequencies are experienced in the Boreal forest belt of Manitoba, Ontario and Quebec, where fresh snow falls more often.

In southern Canada there are several belts of higher frequencies. The open plains of southeastern Saskatchewan and southwestern Manitoba are one such belt, the Regina area being worst affected (9.5% in January at Regina airport). Another is the Georgian Bay coast of southern Ontario, the belt extending inland about 50 miles. A third is the windy St. Lawrence Valley from Montreal to below the Saguenay (whose valley is notorious for blowing snow, with 11.0% at Bagotville). Other centers include much of Nova Scotia and Newfoundland (especially the west coast).

The distribution clearly indicates the controlling influences of strong wind and frequent snowfall. Winds of less than 15 km h$^{-1}$ rarely cause blowing snow, but above that speed blowing becomes progressively more likely. A careful analysis by FRASER (1964) of the relation between wind and visibility in blowing snow in the Arctic yielded the data from which we have constructed Table VII.

TABLE VII

VISIBILITY (%) IN BLOWING SNOW AS A FUNCTION OF WIND SPEED (WHEN THE SPEED EXCEEDS 15 KM H⁻¹)

| Wind speeds (km h⁻¹) | Visibility range | | |
|---|---|---|---|
| | <0.8 km | 0.8–4 km | over 4 km |
| *Baker Lake:* | | | |
| 16–31 | 4 | 51 | 45 |
| 32–47 | 18 | 51 | 31 |
| >48 | 70 | 24 | 6 |
| *Churchill:* | | | |
| 16–31 | 2 | 62 | 36 |
| 32–47 | 8 | 69 | 23 |
| >48 | 32 | 58 | 10 |
| *Isachsen:* | | | |
| 16–31 | 8 | 65 | 27 |
| 32–47 | 29 | 59 | 12 |
| >48 | 69 | 26 | 5 |

Source: FRASER (1964).

No simple relationship can be read from this table, nor can it from more complete presentations. Much depends on local site, and on the surrounding terrain. Open tundra, forest tundra and prairie areas have high surface wind speeds for a given pressure gradient, and hence high blowing snow incidence. Freshly fallen dry snow lifts readily, wind-crusted or wet snow much less so. Densely forested areas are to a large extent protected. For these and other reasons Fig.33 should be taken only as a very rough out-line of the relative frequency of blowing snow.

The high frequency of blowing snow over the tundra surfaces of the eastern Canadian Arctic and the adjacent forest tundra is one of the remarkable elements of the region's climate. The snow is heavily drifted, and many areas of the surface are blown largely clear. The result is a patchwork of highly contrasted ecological sites, some protected from winter desiccation and snow-blasting, others naked. It is also virtually impossible to attempt accurate measurement of falling snow, as has been shown in pp.118–120. The distance between the original formation of a snowflake, and its final incorporation into a stable snowpack, must often be very large.

**Freezing precipitation**

Freezing precipitation falls from warmer air as rain, drizzle or wet snow, and then freezes on contact with the cold ground, vegetation, overhead wires or structures of all kinds. It is a major hazard in many parts of northern North America during the season of freezing temperatures. If accompanied by strong winds it may cause immense damage and disruption.

Like blowing snow freezing precipitation is a hazard mainly in the eastern part of Canada. It occurs occasionally along the Pacific and Bering Sea coasts of Alaska, including the Aleutian Islands, but is rare in interior Alaska, the Yukon and much of coastal and northern British Columbia. In the rest of Canada, however, it is frequent and often

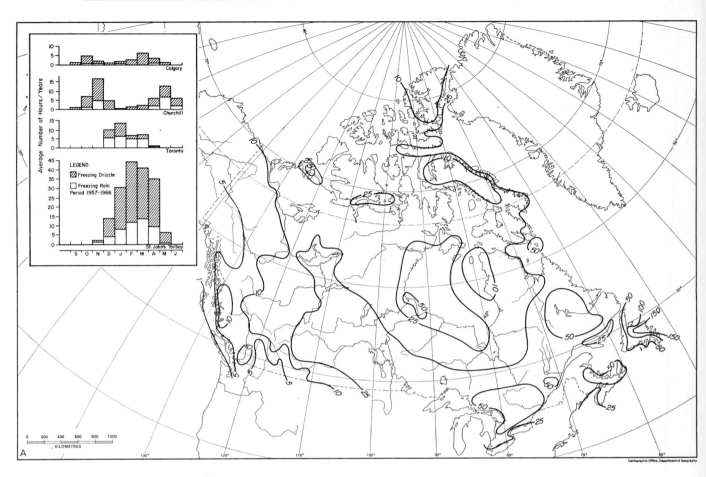

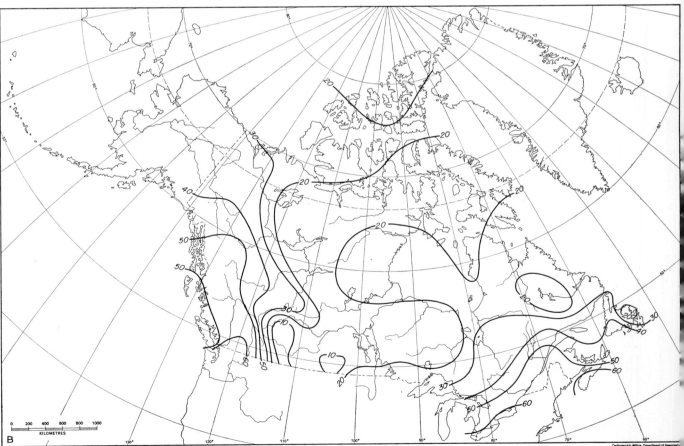

Fig.34. Mean annual hours of freezing precipitation, Canada, 1957–66 (A), with percentage falling as freezing rain (B). Inset shows month-by-month distribution at representative stations. (After McKay and Thompson, 1969.)

destructive. Fig.34 shows the number of days per annum on which it has recently occurred, together with the proportion made up by freezing rain; the chart also shows the characteristic seasonal distribution.

A recent exhaustive study by McKay and Thompson (1969), on which Fig.34 is based, has established the frequency with which heavy freezing rain (ice storms) occurs in the urban areas of eastern Canada. An accumulation of 2 cm of ice on a level surface has an average return period of about 5 years in Toronto and St. Johns, Newfoundland, though the frequency of lighter falls is much higher in St. Johns.

Freezing rain is destructive because it characteristically deposits glaze (clear ice, with density near 0.9 g cm$^{-3}$) on wires, overhead structures, road surfaces and roofs. It falls from warm, overriding air in deep cyclones, in which a shallow wedge of sub-freezing air is trapped near the ground. Surface and ground temperatures are usually in the range 0 to $-5°$C, but one of the writers recalls heavy freezing rain in the Montreal district at $-8°$C. At lower temperatures still in the cold wedge the rain falls as ice pellets, which are less destructive but still a hazard to traffic. With high winds, common in the cyclones concerned, the load on power lines, all kinds of light structures and even roofs may become unbearable, and a severe ice storm produces chaos in the modern urban community. Parts of Montreal, for example, were without power (and hence heat) for over a week in February, 1961.

Freezing rain occasionally falls in the Vancouver district and in the southern interior of British Columbia. Thence eastwards to the St. Lawrence Valley most freezing precipitation is freezing drizzle, which tends to deposit only thin layers of rime (density between 0.3 and 0.8 g cm$^{-3}$). The major belt of high risk of freezing rain extends across southern Ontario and Quebec into New Brunswick, Nova Scotia and southern Newfoundland.

Wet snow may also be destructive, especially when it falls at times when leaves or fruit are on the trees, or crops in the ground. It is very common early and late in the freezing season all the way from the Aleutians to Newfoundland. McKay and Thompson (1969) cite a case at Sedgewick, Alberta, when wet snow formed rime deposits up to 15 cm thickness on telephone poles, many of which failed, at a temperature of 2°C.

### Tornadoes and hail

We have already treated the incidence of thunderstorms in Canada and Alaska (pp.97–99). They are a normal and not very hazardous element of the summer months in all but the Arctic and Pacific coastal districts. In two areas, however, special modifications of thunderstorm development introduce a real risk of damage and loss of life. These areas are the prairies and southwestern Ontario, in both of which tornadoes and severe hailstorms sometimes develop within thunderstorm areas, chiefly in the afternoon and evening.

Fig.35, modified from McKay and Lowe (1960), shows the distribution of tornado frequency for the period 1916–1955 over part of Canada and the United States. In the Canadian prairies the risk is high only in June, July and August (insert on Fig.35), and is largely confined to southern Saskatchewan and southwestern Manitoba, frequencies being highest at the international boundary. A few tornadoes have been reported in Alberta, chiefly near Calgary. Heavy damage and 28 deaths occurred in Regina in 1912, and several tornadoes have occurred in or near Winnipeg. In southwestern Ontario the

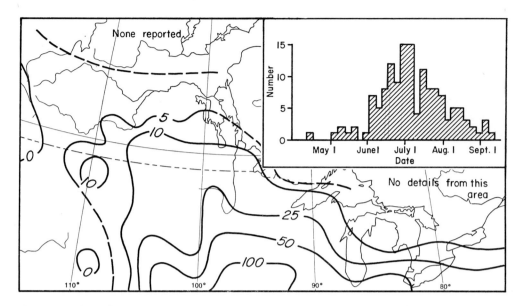

Fig.35. Number of tornadoes, 1916–55, over central North America. Inset shows month-by-month distribution over the Canadian prairie provinces. (After MCKAY and LOWE, 1960.)

risk becomes high during May and remains high into September. Most of the storms in this area also affect Michigan, crossing into Canada over Georgian Bay, Lake St. Clair or the Detroit and St. Clair rivers. The risk falls off very rapidly north and east of the Niagara cuesta, though small twisters occasionally do local damage in eastern Ontario and southern Quebec.

Tornadoes occur in Canada primarily near the mid-westerly jet axis. They commonly move from between west-northwest and south-southwest, near the crest of frontal waves, or ahead of cold fronts. Heavy rainfall often accompanies the storms.

Hail is frozen precipitation in pellet form from cumulonimbus cloud masses, and is not to be confused with the frozen rain—"ice pellets"—of winter, which is rarely destructive. "Soft hail", or graupel, is very common in spring showers in all parts of Canada and Alaska, but is too light and slow in descent to cause damage. True hail falls largely in spring and summer, and consists of dense pellets from a few millimeters to a few centimeters in diameter, with a high terminal velocity of fall, and hence capable of damage to crops, persons, and mechanical or electrical equipment.

Light hail, almost always nondestructive, is very common in showers falling from polar maritime airstreams from both oceans. It occurs widely in the warmer months in coastal Alaska and British Columbia, and less frequently in the interior northern plateaus. Hail of this inoffensive sort is also common in Labrador, Newfoundland, eastern Quebec and parts of the maritime provinces, especially in May and June.

Hail as a real hazard to crops, persons and structures occurs chiefly in a belt from eastern interior British Columbia to Manitoba, and has high economic importance on the prairies, where up to 1,000 km² of standing crops may be damaged or destroyed by wind-driven hail in a bad summer. Hail on the prairies has been studied climatologically by CURRIE (no date) and KENDREW and CURRIE (1955), and has been subjected to detailed research as to physical process and possibility of artificial suppression, especially in Alberta (e.g., DOUGLAS and HITSCHFELD, 1959). It is widely believed in the area that the risk of destructive hail varies rapidly from point to point, and insurance rates reflect this belief. It is not yet clear whether this reflects the partial control of mesoscale convection by topography, or the chance paths of a highly erratic mechanism. What is certain is that hail is much more frequent than tornado activity, and has a different incidence. It is greatly feared by the farmers.

Fig.36 (after CURRIE, no date) shows the number of destructive storms (causing over

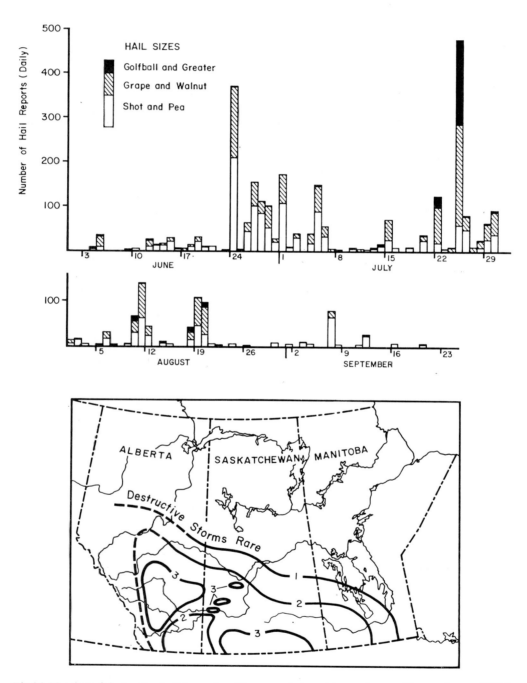

Fig.36. Number of destructive hailstorms in a 35-year period, prairie provinces, with sample year (1957) near Penhold, Alberta. Upper diagrams show day-by-day incidence in 1957 near center of area of maximum frequency. (After CURRIE, no date; DOUGLAS and HITSCHFELD, 1959.)

75% crop loss) occurring in a 35-year period, roughly the working life of a farmer. Two areas of high risk—over three storms in 35 years—are visible. One lies over the central Alberta prairies, east of the foothills, but north of the dry southeast. The other lies over southernmost Saskatchewan, east of the Cypress Hills. But the frequency exceeds 2 in all the grain-growing areas. It will be noted that the distribution is unlike that for tornadoes. We have added to Fig.36, to illustrate the erratic nature of hail incidence and the high frequency of large hailstones, the complete record of storms for the summer of 1957 over an area of 24,000 km² near Calgary and Penhold, Alberta (after DOUGLAS and HITSCH-FELD, 1959).

**Acknowledgement for illustrations**

All the illustrations are based on published data, or on M.S. data supplied by Canada, Atmospheric Environment Service, and by United States Environmental Data Service and U.S. Weather Bureau (now National Weather Service). Permission to publish is gratefully acknowledged. Data sources, or sources of all transcribed charts, are given in the legends. Fig.1–4, 6, 9, 11, 13, 22, 23, 25, 27, 28, 29 and 33 are originals. Fig.37–39 were drawn by Dr. Wolfgang Baier.

**Appendix I—Climatic tables with averages and extremes for representative stations**

**Alaska** (pp.135–144)

The recording period (Rec. yrs.) is between 1931 and 1960 for most cases. Accumulated degree-days of temperature are originally computed from 65°F threshold (= 18.3°C). "Heavy fog" means visibility below one-quarter mile at some hour of the day. Cloudiness is for daylight hours.
The data were provided by the State Climatologist, U.S. Weather Bureau. The tables are arranged from south to north.

**Canada** (pp.145–176)

Vapor pressures not available for sub-freezing period. "Fog" means a day with visibility below 1 km at some hour of observation. Cloudiness is for full 24 h. Recording period: 1931–1960. The tables are grouped per province and arranged from south to north. Otherwise as for Alaska.

TABLE VIII

CLIMATIC TABLE FOR KODIAK
Latitude 57°30′N, longitude 152°45′W

| Month | Temperature (°C) | | | | Relat. humid. (%) | Precipitation (mm) | | Snowfall (mm) |
|---|---|---|---|---|---|---|---|---|
| | daily mean | mean daily range | extremes | | | mean | max. in 24 h | |
| | | | max. | min. | | | | |
| Jan. | −1.2 | 4.6 | 10.6 | −20.6 | 80 | 130 | 81 | 379 |
| Feb. | −0.2 | 4.6 | 13.3 | −18.3 | 79 | 128 | 115 | 343 |
| Mar. | −0.1 | 5.2 | 11.7 | −19.4 | 78 | 96 | 45 | 472 |
| Apr. | 2.6 | 4.9 | 15.0 | −12.2 | 77 | 100 | 70 | 188 |
| May | 6.0 | 4.6 | 24.4 | −6.7 | 79 | 141 | 48 | 8 |
| June | 9.7 | 5.3 | 30.0 | 1.1 | 80 | 100 | 55 | tr. |
| July | 12.1 | 5.2 | 26.7 | 4.4 | 82 | 97 | 63 | 0 |
| Aug. | 12.8 | 5.6 | 27.2 | 2.2 | 82 | 95 | 51 | 0 |
| Sept. | 9.8 | 5.2 | 24.4 | −0.6 | 82 | 151 | 87 | tr. |
| Oct. | 5.2 | 5.1 | 15.6 | −7.2 | 78 | 169 | 92 | 69 |
| Nov. | 1.7 | 4.4 | 12.2 | 12.8 | 80 | 153 | 59 | 140 |
| Dec. | −1.3 | 4.9 | 10.0 | −16.1 | 78 | 118 | 47 | 302 |
| Annual | 4.8 | 4.9 | 30.0 | −20.6 | 79 | 1,477 | 115 | 1,900 |
| Rec.(yrs.) | 30 | 30 | 30 | 18 | 18 | 18 | 18 | 18 |

| Month | Number of days | | | Mean cloud-iness (tenths) | Wind | | 18°C degree-days |
|---|---|---|---|---|---|---|---|
| | precip. >0.25 mm | thunder-storm | heavy fog | | most freq. dir. | mean speed (m/sec) | |
| Jan. | 17 | 0 | 9 | 7 | NW | 5.4 | 583 |
| Feb. | 15 | 0 | 7 | 7 | NW | 5.2 | 536 |
| Mar. | 16 | tr. | 7 | 7 | NW | 5.2 | 576 |
| Apr. | 15 | 0 | 6 | 7 | NW | 4.8 | 472 |
| May | 16 | 0 | 11 | 8 | E | 4.4 | 381 |
| June | 15 | 0 | 14 | 8 | E | 3.7 | 258 |
| July | 15 | tr. | 14 | 8 | E | 2.9 | 190 |
| Aug. | 15 | tr. | 13 | 8 | NW | 3.2 | 174 |
| Sept. | 17 | 0 | 11 | 7 | NW | 3.9 | 357 |
| Oct. | 17 | 0 | 7 | 7 | NW | 4.9 | 420 |
| Nov. | 16 | 0 | 6 | 7 | NW | 5.4 | 496 |
| Dec. | 16 | 0 | 8 | 6 | NW | 5.4 | 597 |
| Annual | 190 | tr. | 113 | 7 | NW | 4.5 | 4,910 |
| Rec.(yrs.) | 18 | 18 | 18 | 18 | 18 | 18 | 18 |

TABLE IX

CLIMATIC TABLE FOR JUNEAU
Latitude 58°22′N, long titude 134°35′W, elevation 5 m

| Month | Mean sta. press. (mbar) | Temperature (°C) | | | | Relat. humid. (%) | Precipitation (mm) | | Snowfall (mm) |
|---|---|---|---|---|---|---|---|---|---|
| | | mean daily | mean daily range | extremes | | | mean | max. in 24 h | |
| | | | | max. | min. | | | | |
| Jan. | 1010.2 | −3.8 | 5.1 | 8.9 | −28.3 | 78 | 102 | 70 | 446 |
| Feb. | 1010.6 | −2.9 | 5.9 | 8.9 | −24.4 | 78 | 78 | 60 | 521 |
| Mar. | 1010.1 | −0.9 | 6.8 | 10.6 | −23.9 | 78 | 83 | 46 | 399 |
| Apr. | 1011.9 | 3.3 | 8.3 | 18.9 | −13.9 | 75 | 73 | 40 | 112 |
| May | 1014.3 | 7.6 | 8.9 | 27.8 | −3.9 | 76 | 82 | 35 | 25 |
| June | 1013.6 | 11.3 | 9.5 | 28.3 | −0.6 | 73 | 86 | 49 | — |
| July | 1016.4 | 12.9 | 8.3 | 28.9 | 2.2 | 78 | 114 | 42 | — |
| Aug. | 1014.5 | 12.3 | 8.3 | 28.3 | −2.8 | 81 | 128 | 61 | — |
| Sept. | 1012.7 | 9.4 | 7.1 | 22.2 | −3.3 | 85 | 169 | 81 | tr. |
| Oct. | 1007.8 | 5.3 | 5.4 | 16.1 | −8.9 | 85 | 212 | 118 | 33 |
| Nov. | 1007.7 | 1.3 | 5.4 | 13.3 | −19.4 | 84 | 154 | 85 | 244 |
| Dec. | 1007.0 | −2.0 | 4.8 | 12.2 | −29.4 | 81 | 107 | 90 | 584 |
| Annual | 1011.4 | 4.5 | 7.1 | 28.9 | −29.4 | 80 | 1,387 | 118 | 12,339 |
| Rec.(yrs.) | 10 | 30 | 30 | 14 | 14 | 14 | 30 | 14 | 30 |

| Month | Number of days | | | Mean cloud-iness (tenths) | Mean sun-shine (h) | Wind | | 18°C degree-days |
|---|---|---|---|---|---|---|---|---|
| | precip. >0.25 mm | thunder-storm | heavy fog | | | most frequ. direct. | mean speed (m/sec) | |
| Jan. | 18 | 0 | 2 | 7.8 | 71 | N | 3.8 | 687 |
| Feb. | 17 | 0 | 2 | 8.1 | 102 | ESE | 3.8 | 594 |
| Mar. | 18 | 0 | 2 | 7.9 | 171 | ESE | 3.9 | 596 |
| Apr. | 17 | 0 | 1 | 8.2 | 200 | ESE | 3.9 | 450 |
| May | 18 | 0 | 1 | 7.8 | 230 | ESE | 3.4 | 336 |
| June | 15 | tr. | tr. | 7.8 | 251 | N | 3.4 | 211 |
| July | 17 | 0 | tr. | 8.2 | 193 | N | 3.2 | 167 |
| Aug. | 18 | 0 | 1 | 7.9 | 161 | N | 3.2 | 188 |
| Sept. | 20 | tr. | 3 | 8.4 | 123 | N | 3.4 | 268 |
| Oct. | 23 | 0 | 3 | 8.6 | 67 | ESE | 4.0 | 402 |
| Nov. | 20 | 0 | 4 | 8.5 | 60 | ESE | 3.7 | 511 |
| Dec. | 21 | 0 | 2 | 8.5 | 51 | ESE | 3.9 | 630 |
| Annual | 222 | tr. | 21 | 8.1 | 1,680 | ESE | 3.7 | 5,037 |
| Rec.(yrs.) | 14 | 14 | 14 | 14 | 29 | 19 | 14 | |

## TABLE X

CLIMATIC TABLE FOR YAKUTAT
Latitude 59°31′N, longitude 139°40′W, elevation 9 m

| Month | Mean sta. press. (mbar) | Temperature (°C) | | | | Relat. humid. (%) | Precipitation (mm) | | Snowfall (mm) |
|---|---|---|---|---|---|---|---|---|---|
| | | mean daily | mean daily range | extremes | | | mean | max. in 24 h | |
| | | | | max. | min. | | | | |
| Jan. | 1006.1 | −2.6 | 7.9 | 6.1 | −30.0 | 81 | 276 | 130 | 1,026 |
| Feb. | 1006.9 | −1.9 | 8.0 | 8.9 | −28.3 | 83 | 208 | 52 | 1,115 |
| Mar. | 1007.7 | −0.3 | 8.2 | 12.8 | −25.0 | 83 | 221 | 85 | 1,008 |
| Apr. | 1010.6 | 2.8 | 8.4 | 18.3 | −16.1 | 83 | 184 | 68 | 455 |
| May | 1013.8 | 7.0 | 8.4 | 24.4 | −5.6 | 85 | 203 | 86 | 18 |
| June | 1014.0 | 10.3 | 7.9 | 25.6 | −1.1 | 85 | 129 | 64 | — |
| July | 1016.5 | 12.3 | 7.1 | 28.9 | 2.2 | 87 | 214 | 63 | — |
| Aug. | 1013.7 | 12.1 | 7.0 | 30.0 | −1.1 | 89 | 277 | 117 | — |
| Sept. | 1011.1 | 9.6 | 8.3 | 25.0 | −3.9 | 89 | 420 | 118 | tr. |
| Oct. | 1004.8 | 5.5 | 7.8 | 15.6 | −11.1 | 85 | 498 | 108 | 130 |
| Nov. | 1003.6 | 1.0 | 7.9 | 15.0 | −23.3 | 87 | 407 | 181 | 417 |
| Dec. | 1001.7 | −2.2 | 6.8 | 7.8 | −26.1 | 86 | 312 | 96 | 1,189 |
| Annual | 1009.2 | 4.4 | 7.9 | 30.0 | −30.0 | 85 | 3,348 | 181 | 5,357 |
| Rec.(yrs.) | 10 | 30 | 30 | 11 | 11 | | 30 | 11 | 30 |

| Month | Number of days | | | Mean cloud-iness (tenths) | Wind | |
|---|---|---|---|---|---|---|
| | precip. >0.25 mm | thunder-storm | heavy fog | | most frequ. direct. | mean speed (m/sec) |
| Jan. | 18 | 0 | 2 | 7.3 | E | 4.1 |
| Feb. | 18 | 0 | 2 | 8.1 | E | 4.4 |
| Mar. | 17 | 0 | 3 | 7.7 | E | 3.6 |
| Apr. | 17 | 0 | 2 | 8.1 | E | 3.7 |
| May | 18 | 0 | 3 | 8.6 | ESE | 3.8 |
| June | 15 | 0 | 4 | 8.5 | ESE | 3.6 |
| July | 15 | 0 | 4 | 8.5 | E | 3.1 |
| Aug. | 17 | 0 | 5 | 8.5 | ESE | 3.3 |
| Sept. | 21 | 0 | 4 | 8.6 | ESE | 3.6 |
| Oct. | 22 | tr. | 1 | 8.4 | E | 4.1 |
| Nov. | 22 | tr. | 1 | 8.4 | E | 4.3 |
| Dec. | 23 | tr. | 1 | 8.5 | E | 4.4 |
| Annual | 223 | | 32 | 8.3 | E | 3.8 |
| Rec.(yrs.) | 11 | 9 | 9 | 9 | 9 | 9 |

TABLE XI

CLIMATIC TABLE FOR BETHEL
Latitude 60°47′N, longitude 161°43′W, elevation 3 m

| Month | Mean sta. press. (mbar) | Temperature (°C) | | | | Relat. humid. (%) | Precipitation (mm) | | Snowfall (mm) |
|---|---|---|---|---|---|---|---|---|---|
| | | mean daily | mean daily range | extremes | | | mean | max. in 24 h | |
| | | | | max. | min. | | | | |
| Jan. | 1007.6 | −15.8 | 8.1 | 6.1 | −46.7 | 80 | 28 | 48 | 295 |
| Feb. | 1006.6 | −13.2 | 6.6 | 8.3 | −42.8 | 79 | 28 | 24 | 274 |
| Mar. | 1009.8 | −11.3 | 8.7 | 8.9 | −41.1 | 80 | 26 | 25 | 318 |
| Apr. | 1008.1 | −3.4 | 8.9 | 13.9 | −35.0 | 82 | 15 | 10 | 112 |
| May | 1007.6 | 3.9 | 8.9 | 23.3 | −20.6 | 78 | 24 | 28 | 28 |
| June | 1010.6 | 10.9 | 8.9 | 28.9 | −1.1 | 76 | 30 | 21 | tr. |
| July | 1011.9 | 12.6 | 7.9 | 30.0 | 1.7 | 82 | 52 | 49 | — |
| Aug. | 1008.3 | 11.3 | 6.7 | 27.2 | −1.1 | 89 | 107 | 61 | — |
| Sept. | 1006.5 | 7.0 | 7.2 | 21.1 | −7.8 | 88 | 66 | 39 | 10 |
| Oct. | 1002.1 | −0.3 | 6.9 | 18.3 | −18.9 | 88 | 39 | 48 | 91 |
| Nov. | 1000.3 | −8.2 | 7.8 | 8.9 | −32.8 | 86 | 27 | 22 | 175 |
| Dec. | 1003.5 | −15.1 | 8.8 | 5.6 | −42.2 | 81 | 26 | 21 | 277 |
| Annual | 1006.9 | −1.8 | 8.0 | 30.0 | −46.7 | 83 | 468 | 61 | 1,580 |
| Rec.(yrs.) | 10 | 30 | 30 | 15 | 15 | 9 | 30 | 15 | 30 |

| Month | Number of days | | | Mean cloudiness (tenths) | Wind | | 18°C degree-days |
|---|---|---|---|---|---|---|---|
| | precip. >0.25 mm | thunder-storm | heavy fog | | most frequ. direct. | mean speed (m/sec) | |
| Jan. | 11 | 0 | 2 | 6.4 | NNE | 5.0 | 1,056 |
| Feb. | 12 | 0 | 1 | 6.2 | NNE | 5.5 | 882 |
| Mar. | 13 | 0 | 1 | 6.5 | NW | 5.1 | 919 |
| Apr. | 10 | 0 | 1 | 6.9 | NW | 4.6 | 651 |
| May | 11 | tr. | 2 | 7.9 | ESE | 4.3 | 447 |
| June | 11 | 1 | 1 | 8.0 | NW | 4.2 | 223 |
| July | 15 | 1 | 2 | 8.5 | SSW | 4.1 | 177 |
| Aug. | 21 | tr. | 2 | 9.0 | SSW | 4.4 | 219 |
| Sept. | 16 | 0 | 2 | 8.2 | N | 4.3 | 340 |
| Oct. | 14 | 0 | 2 | 7.9 | NE | 4.6 | 578 |
| Nov. | 13 | 0 | 2 | 7.1 | NNE | 4.9 | 796 |
| Dec. | 13 | 0 | 2 | 6.8 | N | 4.8 | 1,036 |
| Annual | 160 | 2 | 20 | 7.5 | NNE | 4.6 | 7,324 |
| Rec.(yrs.) | 15 | 14 | 14 | 14 | 8 | 14 | 15 |

## TABLE XII

CLIMATIC TABLE FOR ANCHORAGE
Latitude 61°10′N, longitude 199°59′W, elevation 27 m

| Month | Mean sta. press. (mbar) | Temperature (°C) | | | | Relat. humid. (%) | Precipitation (mm) | | Snowfall (mm) |
|---|---|---|---|---|---|---|---|---|---|
| | | mean daily | mean daily range | extremes | | | mean | max. in 24 h | |
| | | | | max. | min. | | | | |
| Jan. | 1003.1 | −10.9 | 9.0 | 13.3 | −37.2 | 74 | 20 | 24 | 274 |
| Feb. | 1002.6 | −7.8 | 9.4 | 13.9 | −38.9 | 70 | 18 | 30 | 246 |
| Mar. | 1004.3 | −4.8 | 10.4 | 13.3 | −30.0 | 69 | 13 | 20 | 208 |
| Apr. | 1005.0 | 2.1 | 9.1 | 17.2 | −29.4 | 66 | 11 | 23 | 84 |
| May | 1006.8 | 7.7 | 8.9 | 27.8 | −17.2 | 64 | 13 | 20 | 10 |
| June | 1008.2 | 12.5 | 9.1 | 30.0 | −0.6 | 66 | 25 | 28 | — |
| July | 1010.8 | 13.9 | 8.4 | 28.3 | 1.7 | 73 | 47 | 52 | — |
| Aug. | 1007.5 | 13.1 | 8.4 | 27.8 | −0.6 | 78 | 65 | 43 | — |
| Sept. | 1004.6 | 8.8 | 8.4 | 22.8 | −7.2 | 78 | 64 | 46 | tr. |
| Oct. | 998.7 | 1.7 | 7.6 | 17.2 | −21.1 | 78 | 47 | 40 | 140 |
| Nov. | 998.4 | −5.4 | 7.4 | 15.6 | −29.4 | 76 | 26 | 20 | 246 |
| Dec. | 997.0 | −9.8 | 8.3 | 11.7 | −36.1 | 75 | 24 | 41 | 307 |
| Annual | 1003.9 | 1.8 | 8.7 | 30.0 | −38.9 | 72 | 374 | 52 | 1,516 |
| Rec.(yrs.) | 10 | 30 | 30 | 35 | 35 | 18 | 30 | 35 | 30 |

| Month | Number of days | | | Mean cloud-iness (tenths) | Mean sun-shine (h) | Wind | | 18°C degree-days |
|---|---|---|---|---|---|---|---|---|
| | precip. >0.25 mm | thunder-storm | heavy fog | | | most frequ. direct. | mean speed (m/sec) | |
| Jan. | 7 | 0 | 5 | 6.7 | 78 | NE | 2.3 | 905 |
| Feb. | 6 | 0 | 3 | 6.8 | 114 | N | 2.6 | 730 |
| Mar. | 7 | 0 | 1 | 6.7 | 210 | N | 2.6 | 718 |
| Apr. | 4 | 0 | 1 | 6.9 | 254 | N | 2.5 | 488 |
| May | 5 | tr. | tr. | 7.6 | 268 | S | 2.9 | 329 |
| June | 7 | tr. | tr. | 7.6 | 288 | S | 2.8 | 175 |
| July | 11 | tr. | tr. | 7.5 | 255 | S | 2.5 | 136 |
| Aug. | 15 | tr. | 1 | 7.8 | 184 | NW | 2.3 | 162 |
| Sept. | 15 | tr. | 2 | 7.8 | 128 | NNE | 2.3 | 286 |
| Oct. | 11 | 0 | 2 | 7.8 | 96 | N | 2.4 | 516 |
| Nov. | 8 | 0 | 4 | 7.2 | 68 | N | 2.3 | 713 |
| Dec. | 7 | 0 | 6 | 7.3 | 49 | NE | 2.2 | 872 |
| Annual | 103 | 1 | 25 | 7.3 | 1,992 | N | 2.5 | 6,030 |
| Rec.(yrs.) | 35 | 35 | 19 | 17 | 19 | 8 | 8 | 35 |

TABLE XIII

CLIMATIC TABLE FOR MCGRATH
Latitude 62°58′N, longitude 155°37′W, elevation 102 m

| Month | Mean sta. press. (mbar) | Temperature (°C) | | extremes | | Relat. humid. (%) | Precipitation (mm) | | Snowfall (mm) |
|---|---|---|---|---|---|---|---|---|---|
| | | mean daily | mean daily range | max. | min. | | mean | max. in 24 h | |
| Jan. | 999.7 | −22.8 | 10.8 | 12.2 | −53.3 | 71 | 32 | 26 | 424 |
| Feb. | 999.7 | −17.7 | 12.7 | 12.8 | −53.3 | 71 | 29 | 29 | 361 |
| Mar. | 999.9 | −13.2 | 15.4 | 10.6 | −46.1 | 71 | 24 | 27 | 356 |
| Apr. | 997.5 | −2.4 | 13.6 | 17.2 | −33.3 | 67 | 12 | 17 | 107 |
| May | 997.6 | 6.7 | 11.9 | 26.7 | −18.9 | 65 | 22 | 23 | 10 |
| June | 999.1 | 13.6 | 12.3 | 31.1 | 0.0 | 65 | 42 | 61 | tr. |
| July | 1000.7 | 14.8 | 11.1 | 30.6 | 1.1 | 71 | 62 | 32 | — |
| Aug. | 998.1 | 12.3 | 9.9 | 28.3 | −2.2 | 81 | 96 | 53 | tr. |
| Sept. | 996.8 | 6.6 | 9.3 | 24.4 | −14.4 | 81 | 66 | 42 | 18 |
| Oct. | 993.1 | −2.9 | 8.2 | 15.0 | −30.0 | 81 | 34 | 39 | 206 |
| Nov. | 992.9 | −14.8 | 9.1 | 8.3 | −45.0 | 77 | 27 | 33 | 348 |
| Dec. | 994.4 | −21.7 | 9.6 | 6.7 | −51.7 | 73 | 26 | 25 | 424 |
| Annual | 997.4 | −3.4 | 11.2 | 31.1 | −53.3 | 73 | 472 | 61 | 2,253 |
| Rec.(yrs.) | 10 | 30 | 30 | 15 | 15 | 15 | 30 | 15 | 30 |

| Month | Number of days | | | Mean cloud-iness (tenths) | Wind | | 18°C degree-days |
|---|---|---|---|---|---|---|---|
| | precip. >0.25 mm | thunder-storm | heavy fog | | most frequ. direct. | mean speed (m/sec) | |
| Jan. | 10 | 0 | 2 | 6.5 | NW | 1.2 | 1,274 |
| Feb. | 10 | 0 | tr. | 6.4 | NW | 1.6 | 1,009 |
| Mar. | 11 | 0 | tr. | 6.5 | NW | 1.8 | 977 |
| Apr. | 6 | 0 | 0 | 6.2 | N | 2.2 | 623 |
| May | 9 | 1 | tr. | 7.5 | E | 2.5 | 360 |
| June | 12 | 3 | tr. | 7.8 | S | 2.5 | 143 |
| July | 14 | 2 | 1 | 7.8 | S | 2.4 | 116 |
| Aug. | 20 | tr. | 1 | 8.5 | S | 2.3 | 188 |
| Sept. | 15 | tr. | 1 | 8.0 | N | 2.2 | 352 |
| Oct. | 13 | 0 | 1 | 8.6 | N | 1.7 | 658 |
| Nov. | 12 | 0 | 1 | 7.4 | ESE | 1.1 | 995 |
| Dec. | 12 | 0 | 2 | 6.9 | NW | 1.1 | 1,240 |
| Annual | 144 | 6 | 9 | 7.3 | NW | 1.9 | 7,934 |
| Rec.(yrs.) | 15 | 15 | 15 | 15 | 9 | 8 | |

TABLE XIV

CLIMATIC TABLE FOR NOME
Latitude 64°30′N, longitude 165°20′W, elevation 4 m

| Month | Mean sta. press. (mbar) | Temperature (°C) | | | | Relat. humid. (%) | Precipitation (mm) | | Snowfall (mm) |
|---|---|---|---|---|---|---|---|---|---|
| | | mean daily | mean daily range | extremes | | | mean | max. in 24 h | |
| | | | | max. | min. | | | | |
| Jan. | 1012.7 | −15.3 | 7.9 | 3.9 | −39.4 | 80 | 26 | 20 | 290 |
| Feb. | 1011.0 | −14.7 | 8.7 | 8.3 | −41.1 | 76 | 24 | 11 | 170 |
| Mar. | 1013.1 | −13.4 | 9.3 | 5.6 | −38.9 | 79 | 22 | 17 | 254 |
| Apr. | 1011.1 | −6.0 | 8.1 | 10.6 | −32.8 | 81 | 20 | 14 | 163 |
| May | 1010.6 | 1.7 | 6.9 | 20.0 | −23.9 | 79 | 18 | 18 | 41 |
| June | 1011.9 | 7.7 | 7.3 | 27.2 | −3.9 | 81 | 24 | 52 | 2.5 |
| July | 1012.1 | 9.7 | 5.7 | 23.9 | 0.6 | 87 | 58 | 45 | — |
| Aug. | 1008.0 | 9.4 | 5.8 | 22.8 | −1.1 | 87 | 97 | 61 | tr. |
| Sept. | 1008.2 | 5.5 | 6.6 | 17.2 | −8.3 | 82 | 68 | 32 | 13 |
| Oct. | 1004.9 | −1.3 | 6.3 | 15.0 | −19.4 | 80 | 43 | 58 | 117 |
| Nov. | 1003.9 | −8.6 | 6.8 | 6.7 | −39.4 | 83 | 29 | 12 | 277 |
| Dec. | 1007.3 | −14.3 | 7.4 | 1.7 | −40.6 | 78 | 25 | 28 | 231 |
| Annual | 1009.6 | −3.3 | 7.3 | 27.2 | −41.1 | 81 | 454 | 61 | 1,557 |
| Rec.(yrs.) | 10 | 30 | 30 | 11 | 11 | 11 | 30 | 11 | 30 |

| Month | Number of days | | | Mean cloud- iness (tenths) | Mean sun- shine (h) | Wind | | 18°C degree- days |
|---|---|---|---|---|---|---|---|---|
| | precip. >0.25 mm | thunder- storm | heavy fog | | | most frequ. direct. | mean speed (m/sec) | |
| Jan. | 10 | 0 | 3 | 6.0 | 72 | E | 5.4 | 1,044 |
| Feb. | 9 | 0 | 1 | 5.3 | 109 | ENE | 5.0 | 925 |
| Mar. | 11 | 0 | 2 | 6.0 | 193 | E | 5.0 | 983 |
| Apr. | 9 | 0 | 2 | 6.5 | 226 | ENE | 4.9 | 730 |
| May | 8 | 0 | 3 | 6.8 | 285 | NE | 4.5 | 517 |
| June | 9 | tr. | 4 | 7.1 | 297 | SW | 4.2 | 318 |
| July | 14 | tr. | 4 | 8.1 | 204 | WSW | 4.4 | 267 |
| Aug. | 18 | 0 | 2 | 8.7 | 146 | SW | 4.8 | 276 |
| Sept. | 14 | 0 | tr. | 7.7 | 142 | N | 5.1 | 385 |
| Oct. | 9 | 0 | tr. | 7.1 | 101 | NE | 4.9 | 608 |
| Nov. | 12 | 0 | 1 | 7.2 | 67 | N | 5.2 | 808 |
| Dec. | 10 | 0 | 2 | 6.3 | 42 | E | 4.4 | 1,011 |
| Annual | 133 | tr. | 24 | 6.9 | 1,884 | E | 4.8 | 7,872 |
| Rec.(yrs.) | 11 | 11 | 11 | 11 | 27 | 9 | 11 | 11 |

TABLE XV

CLIMATIC TABLE FOR FAIRBANKS
Latitude 64°49′N, longitude 147°52′W, elevation 133 m

| Month | Mean sta. press. (mbar) | Temperature (°C) | | | | Relat. humid. (%) | Precipitation (mm) | | Snowfall (mm) |
| | | mean daily | mean daily range | extremes | | | mean | max. in 24 h | |
| | | | | max. | min. | | | | |
|---|---|---|---|---|---|---|---|---|---|
| Jan. | 997.8 | −23.9 | 11.4 | 5.6 | −54.4 | 69 | 23 | 47 | 351 |
| Feb. | 995.6 | −19.4 | 13.8 | 10.0 | −50.0 | 68 | 13 | 20 | 208 |
| Mar. | 996.6 | −12.8 | 16.2 | 12.8 | −45.0 | 66 | 10 | 17 | 178 |
| Apr. | 994.6 | −1.4 | 14.2 | 20.6 | −35.6 | 61 | 6 | 17 | 74 |
| May | 993.7 | 8.4 | 13.3 | 32.2 | −17.8 | 58 | 18 | 22 | 13 |
| June | 993.4 | 14.7 | 14.2 | 32.8 | −1.1 | 62 | 35 | 39 | tr. |
| July | 995.9 | 15.4 | 13.1 | 33.9 | +1.1 | 70 | 47 | 55 | tr. |
| Aug. | 993.9 | 12.4 | 12.3 | 30.6 | −5.0 | 76 | 56 | 59 | tr. |
| Sept. | 993.1 | 6.4 | 11.4 | 28.9 | −11.1 | 77 | 28 | 31 | 18 |
| Oct. | 989.9 | −3.2 | 10.2 | 17.8 | −33.3 | 79 | 22 | 30 | 236 |
| Nov. | 989.9 | −15.6 | 10.6 | 12.2 | −40.6 | 73 | 15 | 24 | 226 |
| Dec. | 991.4 | −22.1 | 10.9 | 14.4 | −50.6 | 71 | 14 | 19 | 231 |
| Annual | 993.8 | −3.4 | 12.7 | 33.9 | −54.4 | 69 | 287 | 59 | 1,534 |
| Rec.(yrs.) | 10 | 30 | 30 | 28 | 28 | 17 | 30 | 28 | 30 |

| Month | Number of days | | | Mean cloud- iness (tenths) | Mean sun- shine (h) | Wind | | 18°C degree- days |
| | precip. > 0.25 mm | thunder- storm | heavy fog | | | most frequ. direct. | mean speed (m/sec) | |
|---|---|---|---|---|---|---|---|---|
| Jan. | 9 | 0 | 4 | 6.7 | 54 | N | 1.4 | 1,309 |
| Feb. | 7 | 0 | 4 | 6.4 | 120 | N | 1.7 | 1,055 |
| Mar. | 7 | 0 | tr. | 6.3 | 224 | N | 2.1 | 965 |
| Apr. | 4 | 0 | tr. | 6.2 | 302 | N | 2.6 | 593 |
| May | 8 | 1 | tr. | 7.0 | 319 | N | 3.0 | 308 |
| June | 11 | 3 | tr. | 7.3 | 334 | SW | 2.9 | 123 |
| July | 13 | 3 | 1 | 7.2 | 274 | SW | 2.6 | 95 |
| Aug. | 15 | 1 | 1 | 7.7 | 164 | N | 2.5 | 184 |
| Sept. | 10 | tr. | 2 | 7.9 | 122 | N | 2.5 | 356 |
| Oct. | 11 | tr. | 2 | 8.1 | 85 | N | 2.2 | 668 |
| Nov. | 9 | 0 | 2 | 7.0 | 71 | N | 1.7 | 1,017 |
| Dec. | 8 | 0 | 4 | 7.1 | 36 | N | 1.3 | 1,251 |
| Annual | 112 | 8 | 20 | 7.1 | 2,105 | N | 2.2 | 7,925 |
| Rec.(yrs.) | 28 | 28 | 28 | 14 | 20 | 8 | 27 | 28 |

TABLE XVI

CLIMATIC TABLE FOR KOTZEBUE
Latitude 66°52′N, longitude 160°38′W, elevation 3 m

| Month | Mean sta. press. (mbar) | Temperature (°C) | | | | Relat. humid. (%) | Precipitation (mm) | | Snowfall (mm) |
|---|---|---|---|---|---|---|---|---|---|
| | | mean daily | mean daily range | extremes | | | mean | max. in 24 h | |
| | | | | max. | min. | | | | |
| Jan. | 1016.6 | −20.9 | 7.8 | 2.2 | −41.7 | 71 | 10 | 6 | 140 |
| Feb. | 1014.3 | −20.0 | 8.0 | 1.7 | −44.4 | 72 | 8 | 17 | 125 |
| Mar. | 1015.9 | −18.9 | 9.6 | 1.1 | −44.4 | 73 | 7 | 12 | 152 |
| Apr. | 1013.9 | −10.4 | 10.6 | 5.6 | −42.2 | 79 | 8 | 5 | 86 |
| May | 1012.6 | −0.6 | 7.4 | 23.3 | −27.8 | 84 | 8 | 12 | 25 |
| June | 1012.1 | 6.6 | 6.9 | 27.2 | −6.7 | 85 | 12 | 20 | 3 |
| July | 1011.6 | 11.5 | 6.6 | 27.8 | 1.1 | 83 | 37 | 45 | — |
| Aug. | 1008.4 | 10.3 | 5.7 | 23.9 | −0.6 | 86 | 55 | 38 | tr. |
| Sept. | 1009.5 | 4.9 | 5.6 | 17.2 | −8.3 | 85 | 31 | 22 | 25 |
| Oct. | 1007.1 | −4.1 | 5.4 | 10.6 | −22.2 | 84 | 15 | 12 | 142 |
| Nov. | 1007.5 | −13.7 | 6.1 | 3.3 | −37.8 | 78 | 9 | 7 | 198 |
| Dec. | 1010.1 | −19.8 | 6.9 | 1.1 | −43.9 | 72 | 7 | 10 | 155 |
| Annual | 1011.5 | −6.3 | 7.2 | 27.8 | −44.4 | 79 | 208 | 45 | 52 |
| Rec.(yrs.) | 10 | 30 | 30 | 15 | 15 | 14 | 30 | 15 | 30 |

| Month | Number of days | | | Mean cloud-iness (tenths) | Wind | | 18°C degree-days |
|---|---|---|---|---|---|---|---|
| | precip. >0.25 mm | thunder-storm | heavy fog | | most frequ. direct. | mean speed (m/sec) | |
| Jan. | 7 | 0 | 1 | 5.7 | E | 6.8 | 1,218 |
| Feb. | 7 | 0 | 1 | 5.5 | E | 6.4 | 1,073 |
| Mar. | 9 | 0 | 1 | 6.1 | E | 5.9 | 1,155 |
| Apr. | 6 | 0 | 2 | 5.8 | ESE | 6.0 | 863 |
| May | 6 | 0 | 4 | 6.5 | W | 4.6 | 587 |
| June | 8 | 0 | 5 | 7.0 | W | 5.5 | 353 |
| July | 11 | tr. | 3 | 7.7 | W | 5.8 | 212 |
| Aug. | 16 | tr. | 1 | 8.3 | W | 6.2 | 248 |
| Sept. | 13 | 0 | 1 | 7.7 | ESE | 5.8 | 402 |
| Oct. | 11 | 0 | 1 | 7.2 | NE | 6.1 | 694 |
| Nov. | 10 | 0 | 1 | 6.6 | ESE | 6.2 | 960 |
| Dec. | 9 | 0 | 1 | 6.2 | NE | 5.8 | 1,181 |
| Annual | 113 | tr. | 22 | 6.7 | W | 5.9 | 8,946 |
| Rec.(yrs.) | | | | | | | |

TABLE XVII

CLIMATIC TABLE FOR BARROW
Latitude 71°18′N, longitude 156°47′W, elevation 7 m

| Month | Mean sta. press. (mbar) | Temperature (°C) | | | | Relat. humid. (%) | Precipitation (mm) | | Snowfall (mm) |
|---|---|---|---|---|---|---|---|---|---|
| | | mean daily | mean daily range | extremes | | | mean | max. in 24 h | |
| | | | | max. | min. | | | | |
| Jan. | 1021.7 | −26.8 | 7.6 | 0.6 | −47.2 | 65 | 5 | 18 | 56 |
| Feb. | 1019.7 | −27.9 | 6.8 | −0.6 | −48.9 | 61 | 4 | 8 | 53 |
| Mar. | 1022.0 | −25.9 | 7.2 | −1.1 | −46.7 | 64 | 3 | 7 | 41 |
| Apr. | 1018.9 | −17.7 | 8.0 | 5.6 | −41.1 | 74 | 3 | 5 | 41 |
| May | 1019.6 | −7.6 | 5.7 | 7.2 | −27.8 | 87 | 3 | 8 | 38 |
| June | 1014.8 | 0.6 | 3.8 | 21.1 | −13.3 | 93 | 9 | 21 | 10 |
| July | 1012.1 | 3.9 | 6.4 | 25.6 | −5.6 | 92 | 20 | 22 | 18 |
| Aug. | 1011.4 | 3.3 | 5.3 | 22.8 | −6.7 | 93 | 23 | 16 | 15 |
| Sept. | 1013.0 | −0.8 | 3.7 | 16.7 | −17.2 | 92 | 16 | 13 | 71 |
| Oct. | 1011.6 | −8.6 | 5.3 | 6.1 | −28.3 | 85 | 13 | 25 | 63 |
| Nov. | 1015.0 | −18.2 | 6.7 | 3.9 | −40.0 | 76 | 6 | 10 | 89 |
| Dec. | 1017.7 | −24.0 | 6.9 | 1.1 | −48.3 | 66 | 4 | 7 | 69 |
| Annual | 1016.5 | −12.4 | 6.2 | 25.6 | −48.9 | 79 | 110 | 25 | 663 |
| Rec.(yrs.) | 10 | 30 | 30 | 37 | 37 | 13 | 30 | 30 | 30 |

| Month | Number of days | | | Mean cloud-iness (tenths) | Wind | | 18°C degree-days |
|---|---|---|---|---|---|---|---|
| | precip. >0.25 mm | thunder-storm | heavy fog | | most frequ. direct. | mean speed (m/sec) | |
| Jan. | 4 | 0 | 2 | – | ESE | 4.9 | 1,397 |
| Feb. | 4 | 0 | 1 | 5.3 | ENE | 5.1 | 1,294 |
| Mar. | 3 | 0 | 1 | 5.1 | NE | 4.9 | 1,370 |
| Apr. | 3 | 0 | 4 | 5.9 | E | 5.1 | 1,079 |
| May | 3 | 0 | 7 | 8.4 | NE | 5.3 | 802 |
| June | 4 | tr. | 13 | 8.0 | E | 5.1 | 531 |
| July | 8 | tr. | 13 | 8.2 | SW | 5.3 | 446 |
| Aug. | 10 | tr. | 9 | 9.0 | E | 5.7 | 466 |
| Sept. | 9 | 0 | 4 | 9.2 | ENE | 6.1 | 574 |
| Oct. | 9 | 0 | 4 | 8.7 | NE | 6.3 | 833 |
| Nov. | 6 | 0 | 2 | – | NE | 5.6 | 1,094 |
| Dec. | 4 | 0 | 2 | – | ENE | 4.9 | 1,311 |
| Annual | | tr. | 62 | | NE | 5.4 | 11,197 |
| Rec.(yrs.) | 37 | 37 | 16 | 16 | 7 | 30 | 37 |

TABLE XVIII

CLIMATIC TABLE FOR LETHBRIDGE, ALTA. [1]
Latitude 49°38′N, longitude 112°48′W, elevation 280 m

| Month | Mean sta. press. (mbar) | Temperature (°C) | | | | Mean vapor press. (mbar) | Precipitation (mm) | | Snowfall (mm) |
|---|---|---|---|---|---|---|---|---|---|
| | | mean daily | mean daily range | extremes | | | mean | max. in 24 h | |
| | | | | max. | min. | | | | |
| Jan. | 907.8 | −8.2 | 11.2 | 18 | −43 | 2.2 | 22 | 18.8 | 221 |
| Feb. | 907.3 | −7.1 | 11.3 | 19 | −42 | 2.7 | 27 | 24.4 | 264 |
| Mar. | 906.7 | −2.4 | 11.1 | 23 | −38 | 3.5 | 27 | 31.8 | 249 |
| Apr. | 907.3 | 5.4 | 13.0 | 31 | −25 | 5.6 | 35 | 53.1 | 173 |
| May | 907.8 | 11.2 | 13.6 | 33 | −12 | 7.5 | 53 | 48.0 | 64 |
| June | 907.6 | 14.7 | 12.8 | 38 | −2 | 10.6 | 81 | 61.7 | 13 |
| July | 909.5 | 18.9 | 15.5 | 39 | 2 | 12.3 | 43 | 40.6 | tr. |
| Aug. | 909.3 | 17.4 | 15.3 | 37 | 0 | 11.4 | 42 | 54.6 | 3 |
| Sept. | 909.5 | 12.7 | 13.8 | 37 | −16 | 8.4 | 35 | 36.3 | 76 |
| Oct. | 908.4 | 7.4 | 12.8 | 32 | −22 | 6.6 | 27 | 39.1 | 163 |
| Nov. | 907.8 | −0.4 | 11.0 | 23 | −34 | 4.0 | 27 | 37.9 | 254 |
| Dec. | 906.6 | −4.6 | 10.7 | 18 | −35 | 2.9 | 20 | 22.9 | 191 |
| Annual | 908.0 | 5.4 | 12.6 | 39 | −43 | 6.5 | 439 | 61.7 | 1,671 |

| Month | Number of days | | | Mean cloud-iness (tenths) | Mean sun-shine (h) | Wind | | 18°C degree-days |
|---|---|---|---|---|---|---|---|---|
| | precip. >0.25 mm | thunder-storm | heavy fog | | | most frequ. direct. | mean speed (m/sec) | |
| Jan. | 9 | 0 | 2.1 | 6.3 | 103 | W | 7.1 | 832 |
| Feb. | 9 | 0 | 2.0 | 6.4 | 126 | W | 6.6 | 717 |
| Mar. | 10 | 0.1 | 3.1 | 6.4 | 162 | W | 6.3 | 644 |
| Apr. | 8 | 0.1 | 1.0 | 6.6 | 210 | W | 6.8 | 387 |
| May | 10 | 1.8 | 0.6 | 6.4 | 265 | W | 6.2 | 224 |
| June | 12 | 4.7 | 0.7 | 6.4 | 271 | W | 5.8 | 118 |
| July | 8 | 6.0 | 0.2 | 4.2 | 345 | W | 5.2 | 31 |
| Aug. | 8 | 5.3 | 0.8 | 4.8 | 300 | W | 5.0 | 62 |
| Sept. | 7 | 1.1 | 0.7 | 5.3 | 216 | W | 5.8 | 177 |
| Oct. | 6 | 0.1 | 0.9 | 5.5 | 177 | W | 6.6 | 339 |
| Nov. | 8 | 0 | 2.1 | 6.1 | 113 | WSW | 6.6 | 562 |
| Dec. | 8 | 0 | 2.0 | 6.1 | 96 | W | 7.0 | 709 |
| Annual | 103 | 19.2 | 16.2 | 5.9 | 2,384 | W | 6.3 | 4,802 |

TABLE XIX

CLIMATIC TABLE FOR CALGARY, ALTA.
Latitude 51°06′N, longitude 114°01′W, elevation 329 m

| Month | Mean sta. press. (mbar) | Temperature (°C) | | | | Mean vapor press. (mbar) | Precipitation (mm) | | Snowfall (mm) |
|---|---|---|---|---|---|---|---|---|---|
| | | mean daily | mean daily range | extremes | | | mean | max. in 24 h | |
| | | | | max. | min. | | | | |
| Jan. | 888.3 | −9.9 | 11.3 | 16 | −43 | 1.9 | 17 | 20.3 | 170 |
| Feb. | 888.4 | −8.8 | 11.5 | 19 | −41 | 2.2 | 20 | 27.7 | 198 |
| Mar. | 888.0 | −4.4 | 10.6 | 19 | −37 | 3.2 | 26 | 20.3 | 249 |
| Apr. | 889.3 | 3.6 | 12.3 | 28 | −30 | 5.1 | 35 | 32.3 | 251 |
| May | 890.4 | 9.8 | 13.1 | 31 | −17 | 6.9 | 52 | 40.6 | 84 |
| June | 890.4 | 13.0 | 12.2 | 34 | −2 | 9.4 | 88 | 79.3 | 20 |
| July | 936.4 | 16.7 | 14.2 | 36 | 0 | 11.8 | 58 | 67.6 | 0 |
| Aug. | 891.8 | 15.1 | 14.2 | 34 | 0 | 10.9 | 59 | 80.8 | tr. |
| Sept. | 891.9 | 10.9 | 13.1 | 31 | −12 | 7.7 | 35 | 30.7 | 66 |
| Oct. | 890.1 | 5.4 | 12.7 | 29 | −22 | 5.8 | 23 | 23.9 | 140 |
| Nov. | 888.5 | −2.2 | 11.0 | 22 | −32 | 3.3 | 16 | 28.2 | 155 |
| Dec. | 887.4 | −6.6 | 11.1 | 19 | −36 | 2.4 | 15 | 14.0 | 152 |
| Annual | 889.8 | 3.6 | 12.3 | 36 | −43 | 5.9 | 444 | 80.8 | 1,485 |

| Month | Number of days | | | Mean cloud-iness (tenths) | Mean sun-shine (h) | Wind | | 18°C degree-days |
|---|---|---|---|---|---|---|---|---|
| | precip. >0.25 mm | thunder-storm | heavy fog | | | most frequ. direct. | mean speed (m/sec) | |
| Jan. | 9 | 0 | 3.4 | 6.2 | 101 | W | 4.5 | 875 |
| Feb. | 10 | 0.1 | 2.3 | 6.4 | 117 | W | 4.3 | 766 |
| Mar. | 11 | 0 | 3.2 | 6.5 | 146 | N,S | 4.3 | 704 |
| Apr. | 10 | 0.2 | 2.4 | 6.6 | 188 | SE | 5.0 | 443 |
| May | 11 | 1.4 | 1.5 | 6.6 | 240 | NW | 5.0 | 265 |
| June | 14 | 5.8 | 1.6 | 6.8 | 234 | N,NW | 4.6 | 162 |
| July | 11 | 8.2 | 1.1 | 5.1 | 318 | NW | 4.1 | 61 |
| Aug. | 12 | 5.3 | 1.4 | 5.5 | 275 | N | 4.0 | 103 |
| Sept. | 8 | 0.9 | 1.6 | 5 6 | 186 | NW | 4.4 | 223 |
| Oct. | 6 | 0 | 1.5 | 5.7 | 159 | S,W | 4.4 | 399 |
| Nov. | 7 | 0 | 3.2 | 6.1 | 111 | W | 4.3 | 617 |
| Dec. | 8 | 0 | 2.6 | 5.9 | 91 | W | 4.4 | 772 |
| Annual | 117 | 21.9 | 25.8 | 6.1 | 2,166 | N | 4.4 | 5,390 |

TABLE XX

CLIMATIC TABLE FOR EDMONTON, ALTA.
Latitude 53°34′N, longitude 113°31′W, elevation 206 m

| Month | Mean sta. press. (mbar) | Temperature (°C) | | | | Mean vapor press. (mbar) | Precipitation (mm) | | Snowfall (mm) |
|---|---|---|---|---|---|---|---|---|---|
| | | mean daily | mean daily range | extremes | | | mean | max. in 24 h | |
| | | | | max. | min. | | | | |
| Jan. | 935.2 | −14.1 | 9.6 | 13 | −47 | 1.7 | 24 | 15.2 | 239 |
| Feb. | 934.7 | −11.6 | 10.2 | 14 | −46 | 1.8 | 20 | 12.5 | 193 |
| Mar. | 933.9 | −5.5 | 9.8 | 21 | −36 | 3.0 | 21 | 18.0 | 198 |
| Apr. | 934.6 | 4.2 | 11.7 | 32 | −25 | 5.1 | 28 | 38.1 | 152 |
| May | 934.4 | 11.2 | 12.9 | 34 | −12 | 7.2 | 46 | 40.9 | 30 |
| June | 933.7 | 14.3 | 12.0 | 37 | −2 | 10.2 | 80 | 64.0 | 0 |
| July | 935.2 | 17.3 | 12.7 | 34 | 1 | 13.2 | 85 | 114.1 | 0 |
| Aug. | 935.3 | 15.6 | 12.6 | 36 | −2 | 12.3 | 65 | 51.6 | 0 |
| Sept. | 935.0 | 10.8 | 12.3 | 32 | −12 | 8.7 | 34 | 48.8 | 23 |
| Oct. | 933.5 | 5.1 | 11.6 | 28 | −25 | 5.8 | 23 | 18.0 | 104 |
| Nov. | 933.7 | −4.2 | 9.5 | 24 | −34 | 3.2 | 22 | 39.9 | 188 |
| Dec. | 932.7 | −10.4 | 8.5 | 15 | −48 | 1.9 | 25 | 21.1 | 239 |
| Annual | 934.3 | 2.7 | 11.0 | 37 | −48 | 6.2 | 473 | 114.1 | 1,366 |

| Month | Number of days | | | Mean cloud-iness (tenths) | Mean sun-shine (h) | Wind | | 18°C degree-days |
|---|---|---|---|---|---|---|---|---|
| | precip. >0.25 mm | thunder-storm | heavy fog | | | most frequ. direct. | mean speed (m/sec) | |
| Jan. | 12 | 0 | 3.6 | 6.5 | 86 | S | 3.5 | 1,005 |
| Feb. | 10 | 0 | 2.1 | 6.5 | 119 | S | 3.6 | 844 |
| Mar. | 10 | 0 | 1.8 | 6.4 | 163 | S | 4.0 | 739 |
| Apr. | 7 | 0.3 | 1.2 | 6.4 | 221 | S | 4.8 | 425 |
| May | 9 | 1.3 | 0.8 | 6.4 | 258 | NW | 4.7 | 222 |
| June | 13 | 4.4 | 0.7 | 6.7 | 251 | NW | 4.4 | 123 |
| July | 13 | 8.5 | 1.1 | 5.8 | 315 | NW | 4.0 | 41 |
| Aug. | 12 | 5.7 | 2.3 | 5.7 | 269 | NW | 3.7 | 100 |
| Sept. | 9 | 1.1 | 1.1 | 5.8 | 186 | NW | 4.0 | 228 |
| Oct. | 7 | 0 | 1.2 | 5.9 | 157 | S | 4.0 | 410 |
| Nov. | 8 | 0 | 2.3 | 6.3 | 100 | S | 3.6 | 675 |
| Dec. | 11 | 0 | 1.6 | 6.4 | 78 | S | 3.3 | 890 |
| Annual | 121 | 21.3 | 19.8 | 6.2 | 2,203 | S | 4.0 | 5,704 |

TABLE XXI

CLIMATIC TABLE FOR VANCOUVER, B.C.
Latitude 49°11′N, longitude 123°10′W, elevation 2 m

| Month | Mean sta. press. (mbar) | Temperature (°C) | | | | Mean vapor press. (mbar) | Precipitation (mm) | | Snowfall (mm) |
|---|---|---|---|---|---|---|---|---|---|
| | | mean daily | mean daily range | extremes | | | mean | max. in 24 h | |
| | | | | max. | min. | | | | |
| Jan. | 1017.2 | 2.9 | 5.1 | 14 | −18 | 6.1 | 140 | 94.0 | 188 |
| Feb. | 1016.8 | 4.1 | 6.8 | 15 | −16 | 7.2 | 120 | 60.7 | 112 |
| Mar. | 1016.0 | 6.2 | 7.0 | 19 | −9 | 7.7 | 96 | 56.6 | 38 |
| Apr. | 1016.8 | 9.1 | 8.3 | 24 | −3 | 9.1 | 58 | 48.3 | tr. |
| May | 1016.8 | 12.8 | 8.8 | 28 | 1 | 11.4 | 49 | 35.6 | tr. |
| June | 1016.6 | 15.8 | 8.1 | 28 | 4 | 13.2 | 47 | 49.5 | 0 |
| July | 1017.5 | 17.7 | 9.5 | 31 | 7 | 15.3 | 26 | 37.3 | 0 |
| Aug. | 1016.7 | 17.6 | 8.9 | 33 | 6 | 15.3 | 35 | 46.2 | 0 |
| Sept. | 1016.7 | 14.3 | 8.2 | 29 | 1 | 13.2 | 54 | 51.8 | 0 |
| Oct. | 1016.8 | 10.2 | 6.9 | 22 | −6 | 10.6 | 117 | 67.3 | tr. |
| Nov. | 1017.3 | 6.2 | 6.1 | 17 | −12 | 8.4 | 138 | 80.3 | 25 |
| Dec. | 1016.7 | 4.2 | 5.3 | 14 | −13 | 7.1 | 164 | 87.1 | 89 |
| Annual | 1016.8 | 10.2 | 7.1 | 33 | −18 | 10.4 | 1,044 | 94.0 | 452 |

| Month | Number of days | | | Mean cloud-iness (tenths) | Mean sun-shine (h) | Wind | | 18°C degree-days |
|---|---|---|---|---|---|---|---|---|
| | precip. >0.25 mm | thunder-storm | heavy fog | | | most frequ. direct. | mean speed (m/sec) | |
| Jan. | 19 | 0 | 8.6 | 8.1 | 58 | E | 3.7 | 479 |
| Feb. | 16 | 0 | 6.7 | 7.6 | 89 | E | 3.8 | 402 |
| Mar. | 16 | 0.2 | 3.3 | 7.1 | 124 | E | 4.0 | 376 |
| Apr. | 13 | 0.4 | 1.4 | 6.8 | 195 | E | 3.9 | 278 |
| May | 10 | 0.4 | 1.0 | 6.4 | 250 | E | 3.7 | 172 |
| June | 11 | 0.6 | 0.5 | 6.8 | 229 | E | 3.6 | 87 |
| July | 6 | 0.6 | 0.8 | 4.8 | 311 | E | 3.4 | 45 |
| Aug. | 8 | 0.7 | 4.4 | 5.1 | 250 | E | 3.4 | 48 |
| Sept. | 9 | 0.3 | 10.8 | 5.5 | 190 | E | 3.3 | 122 |
| Oct. | 15 | 0.3 | 13.1 | 7.1 | 114 | E | 3.5 | 253 |
| Nov. | 17 | 0.3 | 9.8 | 7.9 | 71 | E | 3.6 | 365 |
| Dec. | 19 | 0 | 10.6 | 8.1 | 44 | E | 3.8 | 437 |
| Annual | 159 | 3.8 | 71.0 | 6.8 | 1,925 | E | 3.6 | 3,064 |

## TABLE XXII

CLIMATIC TABLE FOR PRINCE GEORGE, B.C.
Latitude 53°53′N, longitude 122°41′W, elevation 206 m

| Month | Mean sta. press. (mbar) | Temperature (°C) | | | | Mean vapor press. (mbar) | Precipitation (mm) | | Snowfall (mm) |
|---|---|---|---|---|---|---|---|---|---|
| | | mean daily | mean daily range | extremes | | | mean | max. in 24 h | |
| | | | | max. | min. | | | | |
| Jan. | 935.4 | −11.3 | 9.5 | 8 | −50 | 2.2 | 56 | 29 0 | 500 |
| Feb. | 934.2 | −7.5 | 10.8 | 13 | −45 | 3.0 | 44 | 22.9 | 396 |
| Mar. | 932.8 | −2.3 | 11.3 | 18 | −38 | 3.8 | 36 | 19.8 | 226 |
| Apr. | 934.2 | 4.3 | 11.9 | 23 | −25 | 5.3 | 28 | 15.5 | 66 |
| May | 935.5 | 9.7 | 14.5 | 30 | −8 | 7.7 | 43 | 21.3 | 5 |
| June | 935.2 | 12.9 | 13.1 | 34 | −3 | 10.1 | 62 | 38.9 | tr. |
| July | 937.1 | 14.9 | 14.6 | 33 | −2 | 11.8 | 64 | 26.4 | tr. |
| Aug. | 936.8 | 13.7 | 13.9 | 33 | −4 | 11.4 | 65 | 50.0 | 0 |
| Sept. | 936.9 | 10.1 | 13.2 | 27 | −12 | 9.4 | 56 | 27.7 | 5 |
| Oct. | 934.7 | 4.8 | 9.6 | 24 | −25 | 6.6 | 59 | 21.3 | 74 |
| Nov. | 934.6 | −2.5 | 7.8 | 16 | −42 | 4.2 | 57 | 22.1 | 305 |
| Dec. | 933.4 | −6.6 | 7.6 | 12 | −43 | 3.0 | 56 | 29.0 | 445 |
| Annual | 935.1 | 3.3 | 11.5 | 34 | −50 | 6.5 | 626 | 50.0 | 2,022 |

| Month | Number of days | | | Mean cloud-iness (tenths) | Mean sun-shine (h) | Wind | | 18°C degree-days |
|---|---|---|---|---|---|---|---|---|
| | precip. >0.25 mm | thunder-storm | heavy fog | | | most frequ. direct. | mean speed (m/sec) | |
| Jan. | 17 | 0 | 4.9 | 7.5 | 54 | S | 3.6 | 895 |
| Feb. | 14 | 0 | 2.7 | 7.2 | 89 | S | 3.5 | 732 |
| Mar. | 13 | 0.2 | 2.1 | 6.7 | 130 | S | 3.8 | 623 |
| Apr. | 11 | 0.5 | 0.9 | 6.4 | 183 | S | 3.8 | 415 |
| May | 11 | 1.5 | 1.3 | 6.4 | 248 | S | 3.3 | 260 |
| June | 14 | 4.1 | 1.2 | 7.3 | 240 | S | 3.1 | 155 |
| July | 14 | 4.7 | 3.0 | 6.5 | 267 | S | 2.7 | 131 |
| Aug. | 13 | 3.8 | 4.6 | 6.1 | 243 | S | 2.7 | 139 |
| Sept. | 13 | 1.1 | 6.6 | 5.9 | 166 | S | 2.8 | 247 |
| Oct. | 15 | 0.3 | 5.2 | 7.1 | 101 | S | 3.4 | 415 |
| Nov. | 15 | 0.1 | 4.4 | 7.5 | 56 | S | 3.5 | 617 |
| Dec. | 17 | 0 | 4.7 | 7.6 | 41 | S | 3.6 | 789 |
| Annual | 167 | 16.3 | 41.6 | 6.9 | 1,816 | S | 3.3 | 5,419 |

TABLE XXIII

CLIMATIC TABLE FOR PRINCE RUPERT, B.C.
Latitude 54°17′N, longitude 130°23′W, elevation 16 m

| Month | Mean sta. press. (mbar) | Temperature (°C) | | | | Mean vapor press. (mbar) | Precipitation (mm) | | Snowfall (mm) |
|---|---|---|---|---|---|---|---|---|---|
| | | mean daily | mean daily range | extremes | | | mean | max. in 24 h | |
| | | | | max. | min. | | | | |
| Jan. | 1004.8 | 1.8 | 5.1 | 18 | −19 | 5.9 | 225 | 125.0 | 191 |
| Feb. | 1005.4 | 2.4 | 5.6 | 19 | −17 | 6.4 | 177 | 64.5 | 201 |
| Mar. | 1005.3 | 3.8 | 6.4 | 19 | −15 | 6.6 | 196 | 69.6 | 160 |
| Apr. | 1007.2 | 6.3 | 7.5 | 23 | −6 | 7.5 | 173 | 71.1 | 28 |
| May | 1010.1 | 9.5 | 8.0 | 26 | −1 | 9.8 | 130 | 49.0 | 0 |
| June | 1010.3 | 11.7 | 7.2 | 32 | 2 | 11.8 | 108 | 57.2 | 0 |
| July | 1012.0 | 13.4 | 7.0 | 31 | 1 | 13.2 | 117 | 55.4 | 0 |
| Aug. | 1010.8 | 13.9 | 7.0 | 30 | 4 | 13.7 | 149 | 49.3 | 0 |
| Sept. | 1009.1 | 12.1 | 6.7 | 26 | 1 | 12.3 | 217 | 57.9 | 0 |
| Oct. | 1004.7 | 8.7 | 5.3 | 22 | −5 | 9.8 | 336 | 141.0 | 3 |
| Nov. | 1003.9 | 5.2 | 5.1 | 20 | −12 | 7.5 | 293 | 137.9 | 33 |
| Dec. | 1002.3 | 2.8 | 4.7 | 16 | −17 | 6.4 | 278 | 125.5 | 216 |
| Annual | 1007.1 | 7.6 | 6.4 | 32 | −19 | 9.2 | 2,399 | 141.0 | 832 |

| Month | Number of days | | | Mean cloud-iness (tenths) | Mean sun-shine (h) | Wind | | 18°C degree-days |
|---|---|---|---|---|---|---|---|---|
| | precip. >0.25 mm | thunder-storm | heavy fog | | | most frequ. direct. | mean speed (m/sec) | |
| Jan. | 20 | 0 | 0.5 | 6.8 | 37 | SE | 3.8 | 520 |
| Feb. | 18 | 0 | 0.9 | 7.4 | 58 | SE | 3.4 | 449 |
| Mar. | 20 | 0.1 | 0.5 | 7.1 | 80 | SE | 3.3 | 451 |
| Apr. | 19 | 0 | 0.3 | 7.1 | 106 | SE | 3.3 | 360 |
| May | 17 | 0.1 | 1.9 | 7.2 | 140 | SE | 2.2 | 274 |
| June | 16 | 0.1 | 2.4 | 8.1 | 107 | SE | 1.8 | 198 |
| July | 17 | 0.1 | 4.5 | 8.0 | 108 | SE | 1.6 | 152 |
| Aug. | 16 | 0.1 | 6.0 | 7.6 | 112 | SE | 1.7 | 138 |
| Sept. | 18 | 0.2 | 4.9 | 7.3 | 85 | SE | 2.3 | 188 |
| Oct. | 24 | 0.2 | 1.7 | 7.9 | 50 | SE | 3.5 | 299 |
| Nov. | 22 | 0.1 | 0.5 | 7.8 | 35 | SE | 3.7 | 393 |
| Dec. | 23 | 0 | 0.3 | 7.8 | 25 | SE | 3.9 | 482 |
| Annual | 230 | 1.0 | 24.4 | 7.5 | 943 | SE | 2.9 | 3,905 |

TABLE XXIV

CLIMATIC TABLE FOR FORT NELSON, B.C.
Latitude 58°50′N, longitude 122°35′W, elevation 114 m

| Month | Mean sta. press. (mbar) | Temperature (°C) | | | | Mean vapor press. (mbar) | Precipitation (mm) | | Snowfall (mm) |
|---|---|---|---|---|---|---|---|---|---|
| | | mean daily | mean daily range | extremes | | | mean | max. in 24 h | |
| | | | | max. | min. | | | | |
| Jan. | 972.0 | −22.4 | 8.7 | 7 | −52 | 0.8 | 24 | 18.5 | 241 |
| Feb. | 971.4 | −17.6 | 10.9 | 14 | −48 | 0.9 | 26 | 19.6 | 262 |
| Mar. | 969.1 | −8.7 | 13.2 | 17 | −39 | 2.2 | 26 | 21.1 | 254 |
| Apr. | 968.5 | 1.5 | 12.8 | 24 | −34 | 4.0 | 19 | 20.3 | 130 |
| May | 969.1 | 10.0 | 13.1 | 32 | −31 | 7.2 | 39 | 34.5 | 38 |
| June | 967.5 | 14.3 | 12.6 | 34 | −1 | 10.6 | 66 | 52.1 | tr. |
| July | 968.1 | 16.8 | 13.0 | 37 | 1 | 12.8 | 65 | 47.5 | 0 |
| Aug. | 967.9 | 14.7 | 13.2 | 34 | −2 | 11.4 | 51 | 62.7 | tr. |
| Sept. | 968.1 | 9.3 | 12.2 | 33 | −11 | 8.7 | 34 | 28.5 | 43 |
| Oct. | 965.4 | 1.2 | 10.1 | 26 | −28 | 5.3 | 26 | 16.0 | 163 |
| Nov. | 968.2 | −12.1 | 7.6 | 14 | −41 | 1.9 | 31 | 28.5 | 307 |
| Dec. | 968.4 | −20.4 | 7.5 | 9 | −48 | 0.9 | 28 | 35.3 | 282 |
| Annual | 968.6 | −1.1 | 11.3 | 37 | −52 | 5.6 | 435 | 62.7 | 1,720 |

| Month | Number of days | | | Mean cloud-iness (tenths) | Wind | | 18°C degree-days |
|---|---|---|---|---|---|---|---|
| | precip. >0.25 mm | thunder-storm | heavy fog | | most frequ. direct. | mean speed (m/sec) | |
| Jan. | 11 | 0 | 1.1 | 6.1 | S | 1.8 | 1,264 |
| Feb. | 11 | 0 | 0.4 | 6.5 | N | 2.1 | 1,015 |
| Mar. | 11 | 0 | 0.5 | 6.2 | N | 2.5 | 839 |
| Apr. | 7 | 0 | 0.7 | 6.4 | N | 2.9 | 505 |
| May | 9 | 1.0 | 1.4 | 6.9 | N | 2.9 | 258 |
| June | 12 | 3.5 | 3.3 | 7.1 | N | 2.7 | 125 |
| July | 12 | 5.4 | 4.8 | 6.8 | S | 2.5 | 57 |
| Aug. | 11 | 2.1 | 5.4 | 6.2 | S | 2.4 | 119 |
| Sept. | 9 | 0.3 | 6.7 | 6.5 | S | 2.3 | 270 |
| Oct. | 8 | 0 | 3.2 | 6.6 | S | 2.2 | 532 |
| Nov. | 11 | 0 | 2.3 | 6.7 | S | 1.8 | 913 |
| Dec. | 13 | 0 | 1.9 | 6.5 | S | 1.6 | 1,202 |
| Annual | 125 | 12.3 | 31.7 | 6.5 | S | 2.3 | 7,098 |

TABLE XXV

CLIMATIC TABLE FOR WINNIPEG, MAN.
Latitude 49°54′N, longitude 97°15′W, elevation 254 m

| Month | Mean sta. press. (mbar) | Temperature (°C) | | | | Mean vapor press. (mbar) | Precipitation (mm) | | Snowfall (mm) |
|---|---|---|---|---|---|---|---|---|---|
| | | mean daily | mean daily range | extremes | | | mean | max. in 24 h | |
| | | | | max. | min. | | | | |
| Jan. | 989.6 | −17.7 | 9.7 | 7.5 | −44.2 | 1.4 | 26.2 | 19.1 | 259.1 |
| Feb. | 989.7 | −15.5 | 10.4 | 11.7 | −45.0 | 1.8 | 20.8 | 23.6 | 200.7 |
| Mar. | 987.7 | −7.9 | 10.1 | 23.4 | −38.9 | 3.2 | 27.4 | 38.1 | 205.7 |
| Apr. | 986.8 | 3.3 | 10.5 | 34.0 | −29.9 | 5.5 | 29.7 | 33.0 | 99.0 |
| May | 985.3 | 11.3 | 13.0 | 37.5 | −11.7 | 7.7 | 50.0 | 48.5 | 25.4 |
| June | 983.2 | 16.5 | 12.2 | 38.1 | −6.1 | 12.1 | 81.0 | 49.0 | 0.0 |
| July | 984.7 | 20.2 | 12.7 | 42.3 | 1.6 | 15.7 | 68.8 | 69.1 | 0.0 |
| Aug. | 985.2 | 18.9 | 12.8 | 40.4 | −0.9 | 14.0 | 70.1 | 65.3 | 0.0 |
| Sept. | 987.0 | 12.8 | 11.7 | 37.2 | −8.3 | 10.0 | 54.9 | 65.0 | 5.1 |
| Oct. | 985.9 | 6.2 | 10.5 | 30.0 | −20.7 | 7.1 | 36.6 | 74.4 | 68.6 |
| Nov. | 986.5 | −4.8 | 7.8 | 21.8 | −36.7 | 3.8 | 29.0 | 27.7 | 223.5 |
| Dec. | 987.5 | −12.9 | 8.3 | 11.4 | −47.5 | 2.2 | 22.4 | 24.9 | 215.9 |
| Annual | 986.6 | 2.5 | 10.8 | 42.3 | −47.5 | | 516.9 | 74.4 | 1,303.0 |

| Month | Number of days | | | Mean cloud-iness (tenths) | Mean sun-shine (h) | Wind | | 18°C degree-days |
|---|---|---|---|---|---|---|---|---|
| | precip. >0.25 mm | thunder-storm | heavy fog | | | most frequ. direct. | mean speed (m/sec) | |
| Jan. | 12 | 0.0 | 1.7 | 5.8 | 101 | NW | 5.5 | 1,117 |
| Feb. | 10 | 0.0 | 2.9 | 5.4 | 133 | S | 5.4 | 955 |
| Mar. | 10 | 0.1 | 2.1 | 6.0 | 167 | NW,S | 5.8 | 814 |
| Apr. | 8 | 0.7 | 1.2 | 6.0 | 207 | N | 6.3 | 452 |
| May | 10 | 2.0 | 0.9 | 6.1 | 244 | N,NNE,S | 6.2 | 225 |
| June | 13 | 5.3 | 0.9 | 6.4 | 248 | S | 5.4 | 82 |
| July | 11 | 6.5 | 1.2 | 5.1 | 310 | S | 4.6 | 21 |
| Aug. | 11 | 6.0 | 1.3 | 5.3 | 270 | S | 4.8 | 40 |
| Sept. | 10 | 2.2 | 2.4 | 6.0 | 181 | S | 5.3 | 179 |
| Oct. | 8 | 0.5 | 1.8 | 5.7 | 153 | S | 5.7 | 379 |
| Nov. | 11 | 0.0 | 2.1 | 7.2 | 82 | S | 6.0 | 695 |
| Dec. | 11 | 0.0 | 3.1 | 6.3 | 81 | S | 5.5 | 976 |
| Annual | 125 | 23.3 | 21.6 | 5.9 | 2,177 | S | 5.5 | 5,933 |

TABLE XXVI

CLIMATIC TABLE FOR THE PAS, MAN.
Latitude 53°58′N, longitude 101°06′W, elevation 83 m

| Month | Mean sta. press. (mbar) | Temperature (°C) | | | | Mean vapor press. (mbar) | Precipitation (mm) | | Snowfall (mm) |
|-------|-------|-------|-------|-------|-------|-------|-------|-------|-------|
| | | mean daily | mean daily range | extremes | | | mean | max. in 24 h | |
| | | | | max. | min. | | | | |
| Jan. | 984.7 | −21.7 | 9.5 | 4 | −45 | | 20 | 12.2 | 201 |
| Feb. | 985.1 | −18.2 | 11.6 | 8 | −45 | | 17 | 10.2 | 165 |
| Mar. | 984.0 | −11.4 | 12.9 | 12 | −38 | | 21 | 26.2 | 211 |
| Apr. | 983.3 | −0.3 | 11.2 | 30 | −27 | 5.1 | 26 | 22.1 | 155 |
| May | 982.0 | 8.7 | 11.8 | 33 | −13 | 7.2 | 45 | 41.4 | 43 |
| June | 979.1 | 13.9 | 11.4 | 36 | −3 | 10.9 | 60 | 48.0 | 3 |
| July | 979.3 | 18.2 | 11.3 | 37 | 3 | 15.9 | 68 | 56.6 | 0 |
| Aug. | 980.2 | 16.7 | 10.9 | 34 | 2 | 13.7 | 59 | 68.3 | 0 |
| Sept. | 980.0 | 10.3 | 9.7 | 30 | −4 | 9.4 | 55 | 53.3 | 18 |
| Oct. | 980.3 | 3.3 | 9.0 | 26 | −13 | 6.6 | 28 | 29.0 | 91 |
| Nov. | 981.6 | −7.9 | 6.6 | 13 | −33 | | 29 | 14.2 | 277 |
| Dec. | 983.0 | −16.8 | 8.9 | 5 | −43 | | 23 | 14.0 | 226 |
| Annual | 981.9 | −0.4 | 10.5 | 37 | −45 | | 451 | 68.3 | 1,390 |

| Month | Number of days | | | Mean cloud-iness (tenths) | Mean sun-shine (h) | Wind | | 18°C degree-days |
|-------|-------|-------|-------|-------|-------|-------|-------|-------|
| | precip. >0.25 mm | thunder-storm | heavy fog | | | most frequ. direct. | mean speed (m/sec) | |
| Jan. | 12 | 0 | 1.1 | 5.1 | 94 | W | 4.4 | 1,240 |
| Feb. | 10 | 0 | 0.5 | 5.1 | 123 | E | 4.3 | 1,033 |
| Mar. | 9 | 0 | 1.1 | 5.2 | 165 | W | 4.5 | 9.216 |
| Apr. | 8 | 0.2 | 1.0 | 5.8 | 211 | W | 4.8 | 558 |
| May | 8 | 1.4 | 0.9 | 6.0 | 258 | N | 4.9 | 299 |
| June | 10 | 3.8 | 1.1 | 6.2 | 244 | N | 4.5 | 142 |
| July | 12 | 7.2 | 1.6 | 5.5 | 300 | W | 4.8 | 36 |
| Aug. | 12 | 6.0 | 2.8 | 5.8 | 259 | E,W | 4.6 | 71 |
| Sept. | 11 | 1.7 | 1.7 | 6.4 | 158 | W | 5.0 | 240 |
| Oct. | 9 | 0.4 | 1.9 | 6.4 | 125 | W | 5.4 | 465 |
| Nov. | 13 | 0 | 2.4 | 7.3 | 64 | W | 5.2 | 787 |
| Dec. | 13 | 0 | 1.6 | 5.8 | 66 | W | 4.7 | 1,088 |
| Annual | 127 | 20.7 | 17.7 | 5.9 | 2,067 | W | 4.8 | 6,880 |

TABLE XXVII

CLIMATIC TABLE FOR CHURCHILL, MAN.
Latitude 85°45′N, longitude 94°04′W, elevation 11 m

| Month | Mean sta. press. (mbar) | Temperature (°C) | | | | Mean vapor press. (mbar) | Precipitation (mm) | | Snowfall (mm) |
|---|---|---|---|---|---|---|---|---|---|
| | | mean daily | mean daily range | extremes | | | mean | max. in 24 h | |
| | | | | max. | min. | | | | |
| Jan. | 1013.0 | −27.5 | 7.6 | 0 | −45 | | 13 | 8.9 | 127 |
| Feb. | 1015.2 | −26.4 | 8.4 | 1 | −45 | 0.4 | 14 | 12.7 | 140 |
| Mar. | 1015.8 | −19.8 | 9.2 | 4 | −44 | 1.0 | 17 | 17.0 | 165 |
| Apr. | 1014.3 | −10.7 | 8.9 | 18 | −33 | 1.6 | 26 | 27.9 | 251 |
| May | 1013.7 | −2.3 | 6.7 | 27 | −21 | 4.2 | 30 | 29.5 | 165 |
| June | 1008.6 | 5.8 | 8.6 | 31 | −9 | 7.2 | 41 | 32.5 | 18 |
| July | 1006.6 | 12.0 | 10.2 | 33 | −2 | 10.9 | 52 | 52.6 | tr. |
| Aug. | 1007.9 | 11.6 | 8.1 | 33 | 1 | 10.9 | 61 | 40.6 | 0 |
| Sept. | 1007.9 | 5.7 | 5.8 | 24 | −7 | 8.1 | 53 | 42.2 | 36 |
| Oct. | 1007.6 | −1.1 | 5.5 | 21 | −22 | 5.1 | 38 | 23.9 | 246 |
| Nov. | 1009.9 | −11.7 | 7.2 | 5 | −36 | 2.0 | 39 | 21.6 | 381 |
| Dec. | 1011.1 | −21.9 | 8.2 | 2 | −39 | 0.8 | 23 | 20.3 | 226 |
| Annual | 1011.0 | −7.2 | 7.8 | 33 | −45 | | 407 | 52.6 | 1,755 |

| Month | Number of days | | | Mean cloud-iness (tenths) | Mean sun-shine (h) | Wind | | 18°C degree-days |
|---|---|---|---|---|---|---|---|---|
| | precip. >0.25 mm | thunder-storm | heavy fog | | | most frequ. direct. | mean speed (m/sec) | |
| Jan. | 9 | 0 | 1.9 | 4.3 | 79 | NW | 6.3 | 1,421 |
| Feb. | 9 | 0 | 1.3 | 4.4 | 127 | NW | 6.4 | 1,265 |
| Mar. | 10 | 0 | 1.7 | 5.1 | 180 | NW | 6.2 | 1,183 |
| Apr. | 12 | 0 | 3.6 | 6.2 | 186 | NW | 6.5 | 872 |
| May | 11 | 0.2 | 5.2 | 7.7 | 166 | N | 5.9 | 640 |
| June | 9 | 1.2 | 7.3 | 7.2 | 212 | N | 5.4 | 375 |
| July | 12 | 2.1 | 6.4 | 6.4 | 285 | N | 5.5 | 200 |
| Aug. | 13 | 1.7 | 6.1 | 6.7 | 232 | NW | 5.6 | 208 |
| Sept. | 14 | 0.3 | 3.4 | 8.2 | 100 | N | 6.6 | 378 |
| Oct. | 15 | 0.1 | 2.9 | 8.2 | 67 | NW | 7.2 | 601 |
| Nov. | 17 | 0 | 1.9 | 7.5 | 44 | NW | 6.8 | 900 |
| Dec. | 13 | 0 | 0.7 | 5.4 | 54 | NW | 7.2 | 1,249 |
| Annual | 144 | 5.6 | 42.4 | 6.4 | 1,732 | NW | 6.3 | 9,292 |

TABLE XXVIII

CLIMATIC TABLE FOR SAINT JOHN, N.B.
Latitude 45°19′N, longitude 65°53′W, elevation 326 m

| Month | Mean sta. press. (mbar) | Temperature (°C) | | | | Mean vapor press. (mbar) | Precipitation (mm) | | Snowfall (mm) |
|---|---|---|---|---|---|---|---|---|---|
| | | mean daily | mean daily range | extremes | | | mean | max. in 24 h | |
| | | | | max. | min. | | | | |
| Jan. | 998.8 | −6.9 | 9.5 | 13 | −32 | 2.5 | 144 | 63.5 | 678 |
| Feb. | 998.2 | −6.4 | 9.4 | 11 | −33 | 2.7 | 122 | 72.1 | 645 |
| Mar. | 998.4 | −2.0 | 8.4 | 19 | −26 | 3.8 | 106 | 60.5 | 422 |
| Apr. | 1000.6 | 3.8 | 9.1 | 22 | −11 | 5.8 | 105 | 99.8 | 170 |
| May | 1001.0 | 9.4 | 10.6 | 30 | −5 | 9.1 | 98 | 62.0 | 5 |
| June | 1000.0 | 13.9 | 10.4 | 34 | −1 | 12.3 | 94 | 71.9 | 0 |
| July | 1001.2 | 17.2 | 10.8 | 31 | 4 | 15.3 | 87 | 72.6 | tr. |
| Aug. | 1001.5 | 17.1 | 10.3 | 34 | 3 | 15.3 | 108 | 122.7 | 0 |
| Sept. | 1004.1 | 13.4 | 9.4 | 34 | −3 | 12.3 | 104 | 85.6 | 0 |
| Oct. | 1003.5 | 8.3 | 8.5 | 27 | −9 | 8.7 | 108 | 72.4 | 15 |
| Nov. | 1001.9 | 2.7 | 7.7 | 22 | −14 | 6.1 | 153 | 99.6 | 112 |
| Dec. | 1001.2 | −4.3 | 8.5 | 16 | −30 | 3.3 | 133 | 68.6 | 434 |
| Annual | 1000.9 | 5.4 | 9.5 | 34 | −33 | 8.1 | 1,362 | 122.7 | 2,481 |

| Month | Number of days | | | Mean cloud-iness (tenths) | Mean sun-shine (h) | Wind | | 18°C degree-days |
|---|---|---|---|---|---|---|---|---|
| | precip. >0.25 mm | thunder-storm | heavy fog | | | most frequ. direct. | mean speed (m/sec) | |
| Jan. | 16 | 0.2 | 3.4 | 6.2 | 93 | N | 4.8 | 784 |
| Feb. | 14 | 0.1 | 3.3 | 6.0 | 117 | N | 5.1 | 698 |
| Mar. | 14 | 0.1 | 4.3 | 6.1 | 150 | N | 4.9 | 630 |
| Apr. | 14 | 0.7 | 6.1 | 6.6 | 157 | NW | 4.7 | 436 |
| May | 14 | 0.8 | 10.2 | 6.6 | 203 | S,SW | 4.3 | 277 |
| June | 13 | 2.7 | 11.9 | 7.0 | 194 | S | 3.8 | 133 |
| July | 13 | 3.0 | 16.7 | 6.4 | 231 | S | 3.6 | 50 |
| Aug. | 13 | 2.5 | 13.0 | 6.0 | 213 | S | 3.5 | 55 |
| Sept. | 12 | 1.1 | 12.5 | 5.8 | 168 | S | 4.0 | 148 |
| Oct. | 12 | 0.6 | 7.7 | 5.8 | 149 | SW | 4.4 | 310 |
| Nov. | 15 | 0.2 | 4.8 | 6.8 | 91 | N,NW | 4.2 | 465 |
| Dec. | 15 | 0.1 | 2.7 | 6.0 | 92 | NW | 4.9 | 702 |
| Annual | 165 | 12.1 | 96.6 | 6.3 | 1,858 | SW | 4.4 | 4,691 |

TABLE XXIX

CLIMATIC TABLE FOR ST. JOHNS TORBAY, NFLD.
Latitude 47°37′N, longitude 52°45′W, elevation 43 m

| Month | Mean sta. press. (mbar) | Temperature (°C) | | | | Mean vapor press. (mbar) | Precipitation (mm) | | Snowfall (mm) |
| | | mean daily | mean daily range | extremes | | | mean | max. in 24 h | |
| | | | | max. | min. | | | | |
|---|---|---|---|---|---|---|---|---|---|
| Jan. | 991.9 | −4.3 | 6.2 | 13 | −23 | 3.7 | 153 | 84.6 | 762 |
| Feb. | 991.8 | −4.7 | 6.5 | 13 | −23 | 3.5 | 163 | 54.9 | 927 |
| Mar. | 991.5 | −2.9 | 5.9 | 14 | −21 | 4.2 | 135 | 45.5 | 726 |
| Apr. | 995.1 | 1.2 | 6.9 | 22 | −14 | 5.6 | 121 | 91.7 | 345 |
| May | 997.8 | 5.6 | 8.6 | 24 | −6 | 7.2 | 99 | 50.8 | 91 |
| June | 996.9 | 10.3 | 9.3 | 29 | −3 | 9.4 | 94 | 75.2 | 10 |
| July | 998.4 | 15.4 | 9.6 | 29 | 1 | 14.2 | 89 | 121.2 | 0 |
| Aug. | 998.2 | 15.4 | 8.2 | 30 | 3 | 14.8 | 102 | 59.4 | 0 |
| Sept. | 999.4 | 12.0 | 8.2 | 28 | −1 | 12.3 | 120 | 66.0 | 0 |
| Oct. | 997.6 | 6.7 | 7.0 | 22 | −6 | 8.7 | 138 | 100.8 | 20 |
| Nov. | 996.4 | 2.9 | 6.2 | 19 | −10 | 6.9 | 163 | 64.0 | 185 |
| Dec. | 993.0 | −1.6 | 6.2 | 16 | −17 | 4.6 | 174 | 85.1 | 732 |
| Annual | 995.7 | 4.7 | 7.3 | 30 | −23 | 7.9 | 1,551 | 121.2 | 3,798 |

| Month | Number of days | | | Mean cloud-iness (tenths) | Mean sun-shine (h) | Wind | | 18°C degree-days |
| | precip. >0.25 mm | thunder-storm | heavy fog | | | most frequ. direct. | mean speed (m/sec) | |
|---|---|---|---|---|---|---|---|---|
| Jan. | 22 | 0 | 7.5 | 8.2 | 57 | W | 8.2 | 701 |
| Feb. | 20 | 0.1 | 7.6 | 7.8 | 75 | W | 7.8 | 649 |
| Mar. | 20 | 0.2 | 8.7 | 7.7 | 93 | W | 7.8 | 659 |
| Apr. | 18 | 0.1 | 12.6 | 7.8 | 116 | W | 6.8 | 515 |
| May | 16 | 0.2 | 15.8 | 7.7 | 152 | SW | 6.5 | 394 |
| June | 13 | 0.6 | 14.1 | 7.5 | 176 | SW.W | 6.4 | 240 |
| July | 13 | 1.2 | 13.8 | 7.0 | 212 | SW | 6.2 | 103 |
| Aug. | 14 | 1.0 | 10.7 | 7.2 | 176 | SW | 6.1 | 100 |
| Sept. | 14 | 0.5 | 8.9 | 6.7 | 149 | W | 6.6 | 190 |
| Oct. | 17 | 0.4 | 8.1 | 7.0 | 117 | W | 7.5 | 361 |
| Nov. | 19 | 0.1 | 9.1 | 8.2 | 59 | W | 7.2 | 462 |
| Dec. | 22 | 0 | 6.4 | 7.9 | 53 | W | 7.7 | 618 |
| Annual | 208 | 4.4 | 123.3 | 7.6 | 1,432 | W | 7.0 | 4,995 |

TABLE XXX

CLIMATIC TABLE FOR GOOSE, NFLD.
Latitude 53°19′N, longitude 60°25′W, elevation 13 m

| Month | Mean sta. press. (mbar) | Temperature (°C) | | | | Mean vapor press. (mbar) | Precipitation (mm) | | Snowfall (mm) |
|---|---|---|---|---|---|---|---|---|---|
| | | mean daily | mean daily range | extremes max. | min. | | mean | max. in 24 h | |
| Jan. | 1005.0 | −16.6 | 8.1 | 8 | −39 | 1.0 | 72 | 30.2 | 699 |
| Feb. | 1005.3 | −14.9 | 9.6 | 11 | −37 | 1.3 | 63 | 39.6 | 607 |
| Mar. | 1006.1 | −8.4 | 9.6 | 12 | −36 | 2.0 | 68 | 37.1 | 638 |
| Apr. | 1007.1 | −1.6 | 8.4 | 19 | −26 | 3.3 | 62 | 42.9 | 483 |
| May | 1007.6 | 5.1 | 9.4 | 32 | −12 | 5.6 | 56 | 33.8 | 178 |
| June | 1005.2 | 11.9 | 10.6 | 34 | −2 | 8.4 | 72 | 29.2 | 20 |
| July | 1003.6 | 16.3 | 10.0 | 38 | 3 | 12.3 | 84 | 35.1 | 0 |
| Aug. | 1004.2 | 14.7 | 9.5 | 33 | 0 | 11.4 | 91 | 65.5 | 0 |
| Sept. | 1005.7 | 10.1 | 8.7 | 30 | −7 | 8.7 | 76 | 43.2 | 28 |
| Oct. | 1006.7 | 3.2 | 6.8 | 23 | −12 | 5.8 | 63 | 30.2 | 246 |
| Nov. | 1006.4 | −4.4 | 7.0 | 16 | −23 | 3.5 | 67 | 40.6 | 511 |
| Dec. | 1003.5 | −12.9 | 8.0 | 12 | −36 | 1.8 | 63 | 32.0 | 594 |
| Annual | 1005.5 | 0.2 | 8.8 | 38 | −39 | 5.4 | 837 | 65.5 | 4,004 |

| Month | Number of days | | | Mean cloud-iness (tenths) | Mean sun-shine (h) | Wind | | 18°C degree-days |
|---|---|---|---|---|---|---|---|---|
| | precip. >0.25 mm | thunder-storm | heavy fog | | | most frequ. direct. | mean speed (m/sec) | |
| Jan. | 16 | 0 | 0.6 | 6.3 | 90 | W | 4.8 | 1,082 |
| Feb. | 14 | 0 | 0.8 | 6.3 | 111 | W | 4.4 | 938 |
| Mar. | 14 | 0 | 1.0 | 6.3 | 143 | W | 4.5 | 830 |
| Apr. | 14 | 0 | 1.0 | 7.3 | 136 | NE | 4.4 | 597 |
| May | 13 | 0 | 1.5 | 7.5 | 176 | NE | 4.2 | 412 |
| June | 15 | 1.0 | 1.3 | 7.6 | 198 | NE | 3.9 | 193 |
| July | 15 | 2.6 | 1.6 | 7.5 | 194 | SW | 3.8 | 72 |
| Aug. | 15 | 1.5 | 1.5 | 7.1 | 187 | SW,W | 3.8 | 114 |
| Sept. | 14 | 0.2 | 1.1 | 7.1 | 124 | W | 4.2 | 247 |
| Oct. | 14 | 0.1 | 1.1 | 7.4 | 91 | W | 4.5 | 468 |
| Nov. | 14 | 0 | 1.4 | 7.3 | 69 | W | 4.2 | 682 |
| Dec. | 15 | 0 | 1.2 | 6.4 | 66 | W | 4.4 | 969 |
| Annual | 173 | 5.4 | 14.1 | 7.0 | 1,585 | W | 4.3 | 6,603 |

TABLE XXXI

CLIMATIC TABLE FOR FORT SMITH, N.W.T.
Latitude 60°01′N, longitude 111°58′W, elevation 62 m

| Month | Mean sta. press. (mbar) | Temperature (°C) | | | | Mean vapor press. (mbar) | Precipitation (mm) | | Snowfall (mm) |
|---|---|---|---|---|---|---|---|---|---|
| | | mean daily | mean daily range | extremes | | | mean | max. in 24 h | |
| | | | | max. | min. | | | | |
| Jan. | 993.6 | −25.4 | 10.3 | 6 | −53 | | 15 | 15.2 | 152 |
| Feb. | 993.4 | −22.2 | 12.0 | 9 | −54 | | 17 | 19.1 | 173 |
| Mar. | 991.7 | −14.4 | 13.6 | 9 | −48 | | 19 | 15.2 | 188 |
| Apr. | 991.2 | −2.8 | 13.2 | 22 | −41 | | 17 | 15.2 | 122 |
| May | 990.4 | 7.6 | 13.2 | 27 | −19 | 6.4 | 26 | 24.9 | 28 |
| June | 987.9 | 12.9 | 14.0 | 33 | −6 | 9.1 | 31 | 28.7 | tr. |
| July | 986.6 | 16.2 | 14.1 | 34 | −3 | 11.4 | 53 | 31.0 | tr. |
| Aug. | 987.1 | 14.2 | 13.5 | 33 | −7 | 10.9 | 35 | 19.6 | tr. |
| Sept. | 987.3 | 7.9 | 10.9 | 32 | −10 | 8.1 | 42 | 22.6 | 41 |
| Oct. | 985.5 | 0.2 | 8.7 | 24 | −25 | | 29 | 20.8 | 183 |
| Nov. | 988.7 | −11.7 | 8.2 | 11 | −41 | | 26 | 23.4 | 251 |
| Dec. | 990.0 | −21.3 | 9.6 | 7 | −51 | | 27 | 13.2 | 264 |
| Annual | 989.5 | −3.2 | 11.7 | 34 | −54 | | 337 | 31.0 | 1,402 |

| Month | Number of days | | | Mean cloud-iness (tenths) | Mean sun-shine (h) | Wind | | 18°C degree-days |
|---|---|---|---|---|---|---|---|---|
| | precip. >0.25 mm | thunder-storm | heavy fog | | | most frequ. direct. | mean speed (m/sec) | |
| Jan. | 11 | 0 | 1.1 | 5.5 | 61 | NW | 2.0 | 1,357 |
| Feb. | 11 | 0 | 0.7 | 5.3 | 117 | SE | 2.9 | 1,145 |
| Mar. | 9 | 0 | 0.6 | 5.6 | 167 | SE | 3.3 | 1,014 |
| Apr. | 7 | 0 | 0.3 | 5.3 | 240 | NW,SE | 3.5 | 633 |
| May | 7 | 0.8 | 0.7 | 6.1 | 282 | SE | 3.3 | 334 |
| June | 8 | 2.6 | 0.7 | 6.2 | 317 | NW | 3.0 | 167 |
| July | 10 | 5.1 | 1.4 | 6.5 | 296 | NW | 3.0 | 76 |
| Aug. | 10 | 3.4 | 2.3 | 6.1 | 268 | NW | 2.9 | 144 |
| Sept. | 12 | 0.2 | 3.1 | 6.8 | 133 | NW | 3.1 | 313 |
| Oct. | 12 | 0 | 3.6 | 7.2 | 86 | NW | 3.4 | 563 |
| Nov. | 13 | 0.1 | 3.3 | 7.4 | 46 | SE | 3.0 | 900 |
| Dec. | 14 | 0 | 1.0 | 6.5 | 27 | SE | 2.3 | 1,228 |
| Annual | 124 | 12.2 | 18.8 | 6.2 | 2,040 | NW | 3.0 | 7,875 |

TABLE XXXII

CLIMATIC TABLE FOR FROBISHER BAY, N.W.T.
Latitude 63°45′N, longitude 68°33′W, elevation 7 m

| Month | Mean sta. press. (mbar) | Temperature (°C) | | | | Mean vapor press. (mbar) | Precipitation (mm) | | Snowfall (mm) |
|---|---|---|---|---|---|---|---|---|---|
| | | mean daily | mean daily range | extremes | | | mean | max. in 24 h | |
| | | | | max. | min. | | | | |
| Jan. | 1009.1 | −26.5 | 7.6 | 4 | −45 | | 22 | 40.6 | 218 |
| Feb. | 1009.3 | −25.5 | 8.4 | 3 | −45 | | 26 | 25.2 | 262 |
| Mar. | 1014.8 | −21.5 | 9.6 | 4 | −42 | | 19 | 23.9 | 193 |
| Apr. | 1014.2 | −13.7 | 9.2 | 5 | −34 | | 21 | 18.0 | 180 |
| May | 1012.5 | −3.1 | 6.6 | 13 | −26 | | 19 | 13.7 | 155 |
| June | 1008.4 | 3.6 | 6.2 | 22 | −8 | 6.4 | 33 | 25.7 | 53 |
| July | 1005.9 | 7.9 | 7.6 | 24 | −1 | 8.4 | 53 | 39.1 | tr. |
| Aug. | 1005.4 | 6.9 | 6.8 | 23 | −1 | 8.1 | 53 | 33.8 | 3 |
| Sept. | 1005.0 | 2.2 | 5.1 | 14 | −15 | 5.8 | 43 | 22.1 | 122 |
| Oct. | 1004.2 | −4.7 | 5.7 | 7 | −21 | | 34 | 18.8 | 292 |
| Nov. | 1006.6 | −12.3 | 6.9 | 6 | −36 | | 33 | 27.9 | 323 |
| Dec. | 1004.9 | −20.5 | 8.6 | 2 | −42 | | 24 | 21.8 | 244 |
| Annual | 1008.4 | −8.9 | 7.3 | 24 | −45 | | 380 | 40.6 | 2,045 |

| Month | Number of days | | | Mean cloud-iness (tenths) | Mean sun-shine (h) | Wind | | 18°C degree-days |
|---|---|---|---|---|---|---|---|---|
| | precip. >0.25 mm | thunder-storm | heavy fog | | | most frequ. direct. | mean speed (m/sec) | |
| Jan. | 10 | 0 | 2.4 | 5.1 | 26 | NW | 3.7 | 1,390 |
| Feb. | 11 | 0 | 1.7 | 5.2 | 75 | NW | 4.3 | 1,238 |
| Mar. | 9 | 0 | 0.5 | 4.6 | 184 | NW | 4.7 | 1,235 |
| Apr. | 9 | 0 | 0.2 | 5.2 | 245 | NW | 5.2 | 960 |
| May | 8 | 0 | 1.0 | 7.6 | 174 | NW | 6.7 | 665 |
| June | 9 | 0 | 4.7 | 7.6 | 142 | SE | 5.4 | 442 |
| July | 11 | 0.1 | 5.6 | 7.5 | 190 | SE | 4.4 | 324 |
| Aug. | 14 | 0 | 4.6 | 7.8 | 130 | SE | 3.8 | 355 |
| Sept. | 12 | 0 | 2.4 | 8.1 | 80 | SE | 4.4 | 483 |
| Oct. | 14 | 0 | 0.4 | 7.9 | 58 | NW | 6.5 | 715 |
| Nov. | 11 | 0 | 0.8 | 6.7 | 38 | NW | 5.2 | 920 |
| Dec. | 12 | 0 | 0.8 | 5.5 | 11 | NW | 4.4 | 1,204 |
| Annual | 130 | 0.1 | 25.1 | 6.6 | 1,353 | NW | 4.9 | 9,930 |

TABLE XXXIII

CLIMATIC TABLE FOR CORAL HARBOUR, N.W.T.
Latitude 64°12′N, longitude 83°22′W, elevation 18 m

| Month | Mean sta. press. (mbar) | Temperature (°C) | | | | Mean vapor press. (mbar) | Precipitation (mm) | | Snowfall (mm) |
|---|---|---|---|---|---|---|---|---|---|
| | | mean daily | mean daily range | extremes | | | mean | max. in 24 h | |
| | | | | max. | min. | | | | |
| Jan. | 1003.8 | −30.4 | 8.5 | −1 | −52 | | 8 | 6.4 | 81 |
| Feb. | 1006.7 | −29.6 | 8.6 | −3 | −48 | | 9 | 7.9 | 91 |
| Mar. | 1010.4 | −23.9 | 10.3 | −1 | −46 | | 9 | 11.7 | 91 |
| Apr. | 1010.9 | −15.7 | 10.7 | 4 | −39 | | 14 | 24.6 | 122 |
| May | 1009.5 | −6.1 | 8.2 | 9 | −38 | 3.3 | 19 | 22.6 | 168 |
| June | 1004.6 | 2.3 | 6.5 | 19 | −12 | 6.1 | 26 | 21.1 | 61 |
| July | 1001.7 | 8.3 | 8.7 | 24 | −1 | 8.4 | 35 | 29.2 | 8 |
| Aug. | 1002.9 | 7.6 | 8.0 | 25 | −3 | 8.4 | 38 | 27.4 | tr. |
| Sept. | 1003.0 | 0.9 | 5.5 | 17 | −13 | 5.6 | 33 | 22.9 | 86 |
| Oct. | 1002.4 | −8.0 | 7.4 | 4 | −29 | | 29 | 17.8 | 246 |
| Nov. | 1004.4 | −16.8 | 8.5 | 2 | −37 | | 17 | 7.6 | 163 |
| Dec. | 1003.4 | −24.8 | 8.5 | −4 | −47 | | 12 | 8.6 | 124 |
| Annual | 1005.3 | −11.3 | 8.3 | 25 | −52 | | 249 | 29.2 | 1,241 |

| Month | Number of days | | | Mean cloudiness (tenths) | Wind | | 18°C degree-days |
|---|---|---|---|---|---|---|---|
| | precip. >0.25 mm | thunder-storm | heavy fog | | most frequ. direct. | mean speed (m/sec) | |
| Jan. | 7 | 0 | 0.9 | 4.2 | N | 5.3 | 1,510 |
| Feb. | 8 | 0 | 0.7 | 4.3 | N | 5.4 | 1,354 |
| Mar. | 7 | 0 | 0.7 | 4.7 | N | 5.6 | 1,309 |
| Apr. | 8 | 0 | 1.4 | 5.3 | N | 5.6 | 1,022 |
| May | 9 | 0 | 3.3 | 7.1 | N | 5.7 | 758 |
| June | 8 | 0.1 | 4.9 | 7.3 | N | 5.3 | 482 |
| July | 10 | 0.4 | 4.6 | 6.9 | N | 5.5 | 310 |
| Aug. | 9 | 0.1 | 4.1 | 7.0 | N | 5.6 | 332 |
| Sept. | 9 | 0 | 4.3 | 7.8 | N | 5.9 | 522 |
| Oct. | 12 | 0 | 4.3 | 7.0 | N | 6.7 | 816 |
| Nov. | 11 | 0 | 2.6 | 6.1 | N | 6.5 | 1,055 |
| Dec. | 9 | 0 | 0.7 | 4.8 | N | 5.8 | 1,337 |
| Annual | 107 | 0.6 | 32.5 | 6.0 | N | 5.7 | 10,806 |

TABLE XXXIV

CLIMATIC TABLE FOR BAKER LAKE, N.W.T.
Latitude 64°18′N, longitude 96°00′W, elevation 4 m

| Month | Mean sta. press. (mbar) | Temperature (°C) | | | | Precipitation (mm) | | Snowfall (mm) |
|---|---|---|---|---|---|---|---|---|
| | | mean daily | mean daily range | extremes | | mean | max. in 24 h | |
| | | | | max. | min. | | | |
| Jan. | 1016.7 | −32.9 | 7.0 | −11 | −49 | 5 | 2.5 | 48 |
| Feb. | 1019.3 | −32.8 | 7.1 | −9 | −50 | 4 | 5.1 | 43 |
| Mar. | 1021.3 | −26.3 | 8.1 | −1 | −50 | 6 | 4.6 | 56 |
| Apr. | 1020.5 | −16.4 | 9.1 | 4 | −36 | 9 | 4.6 | 94 |
| May | 1019.6 | −5.8 | 6.7 | 9 | −26 | 8 | 10.7 | 46 |
| June | 1013.3 | 3.9 | 7.4 | 23 | −13 | 21 | 14.2 | 8 |
| July | 1010.2 | 10.7 | 9.9 | 27 | −2 | 40 | 35.8 | tr. |
| Aug. | 1011.8 | 10.0 | 8.2 | 28 | −2 | 45 | 21.1 | tr. |
| Sept. | 1012.7 | 2.8 | 5.6 | 20 | −9 | 34 | 25.7 | 23 |
| Oct. | 1012.0 | −7.5 | 6.1 | 9 | −28 | 20 | 32.0 | 104 |
| Nov. | 1015.3 | −20.0 | 7.4 | 2 | −40 | 9 | 8.6 | 89 |
| Dec. | 1015.3 | −28.2 | 7.2 | 4 | −46 | 7 | 4.8 | 71 |
| Annual | 1015.7 | −11.9 | 7.4 | 28 | −50 | 208 | 35.8 | 582 |

| Month | Number of days | | | Mean cloud-iness (tenths) | Wind | | 18°C degree-days |
|---|---|---|---|---|---|---|---|
| | precip. >0.25 mm | thunder-storm | heavy fog | | most frequ. direct. | mean speed (m/sec) | |
| Jan. | 7 | 0 | 2.1 | 4.5 | NW | 6.3 | 1,588 |
| Feb. | 6 | 0 | 1.3 | 4.3 | NW | 5.5 | 1,444 |
| Mar. | 7 | 0 | 1.1 | 4.8 | NW | 5.6 | 1,384 |
| Apr. | 9 | 0 | 0.5 | 5.5 | N | 6.3 | 1,041 |
| May | 8 | 0 | 1.6 | 7.0 | N | 5.4 | 749 |
| June | 7 | 0.2 | 2.1 | 7.1 | N | 4.4 | 433 |
| July | 10 | 1.0 | 2.6 | 6.5 | N | 4.7 | 236 |
| Aug. | 10 | 0.3 | 1.2 | 6.7 | N | 5.4 | 258 |
| Sept. | 10 | 0.1 | 0.5 | 7.9 | NW | 5.6 | 465 |
| Oct. | 12 | 0 | 1.1 | 8.0 | N | 6.4 | 801 |
| Nov. | 10 | 0 | 0.8 | 6.0 | N | 6.0 | 1,150 |
| Dec. | 8 | 0 | 0.7 | 4.7 | NW | 6.3 | 1,443 |
| Annual | 104 | 1.6 | 15.6 | 6.1 | N | 5.6 | 10,993 |

TABLE XXXV

CLIMATIC TABLE FOR NORMAN WELLS, N.W.T.
Latitude 65°17′N, longitude 126°48′W, elevation 20 m

| Month | Mean sta. press. (mbar) | Temperature (°C) | | | | Mean vapor press. (mbar) | Precipitation (mm) | | Snowfall (mm) |
|---|---|---|---|---|---|---|---|---|---|
| | | mean daily | mean daily range | extremes | | | mean | max. in 24 h | |
| | | | | max. | min. | | | | |
| Jan. | 1014.4 | −28.2 | 8.3 | 7 | −53 | | 19 | 17.8 | 193 |
| Feb. | 1013.9 | −26.3 | 9.1 | 5 | −54 | | 16 | 12.2 | 157 |
| Mar. | 1011.8 | −18.6 | 12.0 | 11 | −46 | | 12 | 6.9 | 119 |
| Apr. | 1009.6 | −7.6 | 12.6 | 18 | −37 | | 13 | 23.6 | 117 |
| May | 1008.1 | 5.2 | 11.2 | 31 | −17 | 6.1 | 15 | 20.6 | 56 |
| June | 1004.6 | 13.4 | 11.5 | 31 | −3 | 9.8 | 34 | 36.6 | 3 |
| July | 1004.1 | 15.9 | 11.7 | 32 | −1 | 11.4 | 49 | 45.2 | 0 |
| Aug. | 1004.2 | 13.1 | 10.7 | 32 | −6 | 10.6 | 64 | 48.5 | tr. |
| Sept. | 1004.7 | 6.1 | 8.3 | 26 | −14 | 7.5 | 38 | 26.9 | 61 |
| Oct. | 1002.9 | −4.0 | 6.6 | 18 | −27 | | 24 | 11.7 | 213 |
| Nov. | 1007.8 | −17.6 | 6.9 | 8 | −43 | | 24 | 16.5 | 241 |
| Dec. | 1009.4 | −26.0 | 7.7 | 2 | −47 | | 20 | 17.0 | 203 |
| Annual | 1008.0 | −6.2 | 9.8 | 32 | −54 | | 328 | 48.5 | 1,363 |

| Month | Number of days | | | Mean cloud-iness (tenths) | Wind | | 18°C degree-days |
|---|---|---|---|---|---|---|---|
| | precip. >0.25 mm | thunder-storm | heavy fog | | most frequ. direct. | mean speed (m/sec) | |
| Jan. | 13 | 0 | 1.0 | 5.3 | NW | 2.8 | 1,442 |
| Feb. | 12 | 0 | 0.4 | 5.8 | SE | 2.2 | 1,260 |
| Mar. | 10 | 0 | 0.3 | 5.2 | NW | 2.4 | 1,145 |
| Apr. | 7 | 0 | 0.2 | 5.5 | SE | 3.8 | 778 |
| May | 7 | 0.2 | 0.8 | 6.4 | SE | 3.5 | 407 |
| June | 10 | 2.5 | 0.2 | 6.4 | SE | 3.8 | 155 |
| July | 11 | 3.0 | 0.4 | 6.7 | NW | 3.2 | 81 |
| Aug. | 12 | 0.9 | 1.1 | 6.9 | SE | 3.3 | 169 |
| Sept. | 12 | 0.1 | 1.9 | 7.2 | SE | 3.1 | 368 |
| Oct. | 11 | 0 | 3.9 | 7.3 | NW | 3.2 | 692 |
| Nov. | 14 | 0 | 1.3 | 6.7 | NW | 2.5 | 1,078 |
| Dec. | 13 | 0 | 0.3 | 6.0 | NW | 2.3 | 1,374 |
| Annual | 132 | 6.7 | 11.8 | 6.3 | NW | 3.0 | 8,950 |

TABLE XXXVI

CLIMATIC TABLE FOR HALIFAX, N.S.
Latitude 44°39′N, longitude 63°34′W, elevation 8 m

| Month | Mean sta. press. (mbar) | Temperature (°C) | | | | Mean vapor press. (mbar) | Precipitation (mm) | | Snowfall (mm) |
|---|---|---|---|---|---|---|---|---|---|
| | | mean daily | mean daily range | extremes | | | mean | max. in 24 h | |
| | | | | max. | min. | | | | |
| Jan. | 1008.0 | −3.3 | 7.7 | 14 | −25 | 3.3 | 141 | 78.2 | 460 |
| Feb. | 1007.3 | −3.6 | 7.7 | 12 | −24 | 3.2 | 119 | 87.9 | 480 |
| Mar. | 1007.0 | −0.1 | 7.2 | 19 | −19 | 4.4 | 113 | 56.9 | 351 |
| Apr. | 1008.8 | 4.8 | 8.0 | 28 | −11 | 5.8 | 112 | 64.5 | 117 |
| May | 1009.5 | 9.9 | 9.3 | 31 | −2 | 8.7 | 109 | 90.4 | 3 |
| June | 1008.8 | 14.4 | 9.3 | 33 | 2 | 11.8 | 94 | 82.3 | 0 |
| July | 1010.0 | 18.5 | 9.4 | 33 | 4 | 16.4 | 94 | 131.6 | 0 |
| Aug. | 1010.0 | 18.8 | 8.7 | 34 | 7 | 16.4 | 96 | 91.7 | 0 |
| Sept. | 1012.7 | 15.5 | 8.2 | 34 | 2 | 13.7 | 117 | 238.8 | tr. |
| Oct. | 1012.0 | 10.3 | 7.8 | 28 | −4 | 9.8 | 120 | 73.7 | tr. |
| Nov. | 1010.1 | 5.3 | 6.9 | 24 | −14 | 7.2 | 143 | 69.1 | 64 |
| Dec. | 1009.4 | −0.9 | 6.9 | 16 | −23 | 4.2 | 126 | 74.7 | 328 |
| Annual | 1009.5 | 7.4 | 8.1 | 34 | −25 | 8.7 | 1,384 | 238.8 | 1,803 |

| Month | Number of days | | | Mean cloud-iness (tenths) | Mean sun-shine (h) | Wind | | 18°C degree-days |
|---|---|---|---|---|---|---|---|---|
| | precip. >0.25 mm | thunder-storm | heavy fog | | | most frequ. direct. | mean speed (m/sec) | |
| Jan. | 16 | 0.2 | 3.5 | 7.0 | 91 | NW | 5.3 | 674 |
| Feb. | 14 | 0.1 | 3.7 | 6.7 | 117 | NW | 5.6 | 617 |
| Mar. | 13 | 0.1 | 4.5 | 6.4 | 144 | NW | 5.1 | 572 |
| Apr. | 13 | 0.6 | 5.4 | 6.8 | 162 | NW | 5.2 | 412 |
| May | 13 | 0.6 | 9.1 | 6.8 | 201 | NW,SW | 4.4 | 269 |
| June | 11 | 1.6 | 9.5 | 7.0 | 202 | SW | 4.0 | 132 |
| July | 11 | 2.3 | 12.1 | 6.4 | 219 | SW | 3.5 | 32 |
| Aug. | 10 | 2.3 | 7.4 | 6.2 | 222 | SW | 3.5 | 28 |
| Sept. | 10 | 0.9 | 5.4 | 5.7 | 179 | NW | 3.9 | 100 |
| Oct. | 10 | 0.5 | 4.6 | 5.8 | 157 | NW | 4.5 | 254 |
| Nov. | 14 | 0.5 | 4.3 | 7.4 | 92 | NW | 4.9 | 394 |
| Dec. | 16 | 0.3 | 3.2 | 7.0 | 87 | NW | 5.6 | 597 |
| Annual | 151 | 10.0 | 72.7 | 6.6 | 1,873 | NW | 4.6 | 4,089 |

TABLE XXXVII

CLIMATIC TABLE FOR TORONTO, ONT.
Latitude 43°40′N, longitude 79°24′W, elevation 35 m

| Month | Mean sta. press. (mbar) | Temperature (°C) | | | | Mean vapor press. (mbar) | Precipitation (mm) | | Snowfall (mm) |
|---|---|---|---|---|---|---|---|---|---|
| | | mean daily | mean daily range | extremes | | | mean | max. in 24 h | |
| | | | | max. | min. | | | | |
| Jan. | 1003.6 | −3.9 | 7.1 | 16 | −25 | 3.2 | 67 | 52.3 | 348 |
| Feb. | 1002.4 | −3.8 | 7.3 | 14 | −29 | 3.2 | 59 | 42.2 | 335 |
| Mar. | 1001.5 | 0.2 | 7.3 | 27 | −21 | 4.2 | 67 | 42.9 | 254 |
| Apr. | 1001.5 | 7.0 | 8.6 | 29 | −12 | 5.8 | 66 | 46.7 | 71 |
| May | 1001.0 | 13.2 | 10.2 | 32 | −1 | 9.1 | 70 | 51.8 | tr. |
| June | 1000.8 | 19.0 | 10.4 | 35 | 3 | 14.2 | 63 | 63.5 | 0 |
| July | 1001.6 | 21.9 | 10.7 | 41 | 7 | 17.0 | 74 | 65.0 | 0 |
| Aug. | 1002.1 | 21.1 | 10.2 | 38 | 7 | 16.4 | 61 | 81.8 | 0 |
| Sept. | 1003.9 | 16.6 | 9.8 | 38 | 1 | 12.8 | 65 | 67.6 | tr. |
| Oct. | 1003.9 | 10.6 | 8.6 | 28 | −6 | 9.4 | 60 | 97.0 | 3 |
| Nov. | 1001.8 | 4.3 | 6.7 | 24 | −16 | 6.1 | 63 | 51.3 | 124 |
| Dec. | 1003.0 | −1.8 | 6.5 | 15 | −30 | 3.8 | 61 | 52.1 | 259 |
| Annual | 1002.2 | 8.7 | 8.6 | 41 | −30 | 8.8 | 776 | 81.8 | 1,394 |

| Month | Number of days | | | Mean cloudiness (tenths) | Mean sunshine (h) | Wind | | 18°C degree-days |
|---|---|---|---|---|---|---|---|---|
| | precip. >0.25 mm | thunder-storm | heavy fog | | | most frequ. direct. | mean speed (m/sec) | |
| Jan. | 16 | 0 | 4.2 | 7.3 | 78 | W | 6.6 | 685 |
| Feb. | 13 | 0.2 | 3.1 | 6.9 | 105 | W | 6.5 | 622 |
| Mar. | 13 | 0.5 | 4.0 | 6.3 | 139 | NW | 6.2 | 562 |
| Apr. | 12 | 1.8 | 3.3 | 6.2 | 170 | E,NW | 5.8 | 342 |
| May | 12 | 3.0 | 3.8 | 6.3 | 220 | E | 4.7 | 165 |
| June | 9 | 3.9 | 2.7 | 5.8 | 257 | E,SW | 4.1 | 34 |
| July | 10 | 4.7 | 1.8 | 5.1 | 287 | SW | 3.7 | 4 |
| Aug. | 9 | 3.3 | 1.7 | 5.2 | 259 | SW | 3.8 | 10 |
| Sept. | 9 | 2.2 | 4.0 | 5.3 | 198 | E,SW,NW | 4.3 | 84 |
| Oct. | 9 | 0.8 | 4.8 | 5.6 | 154 | SW | 4.5 | 244 |
| Nov. | 13 | 0.6 | 3.8 | 7.3 | 84 | SW | 5.8 | 422 |
| Dec. | 13 | 0.2 | 4.3 | 7.3 | 75 | SW | 6.5 | 617 |
| Annual | 138 | 21.2 | 41.5 | 6.2 | 2,026 | SW | 5.2 | 3,792 |

TABLE XXXVIII

CLIMATIC TABLE FOR OTTAWA, ONT.
Latitude 45°19′N, longitude 75°40′W, elevation 38 m

| Month | Mean sta. press. (mbar) | Temperature (°C) | | | | Mean vapor press. (mbar) | Precipitation (mm) | | Snowfall (mm) |
|---|---|---|---|---|---|---|---|---|---|
| | | mean daily | mean daily range | extremes | | | mean | max. in 24 h | |
| | | | | max. | min. | | | | |
| Jan. | 1004.6 | −10.8 | 8.7 | 11 | −36 | 1.8 | 60 | 29.5 | 470 |
| Feb. | 1002.7 | −9.8 | 8.8 | 12 | −36 | 1.9 | 58 | 45.2 | 488 |
| Mar. | 1002.0 | −3.6 | 8.1 | 27 | −31 | 3.3 | 66 | 48.3 | 409 |
| Apr. | 1002.0 | 5.3 | 10.2 | 29 | −14 | 5.8 | 67 | 48.3 | 81 |
| May | 1001.4 | 12.8 | 11.7 | 33 | −4 | 9.8 | 75 | 44.7 | tr. |
| June | 1000.5 | 18.2 | 11.2 | 36 | 1 | 14.8 | 76 | 77.5 | 0 |
| July | 1002.1 | 20.7 | 11.7 | 37 | 5 | 17.0 | 78 | 71.6 | 0 |
| Aug. | 1002.4 | 19.5 | 11.6 | 38 | 3 | 15.9 | 76 | 90.4 | 0 |
| Sept. | 1004.6 | 14.7 | 10.7 | 35 | −3 | 12.3 | 77 | 93.2 | tr. |
| Oct. | 1004.6 | 8.2 | 9.7 | 28 | −8 | 8.4 | 67 | 50.6 | 8 |
| Nov. | 1002.6 | 0.8 | 7.2 | 24 | −19 | 5.1 | 67 | 36.6 | 203 |
| Dec. | 1003.7 | −8.0 | 8.0 | 15 | −34 | 2.5 | 83 | 35.3 | 528 |
| Annual | 1002.8 | 5.7 | 9.8 | 38 | −36 | 8.2 | 850 | 93.2 | 2,187 |

| Month | Number of days | | | Mean cloud-iness (tenths) | Mean sun-shine (h) | Wind | | 18°C degree-days |
|---|---|---|---|---|---|---|---|---|
| | precip. >0.25 mm | thunder-storm | heavy fog | | | most frequ. direct. | mean speed (m/sec) | |
| Jan. | 16 | 0 | 3.3 | 6.6 | 93 | NW | 4.9 | 902 |
| Feb. | 14 | 0 | 3.6 | 6.4 | 110 | NW | 5.1 | 794 |
| Mar. | 13 | 0.2 | 3.5 | 6.1 | 140 | NW | 5.2 | 681 |
| Apr. | 12 | 0.9 | 2.5 | 6.3 | 168 | NW | 5.1 | 390 |
| May | 12 | 2.2 | 1.5 | 6.3 | 230 | NW | 4.7 | 181 |
| June | 11 | 5.0 | 1.7 | 6.1 | 250 | SW | 4.3 | 47 |
| July | 11 | 5.9 | 1.6 | 5.5 | 281 | SW | 4.0 | 12 |
| Aug. | 10 | 4.3 | 2.3 | 5.4 | 254 | SW | 3.9 | 43 |
| Sept. | 11 | 2.4 | 3.1 | 5.6 | 173 | SW | 4.3 | 123 |
| Oct. | 11 | 1.0 | 5.0 | 6.0 | 135 | SW | 4.5 | 315 |
| Nov. | 14 | 0.2 | 3.6 | 7.4 | 78 | NW | 4.8 | 524 |
| Dec. | 17 | 0 | 3.8 | 7.0 | 77 | E | 4.8 | 816 |
| Annual | 152 | 22.1 | 35.5 | 6.2 | 1,989 | NW | 4.6 | 4,829 |

TABLE XXXIX

CLIMATIC TABLE FOR NORTH BAY, ONT.
Latitude 46°22′N, longitude 79°25′W, elevation 113 m

| Month | Mean sta. press. (mbar) | Temperature (°C) | | | | Mean vapor press. (mbar) | Precipitation (mm) | | Snowfall (mm) |
| | | mean daily | mean daily range | extremes | | | mean | max. in 24 h | |
| | | | | max. | min. | | | | |
|---|---|---|---|---|---|---|---|---|---|
| Jan. | 971.3 | −12.2 | 9.1 | 12 | −40 | 1.8 | 81 | 31.8 | 704 |
| Feb. | 971.1 | −10.7 | 9.0 | 10 | −39 | 1.9 | 66 | 27.4 | 577 |
| Mar. | 969.5 | −5.7 | 9.0 | 24 | −37 | 3.0 | 72 | 48.3 | 490 |
| Apr. | 972.7 | 3.1 | 9.6 | 27 | −21 | 4.6 | 68 | 73.7 | 178 |
| May | 970.4 | 10.5 | 11.2 | 28 | −8 | 8.7 | 79 | 45.2 | 10 |
| June | 970.1 | 15.9 | 10.4 | 33 | −2 | 13.7 | 93 | 63.8 | tr. |
| July | 971.5 | 18.7 | 10.2 | 32 | 3 | 16.4 | 101 | 59.4 | 0 |
| Aug. | 972.2 | 17.6 | 10.1 | 33 | 1 | 15.3 | 88 | 96.3 | 0 |
| Sept. | 973.1 | 12.8 | 9.2 | 33 | −4 | 11.8 | 114 | 66.8 | 3 |
| Oct. | 971.3 | 6.6 | 8.4 | 27 | −11 | 8.1 | 93 | 44.5 | 53 |
| Nov. | 969.4 | −1.1 | 6.4 | 19 | −24 | 4.6 | 97 | 36.6 | 371 |
| Dec. | 970.0 | −9.5 | 8.4 | 13 | −37 | 2.5 | 82 | 29.2 | 643 |
| Annual | 971.1 | 3.8 | 9.2 | 33 | −40 | 7.7 | 1,034 | 96.3 | 3,029 |

| Month | Number of days | | | Mean cloud-iness (tenths) | Wind | | 18°C degree-days |
| | precip. >0.25 mm | thunder-storm | heavy fog | | most frequ. direct | mean speed (m/sec) | |
|---|---|---|---|---|---|---|---|
| Jan. | 19 | 0 | 5.9 | 6.8 | N | 4.8 | 945 |
| Feb. | 18 | 0.1 | 4.6 | 6.7 | N | 5.1 | 819 |
| Mar. | 14 | 0.2 | 4.8 | 6.2 | N,SW | 5.2 | 745 |
| Apr. | 12 | 0.8 | 4.7 | 6.3 | SW | 5.2 | 457 |
| May | 13 | 1.4 | 5.0 | 6.6 | SW | 4.8 | 246 |
| June | 13 | 4.2 | 5.1 | 6.3 | SW | 4.5 | 88 |
| July | 12 | 5.8 | 6.1 | 5.8 | SW | 4.1 | 31 |
| Aug. | 12 | 3.6 | 3.9 | 5.9 | SW | 4.5 | 66 |
| Sept. | 14 | 3.2 | 5.6 | 6.3 | SW | 4.5 | 168 |
| Oct. | 14 | 1.0 | 5.9 | 6.6 | SW | 4.8 | 363 |
| Nov. | 19 | 0.6 | 5.5 | 8.1 | N | 4.9 | 583 |
| Dec. | 21 | 0 | 6.0 | 7.3 | N | 4.8 | 863 |
| Annual | 181 | 20.9 | 63.1 | 6.6 | SW | 4.8 | 5,376 |

TABLE XL

CLIMATIC TABLE FOR KAPUSKASING, ONT.
Latitude 49°25′N, longitude 82°28′W, elevation 70 m

| Month | Mean sta. press. (mbar) | Temperature (°C) | | | | Mean vapor press. (mbar) | Precipitation (mm) | | Snowfall (mm) |
|---|---|---|---|---|---|---|---|---|---|
| | | mean daily | mean daily range | extremes | | | mean | max. in 24 h | |
| | | | | max. | min. | | | | |
| Jan. | 988.5 | −17.8 | 10.7 | 8 | −42 | 1.2 | 55 | 38.1 | 531 |
| Feb. | 987.6 | −15.4 | 11.9 | 12 | −41 | 1.4 | 45 | 27.9 | 434 |
| Mar. | 987.4 | −9.3 | 12.5 | 19 | −42 | 2.4 | 54 | 35.6 | 462 |
| Apr. | 987.1 | 0.2 | 12.0 | 29 | −26 | 4.2 | 53 | 30.2 | 239 |
| May | 986.7 | 7.9 | 12.6 | 33 | −11 | 7.2 | 80 | 55.9 | 99 |
| June | 985.0 | 14.3 | 12.9 | 36 | −4 | 11.4 | 94 | 46.5 | tr. |
| July | 985.4 | 17.3 | 12.7 | 35 | 1 | 14.2 | 85 | 47.5 | 0 |
| Aug. | 986.7 | 16.0 | 12.4 | 35 | −1 | 13.7 | 87 | 64.5 | 0 |
| Sept. | 987.2 | 10.5 | 10.0 | 33 | −6 | 10.2 | 90 | 57.9 | 18 |
| Oct. | 987.3 | 4.6 | 8.7 | 28 | −16 | 7.2 | 72 | 47.2 | 157 |
| Nov. | 985.1 | −4.9 | 6.6 | 19 | −34 | 3.8 | 83 | 44.5 | 617 |
| Dec. | 986.8 | −14.0 | 9.6 | 10 | −42 | 1.7 | 60 | 35.6 | 569 |
| Annual | 986.7 | 0.8 | 11.0 | 36 | −42 | 6.6 | 858 | 64.5 | 3,126 |

| Month | Number of days | | | Mean cloud-iness (tenths) | Mean sun-shine (h) | Wind | | 18°C degree-days |
|---|---|---|---|---|---|---|---|---|
| | precip. >0.25 mm | thunder-storm | heavy fog | | | most frequ. direct. | mean speed (m/sec) | |
| Jan. | 19 | 0 | 1.2 | 6.5 | 72 | S | 4.4 | 1,121 |
| Feb. | 16 | 0 | 1.2 | 6.4 | 104 | NW | 4.5 | 954 |
| Mar. | 14 | 0.1 | 1.7 | 6.3 | 135 | NW | 4.6 | 858 |
| Apr. | 11 | 0.3 | 1.9 | 6.5 | 167 | NW | 4.6 | 545 |
| May | 14 | 1.3 | 1.3 | 7.0 | 195 | N | 4.4 | 324 |
| June | 14 | 3.3 | 2.4 | 7.1 | 208 | SW | 4.3 | 127 |
| July | 15 | 4.8 | 3.2 | 6.5 | 238 | SW | 4.0 | 39 |
| Aug. | 12 | 3.6 | 3.3 | 6.6 | 202 | S,SW | 3.8 | 93 |
| Sept. | 15 | 2.0 | 4.1 | 7.2 | 126 | S | 4.4 | 235 |
| Oct. | 15 | 0.6 | 2.6 | 7.2 | 90 | S | 4.6 | 427 |
| Nov. | 20 | 0.1 | 3.0 | 8.5 | 44 | S | 4.5 | 697 |
| Dec. | 19 | 0 | 1.6 | 7.2 | 54 | NW | 4.4 | 1,002 |
| Annual | 184 | 16.1 | 27.5 | 6.9 | 1,635 | NW | 4.4 | 6,422 |

TABLE XLI

CLIMATIC TABLE FOR MOOSONEE, ONT.
Latitude 51°16′N, longitude 80°39′W, elevation 3 m

| Month | Mean sta. press. (mbar) | Temperature (°C) | | | | Mean vapor press. (mbar) | Precipitation (mm) | | Snowfall (mm) |
|---|---|---|---|---|---|---|---|---|---|
| | | mean daily | mean daily range | extremes | | | mean | max. in 24 h | |
| | | | | max. | min. | | | | |
| Jan. | 1018.3 | −20.6 | 11.6 | 7 | −47 | 0.9 | 48 | 28.5 | 457 |
| Feb. | 1017.1 | −18.0 | 12.9 | 11 | −46 | 0.9 | 47 | 27.9 | 452 |
| Mar. | 1016.9 | −11.9 | 13.0 | 16 | −42 | 1.9 | 42 | 29.0 | 348 |
| Apr. | 1016.0 | −2.5 | 10.9 | 27 | −32 | 3.3 | 44 | 22.9 | 254 |
| May | 1015.0 | 5.2 | 10.9 | 33 | −17 | 6.4 | 72 | 37.1 | 112 |
| June | 1012.2 | 11.9 | 12.2 | 34 | −6 | 9.8 | 92 | 53.9 | 8 |
| July | 1011.5 | 15.6 | 12.5 | 36 | −2 | 13.2 | 80 | 58.9 | 0 |
| Aug. | 1013.0 | 14.9 | 11.5 | 35 | −1 | 12.8 | 81 | 35.8 | 0 |
| Sept. | 1014.2 | 10.1 | 10.2 | 32 | −6 | 9.8 | 82 | 35.8 | 3 |
| Oct. | 1014.2 | 3.9 | 8.2 | 27 | −17 | 6.9 | 73 | 33.0 | 175 |
| Nov. | 1013.1 | −5.4 | 7.3 | 19 | −34 | 3.7 | 72 | 39.9 | 490 |
| Dec. | 1015.5 | −15.5 | 10.1 | 9 | −42 | 1.5 | 56 | 20.8 | 503 |
| Annual | 1014.8 | −1.1 | 10.9 | 36 | −47 | 5.9 | 789 | 58.9 | 2,802 |

| Month | Number of days | | | Mean cloud-iness (tenths) | Mean sun-shine (h) | Wind | | 18°C degree-days |
|---|---|---|---|---|---|---|---|---|
| | precip. >0.25 mm | thunder-storm | heavy fog | | | most frequ. direct. | mean speed (m/sec) | |
| Jan. | 17 | 0 | 0.7 | 5.7 | 78 | W | 3.0 | 1,207 |
| Feb. | 15 | 0 | 0.4 | 5.7 | 109 | W | 3.2 | 1,027 |
| Mar. | 13 | 0.1 | 1.2 | 5.9 | 135 | NW | 3.6 | 939 |
| Apr. | 13 | 0.2 | 1.8 | 6.5 | 160 | NW | 4.1 | 625 |
| May | 15 | 0.9 | 1.4 | 7.0 | 188 | N | 3.8 | 408 |
| June | 15 | 2.7 | 2.0 | 6.8 | 179 | N | 3.6 | 192 |
| July | 16 | 3.6 | 2.3 | 6.1 | 241 | SW | 3.3 | 98 |
| Aug. | 15 | 2.7 | 2.6 | 6.6 | 195 | SW | 3.2 | 116 |
| Sept. | 17 | 0.7 | 1.8 | 6.9 | 119 | SW | 3.6 | 248 |
| Oct. | 15 | 0.3 | 1.9 | 7.0 | 80 | SW | 3.7 | 448 |
| Nov. | 19 | 0.1 | 1.8 | 8.1 | 43 | W | 3.5 | 713 |
| Dec. | 20 | 0 | 0.4 | 6.5 | 45 | W | 3.2 | 1,047 |
| Annual | 190 | 11.3 | 18.3 | 6.6 | 1,572 | W | 3.5 | 7,068 |

## TABLE XLII

CLIMATIC TABLE FOR TROUT LAKE, ONT.
Latitude 53°50′N, longitude 89°52′W, elevation 67 m

| Month | Mean sta. press. (mbar) | Temperature (°C) | | | | Mean vapor press. (mbar) | Precipitation (mm) | | Snowfall (mm) |
|---|---|---|---|---|---|---|---|---|---|
| | | mean daily | mean daily range | extremes | | | mean | max. in 24 h | |
| | | | | max. | min. | | | | |
| Jan. | 989.9 | −23.9 | 10.9 | 2 | −48 | 0.6 | 25 | 28.5 | 246 |
| Feb. | 990.8 | −21.4 | 12.5 | 5 | −47 | 0.6 | 21 | 15.2 | 211 |
| Mar. | 990.8 | −14.4 | 14.3 | 12 | −42 | 1.5 | 17 | 10.2 | 163 |
| Apr. | 990.1 | −4.6 | 12.4 | 21 | −32 | 3.0 | 26 | 33.0 | 196 |
| May | 989.2 | 3.6 | 11.3 | 30 | −21 | 5.6 | 48 | 25.4 | 196 |
| June | 986.2 | 11.2 | 10.8 | 32 | −7 | 9.8 | 75 | 48.8 | 3 |
| July | 985.5 | 15.9 | 10.0 | 36 | 0 | 13.7 | 100 | 75.2 | tr. |
| Aug. | 986.8 | 14.7 | 9.0 | 30 | −1 | 13.2 | 92 | 84.1 | 0 |
| Sept. | 986.7 | 8.7 | 7.5 | 29 | −8 | 9.4 | 75 | 64.0 | 46 |
| Oct. | 987.2 | 1.8 | 6.5 | 20 | −19 | 6.4 | 53 | 29.2 | 196 |
| Nov. | 987.2 | −8.9 | 6.4 | 10 | −34 | 2.6 | 47 | 30.5 | 439 |
| Dec. | 987.8 | −19.3 | 9.0 | 4 | −44 | 1.0 | 29 | 15.2 | 287 |
| Annual | 988.2 | −3.1 | 10.0 | 36 | −48 | 5.6 | 608 | 84.1 | 1,983 |

| Month | Number of days | | | Mean cloud-iness (tenths) | Wind | | 18°C degree-days |
|---|---|---|---|---|---|---|---|
| | precip. >0.25 mm | thunder-storm | heavy fog | | most frequ. direct. | mean speed (m/sec) | |
| Jan. | 12 | 0 | 1.0 | 56 | NW | 4.0 | 1,309 |
| Feb. | 12 | 0 | 0.6 | 52 | NW | 4.0 | 1,121 |
| Mar. | 9 | 0 | 1.4 | 52 | NW | 4.0 | 1,015 |
| Apr. | 9 | 0.1 | 1.3 | 63 | N,NW | 4.7 | 688 |
| May | 12 | 0.8 | 2.1 | 68 | NW | 4.5 | 458 |
| June | 13 | 2.5 | 2.2 | 69 | NW | 4.5 | 217 |
| July | 15 | 6.1 | 1.7 | 64 | S,SW | 4.3 | 81 |
| Aug. | 15 | 4.4 | 1.4 | 66 | NW | 4.6 | 122 |
| Sept. | 15 | 1.4 | 1.2 | 76 | NW | 5.2 | 290 |
| Oct. | 13 | 0.4 | 1.7 | 75 | NW | 5.1 | 512 |
| Nov. | 17 | 0 | 1.7 | 82 | NW | 5.1 | 817 |
| Dec. | 15 | 0 | 1.1 | 62 | W | 4.2 | 1,168 |
| Annual | 157 | 15.7 | 17.4 | 65 | NW | 4.5 | 7,799 |

TABLE XLIII

CLIMATIC TABLE FOR MONTREAL, QUE.
Latitude 45°30′N, longitude 73°35′W, elevation 17 m

| Month | Mean sta. press. (mbar) | Temperature (°C) | | | | Mean vapor press. (mbar) | Precipitation (mm) | | Snowfall (mm) |
|---|---|---|---|---|---|---|---|---|---|
| | | mean daily | mean daily range | extremes | | | mean | max. in 24 h | |
| | | | | max. | min. | | | | |
| Jan. | 1010.6 | −8.7 | 7.4 | 13 | −30 | 2.1 | 87 | 44.5 | 574 |
| Feb. | 1008.5 | −7.8 | 7.5 | 10 | −34 | 2.1 | 76 | 30.5 | 597 |
| Mar. | 1008.2 | −2.1 | 7.0 | 25 | −27 | 3.7 | 86 | 42.4 | 472 |
| Apr. | 1007.9 | 6.2 | 8.4 | 30 | −14 | 6.4 | 83 | 41.2 | 112 |
| May | 1007.2 | 13.6 | 9.4 | 32 | −3 | 10.6 | 81 | 51.8 | tr. |
| June | 1006.1 | 18.9 | 9.1 | 34 | 1 | 15.3 | 91 | 68.1 | 0 |
| July | 1007.1 | 21.6 | 8.9 | 36 | 9 | 18.3 | 102 | 90.4 | 0 |
| Aug. | 1007.9 | 20.5 | 8.8 | 35 | 8 | 17.0 | 87 | 75.7 | 0 |
| Sept. | 1010.3 | 15.6 | 8.3 | 32 | 1 | 12.8 | 95 | 49.3 | tr. |
| Oct. | 1010.6 | 9.4 | 7.5 | 29 | −7 | 8.7 | 83 | 86.1 | 13 |
| Nov. | 1008.9 | 2.3 | 5.8 | 22 | −17 | 5.6 | 88 | 52.8 | 211 |
| Dec. | 1009.9 | −5.9 | 6.5 | 15 | −34 | 2.9 | 89 | 62.5 | 526 |
| Annual | 1008.6 | 6.9 | 7.9 | 36 | −34 | 8.8 | 1,048 | 90.4 | 2,505 |

| Month | Number of days | | | Mean cloud-iness (tenths) | Mean sun-shine (h) | Wind | | 18°C degree-days |
|---|---|---|---|---|---|---|---|---|
| | precip. >0.25 mm | thunder-storm | heavy fog | | | most frequ. direct. | mean speed (m/sec) | |
| Jan. | 18 | 0.1 | 2.3 | 6.7 | 86 | SW | 5.6 | 839 |
| Feb. | 15 | 0.1 | 1.8 | 6.1 | 103 | W | 5.7 | 738 |
| Mar. | 12 | 0.4 | 1.8 | 6.1 | 140 | SW | 5.6 | 632 |
| Apr. | 13 | 0.7 | 1.8 | 6.3 | 154 | SW | 5.5 | 365 |
| May | 14 | 1.9 | 1.3 | 6.4 | 209 | SW | 5.0 | 160 |
| June | 13 | 4.4 | 1.6 | 6.1 | 230 | SW | 4.4 | 30 |
| July | 13 | 5.7 | 1.2 | 5.7 | 253 | SW | 4.2 | 9 |
| Aug. | 10 | 3.8 | 1.5 | 5.4 | 240 | SW | 4.1 | 16 |
| Sept. | 13 | 2.2 | 2.0 | 5.6 | 167 | SW | 4.4 | 92 |
| Oct. | 13 | 1.0 | 2.9 | 5.9 | 132 | SW | 4.7 | 276 |
| Nov. | 15 | 0.1 | 2.9 | 7.4 | 69 | SW | 5.2 | 480 |
| Dec. | 17 | 0.1 | 2.8 | 7.0 | 69 | W | 5.3 | 753 |
| Annual | 166 | 20.5 | 23.9 | 6.2 | 1,852 | SW | 5.0 | 4,388 |

## TABLE XLIV

CLIMATIC TABLE FOR QUEBEC, QUE.
Latitude 46°48′N, longitude 71°23′W, elevation 23 m

| Month | Mean sta. press. (mbar) | Temperature (°C) | | | | Mean vapor press. (mbar) | Precipitation (mm) | | Snowfall (mm) |
|---|---|---|---|---|---|---|---|---|---|
| | | mean daily | mean daily range | extremes | | | mean | max. in 24 h | |
| | | | | max. | min. | | | | |
| Jan. | 1007.0 | −11.5 | 8.4 | 11 | −36 | 2.0 | 79 | 92.5 | 665 |
| Feb. | 1005.6 | −10.7 | 8.7 | 12 | −32 | 2.0 | 77 | 42.9 | 701 |
| Mar. | 1005.0 | −4.9 | 8.3 | 18 | −29 | 3.5 | 70 | 33.0 | 450 |
| Apr. | 1004.9 | 3.2 | 9.3 | 27 | −18 | 5.6 | 78 | 46.0 | 147 |
| May | 1004.5 | 10.8 | 11.8 | 32 | −7 | 9.1 | 75 | 34.5 | 6 |
| June | 1003.1 | 16.4 | 11.5 | 34 | 1 | 13.2 | 106 | 104.4 | 0 |
| July | 1004.0 | 19.3 | 11.5 | 36 | 4 | 16.4 | 107 | 74.9 | 0 |
| Aug. | 1004.9 | 18.2 | 11.6 | 35 | 4 | 15.9 | 91 | 131.3 | 0 |
| Sept. | 1007.5 | 13.3 | 10.2 | 31 | −1 | 11.8 | 101 | 72.6 | tr. |
| Oct. | 1007.7 | 6.9 | 8.9 | 28 | −7 | 8.1 | 81 | 44.7 | 36 |
| Nov. | 1006.1 | −0.1 | 6.4 | 22 | −20 | 5.3 | 94 | 52.6 | 282 |
| Dec. | 1006.3 | −8.8 | 7.8 | 15 | −36 | 2.6 | 99 | 44.7 | 757 |
| Annual | 1005.6 | 4.4 | 9.5 | 36 | −36 | 8.0 | 1,058 | 131.3 | 3,044 |

| Month | Number of days | | | Mean cloud-iness (tenths) | Mean sun-shine (h) | Wind | | 18°C degree-days |
|---|---|---|---|---|---|---|---|---|
| | precip. >0.25 mm | thunder-storm | heavy fog | | | most frequ. direct. | mean speed (m/sec) | |
| Jan. | 16 | 0 | 3.1 | 6.6 | 82 | W | 4.5 | 925 |
| Feb. | 14 | 0 | 1.5 | 6.3 | 102 | W | 4.6 | 820 |
| Mar. | 13 | 0.2 | 2.6 | 5.9 | 128 | NE | 4.8 | 719 |
| Apr. | 12 | 0.4 | 3.6 | 6.3 | 157 | NE | 4.5 | 455 |
| May | 12 | 1.3 | 1.8 | 6.3 | 199 | NE | 4.6 | 238 |
| June | 13 | 3.2 | 1.7 | 6.4 | 201 | SW | 3.7 | 69 |
| July | 13 | 4.6 | 2.1 | 5.9 | 228 | SW | 3.2 | 31 |
| Aug. | 11 | 3.1 | 2.2 | 5.5 | 162 | SW | 3.2 | 47 |
| Sept. | 12 | 1.2 | 3.5 | 5.8 | 157 | W | 3.4 | 152 |
| Oct. | 12 | 0.4 | 3.2 | 6.1 | 123 | W | 3.7 | 353 |
| Nov. | 15 | 0.2 | 4.7 | 7.4 | 65 | W | 4.4 | 553 |
| Dec. | 16 | 0 | 3.7 | 6.8 | 65 | W | 4.3 | 842 |
| Annual | 159 | 14.6 | 33.7 | 6.3 | 1,669 | SW,W | 4.1 | 5,206 |

TABLE XLV

CLIMATIC TABLE FOR FORT CHIMO, QUE.
Latitude 50°06′N, longitude 60°26′W, elevation 11 m

| Month | Mean sta. press. (mbar) | Temperature (°C) | | | | Mean vapor press. (mbar) | Precipitation (mm) | | Snowfall (mm) |
|---|---|---|---|---|---|---|---|---|---|
| | | mean daily | mean daily range | extremes | | | mean | max. in 24 h | |
| | | | | max. | min. | | | | |
| Jan. | 1009.7 | −23.9 | 8.9 | 6 | −46 | | 21 | 18.3 | 203 |
| Feb. | 1008.7 | −22.7 | 10.1 | 5 | −43 | | 20 | 17.3 | 203 |
| Mar. | 1013.1 | −16.2 | 10.3 | 8 | −42 | | 21 | 18.0 | 206 |
| Apr. | 1011.0 | −9.3 | 10.0 | 11 | −33 | | 19 | 17.8 | 160 |
| May | 1010.1 | 0.4 | 8.2 | 31 | −19 | | 33 | 27.9 | 135 |
| June | 1006.5 | 7.3 | 10.6 | 31 | −6 | | 38 | 29.0 | 41 |
| July | 1003.7 | 11.8 | 11.2 | 32 | −2 | 9.8 | 50 | 22.4 | tr. |
| Aug. | 1003.4 | 10.5 | 9.3 | 28 | −2 | 9.1 | 62 | 25.2 | tr. |
| Sept. | 1004.8 | 5.6 | 7.2 | 26 | −4 | 7.2 | 52 | 22.9 | 25 |
| Oct. | 1005.6 | −0.3 | 5.5 | 18 | −15 | | 39 | 15.2 | 211 |
| Nov. | 1006.1 | −8.4 | 7.0 | 10 | −31 | | 34 | 19.8 | 312 |
| Dec. | 1005.0 | −17.8 | 8.3 | 8 | −40 | | 28 | 15.5 | 269 |
| Annual | 1007.3 | −5.2 | 8.9 | 32 | −46 | | 417 | 29.0 | 1,765 |

| Month | Number of days | | | Mean cloud-iness (tenths) | Wind | | 18°C degree-days |
|---|---|---|---|---|---|---|---|
| | precip. >0.25 mm | thunder-storm | heavy fog | | most frequ. direct. | mean speed (m/sec) | |
| Jan. | 10 | 0 | 0.5 | 5.4 | SW | 4.7 | 1,309 |
| Feb. | 9 | 0 | 0.5 | 5.6 | SW | 4.7 | 1,159 |
| Mar. | 12 | 0 | 0.8 | 5.8 | W | 4.3 | 1,070 |
| Apr. | 10 | 0 | 1.9 | 5.8 | N | 4.5 | 828 |
| May | 10 | 0 | 0.9 | 7.4 | N | 4.7 | 554 |
| June | 11 | 0.6 | 1.9 | 7.3 | N | 4.6 | 332 |
| July | 13 | 0.5 | 1.8 | 7.1 | N | 4.4 | 202 |
| Aug. | 14 | 0.5 | 1.7 | 7.6 | N | 4.6 | 243 |
| Sept. | 16 | 0.5 | 1.2 | 8.0 | SW | 5.3 | 382 |
| Oct. | 14 | 0 | 0.4 | 8.1 | W | 5.3 | 579 |
| Nov. | 14 | 0.1 | 0.6 | 7.4 | SW | 5.0 | 802 |
| Dec. | 15 | 0 | 0.8 | 6.0 | SW | 4.9 | 1,119 |
| Annual | 148 | 2.2 | 13.0 | 6.8 | SW | 4.8 | 8,580 |

TABLE XLVI

CLIMATIC TABLE FOR PORT HARRISON, QUE.
Latitude 58°27′N, longitude 78°08′W, elevation 6 m

| Month | Mean sta. press. (mbar) | Temperature (°C) | | | | Mean vapor press (mbar) | Precipitation (mm) | | Snowfall (mm) |
|-------|------|------|------|------|------|------|------|------|------|
| | | mean daily | mean daily range | extremes | | | mean | max. in 24 h | |
| | | | | max. | min. | | | | |
| Jan. | 1012.7 | −25.0 | 7.9 | 1 | −44 | | 14 | 13.5 | 137 |
| Feb. | 1013.2 | −25.3 | 8.1 | 5 | −44 | | 9 | 10.4 | 86 |
| Mar. | 1015.8 | −19.8 | 9.2 | 4 | −45 | | 16 | 22.9 | 163 |
| Apr. | 1014.1 | −10.8 | 8.7 | 6 | −34 | | 17 | 14.0 | 142 |
| May | 1013.1 | −2.2 | 6.2 | 23 | −22 | 4.4 | 23 | 25.4 | 122 |
| June | 1009.2 | 4.4 | 7.8 | 26 | −8 | 6.9 | 30 | 35.6 | 38 |
| July | 1007.3 | 8.9 | 8.3 | 28 | −7 | 9.1 | 51 | 24.9 | 3 |
| Aug. | 1006.9 | 8.6 | 7.2 | 24 | −2 | 9.4 | 54 | 23.1 | tr. |
| Sept. | 1007.7 | 5.0 | 5.4 | 23 | −4 | 7.7 | 62 | 30.2 | 33 |
| Oct. | 1008.7 | −0.4 | 5.5 | 17 | −23 | | 49 | 22.6 | 249 |
| Nov. | 1007.6 | −8.1 | 6.4 | 8 | −34 | | 47 | 22.9 | 434 |
| Dec. | 1008.4 | −18.3 | 7.6 | 2 | −43 | | 23 | 10.7 | 231 |
| Annual | 1010.4 | −6.9 | 7.3 | 28 | −45 | | 395 | 35.6 | 1,638 |

| Month | Number of days | | | Mean cloud-iness (tenths) | Mean sun-shine (h) | Wind | | 18°C degree-days |
|-------|------|------|------|------|------|------|------|------|
| | precip. >0.25 mm | thunder-storm | heavy fog | | | most frequ. direct. | mean speed (m/sec) | |
| Jan. | 7 | 0 | 2.4 | 4.7 | 60 | W | 5.7 | 1,343 |
| Feb. | 6 | 0 | 1.2 | 4.2 | 110 | NE | 5.9 | 1,232 |
| Mar. | 8 | 0 | 2.0 | 4.8 | 155 | N | 6.1 | 1,183 |
| Apr. | 9 | 0 | 2.1 | 6.3 | 162 | N | 6.2 | 8,732 |
| May | 11 | 0.1 | 3.6 | 7.9 | 135 | N | 6.6 | 635 |
| June | 8 | 0.4 | 7.0 | 7.2 | 186 | N | 5.9 | 417 |
| July | 11 | 0.7 | 10.2 | 7.0 | 198 | S | 6.1 | 293 |
| Aug. | 13 | 0.4 | 7.8 | 7.3 | 156 | N | 6.3 | 303 |
| Sept. | 16 | 0.1 | 4.8 | 8.1 | 76 | N | 6.8 | 400 |
| Oct. | 16 | 0.1 | 3.0 | 8.3 | 50 | N | 7.0 | 582 |
| Nov. | 16 | 0.1 | 1.1 | 7.9 | 27 | NE | 7.4 | 793 |
| Dec. | 12 | 0 | 1.5 | 6.6 | 27 | N | 6.4 | 1,137 |
| Annual | 133 | 1.9 | 46.7 | 6.7 | 1,342 | N | 6.4 | 9,193 |

TABLE XLVII

CLIMATIC TABLE FOR REGINA, SASK.
Latitude 50°26′N, longitude 104°40′W, elevation 175 m

| Month | Mean sta. press. (mbar) | Temperature (°C) | | | | Mean vapor press. (mbar) | Precipitation (mm) | | Snowfall (mm) |
|---|---|---|---|---|---|---|---|---|---|
| | | mean daily | mean daily range | extremes | | | mean | max. in 24 h | |
| | | | | max. | min. | | | | |
| Jan. | 948.1 | −16.9 | 10.6 | 7 | −46 | 1.5 | 19 | 17.8 | 188 |
| Feb. | 947.9 | −14.8 | 10.9 | 9 | −43 | 1.5 | 17 | 12.5 | 165 |
| Mar. | 947.1 | −8.1 | 10.6 | 21 | −41 | 2.7 | 21 | 19.3 | 183 |
| Apr. | 946.8 | 3.4 | 12.8 | 33 | −24 | 5.3 | 21 | 30.2 | 89 |
| May | 946.3 | 11.2 | 14.9 | 37 | −12 | 7.5 | 40 | 45.5 | 23 |
| June | 945.4 | 15.3 | 13.9 | 38 | −3 | 11.4 | 83 | 95.3 | tr. |
| July | 946.8 | 19.3 | 15.1 | 40 | −2 | 14.2 | 55 | 65.3 | 0 |
| Aug. | 946.8 | 17.8 | 15.2 | 41 | −3 | 13.2 | 49 | 78.7 | 0 |
| Sept. | 946.9 | 11.9 | 14.4 | 37 | −12 | 8.7 | 34 | 62.5 | 28 |
| Oct. | 946.7 | 5.1 | 13.7 | 31 | −19 | 6.1 | 18 | 23.6 | 74 |
| Nov. | 946.6 | −5.4 | 10.3 | 21 | −33 | 3.2 | 20 | 23.9 | 180 |
| Dec. | 946.3 | −12.3 | 10.1 | 15 | −40 | 1.8 | 17 | 9.7 | 163 |
| Annual | 946.8 | 2.2 | 12.7 | 41 | −46 | 6.4 | 394 | 95.3 | 1,093 |

| Month | Number of days | | | Mean cloud-iness (tenths) | Mean sun-shine (h) | Wind | | 18°C degree-days |
|---|---|---|---|---|---|---|---|---|
| | precip. >0.25 mm | thunder-storm | heavy fog | | | most frequ. direct. | mean speed (m/sec) | |
| Jan. | 12 | 0 | 4.3 | 6.1 | 98 | NW | 5.7 | 1,092 |
| Feb. | 10 | 0 | 4.6 | 6.0 | 118 | SE | 5.8 | 937 |
| Mar. | 11 | 0 | 4.7 | 6.2 | 152 | SE | 5.8 | 818 |
| Apr. | 8 | 0.5 | 3.1 | 6.1 | 215 | SE | 6.3 | 447 |
| May | 8 | 2.2 | 0.9 | 5.9 | 266 | SE | 6.2 | 227 |
| June | 13 | 5.3 | 1.5 | 6.4 | 249 | SE | 5.9 | 112 |
| July | 10 | 6.1 | 1.5 | 4.6 | 334 | SE,W | 5.3 | 43 |
| Aug. | 9 | 5.9 | 1.7 | 5.2 | 286 | SE | 5.4 | 52 |
| Sept. | 8 | 1.2 | 1.4 | 5.4 | 197 | SE,NW | 5.6 | 200 |
| Oct. | 6 | 0.1 | 2.4 | 5.3 | 170 | SE | 5.9 | 412 |
| Nov. | 10 | 0 | 3.2 | 6.3 | 96 | SE | 5.8 | 713 |
| Dec. | 10 | 0.1 | 5.3 | 6.2 | 85 | SE | 5.4 | 950 |
| Annual | 115 | 21.4 | 34.6 | 5.8 | 2,266 | SE | 5.8 | 6,003 |

## TABLE XLVIII

CLIMATIC TABLE FOR SASKATOON, SASK.
Latitude 52°08′N, longitude 106°38′W, elevation 157 m

| Month | Mean sta. press. (mbar) | Temperature (°C) | | | | Mean vapor press. (mbar) | Precipitation (mm) | | Snowfall (mm) |
|-------|------|------|------|------|------|------|------|------|------|
| | | mean daily | mean daily range | extremes | | | mean | max. in 24 h | |
| | | | | max. | min. | | | | |
| Jan. | 956.5 | −17.6 | 9.5 | 9 | −46 | 1.2 | 15 | 25.4 | 147 |
| Feb. | 955.2 | −14.9 | 10.5 | 13 | −44 | 1.3 | 16 | 24.4 | 160 |
| Mar. | 955.2 | −7.9 | 10.3 | 19 | −35 | 2.6 | 15 | 19.6 | 140 |
| Apr. | 954.8 | 3.6 | 12.2 | 33 | −27 | 5.1 | 21 | 21.8 | 97 |
| May | 954.2 | 11.2 | 14.2 | 37 | −8 | 7.2 | 34 | 37.9 | 25 |
| June | 952.5 | 15.4 | 13.4 | 40 | −2 | 10.2 | 58 | 63.0 | tr. |
| July | 954.1 | 19.3 | 14.5 | 40 | 2 | 13.2 | 60 | 55.1 | 0 |
| Aug. | 954.4 | 17.6 | 14.4 | 38 | −1 | 12.8 | 45 | 84.3 | 0 |
| Sept. | 954.5 | 11.6 | 13.3 | 35 | −11 | 8.4 | 34 | 43.7 | 18 |
| Oct. | 954.1 | 4.9 | 12.4 | 32 | −22 | 6.1 | 19 | 22.6 | 69 |
| Nov. | 954.6 | −5.8 | 8.8 | 20 | −34 | 3.2 | 18 | 19.1 | 142 |
| Dec. | 954.7 | −13.2 | 8.8 | 13 | −40 | 1.5 | 17 | 28.5 | 168 |
| Annual | 954.6 | 2.0 | 11.8 | 40 | −46 | 6.1 | 352 | 84.3 | 966 |

| Month | Number of days | | | Mean cloud-iness (tenths) | Mean sun-shine (h) | Wind | | 18°C degree-days |
|-------|------|------|------|------|------|------|------|------|
| | precip. >0.25 mm | thunder-storm | heavy fog | | | most frequ. direct. | mean speed (m/sec) | |
| Jan. | 11 | 0 | 4.2 | 5.9 | 96 | S | 5.2 | 1,114 |
| Feb. | 9 | 0 | 2.9 | 5.8 | 129 | SW | 4.9 | 9,382 |
| Mar. | 8 | 0.1 | 3.0 | 5.7 | 191 | SE | 4.9 | 813 |
| Apr. | 8 | 0.4 | 1.6 | 6.0 | 226 | NW | 5.6 | 443 |
| May | 7 | 1.3 | 0.3 | 6.1 | 275 | SE | 5.6 | 224 |
| June | 10 | 3.2 | 1.0 | 6.5 | 270 | NW | 5.3 | 103 |
| July | 10 | 6.8 | 1.2 | 5.1 | 340 | NW | 4.9 | 31 |
| Aug. | 10 | 5.3 | 3.0 | 5.3 | 293 | SE | 4.7 | 48 |
| Sept. | 8 | 0.7 | 2.0 | 5.6 | 210 | NW | 5.2 | 206 |
| Oct. | 6 | 0 | 2.2 | 5.5 | 166 | S | 5.2 | 417 |
| Nov. | 9 | 0 | 3.2 | 6.7 | 99 | SW | 4.9 | 723 |
| Dec. | 9 | 0 | 4.1 | 6.2 | 86 | S | 4.5 | 976 |
| Annual | 105 | 17.8 | 28.7 | 5.9 | 2,381 | S | 5.1 | 6,038 |

TABLE XLIX

CLIMATIC TABLE FOR WHITEHORSE, YUKON
Latitude 60°43'N, longitude 135°04'N, elevation 2,128 m

| Month | Mean sta. press. (mbar) | Temperature (°C) | | | | Mean vapor press. (mbar) | Precipitation (mm) | | Snowfall (mm) |
|---|---|---|---|---|---|---|---|---|---|
| | | mean daily | mean daily range | extremes | | | mean | max. in 24 h | |
| | | | | max. | min. | | | | |
| Jan. | 929.1 | −18.1 | 8.4 | 8 | −52 | | 18 | 9.4 | 178 |
| Feb. | 928.5 | −14.1 | 9.5 | 10 | −51 | | 14 | 10.4 | 142 |
| Mar. | 926.7 | −7.6 | 11.1 | 11 | −38 | | 15 | 20.3 | 150 |
| Apr. | 928.4 | −0.2 | 10.5 | 21 | −26 | | 11 | 14.2 | 102 |
| May | 930.7 | 7.5 | 12.2 | 30 | −8 | 5.8 | 13 | 12.2 | 20 |
| June | 931.3 | 12.6 | 12.9 | 32 | −2 | 8.4 | 27 | 20.8 | 0 |
| July | 932.6 | 14.2 | 12.1 | 33 | −2 | 9.8 | 35 | 21.1 | 0 |
| Aug. | 931.8 | 12.4 | 11.4 | 30 | −8 | 9.4 | 37 | 30.7 | tr. |
| Sept. | 929.8 | 7.9 | 9.7 | 27 | −10 | 7.5 | 25 | 21.6 | 33 |
| Oct. | 924.4 | 0.7 | 7.4 | 19 | −24 | | 19 | 11.9 | 119 |
| Nov. | 925.6 | −8.2 | 6.7 | 11 | −42 | | 23 | 11.4 | 216 |
| Dec. | 924.6 | −15.1 | 7.7 | 8 | −48 | | 20 | 10.9 | 198 |
| Annual | 928.6 | −0.7 | 10.0 | 33 | −52 | | 257 | 30.7 | 1,158 |

| Month | Number of days | | | Mean cloud-iness (tenths) | Mean sun-shine (h) | Wind | | 18°C degree-days |
|---|---|---|---|---|---|---|---|---|
| | precip. >0.25 mm | thunder-storm | heavy fog | | | most frequ. direct. | mean speed (m/sec) | |
| Jan. | 12 | 0 | 3.7 | 6.7 | 48 | S | 3.9 | 1,130 |
| Feb. | 10 | 0 | 1.3 | 6.8 | 74 | S | 4,0 | 915 |
| Mar. | 8 | 0 | 1.1 | 6.4 | 164 | S | 4.0 | 804 |
| Apr. | 6 | 0 | 0.3 | 6.8 | 246 | S | 3.9 | 555 |
| May | 5 | 0.2 | 0.3 | 7.0 | 265 | SE | 3.9 | 336 |
| June | 8 | 2.2 | 0.9 | 7.1 | 295 | SE | 3.6 | 183 |
| July | 12 | 2.4 | 0.3 | 7.5 | 241 | SE | 3.3 | 133 |
| Aug. | 10 | 0.8 | 1.6 | 7.1 | 219 | SE | 3.5 | 184 |
| Sept. | 9 | 0 | 2.1 | 7.1 | 148 | S,SE | 4.1 | 312 |
| Oct. | 9 | 0 | 1.8 | 7.0 | 114 | S | 4.7 | 546 |
| Nov. | 12 | 0 | 2.2 | 7.7 | 56 | S | 4.1 | 797 |
| Dec. | 12 | 0 | 3.6 | 7.2 | 28 | S | 3.9 | 1,035 |
| Annual | 113 | 5.6 | 19.2 | 7.0 | 1,898 | S | 3.9 | 6,930 |

### Appendix II—Mean values (1957–64) of the principal radiative fluxes for representative stations in Canada and Alaska

These are *computed* values, using the model described in HAY (1970). The albedo is that considered representative of the region in which the station is located. $I$ = received global radiation at ground; $I(1-\alpha)$ = the absorbed solar radiation; and $R$ = net radiation. The monthly values are in ly day$^{-1}$, annual values in kly year$^{-1}$.

| | Jan. | Feb. | Mar. | Apr. | May | June | July | Aug. | Sep. | Oct. | Nov. | Dec. | Year |
|---|---|---|---|---|---|---|---|---|---|---|---|---|---|
| *Barrow:* | | | | | | | | | | | | | |
| $I$ | — | 27 | 173 | 386 | 453 | 458 | 375 | 225 | 118 | 50 | 2 | — | 69.2 |
| $I(1-\alpha)$ | — | 5 | 38 | 116 | 222 | 344 | 308 | 185 | 86 | 22 | 1 | — | 40.5 |
| $R$ | −50 | −60 | −61 | 11 | 136 | 235 | 203 | 107 | 17 | −31 | −34 | −122 | 10.9 |
| *Fairbanks:* | | | | | | | | | | | | | |
| $I$ | 10 | 71 | 215 | 352 | 460 | 471 | 421 | 293 | 187 | 91 | 21 | 2 | 79.2 |
| $I(1-\alpha)$ | 5 | 36 | 116 | 225 | 368 | 400 | 356 | 249 | 155 | 66 | 12 | 1 | 60.8 |
| $R$ | −36 | −25 | 3 | 104 | 205 | 228 | 202 | 122 | 37 | −19 | −57 | − 43 | 22.1 |
| *Anchorage:* | | | | | | | | | | | | | |
| $I$ | 32 | 115 | 269 | 389 | 439 | 462 | 414 | 317 | 203 | 111 | 44 | 16 | 85.7 |
| $I(1-\alpha)$ | 7 | 25 | 86 | 198 | 356 | 388 | 348 | 266 | 171 | 80 | 20 | 5 | 59.6 |
| $R$ | −47 | −47 | −34 | 70 | 206 | 234 | 205 | 137 | 58 | −29 | −61 | − 55 | 19.6 |
| *Whitehorse:* | | | | | | | | | | | | | |
| $I$ | 32 | 101 | 245 | 385 | 470 | 498 | 461 | 348 | 221 | 106 | 39 | 15 | 89.2 |
| $I(1-\alpha)$ | 13 | 41 | 112 | 253 | 382 | 422 | 392 | 296 | 186 | 83 | 24 | 7 | 67.5 |
| $R$ | −42 | −29 | 14 | 122 | 230 | 253 | 222 | 146 | 56 | −24 | −49 | − 52 | 25.9 |
| *Vancouver:* | | | | | | | | | | | | | |
| $I$ | 69 | 124 | 225 | 345 | 469 | 501 | 548 | 430 | 307 | 173 | 91 | 57 | 101.9 |
| $I(1-\alpha)$ | 57 | 104 | 187 | 297 | 403 | 431 | 471 | 370 | 264 | 148 | 78 | 46 | 87.1 |
| $R$ | −13 | 13 | 75 | 164 | 250 | 269 | 300 | 218 | 124 | 36 | −13 | − 20 | 42.9 |
| *Resolute:* | | | | | | | | | | | | | |
| $I$ | — | 9 | 119 | 360 | 557 | 595 | 464 | 263 | 118 | 31 | 1 | — | 76.9 |
| $I(1-\alpha)$ | — | 2 | 24 | 88 | 170 | 326 | 340 | 207 | 65 | 13 | 1 | — | 37.8 |
| $R$ | −83 | −90 | −72 | −24 | 65 | 200 | 206 | 102 | −7 | −64 | −79 | − 78 | 2.5 |
| *Regina:* | | | | | | | | | | | | | |
| $I$ | 123 | 212 | 340 | 449 | 527 | 591 | 595 | 500 | 369 | 233 | 122 | 92 | 126.5 |
| $I(1-\alpha)$ | 56 | 90 | 185 | 324 | 450 | 473 | 464 | 375 | 280 | 178 | 80 | 49 | 91.7 |
| $R$ | −23 | 2 | 71 | 175 | 277 | 285 | 272 | 188 | 115 | 31 | −20 | − 29 | 41.1 |
| *Kenora:* | | | | | | | | | | | | | |
| $I$ | 127 | 231 | 366 | 458 | 538 | 535 | 520 | 441 | 307 | 200 | 107 | 100 | 119.7 |
| $I(1-\alpha)$ | 56 | 102 | 177 | 298 | 445 | 460 | 447 | 380 | 263 | 168 | 73 | 48 | 89.0 |
| $R$ | −23 | − 4 | 54 | 154 | 279 | 286 | 278 | 221 | 121 | 38 | − 8 | − 34 | 41.6 |
| *Frobisher:* | | | | | | | | | | | | | |
| $I$ | 18 | 94 | 253 | 457 | 532 | 504 | 448 | 296 | 190 | 114 | 30 | 6 | 89.7 |
| $I(1-\alpha)$ | 4 | 19 | 51 | 118 | 190 | 353 | 354 | 242 | 131 | 49 | 9 | 1 | 46.4 |
| $R$ | −70 | −61 | −57 | − 2 | 80 | 216 | 211 | 124 | 31 | −34 | −77 | − 71 | 9.0 |
| *Moosonee:* | | | | | | | | | | | | | |
| $I$ | 99 | 180 | 304 | 392 | 410 | 491 | 441 | 357 | 248 | 149 | 76 | 73 | 98.1 |
| $I(1-\alpha)$ | 42 | 76 | 148 | 238 | 333 | 407 | 380 | 307 | 210 | 122 | 52 | 43 | 71.9 |
| $R$ | −32 | −16 | 39 | 116 | 198 | 246 | 221 | 168 | 88 | 19 | −16 | − 34 | 30.5 |
| *Windsor:* | | | | | | | | | | | | | |
| $I$ | 141 | 213 | 311 | 391 | 503 | 539 | 519 | 461 | 359 | 250 | 142 | 116 | 120.2 |
| $I(1-\alpha)$ | 75 | 113 | 193 | 302 | 398 | 420 | 404 | 355 | 276 | 192 | 102 | 72 | 88.5 |
| $R$ | − 5 | 24 | 79 | 152 | 223 | 233 | 220 | 183 | 114 | 39 | −13 | − 16 | 37.6 |

|   | Jan. | Feb. | Mar. | Apr. | May | June | July | Aug. | Sep. | Oct. | Nov. | Dec. | Year |
|---|------|------|------|------|-----|------|------|------|------|------|------|------|------|
| | *Montreal:* | | | | | | | | | | | | |
| I | 116 | 193 | 309 | 389 | 492 | 540 | 498 | 428 | 340 | 216 | 104 | 91 | 113.2 |
| I(1-α) | 79 | 130 | 232 | 303 | 393 | 432 | 398 | 342 | 272 | 172 | 80 | 66 | 88.4 |
| R | 3 | 40 | 112 | 162 | 222 | 249 | 221 | 178 | 115 | 33 | −18 | − 12 | 39.8 |
| | *Goose Bay:* | | | | | | | | | | | | |
| I | 81 | 167 | 280 | 392 | 432 | 455 | 437 | 359 | 259 | 157 | 78 | 62 | 96.2 |
| I(1-α) | 36 | 71 | 140 | 232 | 334 | 391 | 372 | 305 | 220 | 120 | 50 | 30 | 70.2 |
| R | −30 | −14 | 41 | 120 | 201 | 239 | 217 | 164 | 88 | 14 | −28 | − 46 | 29.6 |
| | *Halifax:* | | | | | | | | | | | | |
| I | 120 | 206 | 293 | 369 | 465 | 464 | 458 | 416 | 330 | 223 | 122 | 101 | 108.6 |
| I(1-α) | 73 | 118 | 202 | 289 | 376 | 376 | 371 | 337 | 267 | 180 | 96 | 71 | 84.1 |
| R | −11 | 22 | 96 | 166 | 235 | 228 | 227 | 200 | 132 | 55 | − 2 | − 12 | 40.7 |

## Appendix III—Crop water balance for Canada[1]

The interrelationships between weather, soils and crops are complex and can only be determined from long-term climatic records for various elements and from crop data by means of biomathematical models. Modern electronic data processing facilities make such an analysis possible, but basic knowledge of these physical–biological interrelationships is still lacking.

Two main approaches have been used in developing so-called crop-weather models: (*1*) the climatological approach, in which climatic means or totals are related to yields or similar crop production parameters; and (*2*) the micrometeorological approach, in which selected elements of the microclimate are related to measurable reactions of plants such as transpiration rate, growth or photosynthesis.

In Canada, progress has been made in developing crop–weather models that use standard climatic data as input and provide meaningful parameters related to soils, crop development and production. In particular, computer techniques are available for estimating the following parameters.

(*1*) Potential evapotranspiration from a regression technique using daily maximum and minimum temperatures and tabulated solar energy and day-length data as input (BAIER and ROBERTSON, 1965; BAIER et al., 1970).

(*2*) Soil moisture in six or less divisions in the soil profile from a computerized physical model, called the "versatile soil moisture budget", using as input daily precipitation and estimated potential evapotranspiration (BAIER and ROBERTSON, 1966; BAIER et al., 1970).

(*3*) Rate of crop development towards maturity from a mathematical model using photoperiod and day and night temperatures as input. This provides a biometeorological time scale which synthesizes daily phenological data that are otherwise not available (ROBERTSON, 1968).

### Crop water balance technique

The above techniques were used to compute from standard climatic data (1931–60) the daily water balance of a wheat crop planted every year at the average spring wheat

---

[1] By Wolfgang Baier.

seeding time for each of 10 locations across Canada (Table L). For comparison purposes across the country, it was presumed that the capacity of the soil for plant available water was 20 cm and the initial water content in spring 1931 was 15 cm. The relationship between available water in the soil and the ratio of actual to potential evapotranspiration (*AE/PE*) was such that there was no reduction in the *AE/PE* ratio from 100% to 70% and a linear decrease from 70% to 0% available soil water. Crop coefficients reflecting the consumptive water use of wheat during its growing season were adjusted for soil dryness once the jointing stage had been reached. It was assumed that all the accumulated snow after thawing penetrated into the soil (BAIER et al., 1970).

TABLE L

LOCATION OF SELECTED CLIMATIC STATIONS AND AVERAGE PLANTING DATES OF SPRING WHEAT

| Station | Province | Latitude North | Longitude West | Elevation (m) | Average Planting Date |
|---------|----------|----------------|----------------|---------------|------------------------|
| Agassiz | B.C. | 49°17′ | 121°46′ | 15 | Apr. 11 |
| Ft. Simpson | N.W.T. | 61°52′ | 121°21′ | 129 | May 23 |
| Ft. Vermilion | Alta. | 58°23′ | 116°03′ | 279 | May 13 |
| Lethbridge | Alta. | 49°38′ | 112°48′ | 920 | Apr. 21 |
| Swift Current | Sask. | 50°17′ | 107°45′ | 762 | Apr. 25 |
| Brandon | Man. | 49°52′ | 99°59′ | 366 | May 5 |
| Ottawa | Ont. | 45°24′ | 75°43′ | 79 | May 10 |
| Harrow | Ont. | 42°02′ | 82°53′ | 191 | Apr. 19 |
| Normandin | Que. | 48°51′ | 72°32′ | 137 | May 23 |
| Charlottetown | P.E.I. | 46°15′ | 63°08′ | 23 | May 17 |

The estimates of the daily water balance were summarized over periods based on the following computed dates when selected crop development stages (as defined by ROBERT-SON, 1968) were reached: *P* = average planting date of spring wheat for the location; *E* = emergence; *J* = jointing; *H* = heading; *S* = soft dough; *R* = ripe; *F* = freeze-up of soil surface, assumed when daily maximum temperatures smoothed over 5 days dropped below 0°C.

To interpret the results for the individual stations correctly, it is important to remember that, because of the daily balancing procedure, soil moisture surplus and deficit may occur for periods in individual years as they do in nature. This will not be found in the case of monthly or seasonal water budgets using long-term average values. The present crop water balance approach emphasizes the distribution of daily soil water deficits and surplus in relation to crop development, although the limiting effects of freezing temperatures on growth and production cannot be ignored. Probable dates of the last spring and the first autumn frosts (minimum temperature <0°C) were superimposed on the water balance results. These dates have been published for 59 stations across Canada (COLIGA-DO et al., 1968).

**Water balance and temperature regime**

The results of the daily water balance calculation for the 1931–60 period at 10 selected locations across Canada are illustrated in Fig.37 and summarized over the growing

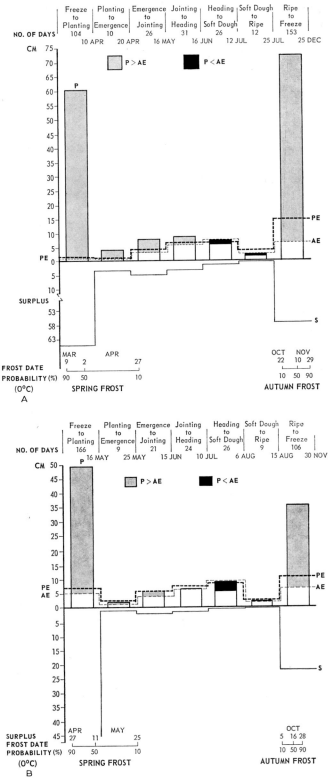

Fig.37. Daily estimates of crop water balance accumulated over developing periods of wheat and frost date probabilities, 1931–60 average: *P* = precipitation; *PE* = potential evapotranspiration; *AE* = actual evapotranspiration; *Surplus* = drainage + runoff. A. Agassiz, B.C.; B. Charlottetown, P.E.I.; C. Normandin, Que.; D. Ottawa, Ont.; E. Harrow, Ont.; F. Fort Simpson, N.W.T.; G. Brandon, Man.; H. Fort Vermilion, Alta.; I. Lethbridge, Alta.; J. Swift Current, Sask.

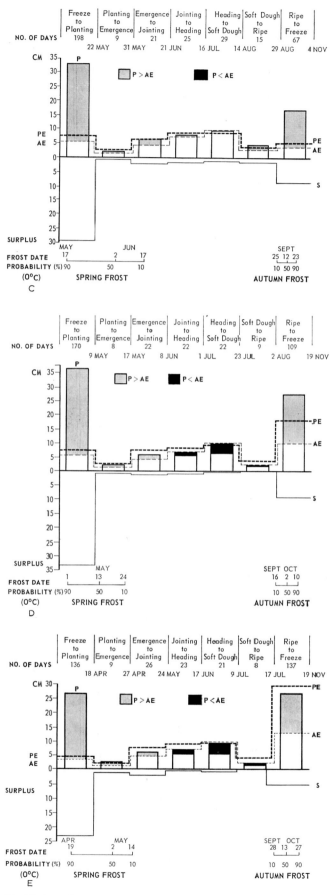

Fig.37 C–E (legend see p.180)

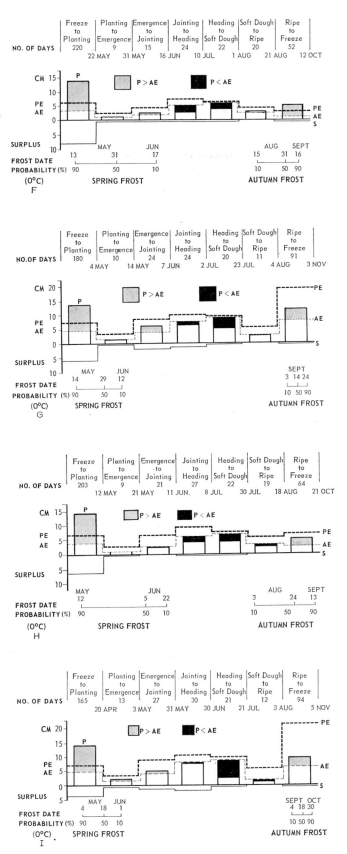

Fig.37F–I (legend see p.180)

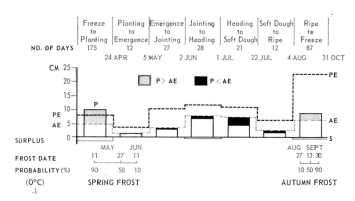

Fig.37J (legend see p.180)

season from planting to ripe of spring wheat in Table L. The locations are listed in the order of decreasing spring soil moisture and increasing stress (*PE–AE*).

Average over-winter precipitation at Agassiz and Charlottetown (Fig.37A, B) is plentiful so that surplus is substantial and the available soil moisture at planting time is usually close to storage capacity (20 cm). Because of this reserve and the adequate summer rainfall, *PE–AE* is small and some surplus occurs almost throughout the growing season of wheat (*P–R*). Early frost in autumn is no threat to wheat ripening but late spring frosts can occur at the time of emergence 10% of the time.

Average spring soil moisture reserves at Normandin, Ottawa and Harrow (Fig.37C–E) are all close to capacity and the *R–P* surplus is moderate. Although some surplus occurs even throughout the growing season, there is at the same time serious stress, particularly from *E* to *H*, increasing towards the south. Autumn frost is no problem but late spring frosts occur after emergence 10% of the time at Ottawa and more frequently (50%) at Normandin because of the colder climate and at Harrow because of the earlier planting date.

Ripening of wheat is endangered by early autumn frost 10% of the time only at Fort Simpson (Fig.37F) of the 10 selected stations. Here, late frosts in spring are also experienced after emergence 50% of the time. On the other hand, average spring soil moisture reserves are still 91% of storage capacity, although surplus is only 8.6 cm from *R* to *P* and 1.2 from *P* to *R*. Some stress during the growing season (7.2 cm) particularly for *P–H* is common.

At Brandon, Fort Vermilion and Lethbridge (Fig.37G–I) the average soil moisture content is 78–88% of capacity in spring and 42–55% at ripe. Surplus is low from *R* to *P* (3–6 cm) as well as from *P* to *R* (1–3 cm). Stress occurs in each of the five crop developing stages and accumulates to 11–14 cm over the growing season. Autumn frosts do not endanger wheat ripening at Brandon and Lethbridge but they do so in almost 50% of the time at Fort Vermilion. Spring frosts, however, are frequent at all three locations, 90% of the time after emergence and 10% after jointing.

The climate at Swift Current (Fig.37J) is unique amongst the 10 selected stations. Soil moisture at planting time is on the average only 52% of capacity and at ripe only 23%. Surplus is negligible throughout the year. Stress is significant during each of the five crop development stages and accumulates to 20 cm from *P* to *R*. Spring frosts must be expected after emergence in 90% of the time and after jointing in almost 50% of the time. In autumn, however, frost hardly ever affects ripening.

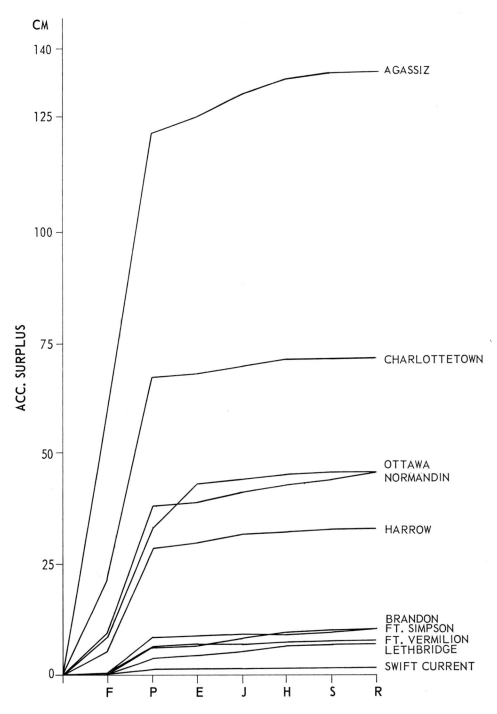

Fig.38. Accumulated daily surplus (drainage + surplus) beginning at estimated ripe date of wheat: 1931–60 average.

TABLE LI

SEASONAL TOTALS OF CROP WATER BALANCE[1] (COMPUTED DAILY) FROM PLANTING TO RIPE OF SPRING WHEAT:
1931–60

| Location | $SM$ planting | $SM$ ripe | $B$ | $P$ | $S$ | $AE$ | Stress: $PE-AE$ |
|---|---|---|---|---|---|---|---|
| | (cm) | (cm) | (cm) | (cm) | (cm) | (cm) | (cm) |
| Agassiz | 19.9 | 15.2 | −4.8 | 27.9 | 13.3 | 19.4 | 3.8 |
| Charlottetown | 19.6 | 14.0 | −5.5 | 22.6 | 4.4 | 23.7 | 5.4 |
| Normandin | 19.3 | 17.0 | −2.1 | 30.0 | 7.3 | 24.8 | 5.9 |
| Ottawa | 19.2 | 12.4 | −6.7 | 21.9 | 3.2 | 25.4 | 6.9 |
| Harrow | 19.6 | 11.5 | −8.1 | 20.7 | 4.0 | 24.8 | 7.2 |
| Ft. Simpson | 18.4 | 12.9 | −5.5 | 13.3 | 1.2 | 17.6 | 7.5 |
| Brandon | 17.9 | 11.1 | −6.8 | 21.7 | 3.5 | 25.0 | 11.0 |
| Ft. Vermilion | 15.8 | 9.2 | −6.5 | 15.4 | 0.8 | 21.1 | 11.8 |
| Lethbridge | 17.5 | 8.6 | −9.0 | 18.6 | 2.9 | 24.7 | 14.4 |
| Swift Current | 10.6 | 4.6 | −6.0 | 16.3 | 0.2 | 22.1 | 20.2 |

[1] Terms in crop-water balance equation (measured or computed daily): $AE$ = computed actual evapotranspiration ($AE = P - S - B$); $P$ = observed precipitation; $S$ = computed surplus (runoff + drainage); $B$ = computed soil moisture balance; $PE$ = estimated potential evapotranspiration; $SM$ planting = computed 30-year average soil moisture content at planting; $SM$ ripe = computed 30-year average soil moisture content at ripe.

The distribution of surplus accumulated over the year starting at ripe (Fig.38) clearly singles out the two coastal locations (Agassiz and Charlottetown), the three in eastern Canada (Ottawa, Normandin and Harrow), and the five in western Canada.

On the other hand, stress accumulated from $P$ to $R$ (Fig.39) shows that Swift Current is the most droughty location, followed by Lethbridge, Fort Vermilion and Brandon. These are followed by a group including all eastern locations. Agassiz is the least droughty location. Superimposed on the accumulated $PE–AE$ curves is the length of time in spring when the probability for late frosts exceeds 50%. This period after emergence is as long as 22 days at Swift Current, 15 days each at Brandon, Fort Vermilion and Lethbridge. It is 5 days or less at Harrow, Normandin and Fort Simpson; no frost occurs on the average after emergence at Agassiz and Charlottetown.

**Verification**

A verification of the water balance terms through comparison with actual measurements is difficult but encouraging results have been obtained. The meteorological water budgeting technique employed in this study gave realistic soil moisture estimates under a variety of soils and climates. Estimates of spring soil moisture contents in fallow and stubble lands at Swift Current over 20 seasons were significantly correlated with the measured values and showed no bias in their means (W. Baier, unpublished manuscript). Daily observed soil moisture variations under soil at Ottawa over 10 years were also used to verify the estimates. The coefficient of determination for 5-day soil moisture means was 0.73 and the standard error of estimate 15% of the mean (BAIER, 1969).

In another verification approach, daily soil moisture estimates, averaged over each of five crop development periods, were found more closely related to wheat yields than

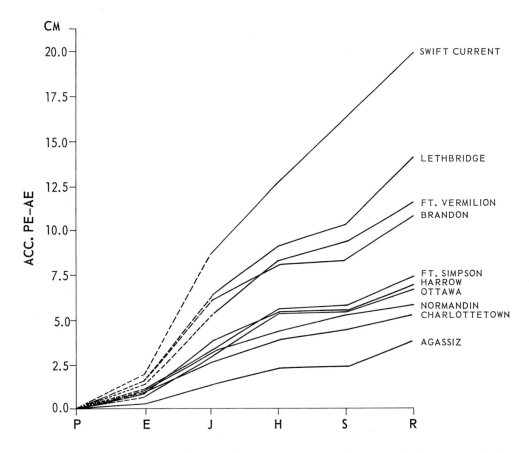

Fig.39. Accumulated daily stress (*PE-AE*) from planting to ripe of wheat: 1931–60 average. (Broken curves indicating periods of frost probability 50 percent).

precipitation totals or temperature means for these periods. Soil moisture accounted for 70% of the total variation in 39 wheat yields at eight selected stations across Canada as compared with maximum temperature (49%), minimum temperature (59%) or rainfall (17%) (BAIER and ROBERTSON, 1968). The same soil moisture estimates explained variations in the number of heads (59%), number of kernels per head (48%) and 1,000-kernel weight (56%) (BAIER and ROBERTSON, 1967).

Although the soil moisture estimates and the parameters derived from the water balance are significantly correlated with crop yields and crop yield components, these correlations are not close enough for realistic crop yield estimations. Seasonal fluctuations of weather elements such as early and late frosts, temperature and radiation influence crops to various degrees. Research is in progress to develop climatic crop–weather models which include soil moisture or derived stress terms and temperature.

Preliminary results are promising in that the yield estimates from 1953 to 1962 at 6 of the 10 selected locations agree satisfactorily with the observed yields which are available from a special crop–weather project (BAIER and ROBERTSON, 1968). With the exception of Fort Simpson, the error in estimating average yields from computed terms of the water balance and observed minimum temperatures was less than 10% of the observed value and even the annual estimates were reasonably correlated with the observations

TABLE LII

OBSERVED AND ESTIMATED WHEAT YIELDS AT SIX SELECTED LOCATIONS IN CANADA: 1953–63

| Location | Number of plantings | Mean[1] obs. yield (kg/ha) | Mean[2] est. yield (kg/ha) | Error (%) | Correlation[3] coefficient |
|---|---|---|---|---|---|
| Fort Simpson | 8 | 2837 | 2295 | −19 | 0.63 |
| Fort Vermilion | 9 | 2720 | 2918 | +7 | 0.67 |
| Swift Current | 10 | 1776 | 1649 | −7 | 0.91 |
| Ottawa | 8 | 2669 | 2594 | −3 | 0.82 |
| Harrow | 7 | 1764 | 1722 | −2 | 0.68 |
| Normandin | 8 | 2669 | 2594 | −3 | 0.82 |
| All plantings used in regression | 78 | 2218 | — | — | 0.67 |

[1] Observed yields available at 6 of the 10 selected locations in this study from special crop-weather project (BAIER and ROBERTSON, 1968).

[2] Estimated yields from regression equation developed from 78 plantings at 8 locations across Canada:

$$\text{Yield} = 8560 - 124\,\text{Min}\,(J\text{-}H) + 2291\,\frac{AE}{PE}\,(J\text{-}H) - 1176\,\frac{AE}{PE}\,(S\text{-}R) + 1296\,\frac{AE}{PE}\,(H\text{-}S) - 46\,\text{Min}\,(E\text{-}J)$$

[3] Correlation between observed and estimated yields.

(Table LII). The important point here is that the primary data input to the simulator model includes only standard daily maximum and minimum temperatures and precipitation. Furthermore, since the model was developed from crop and meteorological data obtained from 78 plantings under a variety of soils, climates and weather situations, the estimates for a particular location and year seemed realistic for climatic classification purposes.

**Applications of crop water balance**

The several parameters obtained from the crop water balance have been used for the interpretation of climates in relation to soils, crops or management systems. For example, spring soil moisture estimates reflect the effectiveness of soil moisture conservation practices simulated in different computer runs; surplus data give background information for drainage studies and soil leaching processes; stress during the five developing stages is related to yields. Depending on the purpose of the climatic evaluations, sites and eventually areas can be classified by one or several of these parameters. No attempt has been made to assign arbitrary class boundaries to the water balance terms presented in this study. Before such a meaningful climatic classification can be developed, it is necessary to determine the relationships between parameters of the water balance and those of the temperature regime on the one hand and crop responses to these elements on the other hand. It is expected that this goal can be achieved in due course. The vast amount of climatological data available in the national meteorological archives can then be exploited through specific climatic interpretations in view of the various agricultural problems which are found all over the world.

## References[1]

ADAMS, W. P. (Editor), 1966. Hydrological studies in Labrador–Ungava. *McGill Univ., Montreal, McGill Sub-Arctic Res. Pap.*, 22: 194 pp.

ALLEN, W. T. R., 1964. *Break-up and Freeze-up Dates in Canada.* Canada, Dept. Transport, Meteorol. Branch, Toronto, Ont., 200 pp.

ANDERSON, R., BOVILLE, B. W. and McCLELLAN, D. E., 1955. An operational frontal contour-analysis model. *Q.J.R. Meteorol. Soc.*, 81: 588–599.

ASHTON, W., 1969. Average soil temperature based on the period 1960 to 1968. *Canada, Dept. Transport, Meteorol. Branch, Climatol. Data Ser.*, 5–69: 8 pp.

BAIER, W., 1969a. Observed and estimated seasonal soil water variations under non-irrigated sod. *Can. J. Soil Sci.*, 49: 181–188.

BAIER, W., 1969b. Concepts of soil moisture availability and their effect on soil moisture estimates from a meteorological budget. *Agric. Meteorol.*, 6: 165–178.

BAIER, W. and ROBERTSON, G. W., 1965. Estimation of latent evaporation from simple weather observations. *Can. J. Plant Sci.*, 45: 276–284.

BAIER, W. and ROBERTSON, G. W., 1966. A new versatile soil moisture budget. *Can. J. Plant Sci.*, 46: 299–315.

BAIER, W. and ROBERTSON, G. W., 1967. Estimating yield components of wheat from calculated soil moisture. *Can. J. Plant Sci.*, 47: 617–630.

BAIER, W. and ROBERTSON, G. W., 1968. The performance of soil moisture estimates as compared with the direct use of climatological data for estimating crop yields. *Agric. Meteorol.*, 5: 17–31.

BAIER, W. and OUELLET, C. E., 1970. Definition of frost versus freeze. *Can. Agric. Serv. Coordinating Comm., 11th Ann. Meet. Can. Comm. Agrometeorol., Ottawa*, pp.22–35 (MS Position Papers).

BAIER, W., CHAPUT, D., RUSSELO, D. and SHARP, W., 1970. *Soil Moisture Estimator Program System. Internal Rept. No.19.* Agromeoterol. Section, Plant Res. Inst., Canada, Dept. of Agriculture, Ottawa, Ont., 63 pp.

BARRY, R. G., 1966. Meteorological aspects of the glacial history of Labrador–Ungava with special reference to atmospheric vapour transport. *Geogr. Bull.*, 8: 319–340.

BARRY, R. G., 1967. Variations in the content and flux of water-vapour over northeastern North America during two winter seasons. *Q.J.R. Meteorol. Soc.*, 93: 535–543.

BARRY, R. G., 1968. Vapour flux divergence and moisture budget calculations for Labrador–Ungava. *Cahiers Géogr. Québec*, 25: 91–102.

BARRY, R. G. and FOGARASI, S., 1968. Climatology studies of Baffin Island, Northwest Territories. *Can., Dept. Energy, Mines Resour., Inland Waters Branch, Ottawa, Tech. Bull.*, 13: 106 pp.

BAUDAT, C. and WRIGHT, J. B., 1969. The unusual winter 1968–69 in British Columbia. *Can., Dept. Transp., Meteorol. Branch, Tech. Memo.*, 730: 23 pp.

BENTON, G. S. and ESTOQUE, M. A., 1954. Water-vapour transfer over the North American continent. *J. Meteorol.*, 11: 462–477.

BERRY, F. A., OWENS, G. V. and WILSON, H. P., 1954. Arctic track charts. *Proc. Meteorol. Conf., Toronto, 1954*, pp.91–102.

BILELLO, M. A., 1964. Method for predicting river and lake ice formation. *J. Appl. Meteorol.*, 3: 38–44.

*BIRD, J. B., 1967. *The Physiography of Arctic Canada.* Johns Hopkins Press, Baltimore, Md., 336 pp.

BOCHKOF, A. P., CHEBOTAREV, A. I. and VOSKRESENSKY, K. P., 1970. Water resources and water balance of the U.S.S.R. *Symp. World Water Balance (Int. Assoc. Sci. Hydrol., Brussels)*, pp.324–330.

BODURTHA, F. T., 1952. An investigation of anticyclogenesis in Alaska. *J. Meteorol.*, 9: 118–125.

BOWLING, S. A., OHTAKE, T. and BENSON, C. S., 1968. Winter pressure systems and ice fog in Fairbanks, Alaska. *J. Appl. Meteorol.*, 7: 961–968.

+BRINKMANN, W. A. R., 1969. *The Definition of Chinook in the Calgary Area.* Ph. D. Thesis, Univ. Calgary, Calgary, Alta., 115 pp.

BRINKMANN, W. A. R. and ASHWELL, I. Y., 1968. The structure and movement of the Chinook in Alberta. *Atmosphere*, 6(2): 1–10.

+BROWN, D. M. and BAIER, W., 1970. Implications of freezing temperatures to Canada's agriculture. *Can. Agric. Serv. Coordinating Comm., 11th Ann. Meet. Can. Comm. Agrometeorol.*, pp.84–124 (MS Position Papers).

---

[1] References marked + are publications with extensive literature references. Those marked * contain more detailed information on the subject being discussed.

*BROWN, D. M., McKAY, G. A. and CHAPMAN, L. J., 1968. The climate of southern Ontario. *Can. Dept. Transp., Meteorol. Branch, Toronto, Climatol. Stud.*, 5: 50 pp.

BRUCE, J. P., 1968. Atlas of rainfall intensity duration frequency data for Canada. *Can., Dept. Transp., Meteorol. Branch, Toronto, Climatol. Stud.*, 8: 30 charts.

BRYSON, R. A., 1966. Air masses, streamlines, and the Boreal forest. *Geogr. Bull.*, 8: 228–269.

BRYSON, R. A. and KUHN, P. M., 1962. Some regional heat budget values for northern Canada. *Geogr. Bull.*, 17: 57–66.

BUDYKO, M. I., 1958. *The Heat Balance of the Earth's Surface*. (Transl. from Russian by Nina A. Stepanova), U.S. Weather Bureau, Washington, D.C., 259 pp. (Specific references, pp.140–151.)

BURBIDGE, F. E., 1951. The modification of continental polar air over Hudson Bay. *Q.J.R. Meteorol. Soc.*, 77: 365–374.

*CANADA, DEPARTMENT OF TRANSPORT, METEOROLOGICAL BRANCH, 1956. *Climatic Summaries, III. Frost Data*. Toronto, 94 pp.

CANADA, DEPARTMENT OF TRANSPORT, METEOROLOGICAL BRANCH, 1964. *Heating Degree-Day Normals below 65°F—Based on the Period 1931–60*. Climatology Division Toronto, CDS 5-64, 12 pp. (see also CDS 8-69: *Supplemental Heating Degree-Day Normals*, 3 pp.)

CANADA, DEPARTMENT OF TRANSPORT, METEOROLOGICAL, BRANCH, 1965. *Average Soil Temperature*. Climatology Division, Toronto, CDS 14-65, 4 pp.

*CANADA, DEPARTMENT OF TRANSPORT, METEOROLOGICAL BRANCH, 1967a. *Climatic Charts*. Climatology Division, Toronto. (A series produced periodically to cover elements of World Climatic Atlas specifications.)

CANADA, DEPARTMENT OF TRANSPORT, METEOROLOGICAL BRANCH, 1967b. *Hourly Data Series* (HDS). Climatology Division, Toronto (covers hourly synoptic reporting stations individually).

*CANADIAN NATIONAL COMMITTEE FOR THE INTERNATIONAL HYDROLOGICAL DECADE, 1969. *Hydrological Atlas of Canada. A Series of Preliminary Charts of Principal Hydrological Parameters*. Ottawa, Department of Energy, Mines and Resources, 5 charts.

CATCHPOLE, A. J. W., 1969. The solar control of diurnal temperature variations at Winnipeg. *Can. Geogr.*, 13: 255–268.

CHAPMAN, L. J., 1970. Occurrence of freezing temperatures in the agricultural areas of Canada. *Can. Agric. Serv. Coordinating Comm., 11th Ann. Meet. Can. Comm. Agrometeorol., Ottawa*, pp.36–41 (MS Position Papers).

*CHAPMAN, L. J. and THOMAS, M. K., 1968. The climate of Northern Ontario. *Can., Dept. Transp., Meteorol. Branch, Toronto, Climatol. Stud.*, 6: 58 pp.

COLIGADO, M. C., BAIER, W. and SLY, W. K., 1968. Risk analyses of weekly climatic data for agricultural and irrigation planning. *Can., Dept. Agric., Plant Res. Inst., Agrometeorol. Sect., Tech. Bull.*, 17–58, 61–77, 8 pp. 26 tables.

COURT, A. and GERSTON, R. D., 1966. Fog frequency in the United States. *Geogr. Rev.*, 60: 543–550.

*CRUTCHER, H. L. and HALLIGAN, D. K., 1967. Upper wind statistics of the northern Western Hemisphere. *ESSA Tech. Rept. EDS-1*, 20 pp., 100 charts.

CUDBIRD, B. S. V., 1964. *Diurnal Averages of Wind Atmospheric Pressure and Temperature at Selected Canadian Stations*. Canada, Department of Transport, Meteorological Branch, Toronto, CIR-4114, CLI-33, 43 pp.

CURRIE, B. W., *Prairie Provinces and Northwest Territories Wind and Storms*. Univ. of Saskatchewan, Saskatoon, Sask, 18pp., 13 charts and tables.

DAVIES, J. A. and BUTTIMOR, P. H., 1969. Reflection coefficients, heating coefficients and net radiation at Simcoe, Southern Ontario. *Agric. Meteorol.*, 6: 373–386.

DAVIES, J. A. and McCAUGHEY, J. H., 1968. Potential evapotranspiration at Simcoe, Southern Ontario. *Arch. Meteorol., Geophys. Bioklimatol.*, B16: 391–417.

DOUGLAS, R. H. and HITSCHFELD, W., 1959. Patterns of hailstorms in Alberta. *Q.J.R. Meteorol. Soc.*, 85: 105–119.

DUNBAR, M. J., 1951. Eastern Arctic waters. *Fisheries Res. Board Can., Ottawa, Bull.*, 88: 131 pp.

EMSLIE, J. H. and SATTERTHWAITE, J., 1966. *Air Pollution–Meteorological Relationships at Vancouver, British Columbia*. Can., Dept. Transp., Meteorol. Branch, Toronto, CIR-4396, TEC-605, 14 pp.

FERGUSON, H. L., O'NEILL, A. D J and CORK, H. F., 1970. Mean evaporation over Canada. *Water Resour. Res.*, 6: 1618–1633.

*FERLAND, M. G. and GAGNON, R. M., 1967. *Climat du Québec méridional*. Québec, Ministère des richesses naturelles, Service de Météorologie, M.P.-13, 93 pp.

FINDLAY, B. F., 1966. The water budget of the Knob Lake area. In: W. P. ADAMS (Editor), *Hydrological Studies in Labrador–Ungava. McGill Sub-Arctic Res. Pap.*, 22: 1–95.

FINDLAY, B. F., 1969. Precipitation in northern Quebec and Labrador: an evaluation of measurement techniques. *Arctic*, 22: 140–150.

FINDLAY, B. F., 1970. Topoclimatological aspects of frost. *Can. Agric. Serv. Coordinating Comm., 11th Ann. Meet. Can. Comm. Agrometeorol., Ottawa*, pp.42–56 (MS Position Papers).

FLOWERS, E. C., McCORMICK, R. A. and KURFIS, K. R., 1969. Atmospheric turbidity over the United States, 1961–1966. *J. Appl. Meteorol.*, 8: 955–962.

FRASER, D. A., 1965. The water cycle over a 16-year period on several forest sites at Chalk River, Ontario, Canada. In: W. E. SOPPER and H. W. LULL (Editors), *Forest Hydrology. Proc. Natl. Sci. Found. Adv. Sci. Semin*. Pennsylvania State University, Pergamon, Oxford and New York, pp.463–475.

FRASER, W. C., 1964. *A Study of Blowing Snow in the Canadian Arctic*. Canada, Dept. Transport, Meteorol. Branch, Toronto, Ont., CIR-4162, TEC-548, 31 pp., + 16 charts.

*GAGNON, R. M. and FERLAND, M., 1967. *Climat du Québec septentrional*. Ministère des Richesses naturelles, Service de Météorologie, Québec, M.P.-10, 107 pp.

GOLD, L. W. and BOYD, D. W., 1965. Annual heat and mass transfer at an Ottawa site. *Can. J. Earth Sci.*, 2: 1–10.

*GRUBBS, B. E. and McCOLLUM, R. D., 1968. *A Climatological Guide to Alaskan Weather*. 11th Weather Squadron, Elmendorf AFB, Alaska, 88 pp.

HAACK, L. C., 1964. *The Frequency of Thunderstorm Occurrence in Ontario*. Canada, Dept. of Transport, Meteorol. Branch, Toronto, Ont., CIR-4160 TEC 546, 5 pp. + 7 charts.

HARE, F. K., 1950a. Climate and zonal divisions of the Boreal forest formation in eastern Canada. *Geograph. Rev.*, 40: 615–635.

HARE, F. K., 1950b. *The Climate of the Eastern Canadian Arctic and Sub-Arctic*. Ph.D. Thesis, Université de Montréal, Montreal, 2 vols., 440 pp.

+HARE, F. K., 1951. Some climatological problems of the Arctic and Sub-Arctic. In: T. F. MALONE (Editor), *Compendium of Meteorology*. Am. Meteorol. Soc., Boston, pp.952–964.

+HARE, F. K., 1968. The Arctic. *Q.J.R. Meteorol. Soc.*, 94: 439–459.

HARE, F. K. and HAY, J. E., 1971. Anomalies in the large-scale annual water balance over northern North America. *Can. Geogr.*, 15: 79–94.

HARE, F. K. and MONTGOMERY, M. R., 1949. Ice, open water and winter climate in the eastern Arctic of North America, 1. *Arctic*, 2: 79–89.

+HARE, F. K. and ORVIG, S., 1958. *The Arctic Circulation*. McGill Univ., Dept. of Meteorol., Montreal, Publ. in Meteorol., 12: 211 pp.

HARRY, J. B. and WRIGHT, K. F., 1967. *The Climate of Vancouver*. Canada, Dept. of Transport, Meteorol. Branch, Toronto, Ont., CIR 2985 TEC 258, 53 pp.

*HAY, J. E., 1970. *Aspects of the Heat and Moisture Balance of Canada*. Ph.D. Thesis, University of London, London, 212 pp.

JOHNSON, C. B., 1948. Anticyclogenesis in eastern Canada during spring. *Bull. Am. Meteorol. Soc.*, 29: 47–55.

KENDALL, G. R. and ANDERSON, S. R., 1966. *Standard Deviations of Monthly and Annual Mean Temperatures*. Canada, Dept. of Transport, Meteorol. Branch, Toronto, Ont., Climatol. Stud. no. 4, 18 pp. + 16 charts.

KENDALL, G. R. and PETRIE, A. G., 1962. *The Frequency of Thunderstorm Days in Canada*. Canada, Dept. of Transport, Meteorol. Branch, Toronto, Ont., CIR-3688, TEC-418, 17 pp. + 9 charts.

KENDREW, W. G. and CURRIE, B. W., 1955. *The Climate of Central Canada*. Queen's Printer, Ottawa, Ont., 194 pp.

KERR, D. P., 1951. The summer-dry climate of Georgia Basin, British Columbia. *Trans. R. Can. Inst.*, 29: 23–31.

KREBS, S. and BARRY, R. G., 1970. The Arctic front and the tundra–taiga boundary. *Geogr. Rev.*, 60: 548–554.

LAYCOCK, A. H., 1960. Drought patterns in the Canadian prairies. *Int. Assoc. Sci. Hydrol., Publ.*, 51: 34–47.

LESLIE, D., 1971. *Estimates of Regional Soil Heat Flux Values for Climatonomic Analysis in Southern Ontario and Quebec*. B.A. Thesis, Univ. of Toronto, Toronto, Ont., 76 pp. (unpublished).

LETTAU, H., 1969. Evapotranspiration climatonomy, 1. A new approach to numerical prediction of monthly evapotranspiration, runoff, and soil moisture storage. *Monthly Weather Rev.*, 97: 691–699.

LONGLEY, R. W., 1959. *The Three-Front Model—a Critical Analysis*. Canada, Dept. Transport, Meteorol. Branch, Toronto, Ont., CIR-3245, TEC-309, 14 pp.

LONGLEY, R. W., 1967. The frequency of winter chinooks in Alberta. *Atmosphere*, 4: 4–16.

MANNING, F. D., 1968. *Winter Snowfall Averages and Extremes at Synoptic Observing Stations*. Canada, Dept. Transport, Meteorol. Branch, Toronto, Ont., CDS 3-68, 21 pp.

MANNING, F. D., 1969. *Monthly Fog Data*. Canada, Dept. Transport, Meteorol. Branch, Toronto, Ont., CDS 1-69, 7 pp.

MARCUS, M., 1964. Climate-glacier studies in the Juneau ice field region, Alaska. *Univ. Chicago, Dept. Geogr., Res. Pap.*, 88: 128 pp.

MATEER, C. L., 1955. A preliminary estimate of the average insolation in Canada. *Can. J. Agric. Sci.*, 35: 579–594.

MCFADDEN, J. D., 1965. The interrelationship of lake ice and climate in central Canada. *Univ. Wisc., Dept. Meteorol., Madison, Tech. Rept.*, 20: 120 pp.

MCINTYRE, D. P., 1950. On the air-mass temperature distribution in the middle and high troposphere in winter. *J. Meteorol.*, 7: 101–107.

MCINTYRE, D. P., 1955. On the barocline structure of the westerlies. *J. Meteorol.*, 12: 201–210.

MCKAY, G. A. and LOWE, A. B., 1960. The tornado in western Canada. *Bull. Am. Meteorol. Soc.*, 41: 1–8.

MCKAY, G. A. and THOMPSON, H. A., 1968. Snowcover in the prairie provinces of Canada. *Trans. Am. Soc. Agric. Eng.*, 11: 812–815.

MCKAY, G. A. and THOMPSON, H. A., 1969. Estimating the hazard of ice accretion in Canada from climatological data. *J. Appl. Meteorol.*, 8: 927–935.

*MCKAY, G. A., MAYBANK, J., MOONEY, O. R. and PELTON, W. L., 1967. The agricultural climate of Saskatchewan. *Canada, Dept. Transport, Meteorol. Branch, Toronto, Climatol. Stud.*, 10: 18 pp.

MOKOSCH, E., 1961. *Location of Lows East of the Rockies*. Canada, Dept. of Transport, Meteorol. Branch, Toronto, Ont., CIR-3575, TEC-386, 16 pp.

MURRAY, W. A., 1964. *Rainfall Intensity–Duration–Frequency Maps for British Columbia*. Canada, Dept. of Transport, Meteorol. Branch, Toronto, Ont., CIR-4031, TEC-318, 8 pp. + 9 charts.

PATRIC, J. H. and BLACK, P. E., 1968. Potential evapotranspiration and climate in Alaska by Thornthwaite's classification. *U.S. Dept. Agric. Forest Serv., Res. Pap.*, PNW-71: 28 pp.

PELTON, W. L. and KORVEN, H. C., 1969. Evapotranspiration estimates from atmometers and pans. *Can. J. Plant Sci.*, 49: 615–621.

PENNER, C. M., 1955. A three-front model for synoptic analysis. *Q.J.R. Meteorol. Soc.*, 81: 89–91.

PETERSON, J. T., 1965. On the distribution of lake temperatures in central Canada as observed from the air. *Univ. Wisc., Dept. Meteorol., Madison, Tech. Rept.*, 22: 35 pp.

PÉWÉ, T. L., 1965. Notes on the physical geography of Alaska. *Proc. Alaskan Sci. Conf., 1965*, pp.103–108.

*POTTER, J. G., 1965a. Snow cover. *Canada, Dept. Transport, Meteorol. Branch, Toronto, Climatol. Stud.*, 3: 69 pp.

POTTER, J. G., 1965b. *Water Content of Freshly Fallen Snow*. Canada, Dept. of Transport, Meteorol. Branch, Toronto, Ont., CIR-4232, TEC-569, 12 pp.

POWE, N. N., 1968. *The Influence of a Broad Valley on the Surface Winds within the Valley*. Canada, Dept. of Transport, Meteorol. Branch, Toronto, Ont., TEC-668, 9 pp.

RAGOTZKIE, R. A. and MCFADDEN, J. D., 1965. Operation freeze-up. An aerial reconnaissance of climate and lake ice in central Canada. *Univ. Wisc., Dept. Meteorol., Madison, Tech. Rept.*, 10: 28 pp.

*RASMUSSON, E. M., 1966. *Atmospheric Water Vapor Transport and the Hydrology of North America*. M.I.T., Dept. Meteorol., Cambridge, Planetary Circulation Proj., Rept. A.1, 1970 pp. + 118 unpaginated chart..

RASMUSSON, E. M., 1968. Atmospheric water vapor transport and the water balance of North America II. Large-scale water balance investigations. *Monthly Weather Rev.*, 96: 720–734.

REED, R. J. and KUNKEL, B., 1960. The Arctic circulation in summer. *J. Meteorol.*, 17: 489–506.

RICHARDS, T. L., 1964. The meteorological aspects of ice cover on the Great Lakes. *Monthy Weather Rev.*, 92: 297–302.

ROBERTSON, G. W., 1968. A biometeorological time scale for a cereal crop involving day and night temperatures and photoperiod. *Int. J. Biometeorol.*, 12: 191–223.

ROBINSON, E., THUMAN, E. C. and WIGGINS, E. J., 1957. Ice fog as a problem of air pollution in the Arctic. *Arctic*, 10: 89–104.

ROUSE, W. R., 1970. Relations between radiant energy supply and evapotranspiration from sloping terrain: an example. *Can. Geogr.*, 14: 27–37.

ROUSE, W. R. and WILSON, R. G., 1969. Time and space variations in the radiant energy fluxes over sloping forested terrain and their influence on seasonal heat and water balances at a middle latitude site. *Geogr. Ann.*, 51A: 160–175.

SABBAGH, M. E. and BRYSON, R. A., 1962. An objective precipitation climatology of Canada. *Dept. Meteorol., Univ. Wisc., Tech. Rept.*, 8: 36 pp., 12 charts, 46 pp. of appendices.

SANDERSON, M. E., 1948. The climates of Canada according to the new Thornthwaite classification. *Sci. Agric.*, 28: 591–617.

SANDERSON, M. E., 1949. Measuring potential evapotranspiration at Norman Wells. *Geogr. Rev.*, 40: 636–645.

SANDERSON, M. E., 1950. Three years of evapotranspiration at Toronto. *Can. J. Res.*, 28C: 482–492.

*SANDERSON, M. E. and PHILLIPS, D. W., 1967. Average annual water surplus in Canada. *Canada, Dept. Transport, Meteorol. Branch, Toronto, Climatol. Stud.*, 9: 76 pp.

*SATER, J. E. (Editor), 1969. *The Arctic Basin.* Arctic Institute of North America, Washington, D.C., 2nd ed., 337 pp.

*SEARBY, H. W., 1969. *Coastal Weather and Marine Data Summary for Gulf of Alaska, Cape Spencer westward to Kodiak Island.* United States, E.S.S.A., E.D.S., Tech. Memor. E.D.S.T.M., 8: 30 pp.

THOM, H. C. S., 1968. *Standard Deviation of Monthly Average Temperature.* E.S.S.A. Tech. Rept., Environmental Data Service, Silver Spring, Md., EDS-3, 10 pp. plus 12 charts.

THOMAS, M. K., 1964. *Snowfall in Canada.* Canada, Dept. of Transport, Meteorol. Branch, Toronto, Ont., CIR-3977, TEC-503, 16 pp.

THOMPSON, H. A., 1963. Freezing and thawing indices in northern Canada. *Proc. Can. Conf. Permafrost, 1st, Nat. Res. Counc., Tech. Memor.*, 76: 18–26.

*THOMPSON, H. A., 1967. The climate of the Canadian Arctic. In: *Can. Year Book.* Queen's Printer, Ottawa, Ont., pp.55–74.

*THOMPSON, H. A., 1968. The climate of Hudson Bay. In: *Science, History and Hudson Bay.* Queen's Printer, Ottawa, Ont., 1: 263–286.

THUMAN, W. C. and ROBINSON, E., 1954. Studies of Alaskan ice fog particles. *J. Meteorol.*, 11: 151–156.

*TITUS, R. L. and TRUHLAR, E. J., 1969. *A New Estimate of Average Global Solar Radiation in Canada.* Canada, Dept. of Transport, Meteorol. Branch, Toronto, Ont., CLI 7-69, 11 pp. + 13 charts.

*U.S. Weather Bureau, 1968. *Climatic Atlas of the United States.* U.S. E.S.S.A., Washington, D.C., 80 pp.

VIERECK, L., 1970. Forest succession and soil development adjacent to the Chena River in interior Alaska. *J. Arctic Alpine Res.*, 2: 1–26.

WALKER, E. R., 1961. *A Synoptic Climatology for Parts of the Western Cordillera.* Publications in Meteorology, Arctic Meteorology Research Group, McGill University, Montreal, 35: 218 pp.

WARKENTIN, J. (Editor), 1968. *Canada, A Geographical Interpretation.* Methuen, London, 608 pp. (Canada's physical geography is reviewed in pp.57–136.)

WEXLER, H., 1937. Formation of polar anticyclones. *Monthly Weather Rev.*, 65: 229–236.

WILLIAMS, G. P., 1965. Correlating freeze-up and break-up with weather conditions. *Can. Geotech. J.*, 2: 313–326.

WILLIAMS, G. P., 1968. Freeze-up and break-up of fresh water lakes. Canada, National Research Council, Division of Building Research, Tech. Pap., 286 (from: *Proceedings of a Conference on Ice Pressures against Structures, Laval University, Quebec*, pp.203–215).

*WILSON, A. and ISERI, K. T., 1969. River discharge to the sea from the shores of the conterminous United States, Alaska, and Puerto Rico. *U.S. Geol. Surv., Washington, D.C., Hydrol. Invest. Atlas*, HA-282, 2 charts.

*WILSON, C. V., 1972. The climate of Quebec. *Canada, Dept. Transport, Meteorol. Branch, Climatol. Stud.*, 11. (Vol. II, "Atlas of Climatic Charts", has been published. The text is in English and in French.)

WRIGHT, J. B. and TRENHOLM, C. H., 1969. *Greater Vancouver Precipitation.* Canada, Dept. of Transport, Meteorol. Branch, Toronto, Ont., TEC-722, 36 pp.

*Chapter 3*

# The Climate of the Conterminous United States

ARNOLD COURT

**Overview**

Air from the south, warm and moist, and air from the north, cold and dry, meet over most of the conterminous United States in constantly varying patterns. Air from the Pacific Ocean on the west affects most of the mountainous western third of the country. It pushes the conflict between the southern and northern flows to the eastern seaboard and out into the Atlantic Ocean. The result is a wide gamut of weather events: blizzards, droughts, hurricanes, ice storms, tornadoes, chinooks, hail, fog, downpours. These alternate with calm, clear periods with bright sunshine and, if they continue, serious air pollution.

Added together, these incessant movements of air provide a wide variety of climates. Although it spans only 25° of latitude and less than 60° of longitude, the conterminous United States has subtropical swamps and deserts, glaciers and frost-free regions, lush forests and vast prairies with extreme annual temperature ranges and precipitation variability.

These various climates are discussed in this contribution, considering first the airflow and the water it carries, storms, precipitation, and water balance. Radiation and temperature, greatly affected by the results of air movement, are described next. Regional and local climates and special topics come last. The discussion is limited to the conterminous United States, excluding the states of Alaska and Hawaii, and various islands under U.S. control, which are described elsewhere in this survey.

Climatic data for 77 stations (Fig.1) are presented in standard format in the Appendix (pp.267–343). Wherever possible, the data are "climatological standard normals" for the period 1931–1960. However, the official weather service uses this period only for mean pressure and temperature, and total precipitation; extremes of temperature, days with rain, thunder, fog, etc., and other elements often are for varying periods of years, usually the longest interval of unchanged instrumental exposure.

In addition, most of the special studies from which maps and conclusions have been extracted were for much shorter periods. This may be an advantage, because "climate", however defined, changes continually, and averages for the last 10 to 15 years are better predictors of future conditions than are "normals" for a longer period ending more than a decade ago (LAMB, 1969).

In the United States, English units are used routinely for measurements of temperature, precipitation, wind speed, area, streamflow, and other quantities. Their conversion to metric units sometimes becomes awkward, as with the base of 65°F (18.3°C) for heating degree days, and with days of precipitation exceeding 0.50 inch (12.7 mm). More impor-

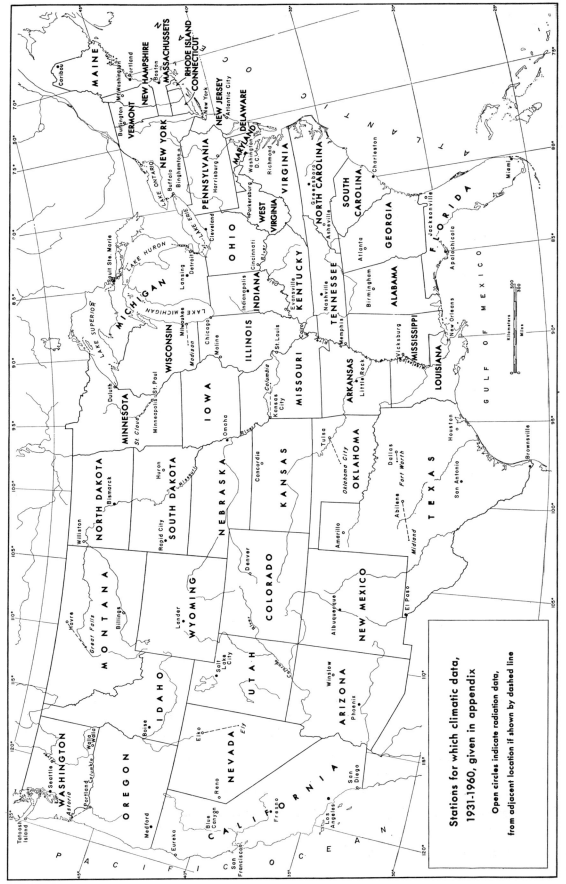

Fig.1. Stations for which climatic data are given in the Appendix. The 48 conterminous states and certain lakes and rivers, discussed in the text, are identified.

tant for comparing aspects of the U.S. climate with those of other countries are the differences in measurement techniques. Temperature midrange, the mean of maximum and minimum, is generally considered the daily mean temperature, although in most places in most seasons it is 1° or 2°C warmer than the mean of 24 equally-spaced observations. Most raingauges have funnels 8 inches (20.3 cm) in diameter, with tops about 1 m above ground or building roof. Thus they catch less than similar gauges closer to the ground. Winds are measured, at widely differing heights, by cup anemometers, either rotating or bridled; the latter measure pressure, and hence are more comparable to Dines and similar pressure anemometers.

Studies of the various aspects of the climate of the United States, or of substantial portions, have become so numerous, detailed, and specific in the past quarter-century that no single survey can hope to do all of them justice. Accordingly, the voluminous results of radar, radiosonde, and rocketsonde observations in the free air are not mentioned, except for the moisture advection part of the water budget (see p.232). Likewise, the detailed hydrologic studies of storm rainfall are not discussed.

From more than a hundred articles and reports, almost two-thirds of them from the past decade, information has been extracted, and maps copied, to provide a survey of the more interesting and important aspects of the climate of the contiguous United States.

## Air movement

In addition to its latitude, from about 25° to 49°N, the major geographic controls on the United States climate are the sources of its airflow: the oceans to the east and west, the warm Gulf of Mexico south of the eastern half and the vast watery Canadian Shield and drier prairie to the north. The general topography of its 7.8 million km² (only 1.6% water) trends north–south. The broad Mississippi Valley, some 1,500 km from the Rocky Mountains on the west to the lower Appalachian Plateau on the east, permits polar and tropical air to ebb and flow through almost 20° of latitude, from the Gulf of Mexico to Canada.

Of the three principal airflows into the United States, that from the south is the most prevalent, at and just above the surface. Sweeping westward across the warm Caribbean Sea, the northern branch of the Atlantic tradewind circulation curves northward over the Gulf of Mexico and pushes across the low coastline. From central Texas to Iowa and the Great Lakes, the average western margin of this flow creates a major climatic transition zone between the humid southeastern third of the country and the semi-arid western plains and plateaus. The grasslands of the Great Plains extend northeastward into the prairies of Iowa and Illinois, along the outer flank of this flow of tropical air and south of the main route of Arctic air (BORCHERT, 1950). (See also pp.19–20.)

Along the outward margin of the tropical airstream, from the southeast corner of New Mexico to the northern border of Illinois, severe hailstorms and tornadoes are most common in summer, and ice storms in winter. This boundary is the region of a "low level jet", a flow of air at 11–14 m/sec from SSW to NNE some 700–900 m above the surface (BONNER, 1968), with much slower speeds above and to the sides of the stream (Fig.2).

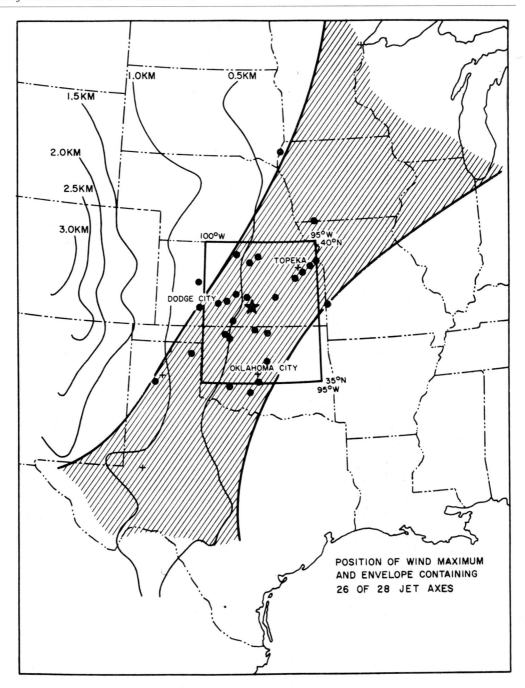

Fig.2. Location of axis of low-level jet in 26 of 28 cases, 1959–1960. Star indicates median latitude and longitude of all 28 cases, centers of which are shown by dots. Land elevations above sea level shown by solid lines. (After BONNER, 1968.)

This jet occurred on 30% of the 6-hourly observations, 1959–60, slightly more frequently in summer; its connection with the maximum incidence of tornadoes, severe thunderstorms, and hail is not yet clear.

The strongly meridional flow through the center of the United States is shown clearly on the map (Fig.3) of the average annual wind frequencies, by directions. [On this and other wind flow maps, the roses are drawn so that the *area* of each sector is proportional

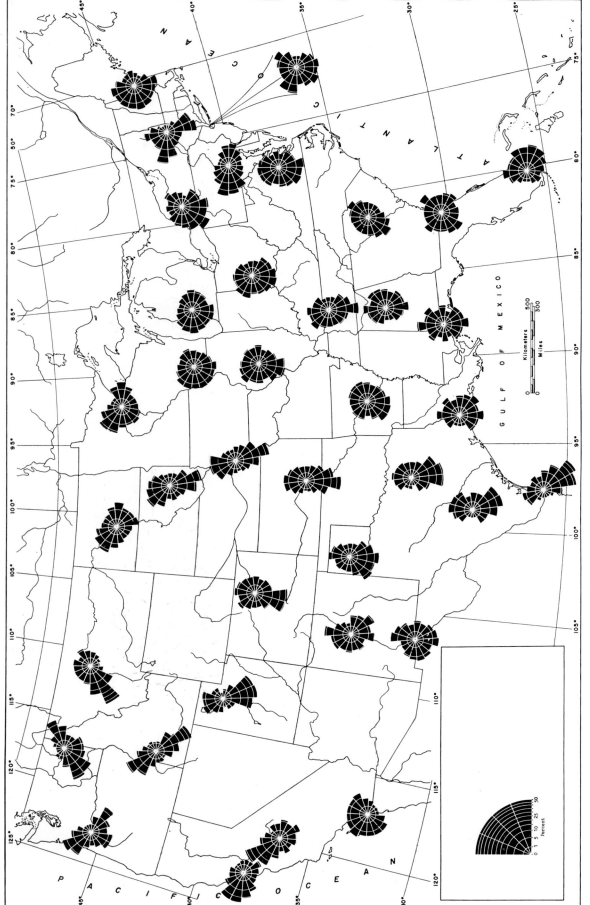

Fig.3. Average annual percentage frequencies of wind direction, 1951–1960. Areas of sectors are proportional to hourly frequencies. (From SLUSSER, 1965.)

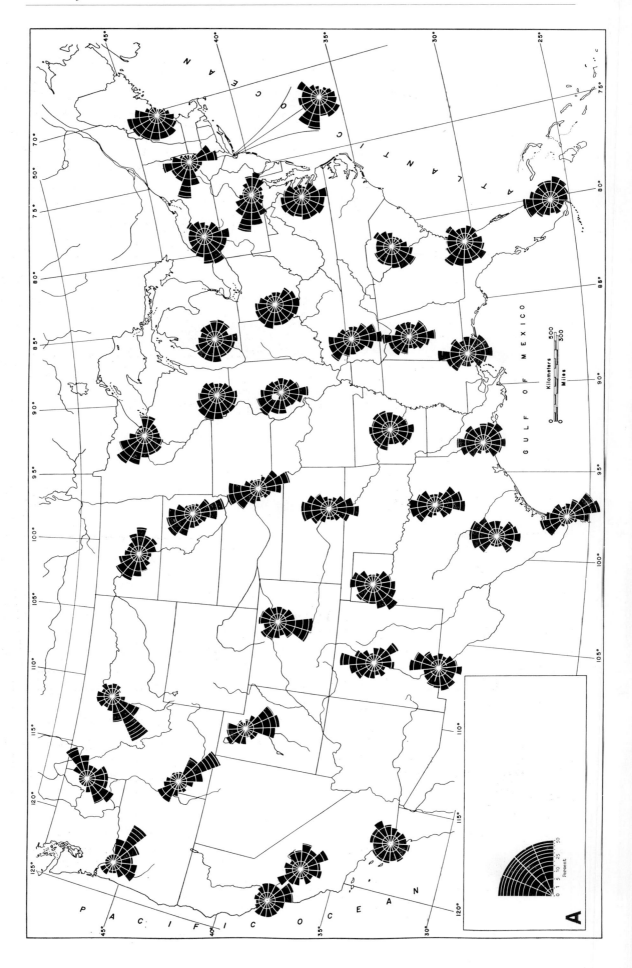

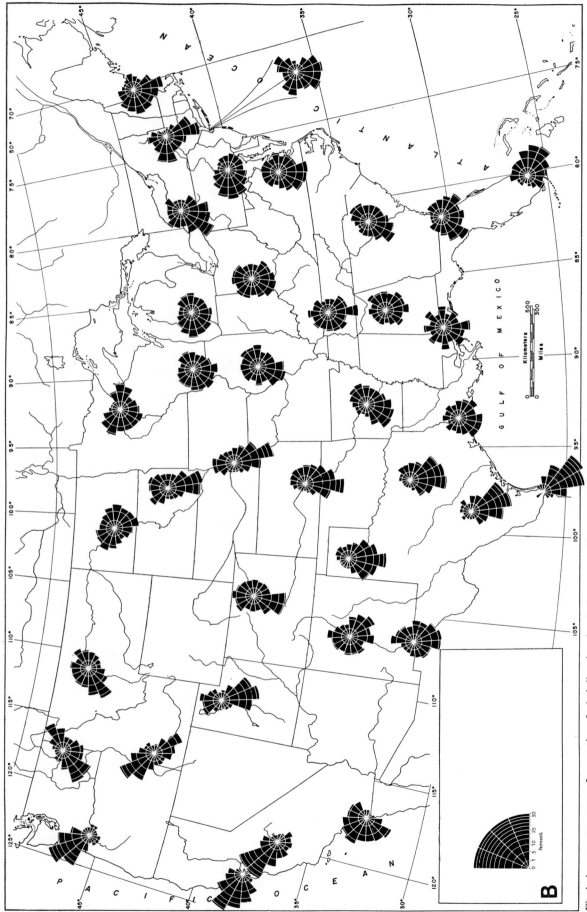

Fig.4. Average percentage frequencies of wind directions in: A. January, and B. July, 1951–1960. Areas of sectors are proportional to hourly frequencies. (From SLUSSER, 1965.)

to the number of hours of wind from that direction (SLUSSER, 1965).] Throughout the western plains, from 95° to 105°W, winds usually are either southerly or northerly. Farther west, many of the roses at inter-mountain stations also show two opposite modes, but these are dictated by the topography around the stations.

On the plains, northerly winds are more common in winter (Fig.4A), southerly in summer (Fig.4B). Wind speeds in this vast region, however, are remarkably uniform from all directions (Fig.5). In northwest Texas and adjacent areas, average annual speeds are more than 5 m/sec.

Other windy places are the northern portions of the Pacific and Atlantic seacoasts. Precise comparisons of windiness are not possible because of the great diversity in anemometer heights and exposures. Of the 77 places for which data are tabulated here, the windiest is the summit of Mt. Washington, N.H., 1,909 m above sea level. Next windiest is Tatoosh Island, Wash., at the extreme northwest corner of the country, with 9 m/sec in December and January. Both Abilene and Amarillo, Texas, where anemometers were only 10 and 18 meters above the ground, reported mean speeds of 7 m/sec in March and April; Buffalo, N.Y., at the east end of Lake Erie, also reported 7 m/sec (Dec.–Mar.), but for an anemometer atop a tall building.

Air movement over the United States is so general that air rarely remains over any one portion long enough to develop distinct air mass characteristics. Instead, the various air masses swirling across the country are modified continually as they move to the north, east, and south. The air masses affecting the United States are chiefly polar maritime air from the North Pacific Ocean, and occasionally from the North Atlantic; tropical maritime from the central Pacific Ocean and, more importantly, from the Caribbean Sea and the Gulf of Mexico; and polar continental air from Canada.

Air mass classification, like climatic classification, has declined in popularity in the past quarter-century, as the impossibility of establishing strict boundaries between types became apparent. Daily weather maps during the 1940's identified a score of different air masses each day, but during the 1960's showed less than half a dozen. From the earlier maps, BRUNNSCHWEILER (1952, 1955) compiled air mass frequencies by seasons. From frequency distributions of July maximum temperatures at 120 stations, 1948–1957, BRYSON (1966) estimated the frequencies of different air masses in July over central and eastern United States and Canada.

In autumn, and sometimes in other seasons, air may stagnate for several days over large portions of the east central United States and on the west coast. In the east, the warm days and cool nights of this "Indian summer" formerly were a delightful reminder of summer, but in recent years this season has been much less welcome. The lack of ventilation, and the inversion which develops in the gently subsiding air, prevent atmospheric pollutants from dispersing. The "air pollution potential" reaches a maximum in October. Stagnation with subsidence also occurs over the west coast and Great Basin (Nevada, Utah and portions of adjacent states), primarily in autumn and winter, similarly increasing the pollution potential. But the resulting outflow of air southwestward through mountain passes across California to the Pacific Ocean may become a hot, dry Santa Ana wind which sweeps away the smog from the Los Angeles Basin and creates extreme danger of hillside brush fires (SERGIUS et al., 1962). Even when the air outflow is not violent, the subsidence can cause extensive warming, anywhere along the Pacific coast and in the intermountain basins. CRAMER and LYNOTT (1970) have documented a

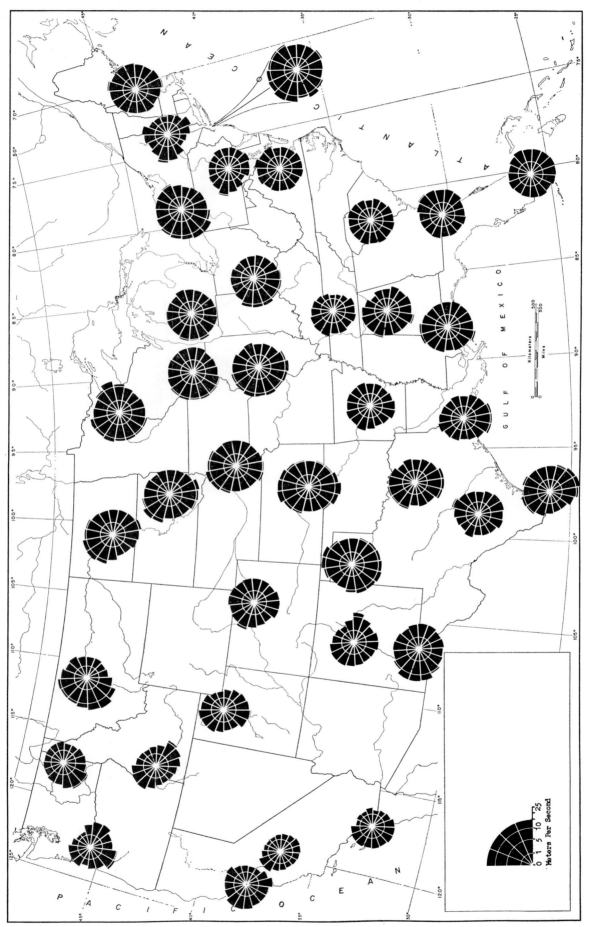

Fig.5. Average annual wind speeds, 1951–1960. Areas of sectors are proportional to hourly speeds. (From SLUSSER, 1965.)

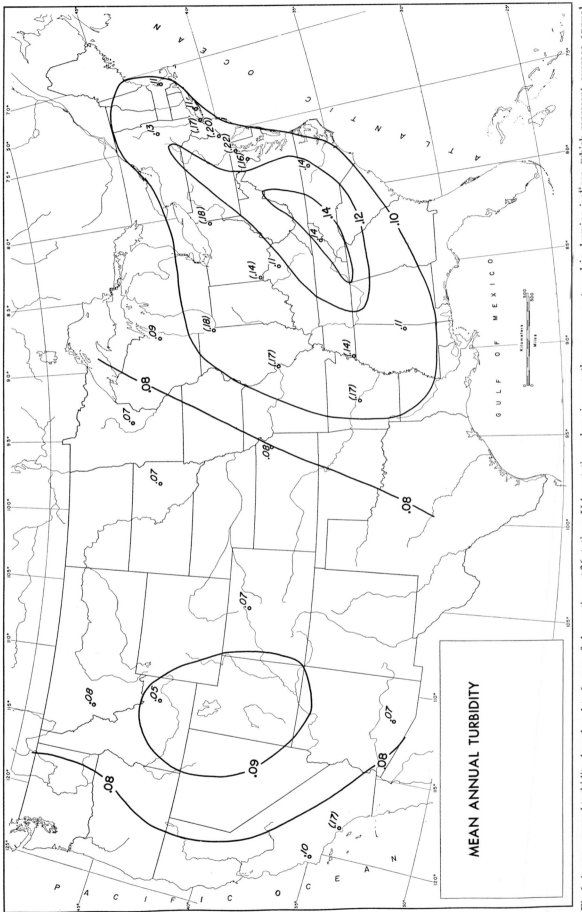

Fig.6. Average annual turbidity, based on 1 to 6 years of observations at 26 stations. Urban station values, in parentheses, were not used in drawing isolines, which represent average annual "background" turbidity. (After FLOWERS et al., 1969.)

20°C warming during four days in July 1960, in Pacific air which had subsided over and just to the east of the Willamette Valley of Oregon.

Atmospheric turbidity, the reduction of direct sunshine by dust, haze, and water vapor in the atmosphere, is greatest over the Appalachian Plateau and the Great Basin. In Fig.6, urban station values are placed in parentheses, and the isolines are based on only the rural and suburban station data so that the pattern represents the average annual background turbidity (FLOWERS et al., 1969).

Strong winds also cause turbidity by raising sand and dust into the troposphere; soil conservation practices have stabilized millions of square kilometers on the Great Plains which, desiccated by decreased precipitation during the mid–1930's, formed the notorious dust bowl. But strong winds blowing over dry fields still create dust storms. Mean monthly dust deposition at 15 sites on the Great Plains and in northeastern United States during 1963–1966 ranged from 17 to 460 kg/ha; the maximum was 4,200 kg/ha during May 1964 at Tribune, Kansas, near the center of the former dust bowl (BROWN et al., 1968).

The atmospheric pollution which forms in stagnant air may have surpassed fog as the primary cause of reduced visibility over parts of the United States. The percent of hours with visibility less than 5 km, based on hourly observations at 53 stations for 10 years ending in 1958 (ELDRIDGE, 1966), is greatest in all four seasons (Fig.7) in western Pennsylvania and around Los Angeles, California. Smoke, haze, and combinations of them were only one-fourth as frequent on January hourly observations at eight stations in 1965 as in 1945, apparently reflecting a reduction in coal usage (BEEBE, 1967). At three airports in east-central U.S. (Akron, Ohio; Lexington, Kentucky; and Memphis, Tennessee), daytime observations showing visibilities of 10 km (6 miles) or less were more than twice as frequent during three years, 1966–1969, as during the preceding three years, 1962–1965 (MILLER et al., 1972).

The mean number of days with fog, which also comes with light winds, based on reports from 251 stations for varying periods ending in 1960 (COURT and GERSTON, 1966), is greatest (60 days/year) over West Virginia, around Los Angeles and elsewhere on the Pacific coast, and also in New England (Fig.8). An essentially equivalent map, based on the same data, was prepared independently by PEACE (1969), who also delineated seven fog climatic regions: Pacific coast, north basin, south basin, Great Plains, Great Lakes, Appalachian, Atlantic and Gulf coasts.

## Moisture

Several thousand cubic kilometers of water are carried into the troposphere over the United States annually in the airflow from the south and west. Estimates of this flow can be made from the studies of BENTON and ESTOQUE (1954) and RASMUSSON (1967), respectively, for the calendar year 1949 and for 24 months, May 1961–April 1963. The difference between the estimated annual net inflow across the western and southern borders and the outflow into Canada and the Atlantic Ocean, in teratons (1 teraton = $10^{12}$ metric tons = $10^{18}$ g = $10^3$ km³ water), is:

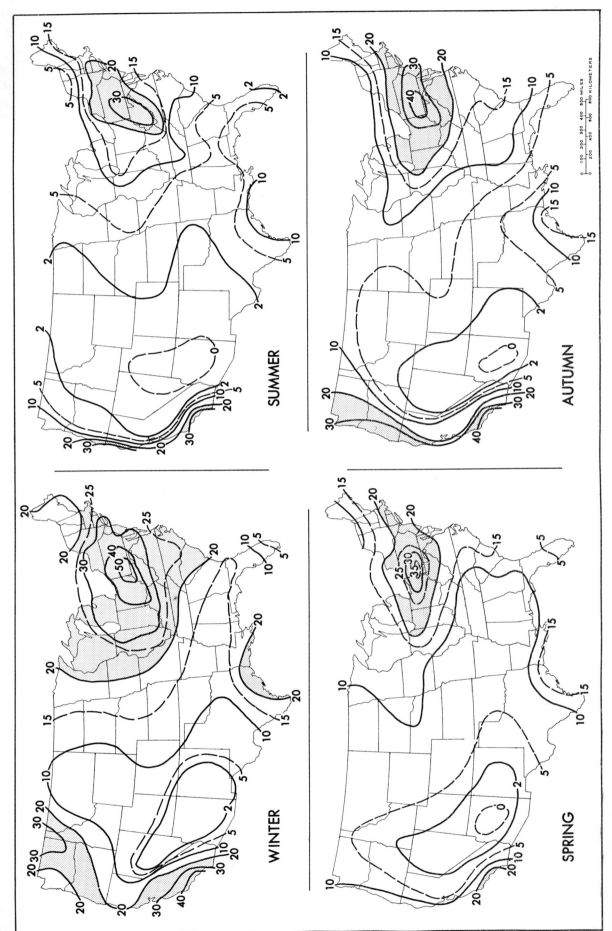

Fig.7. Average percent of hours with visibility less than 5 km, 1948–1958, by seasons. (From ELDRIDGE, 1966.)

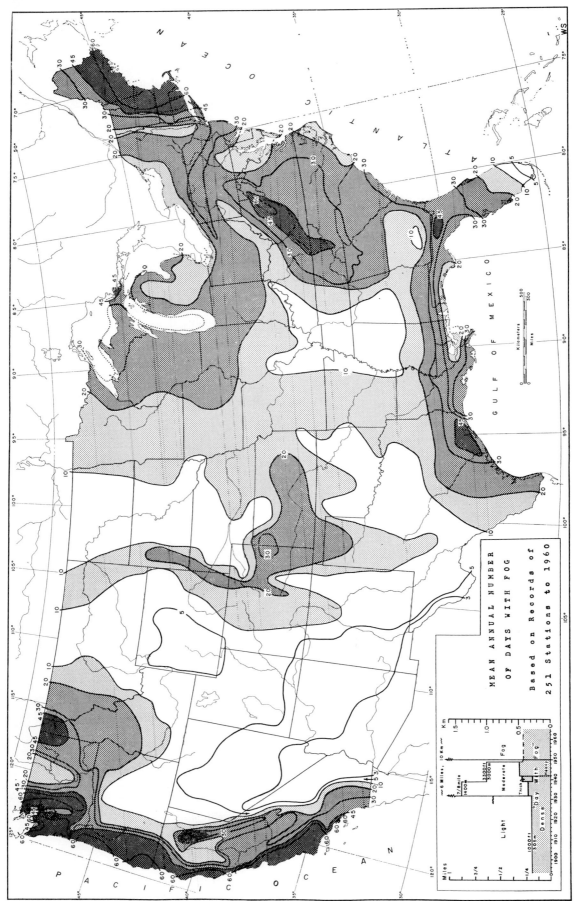

Fig.8. Mean annual number of days with fog (visibility less than 300 m), from records of 251 stations through 1960. (From COURT and GERSTON, 1966.)

1949    :    10 − 8   = 2
1961–62:   9.6 − 7.9 = 1.6
1962–63:   9.0 − 7.6 = 1.4

Thus, somewhat less than 2 teratons, about one-fifth of the total inflow, remains on the ground and must appear as streamflow (see p.232). Most of the inflow is across the Gulf coast: 6.7 teratons in 1949, 4.6 and 3.8 in 1961–62 and 1962–63.

Effects of this flow are shown (Fig.9) in the computed moisture content of the atmosphere up to 325 mbar (8.6 km) over 50 stations during 11 years, 1946–56 (REITAN, 1960a). Annual average of the computed liquid water equivalent, or "precipitable water," over the United States varied little from year to year, from 16.8 mm in 1956 to 18.3 mm in 1946 and 1949, because deficiencies in some regions were compensated by excesses in others. Mean monthly values (mm) for 1946–56 were:

| I | II | III | IV | V | VI | VII | VIII | IX | X | XI | XII |
|---|---|---|---|---|---|---|---|---|---|---|---|
| 9.8 | 9.4 | 10.5 | 13.7 | 18.9 | 24.9 | 29.9 | 29.0 | 22.8 | 17.2 | 12.0 | 11.1 |

The annual mean of 17.5 mm corresponds to 0.13 teratons over the conterminous United States, or as much as enters, on the average, in five days. Slightly lower annual values were found for three later years, 1964–66, from reports of only a dozen U.S. stations as part of a worldwide survey (TULLER, 1968).

This atmospheric moisture content in turn is reflected in the surface dew-points, which decrease northward in all months over the eastern two-thirds of the country, while

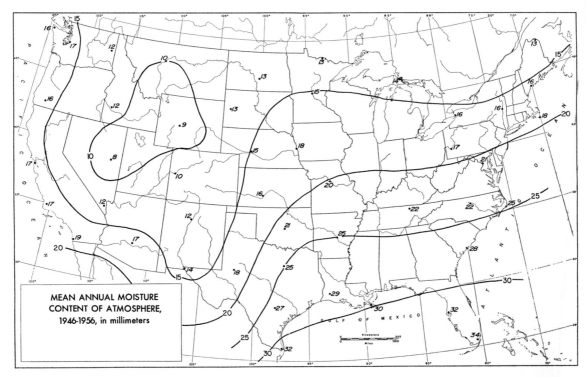

MEAN ANNUAL MOISTURE
CONTENT OF ATMOSPHERE,
1946-1956, in millimeters

Fig.9. Mean annual moisture content of atmosphere up to 325 mbar over 50 stations, 1946–1956, in millimeters of condensed water, or decigrams per square centimeter of earth's surface. (After REITAN, 1960a.)

over the western third any eastward decrease is masked by the rugged terrain (DODD, 1965). In January (Fig.10A) the dew-point decreases from 15°C in south Florida to −15°C and colder on the central Canadian border. In July (Fig.10B) the mean dew-point in southern Florida is almost 25°C, but the driest region is not on the northern border but in the western deserts, with dew-points of 2.5°C—and mean temperatures above 30°C.

About half of the water vapor flowing into the United States is condensed and precipitated as rain, snow, and hail in cyclonic storms, thunderstorms, and hurricanes (p.225). Annual precipitation (Fig.11) is greatest, therefore, along the Gulf coast, and on the western slopes of mountains along the Pacific coast. More than 200 cm falls annually on the mountains of Oregon and Washington, more than 100 cm in California, Idaho, and perhaps other western states, and at the ends of the Appalachian mountain chain near the east coast. At the other extreme, in southeastern California and adjacent parts of Nevada and Arizona the mean annual precipitation is less than 10 cm, but such mean precipitation values can be misleading.

Only in the wetter portions of the United States is the statistical distribution of annual rainfall symmetrical, and close to Gaussian. The coefficient of variation (ratio of standard deviation to mean) of annual precipitation, 1931–1960, at 220 first order stations (Fig.12) varied from 10% at Sault Ste. Marie, between lakes Superior and Huron, to more than 60% at Yuma in the southwest corner of Arizona; it was more than 20% over more than half the country (HERSHFIELD, 1962). Another representation of variability is provided (Fig.13) by the percent of years in which total precipitation was within 25% of the 50-year mean at 206 stations, found by evaluating gamma distributions (EAGLEMAN, 1968). Over the western half of the United States, the distribution of annual rainfall is so non-Gaussian that at many places the median annual rainfall, 1931–1960, is less than 75% of the arithmetic mean (Fig.14), and less than 25% at Yuma and nearby stations (SLUSSER, 1968).

Frequency of precipitation may be more important, for many applications, than amounts of precipitation. In the wettest regions of western Washington, almost 200 days annually have 0.25 mm (0.01 inch) or more, while in the southeast California desert the 10-year total of such days is not as great (Fig.15A).

Standard deviations of the number of days with 0.25 mm, at 169 first order stations, 1939–1958, are shown in Fig.15B. Cooperative (unpaid) observers, reading raingauges only once a day, tend to miss light showers; while their total amounts agree with those at first order stations, their reported frequency of rainy days is lower (VESTAL, 1961). Since 1954 the published records for cooperative stations are of days with more than 2.5 mm (0.10 inch) instead of the old value of 0.25 mm.

Rainfall amounts per rainy day (0.25 mm) may exceed 10 mm along the coasts, but in the interior are only 2.5 in winter and perhaps 7 mm in summer (JORGENSEN, 1967). Days with at least 12.7 mm (0.50 inch) produce 80% of the annual rainfall in wet (> 2,000 mm) regions, but less than 30% in very dry (< 25 mm) places (PAULHUS and MILLER, 1964). The frequency of days with 12.7 mm or more, based on 315 stations (1942–1961), and the ratio of the total rainfall on such days to the grand total, are shown in Fig.16.

The greatest contrast between the maps of days with 0.25 mm and with 12.7 mm is that the isopluves of the former are generally meridional in the Mississippi Valley, while the 12.7-mm day isopluves trend more zonally, as do the isohyets. The Appalachian Plateau,

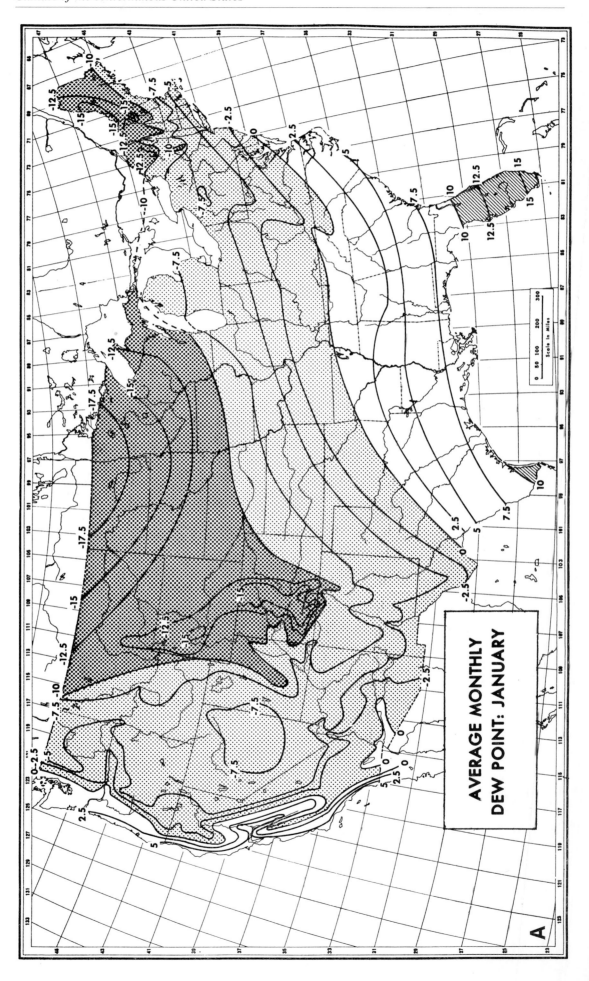

AVERAGE MONTHLY
DEW POINT: JANUARY

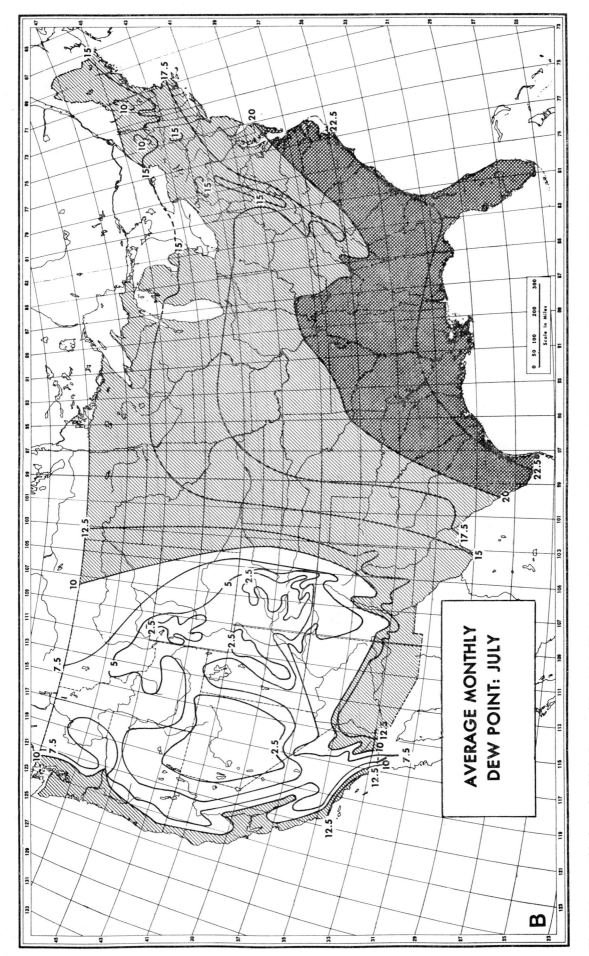

Fig.10. Mean monthly dew point (°C) in January (A) and July (B), based on observations every three hours at 200 stations during 10 years ending in 1959 or 1960. (Especially prepared by Dr. Arthur V. Dodd from data used for DODD, 1965.)

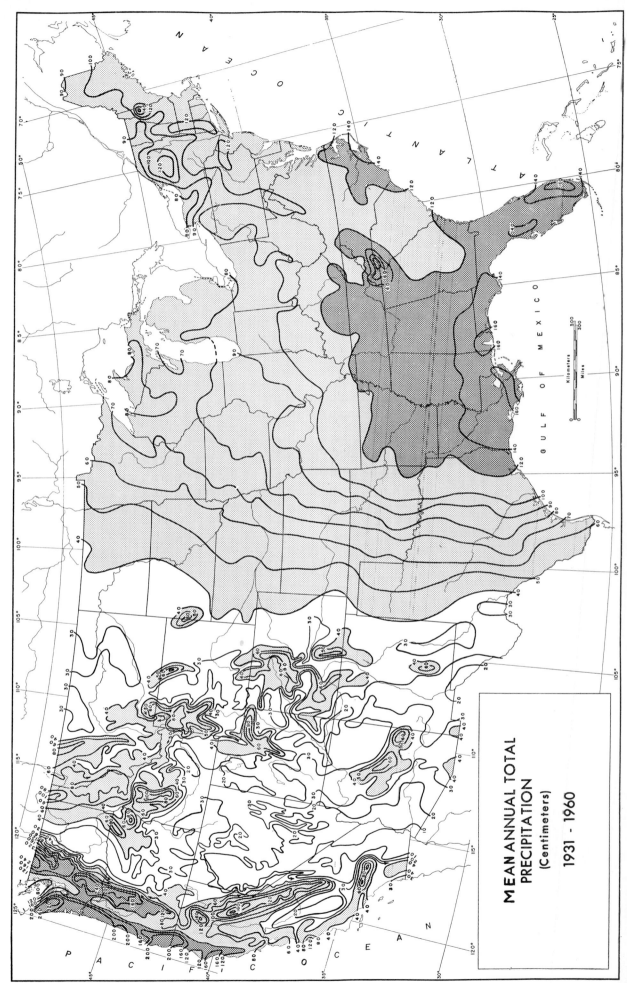

Fig.11. Mean annual total precipitation, based on all available stations adjusted to standard period 1931–1960. (After U.S. Weather Bureau.)

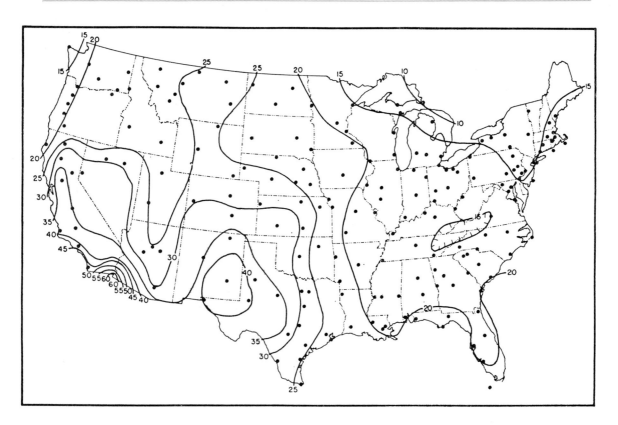

Fig.12. Coefficient of variation (%) of annual precipitation, 1931–1960, at 220 first-order stations. Station locations indicated by dots. (After HERSHFIELD, 1962.)

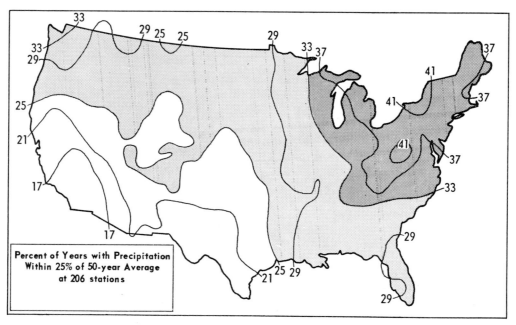

Percent of Years with Precipitation Within 25% of 50-year Average at 206 stations

Fig.13. Percent of years in which precipitation was between 75% and 125% of 50-year mean. Distributions of annual precipitation at each of 206 stations were fitted by gamma distributions, from which interval was obtained. (After EAGLEMAN, 1968.)

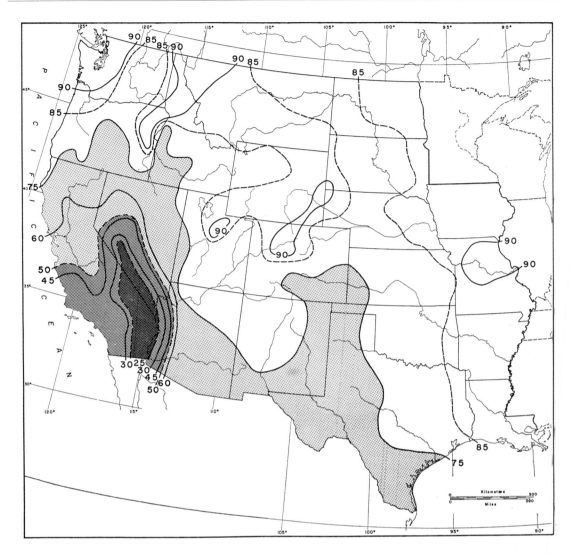

Fig.14. Average ratio (%) of median precipitation to mean precipitation, obtained as average of 12 monthly ratios for 219 stations west of Mississippi River, 1931–1960.(From SLUSSER, 1968.)

in Pennsylvania and West Virginia, has 170 rainy days (0.25 mm) annually, although precipitation there is not much greater (1,000 mm) than in nearby areas with only two-thirds as many rainy days. This is primarily the consequence of humidification of an airstream from the northwest as it passes over the Great Lakes in autumn and winter.

**Precipitation**

June is the wettest month over most of the Great Plains and the upper Mississippi Valley, January is wettest on the Pacific coast, and August has most rain in New Mexico and eastern Arizona, and also in Florida and along the coast to the northeast (Fig.17A). Other portions of the country have no dominant wet season; MARKHAM (1970) expressed the degree of seasonal concentration of precipitation (Fig.17B), analogous to the classic measure of wind constancy, as the ratio of the resultant of 12 monthly precipitation values, considered as vectors on an annual circle, to the (scalar) total annual precipitation. This ratio, varying from 0 to 100%, gives a more graphic representation of the seasonal distribution of rainfall than the classic harmonic analysis, with separate maps of amplitude and phase angle for each harmonic (HORN and BRYSON, 1960).

The winter maximum on the Pacific coast progresses irregularly southward, from early December in Washington to February (in some places late December) in California (PYKE, 1971). It is caused by the southward movement of the band of maximum westerly flow in the troposphere, along which polar maritime air from the Gulf of Alaska encounters moist tropical maritime air from the Pacific Ocean around Hawaii.

Most of the coastal rainfall is released by a train of waves on a front trending SW–NE across the coast. These storms bring rain and snow as far east as the Great Basin, where in 1962–63 they contributed two-thirds or more of the annual rainfall (HOUGHTON, 1969). This percentage decreases to the southeast, and in Arizona and New Mexico winter rain is overshadowed by that from summer thunderstorms. These occur in a flow of moist air from the Caribbean and Gulf of Mexico, arriving around the end of June (BRYSON and LOWRY, 1955); rains are most likely when the easterly flow in the stratosphere decelerates (JETTON et al., 1971).

About one-fourth of the meager rainfall in the southeastern California deserts comes from, or is associated with, tropical storms from the south, usually in August, September or October. Even if the storms move northwestward over the Pacific Ocean off Baja California, the moisture from them flows northward and feeds desert showers (HARRIS, 1969).

Winter rains and snows are the chief water supply to the Colorado River Basin ($6.3 \cdot 10^5$ km$^2$), from which much of southern California and Arizona derive water. But runoff, chiefly in spring and summer, is only about 20% of the winter moisture accumulation computed from the atmospheric water balance. This accumulation is almost twice that estimated from the available raingauges, none of which samples the 3% of the basin above 3.3 km (RASMUSSON, 1970). Raingauges at lower elevations show a linear increase of about 20 cm per kilometer increase in elevation (MARLATT and RIEHL, 1963).

Over the Great Plains and the Mississippi Valley, most of the precipitation comes from cyclonic storms. These form on cold fronts which push southeastward from Canada or cross the mountains from the west and regenerate in their lee, usually in or near southeastern Colorado. Over the 12 states from the Dakotas and Kansas to Ohio, the weeks least likely to be dry and most likely to have 2.5 mm or more of rainfall were in April, May or early June, while the weeks most likely to be dry and least likely to be wet were in late October, mid-November and mid-December (LANDSBERG, 1960a).

At 10 stations from Kansas, Nebraska and South Dakota to Indiana, 83% of the summer precipitation during five years (1952–56) was classified as frontal by RUDD (1961). One-third of all precipitation periods at four Illinois stations, 1944–53, came with cold fronts, another third with warm or stationary fronts (HISER, 1956). The frontal effect was somewhat less pronounced in southern Illinois, and in summer.

Most of the eastern third of the United States has no dominant month or season of precipitation. For the southeastern quarter of the country, from Oklahoma to Virginia and Florida, CROWE (1951) identified three major regions: a Plains borderland, from central Texas to Illinois, with "most rainfall in summer, especially early summer"; the Mississippi lowlands, from east Texas to Alabama and Kentucky, with more than half

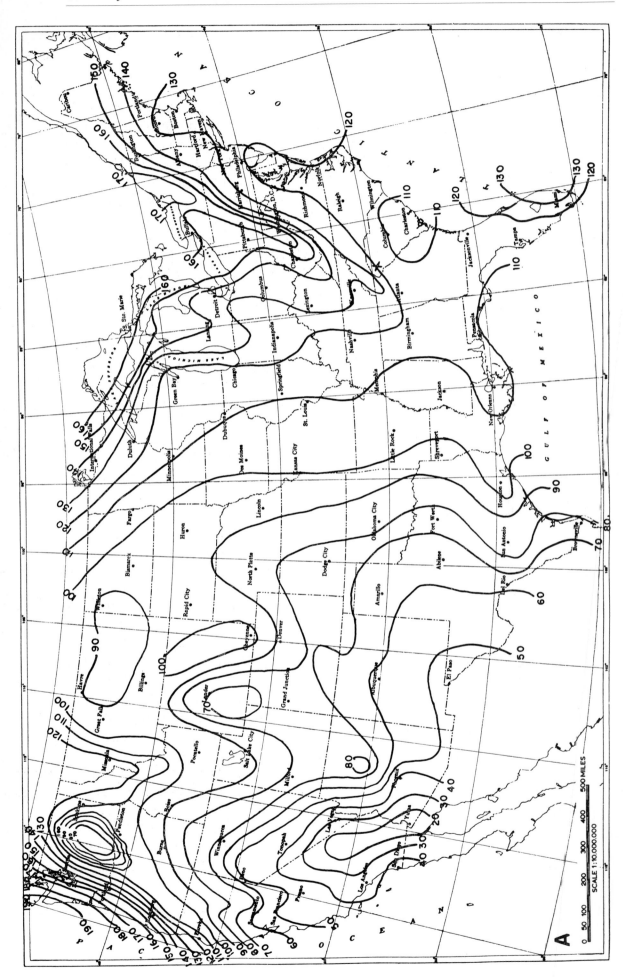

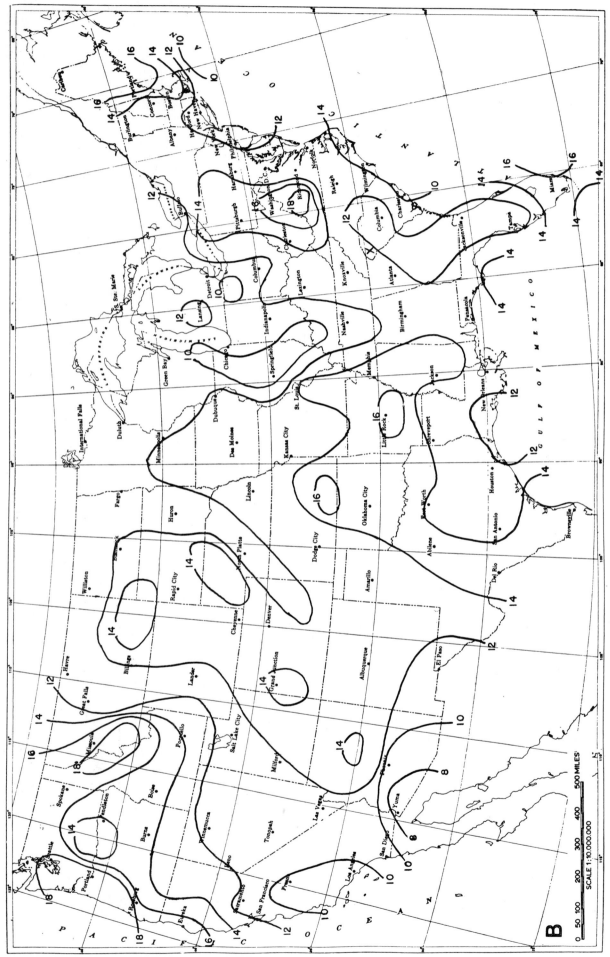

Fig.15. Mean (A) and standard deviation (B) of annual number of days with precipitation of 0.25 mm or more, 1931–1960. (From Environmental Data Service, NOAA.)

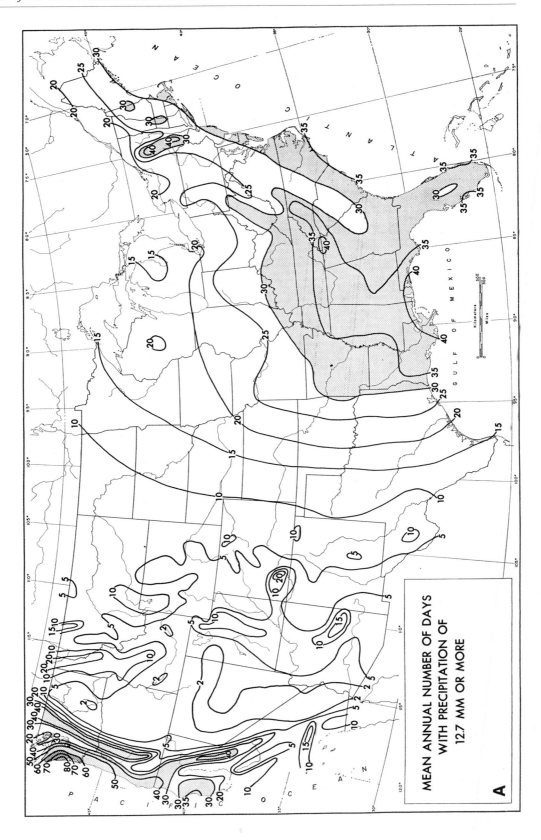

MEAN ANNUAL NUMBER OF DAYS
WITH PRECIPITATION OF
12.7 MM OR MORE

A

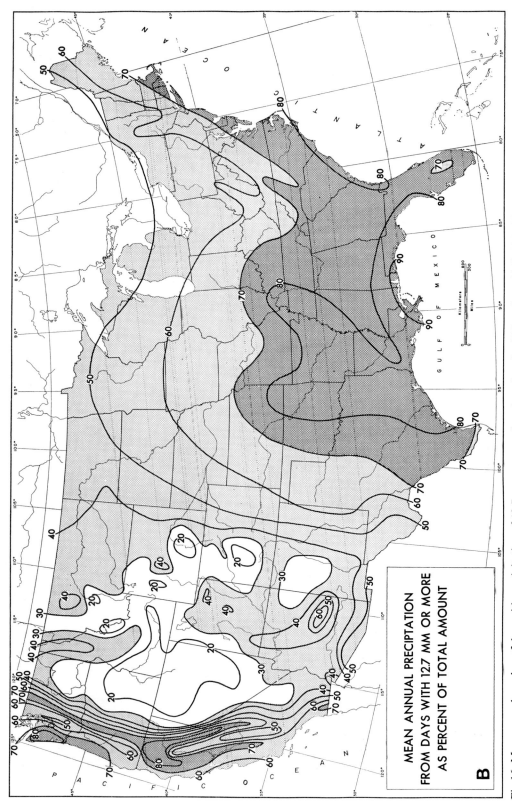

Fig.16. Mean annual number of days with precipitation of 12.7 mm or more, based on data for 315 stations, 1942–1961 (A), and ratio (%) of total annual rainfall on such days to total from all days (B) based on data for 315 stations, 1942–1961. (From PAULHUS and MILLER, 1964.)

MEAN ANNUAL PRECIPTATION
FROM DAYS WITH 12.7 MM OR MORE
AS PERCENT OF TOTAL AMOUNT

B

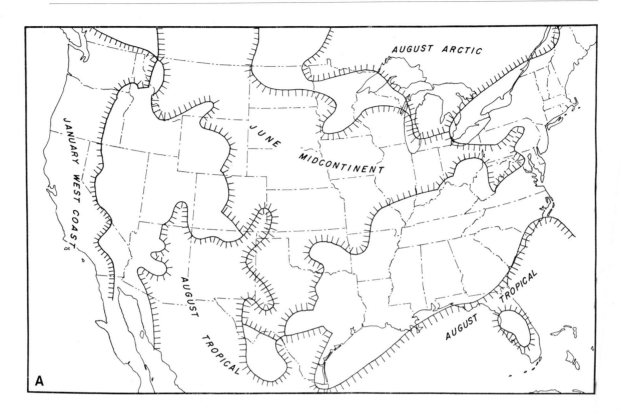

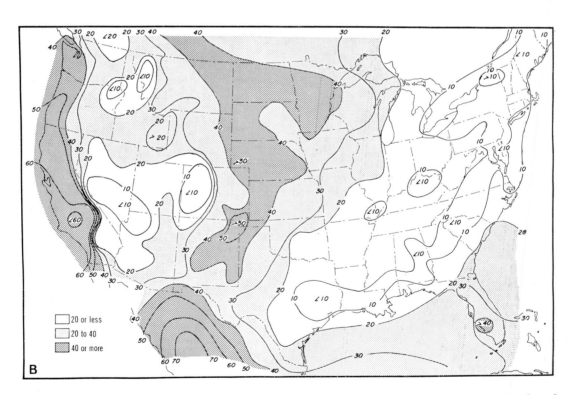

Fig.17. Months of dominant concentration of precipitation. A. Degree of seasonal concentration of precipitation, 1951–1960. B. Figures express vector sum of monthly precipitations (months 30° apart around circle) as percent of scalar sum. (From MARKHAM, 1970.)

218

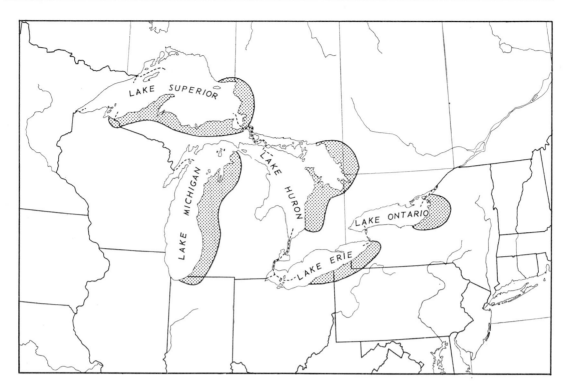

Fig.18. Snowbelts around the Great Lakes. (After EICHENLAUB, 1970.)

the annual precipitation in winter; and the Atlantic coastlands, from southern Louisiana to eastern Virginia and south to Florida, with a midsummer maximum of rainfall. The plains region is the western margin of the tropical air inflow to the country, but causes of the rainfall regimes were not discussed.

The winter preponderance of rainfall in the Mississippi lowland region has been discussed by TREWARTHA (1961). Unusually wet and dry months over the seven southeastern states from Louisiana to Virginia (excluding Florida and Tennessee) are more closely related to anomalous positions and amplitudes of Rossby troughs in the westerly circulation than to surface conditions or tropical storms, FLETCHER (1971) declared.

In northeastern United States, the heaviest rains in all seasons occur in the onshore flow of air in an offshore cyclone, tropical or extratropical. Such cyclones form in the vicinity of Cape Hatteras (35°N), often as new developments on secondary fronts on which the previous cyclone dissipates over the Appalachian Plateau. When such a storm moves northeastward just offshore, or forms farther north, around Cape Cod, it can bring into New England an infamous "back door" cold front and its attendant "northeaster" storm. In winter such storms cause heavy snowfall throughout New England, closing roadways and disrupting communications.

The Great Lakes modify greatly the climate of surrounding regions, especially downwind, to the south and east (CHANGNON and JONES, 1972). The Lakes cover 245,000 km² and contain 24,000 km³ of fresh water (BUE, 1963), the largest such reservoir in the world—four times the annual streamflow of the United States. Over Lake Michigan (59,000 km²) precipitation is about 6% less than that (80 cm) over the surrounding land (CHANGNON, 1968), the deficiency being greatest in spring and summer when the lake is much colder

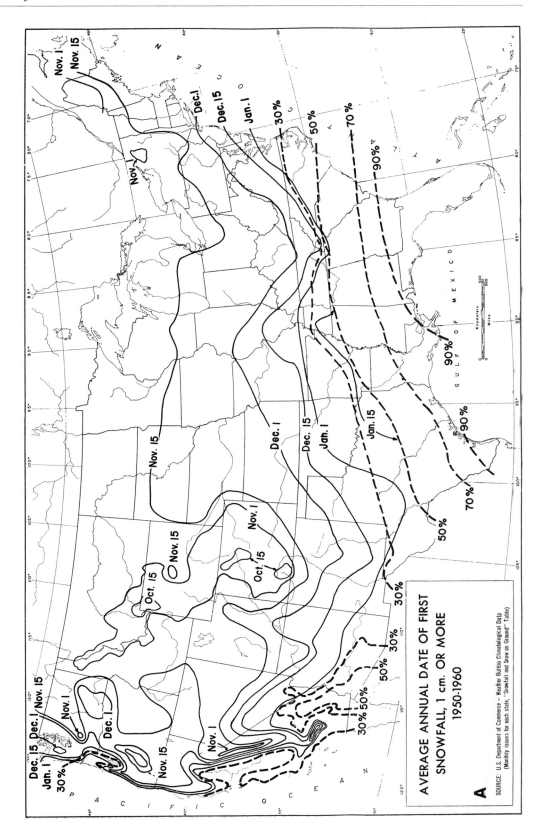

AVERAGE ANNUAL DATE OF FIRST
SNOWFALL, 1 cm. OR MORE
1950-1960

**A**

SOURCE: U.S. Department of Commerce – Weather Bureau Climatological Data
(Monthly issues for each state, "Snowfall and Snow on Ground" Table)

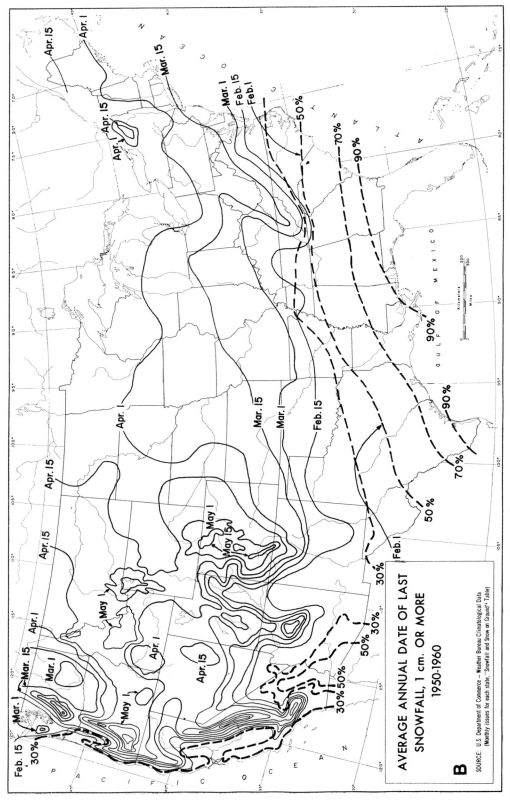

Fig.19. Average annual date of first (A) and last (B) snowfall of 1 cm or more, 1950–1960. Dashed lines give percent of years without snowfall. (Prepared by Donald Larson.)

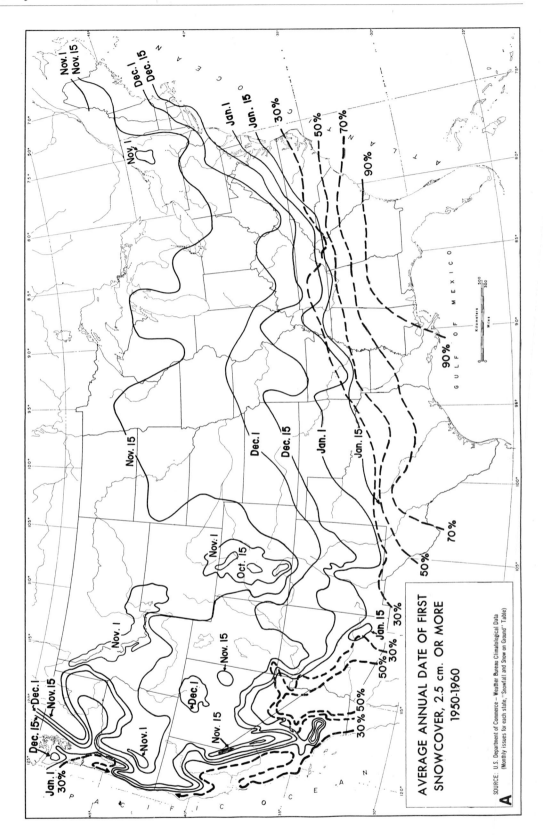

AVERAGE ANNUAL DATE OF FIRST SNOWCOVER, 2.5 cm. OR MORE 1950-1960

SOURCE: U.S. Department of Commerce—Weather Bureau Climatological Data (Monthly issues for each state, "Snowfall and Snow on Ground" Table)

A

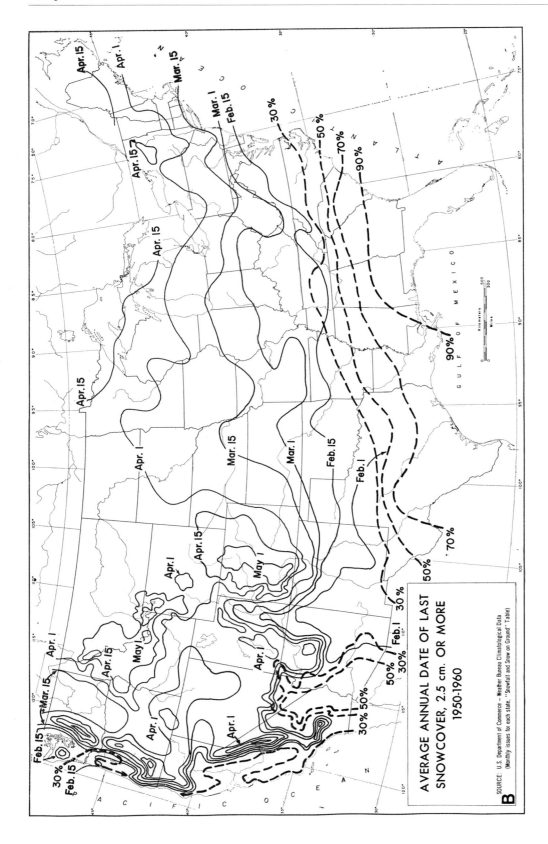

Fig.20. Average annual date of first (A) and last (B) snowcover of 2.5 cm or more, 1950–1960. Dashed lines give percent of years without snowcover. (Prepared by Donald Larson.)

than the land. The reality of this deficiency, and of similar differences over the other lakes, has been argued extensively.

Snowbelts extend along the east sides of the five lakes (Fig.18), with seasonal snowfalls as great as 4 m, three to four times that in areas beyond the lake effect (MULLER, 1966; EICHENLAUB, 1970). On the Atlantic seaboard, the ocean has the reverse effect: snow occurs on 25% of all precipitation days during winter (December–March) on Long Island, east of New York City, but on 40% of such days west and north of the city, only 60 km away (GRILLO and SPAR, 1971). Heaviest snowfalls are in the northwestern mountains, where Paradise Ranger Station, at 1,700 m on the south slopes of Mt. Rainier, received 28.50 m of snow in 1971–72, the greatest total ever reported in the United States.

Average dates of the first and last snowfalls for the entire country (1950–60) are given in Fig.19, which also shows the percentage of years with no snowfall whatever in the southern United States. Similar maps show dates of the first and last snowcover (Fig.20). Across the northern half of the country, snow removal from city streets and rural highways costs several hundred million dollars annually (ROONEY, 1967). Melting of the snow in March, April, and May saturates the soil, replenishes groundwater, and makes streams and rivers bankfull, causing damaging or disastrous floods in some parts of the country almost every year.

Freezing rain, while not of particular importance as water supply to plants or man, is very harmful to man's works, especially communications by road, rail, air, and wire (BENNETT, 1959). Much of northcentral and northeastern United States has an average of more than 6 days per year with freezing rain, and in western Minnesota and southeastern Michigan the 10-year average (1939–48) was more than 15 such days (Fig.21).

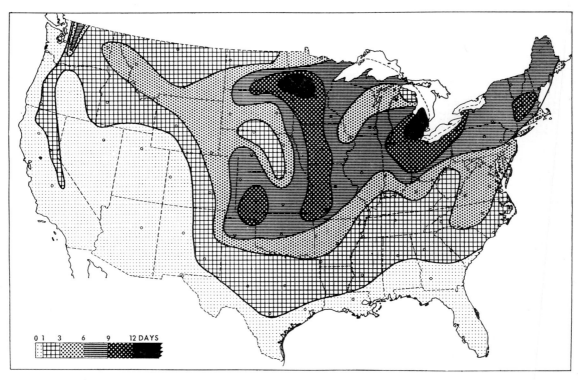

Fig.21. Mean number of days with freezing precipitation, 1939–1948. (From BENNETT, 1959.)

**Storms**

*Tropical storms* cause an August maximum in precipitation in the southeastern United States. Of the 299 tropical cyclones, including hurricanes, identified in the North Atlantic during 30 years (1931–60), a little more than half (167) affected directly the United States (CRY, 1967). In the ensuing decade (1961–70), the U.S. coast was reached by 14 of 48 North Atlantic tropical storms (9 of 30 hurricanes), killing 533 persons (256 in 1969 alone) and causing millions of dollars in damage (CONDON, 1971). Despite their devastation, such storms have beneficial effects, and are necessary parts of the general circulation (LANDSBERG, 1960). (See p.36.)

Along the immediate Gulf and Atlantic coasts, tropical storms were responsible for more than 15% of the total precipitation during the 5-month hurricane season, June–Novem-

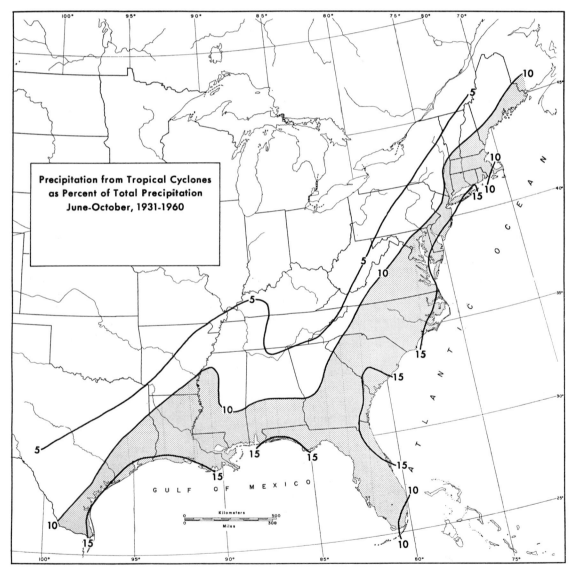

Fig.22. Percent of total precipitation, June–October, attributed to tropical cyclones, 1931–1960. (From CRY, 1967.)

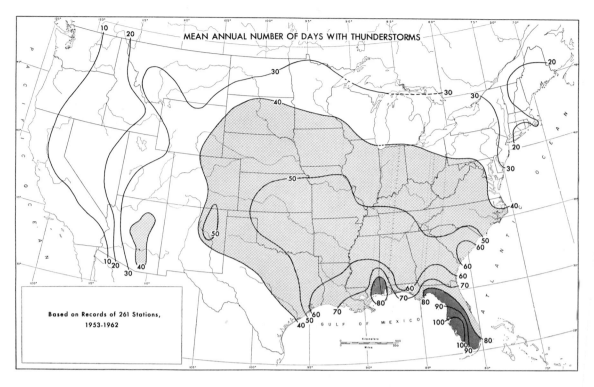

Fig.23. Mean annual number of days with thunderstorms, based on data from 261 stations, 1953–1962. (Prepared by Robert Maples.)

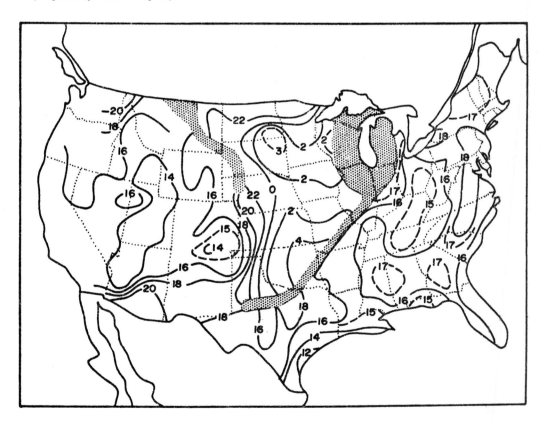

Fig.24. Hour (local standard time) of maximum frequency of thunderstorms, June–August. (After RASMUSSON, 1971.)

ber (Fig.22). Even 500 km inland, the contribution to the seasonal rainfall was about 5%. Most of the areas receiving rainfall from tropical storms were already moist to wet (CRY, 1967), but about once every three years (1925–55) a moderate to severe drought was broken by such rainfall (SUGG, 1968).

*Thunderstorms* supply summer rainfall over much of the United States, especially the southeast, but no assessment of their total contribution can be made. Some thunderstorms are parts of wave cyclones on fronts between moist tropical air and cool Pacific or Arctic air, and some form in the circulation of hurricanes.

Thunderstorms cannot be identified individually to be tabulated climatically, and so are catalogued only by the number of "days with thunder heard" at first-order observing stations. Based on records of 261 such stations, 1953–62, the mean annual number of days with thunderstorms (Fig.23) is 100 or more in southwest Florida; about half the country has more than 40 per year. In the southern Rocky Mountains, the number of days with thunder apparently increases linearly with the altitude of the reporting station, perhaps the result of a greater range to which thunder can be heard. (This bias is responsible for the questionable secondary maximum of thunderstorm activity in northern New Mexico, exaggerated on some maps.)

The largest and most intense thunderstorms appear to occur in and around Kansas and Oklahoma, where they penetrate the tropopause most frequently to bring moisture into the lower stratosphere (LONG, 1966). Over most of the United States, summer thunderstorms occur primarily in the afternoon, but on the Great Plains, from Oklahoma to the Canadian border, they are most frequent (Fig.24) around and after midnight (RASMUSSON, 1971), perhaps associated with a low-level jet. Thunderstorms tend to form in the lee of the Rocky Mountains in late afternoon and propagate eastward during the evening.

*Lightning* killed an average of 137 persons annually during seven years (1959–65), but these deaths were not as concentrated (Fig.25) as the thunderstorms producing the lightning, perhaps because thunderstorms occur mostly at night in their region of greatest frequency. "About 70% of all injuries and fatalities occur in the afternoon, 20% between 6 p.m. and midnight, and approximately 10% between 6 a.m. and noon. Only about 1% occur between midnight and 6 a.m." (ZEGEL, 1967). Annual lightning mortality exceeded 2.5 per million persons in Vermont, Maryland, Arkansas, Wyoming, and New Mexico. In Iowa, lightning strikes average about 10 per square kilometer annually, two-thirds of them from May to August; of 25 persons killed in nine years, 1959–67, 12 were men driving tractors and only two were females (WAITE, 1968).

*Hail* sometimes falls from severe thunderstorms but the correspondence between the two phenomena is difficult to establish because of differences in reporting procedures, and lack of any objective criterion for defining thunderstorm severity other than production of hail and lightning. Hail comes in narrow swaths, and only a tiny fraction of all occurrences occurs and is recorded at regular reporting stations. But hail often damages buildings and crops and so is reported by the populace, especially as insurance claims on crops—until after the harvest.

The "point intensity of hail," derived from the ratio of insurance claims to total insurance in force, 1957–64, "near the Rocky Mountains is much greater than in the northwest, Great Plains, and Middle West" and "decreases rapidly and steadily eastward from the Rocky Mountains"in July and August, the months of peak intensity and

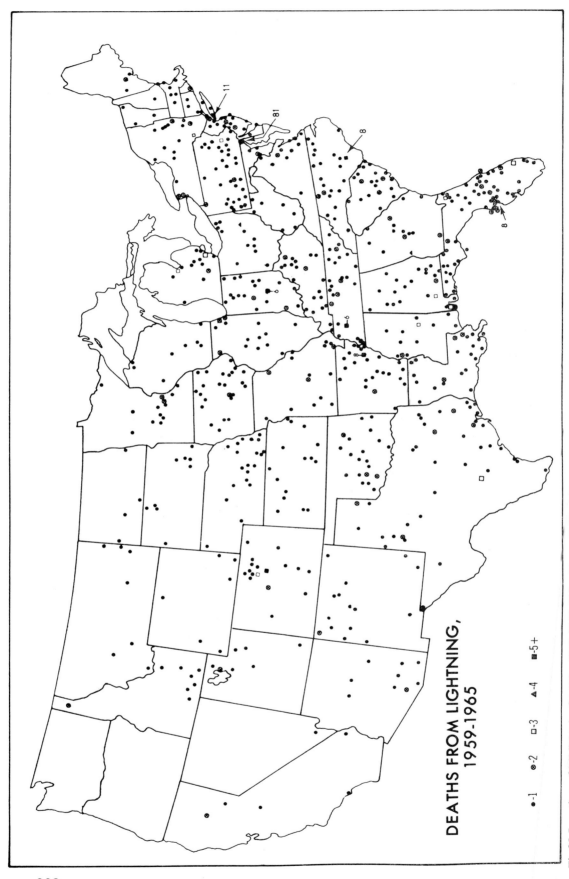

DEATHS FROM LIGHTNING,
1959-1965

Fig.25. Locations of 960 lightning fatalities, 1959–1965. (After ZEGEL, 1967.)

crop damage (CHANGNON and STOUT, 1967). However, "Damaging storm days are most frequent, on the average, in Kansas, Nebraska and the Dakotas… In the mountain states where hail intensity at a point is greatest, storm days are less frequent than in the Great Plains."

These point intensities were not mapped in detail; instead STOUT and CHANGNON (1968) prepared an atlas showing the distribution, for each month and the year, 1901–60, of hail reports from 1,285 stations in 17 states, and 21 first-order stations in 7 additional states to the southeast. Their annual map (Fig.26) shows 27 relative maxima, of which the western 10 were identified as caused by mountain effects, 4 as related to the Great Lakes, and the remaining 12 as arising from "macroscale weather" processes.

This distribution of hail of all sizes is quite different from the map (Fig.27) of the areal

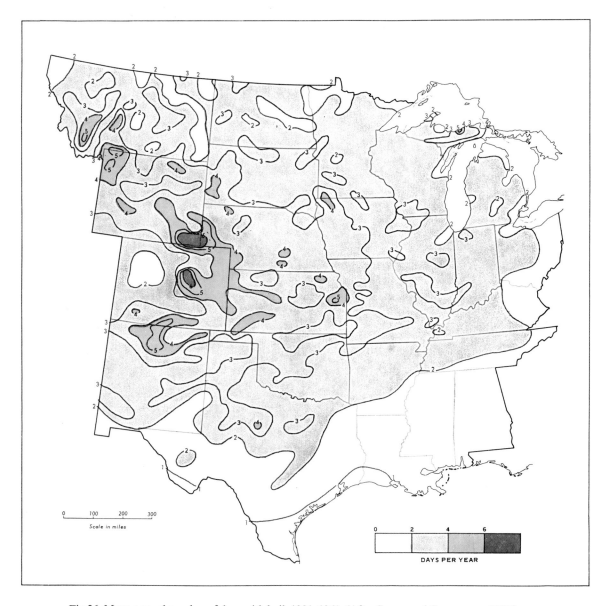

Fig.26. Mean annual number of days with hail, 1901–1960. (After STOUT and CHANGNON, 1968.)

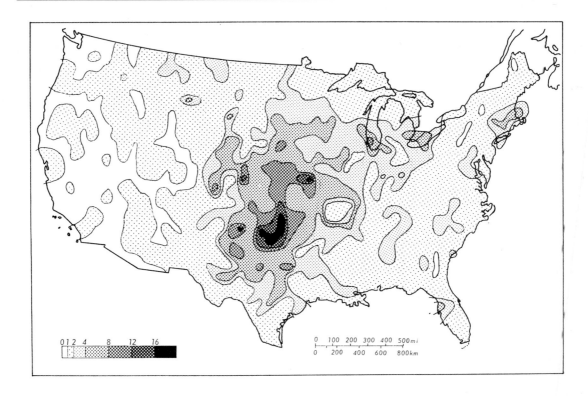

Fig.27. Mean annual incidence of large hail (>19 mm) per 26,000 km², 1955–1967. (Prepared by M.L. Swift from data of PAUTZ, 1969.)

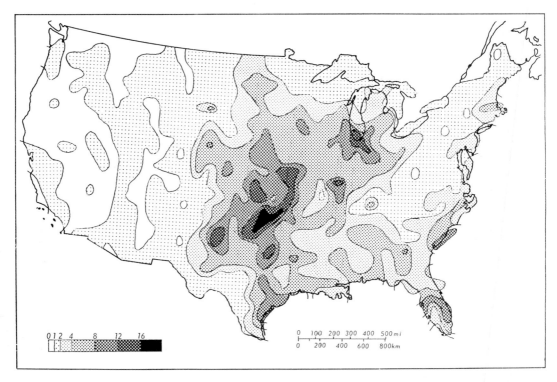

Fig.28. Mean annual incidence of tornadoes per 26,000 km², 1955–1967. (Prepared by M.L. Swift from data of PAUTZ, 1969.)

incidence of hail, 19 mm (3/4 inch) or more in diameter during 15 years, 1953–67, compiled from reports from observers of all kinds, including newspaper accounts and aircraft pilots (PAUTZ, 1969). From Oklahoma and Kansas, more than 10 reports were received annually per 10,000 km² (16 per 10,000 sq. miles).

*Tornadoes*, the most violent of all meteorological phenomena, occur more frequently in the center of the United States than anywhere else in the world. As with hail, tornadoes are tabulated after the fact, on the basis of reports from local residents and damage inspections by meteorologists. Adequate reporting of either phenomenon, therefore, requires fairly dense settlement, and good communications. Until a quarter-century ago, hail and tornado compilations were far from adequate (COURT, 1970).

Tabulations for recent years (PAUTZ, 1969) show "tornado alley" extending diagonally across the country from southwestern Texas to northern Illinois (Fig.28), approximately the northwest side of the prevailing flow of tropical air, and its low-level jet (Fig.2). Maximum annual incidence is 20 tornadoes per 26,000 km²; the average tornado damages about 7.5 km², so even in the middle of tornado alley only about 0.6% of the country-side is damaged annually. During 18 years (1953–70), tornadoes killed about 116 persons annually. Tornadoes are least frequent in winter, when the relatively few that do occur strike the Gulf coast. They increase in number and intensity during spring as the region of maximum incidence moves northwestward.

*Blizzards* are another kind of strong winds, afflicting the northern Plains and upper Mississippi Valley in winter. Blizzards are combinations of cold temperatures, strong winds and blowing or drifting snow: $< -7°C$ and $> 15$ m/sec with moderate snow for ordinary blizzards, $< -12°C$ and $> 20$ m/sec with heavy snow for severe ones. They occur when a mass of Arctic air sweeps southeastward at 20 or 30 m/sec, blowing dust from plowed fields and snow which falls from clouds in the warmer tropical air forced upward by the sharp cold front. Such "cold waves" bring rapid cooling, as much as 25° or even 30°C in less than a day, especially when the previous days have been unseasonably warm because of a persistent flow from the south.

*Chinooks* can cause similar wide oscillations in temperature just east of the Rocky Mountains, and in some intermountain basins. Similar to the classic foehn of the Alps, the chinook is warm and very dry, through addition of latent heat from precipitation on the windward side of the mountains and subsequent compressional heating. It occurs several times each winter at various places in the "chinook belt" 300–400 km wide just east of the Rocky Mountains, from Alberta (Canada) through Montana and eastern Wyoming and Colorado to northeastern New Mexico (GLENN, 1961). In Boulder, Colorado, 76 chinooks occurred in 63 years (1906–68), with 22 in January, but usually a winter has either more than one chinook, or none (JULIAN and JULIAN, 1969).

When the chinook reaches the ground just east of the mountains, temperature rises 10°–30°C, humidity decreases, and snow evaporates quickly. If the winds raise off the ground, cold air returns from the east and temperature drops rapidly. In some cases, chinook and local winds alternate at intervals of minutes or hours, causing spectacular swings of the thermograph.

**Evaporation**

All the rain, snow, and hail produces an annual fall of about 6 teratons of water upon the United States, representing a nation-wide average precipitation of 76 cm. This is more than 40 times the estimated average water content of the atmosphere. Somewhat more than one-fourth (28%) of the precipitation appears as streamflow (MURRAY, 1965). The estimated total annual runoff (1931–60) of 1.6 teratons is gratifyingly close to the net divergence of water vapor flux, estimated at slightly less than 2 teratons (Table I).

TABLE I

ANNUAL WATER BALANCE ESTIMATES FOR THE CONTERMINOUS UNITED STATES ($7.8 \cdot 10^6$ km²)

|  | g/cm² | teratons/ year* | gigatons/day** |
|---|---|---|---|
| Water vapor inflow | 123 | 9.6 | 26 |
| Water vapor outflow | 101 | 7.9 | 22 |
| Water vapor divergence | 22 | 1.7 | 4 |
| Water content of atmosphere | 1.7 | 0.13 | 130 |
| Total precipitation | 76 | 5.9 | 16 |
| Total evaporation | 55 | 4.3 | 12 |
| Net streamflow | 21 | 1.6 | 4 |
| Insolation (400 ly/day) as water evaporated | 252 | 19.6 | 53.8 |
| Lake evaporation rate | 110 | 8.6 | 24 |
| Potential evapotranspiration | 130 | 10.1 | 28 |

* 1 teraton $= 10^{12}$ tons $= 10^{18}$ g $= 10^3$ km³ water
** 1 gigaton $= 10^9$ tons $= 10^{15}$ g $= 1$ km³ water

Water may recycle several times between air and ground in transit across the United States. Total precipitation (6 teratons) is more than half of the total inflow of water vapor (10 teratons), but almost three-fourths of it evaporates, leaving only 1.6 teratons, perhaps a sixth of the vapor inflow, as liquid outflow in rivers.

This streamflow is distributed very unevenly in both time and space. The annual variation in runoff for various regions is shown in Fig.29, based on the standard "water year" beginning with October. Maximum streamflow comes in February and March in southeastern United States, March and April in the northeast and northcentral, April and May in much of the Mississippi Valley, and May and June in the Rocky Mountains and much of the Pacific coast (BUSBY, 1963).

In the Great Basin and the desert southwest, all precipitation evaporates before it can reach the ocean, making the evaporation/precipitation ratio 100%. In the mountains of Washington and New England, the ratio is less than 10%. Ratios computed by RASMUSSON (1971) for a 5-year period (1958–63) were 84% for the Central Plains ($4.2 \cdot 10^6$ km²) and 61% for the eastern region ($2.2 \cdot 10^2$ km⁶), including the Great Lakes.

Evaporation from lakes cannot be measured directly, except in a few cases, and is esti-

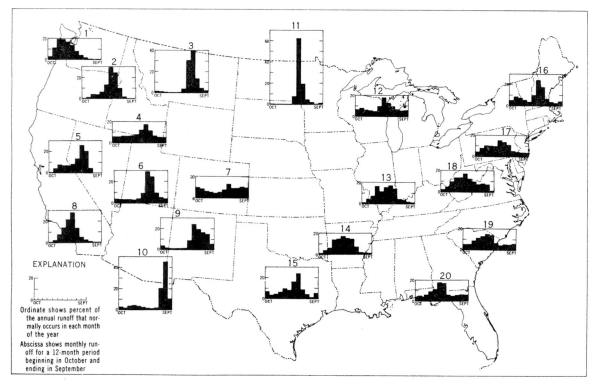

Fig.29. Normal monthly distribution of run-off on 20 representative streams, 1931–1960. (From Busby, 1963.)

mated from the water loss of special evaporation pans, of which several types are in use. Maximum pan evaporation during 12 months (1958–59) was 4 m in Death Valley, California (Robinson and Hunt, 1961). Under average weather conditions, a "Class A" pan loses about 43% more water than a lake or reservoir (Kohler et al., 1955).

Lake evaporation, estimated from observed pan evaporation using corrections derived for each location (Fig.30), is greatest in the southwest deserts (210 cm) and least in the extreme northwest and northeast corners of the country (50 cm). Actual evaporation from Lakes Superior (83,000 km²) and Ontario (19,000 km²) was computed from the difference between inflow and outflow as 52 and 81 cm, respectively (Morton, 1967), slightly more than would be estimated from the map based on evaporation pans. Slightly smaller values were found through a modified mass transfer computation by Richards and Irbe (1969): Superior 46 cm; Huron and Ontario 72 cm; Erie 91 cm.

For the conterminous United States as a whole, lake evaporation averages about 110 cm, just twice the difference between the annual precipitation (76 cm) and runoff (21 cm). The country's water budget, therefore, resembles that of an otherwise completely dry, non-evaporating region half-covered by lakes, which receive all the precipitation. In actuality, lakes and other water bodies are minor contributors to the total evaporation; most of it is from trees and crops as part of their growth mechanism, and represents the major waste output in the production of food and fiber.

Pan evaporation is an indicator of potential evapotranspiration, the maximum water loss from soil and plants adequately supplied with water. From application of a new formula for net radiation to an adaptation of Penman's evaporation formula, Chang and Okimoto (1970) estimate potential evapotranspiration for the country as 134 cm,

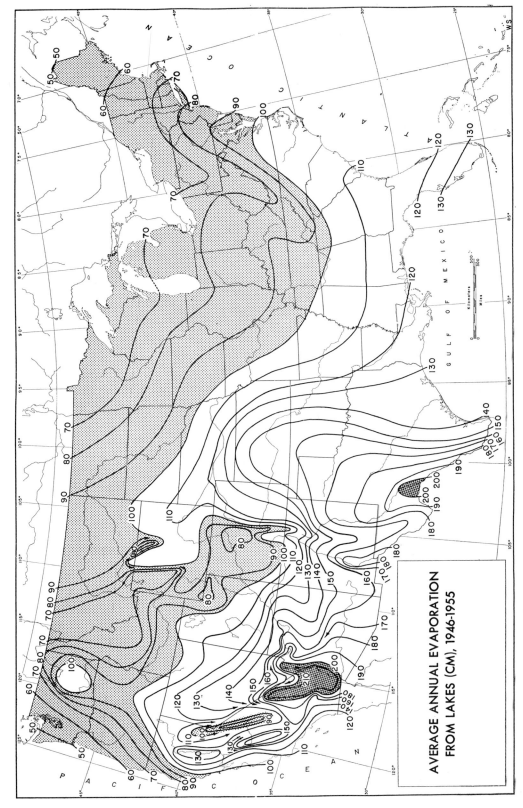

Fig.30. Average annual evaporation (cm) from lakes, 1946–1955, derived from pan evaporation records. (After KOHLER et al., 1955.)

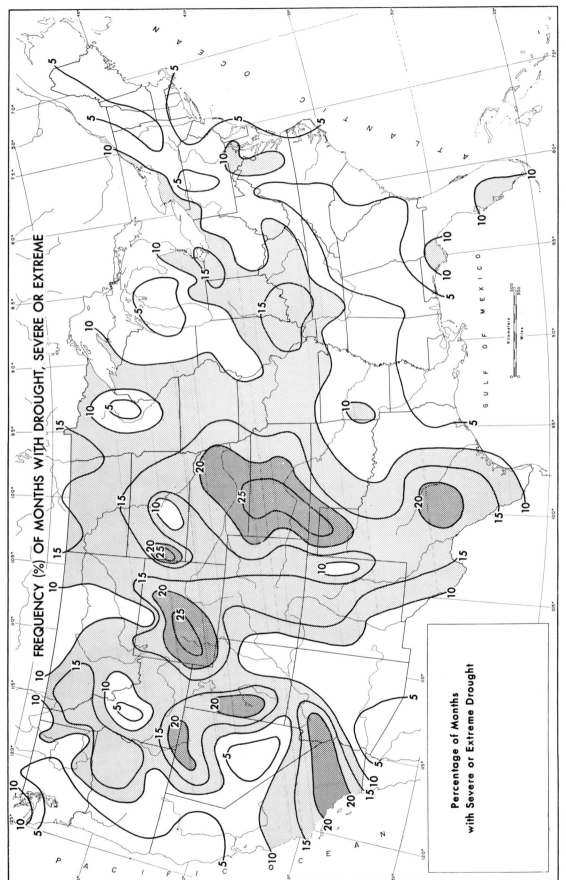

Fig. 31. Percent of months with severe or extreme drought (%), 1931–1960, in eastern United States, and 1931–1967 in 11 western states. (After manuscript map supplied by Dr. Wayne C. Palmer.)

20% more than the average lake evaporation (110 cm), and 53% of the estimated total incident sunshine.

Because potential evaporation varies much less than precipitation, drought occurs every year in some part of the United States (Fig.31). Drought begins after several weeks or months with less than average precipitation (HOFMANN and RANTZ, 1968). Conceivably, even with the usual rainfall, drought could be caused by above-average evaporation, such as might occur in persistent incursions of unusually dry, warm air, with more insolation than usual.

Actually, both these conditions usually occur simultaneously. A westward shift of the Bermuda anticyclone in summer brings moist air over Texas, which then receives showers while the southeastern states receive little rain and are desiccated by descending air. Conversely, an eastward shift brings the moisture over the southeastern states while hot, dry air descends from northern Mexico into New Mexico, Texas and beyond (NAMIAS, 1955).

The Palmer drought index, PDI, used by the government weather and agricultural services (PALMER, 1965), measures drought intensity each week as the difference between actual evapotranspiration and the adjusted normal value. "The long-term mean value is adjusted upward or downward depending on the departure of the week's temperature from normal. Successive weekly values of this computed abnormal evapotranspiration deficit have been combined into a measure of the cumulative severity of agricultural drought."

In about half of the 40 years, 1931–70, droughts had begun by August in one or more parts of New Mexico, Texas and Oklahoma, and 70% of them continued through November (PALMER, 1971). Droughts also occur, albeit less frequently, in New England (SPAR, 1968), the Ohio and Mississippi valleys, and on the northern plains, and even in the western mountains. In the western third of the country, where most agriculture is irrigated, drought is of less concern until it becomes so persistent and widespread as to affect the mountain sources of the irrigation water (THOMAS et al., 1962).

Droughts sometimes are followed by floods. In southwestern Oklahoma, Cotton County was classified as a drought area in June 1971, after a year with little rain; on August 15 it received 30 cm in three hours, and farmers who had been receiving drought assistance now could obtain flood payments also.

**Heat**

Sunshine varies across the United States from more than 4,000 h annually in south-eastern California and southwestern Arizona to less than 1,800 h in northwestern Washington and in northern New Hampshire (Fig.32). Actual solar energy received also varies over the United States by little more than a factor of two, from an annual average of 250 ly/day in northwest Washington to more than 500 ly/day in the desert southwest (Fig.33). Through the year the contrasts change: in January (Fig.34) the southwestern deserts receive three times as much radiation as the Pacific northwest, but in July (Fig.34) the ratio is only 750:500, with the 750-ly average at Inyokern somewhat questionable.

The sunshine duration map is quite detailed, perhaps unjustifiably so; it is based on

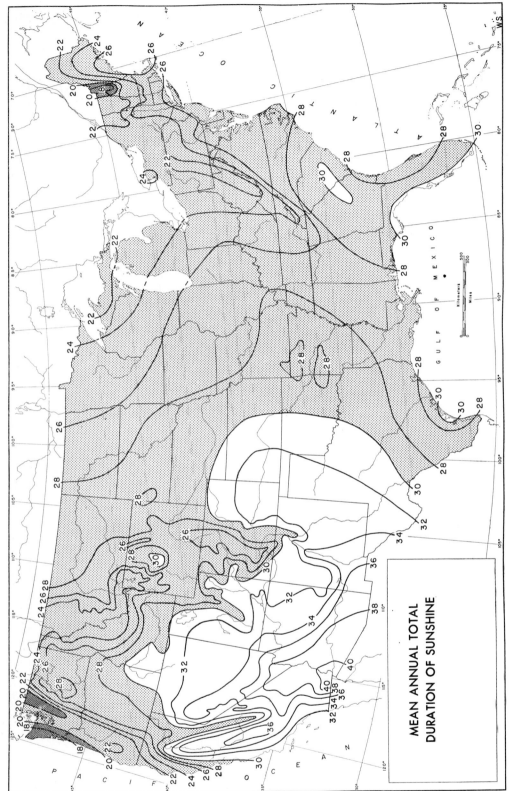

Fig.32. Mean annual duration of sunshine, in hundreds of hours, based on black-bulb sunshine recorders at 125 stations, 1931–1960, and 50 for fewer years. (After U.S. Weather Bureau National Atlas map.)

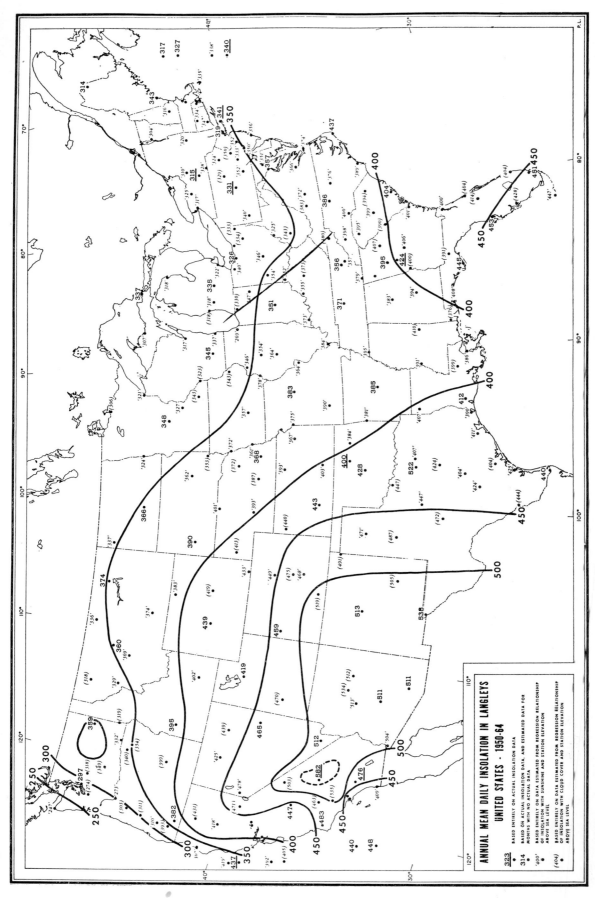

Fig.33. Annual mean daily insolation, in langleys, 1950–1964. (From Dr. Iven Bennett.)

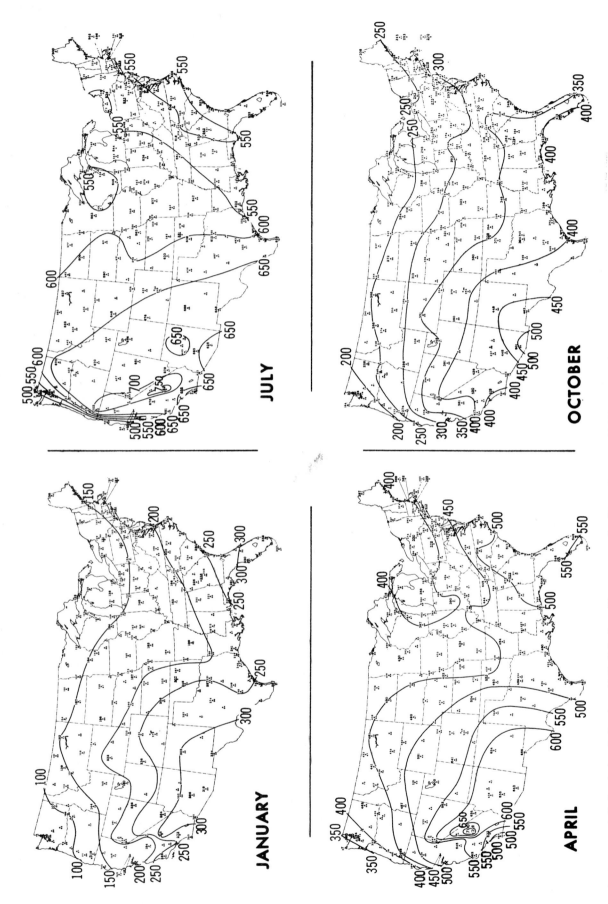

JULY

OCTOBER

JANUARY

APRIL

Fig.34. Mean daily insolation, in langleys, 1950–1964, for mid-season months. (From BENNETT, 1965.)

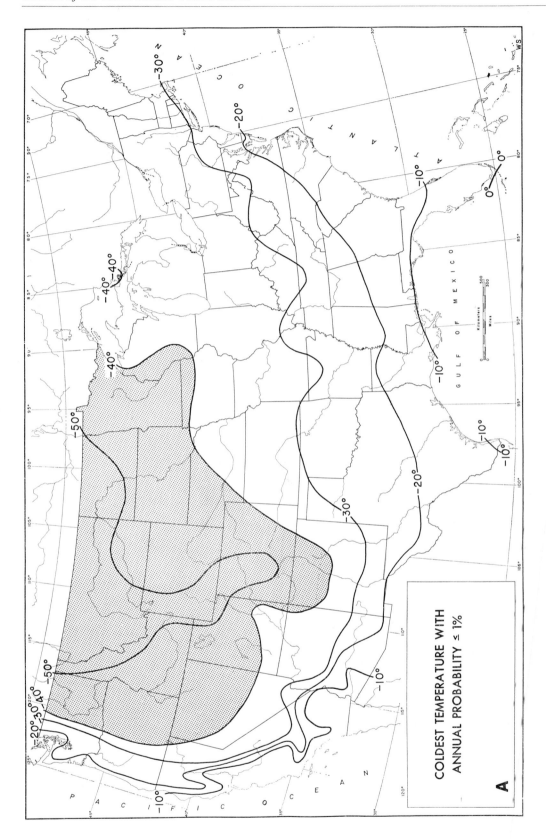

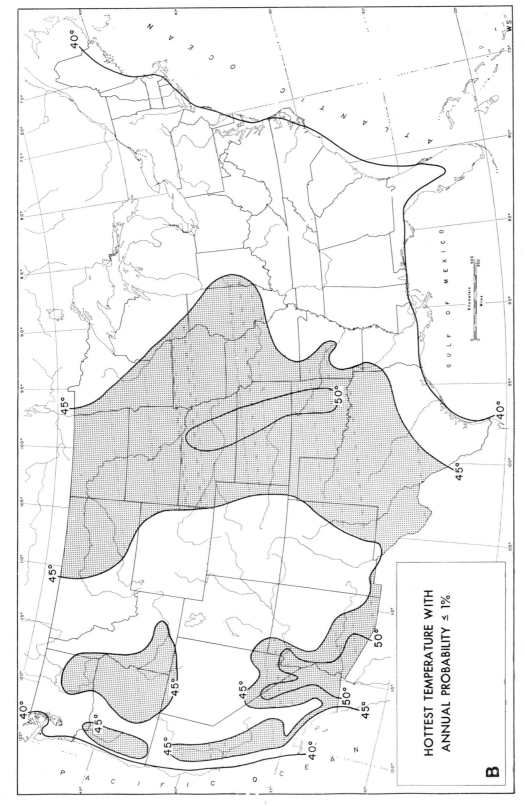

Fig.35. Coldest (A) and hottest (B) temperatures with annual probability of 1% or less, estimated from annual extremes, 1931–1960, at 220 first-order stations. (Prepared by David Waco.)

HOTTEST TEMPERATURE WITH
ANNUAL PROBABILITY ≤ 1%

B

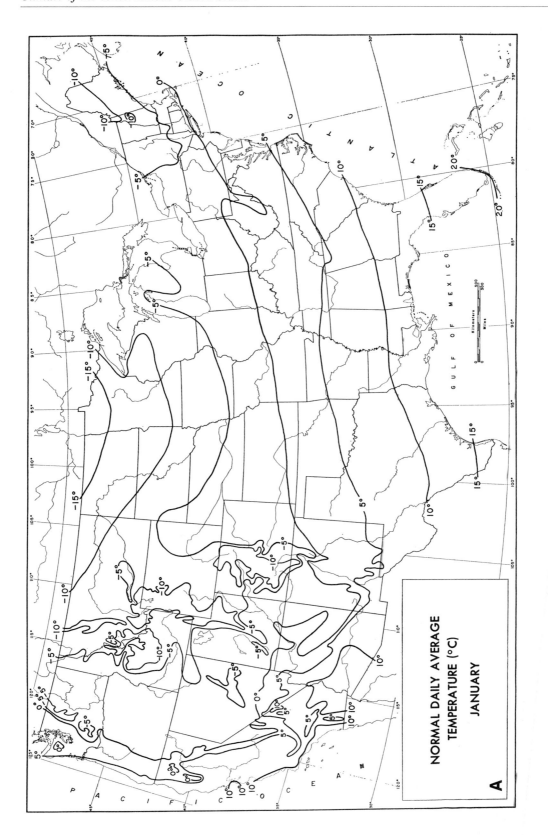

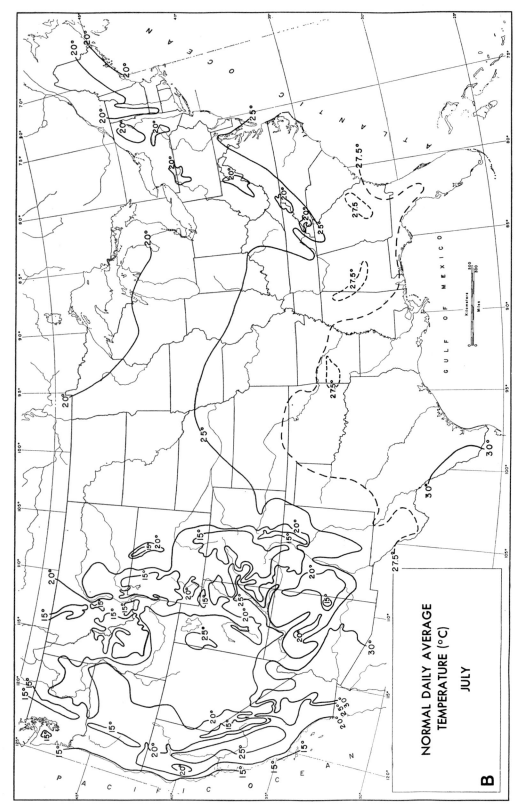

Fig.36. Mean monthly temperatures, 1931–1960, in January (A) and July (B), interpolated from U.S. Weather Bureau National Atlas map.

electric black-bulb records at 175 stations, 125 of them for the full 1931–60 period, the other 50 for shorter periods. The intensity maps are estimates of the 15-year mean, 1950–64, from complete pyrheliometer data for 9 stations, partially interpolated values for 50 others, and estimates for 113 stations from multiple linear regression on sunshine and station elevation (BENNETT, 1965).

The nation-wide average of about 400 ly/day or 150 kly/year would be enough to evaporate, at 580 cal./g, some 250 cm/year from the surface of the United States. This is almost 5 times the actual annual difference of 55 cm between precipitation and runoff, indicating that barely one-fifth of the incoming radiation is used in evaporation. The rest is reflected, radiated, or absorbed as heat and transferred to the air.

Cloudiness can be inferred from sunshine duration, but with caution. The sum of average daytime cloudiness and of sunshine duration, both in percent, varied between 111 and 119%, depending on region, at 132 stations for their entire periods of record to 1958 (FOX, 1961).

Air temperatures in the 48 conterminous United States have been reliably observed as cold as −57°C (−70°F) at Rogers Pass, Montana, on January 20, 1954 (LUDLUM, 1954), and as hot as 54°C (129°F) in Death Valley, California, 18 July, 1960; the widely quoted value of 56°C (134°F) in 1913 (LUDLUM, 1963) apparently was caused by a separation in the maximum thermometer (COURT, 1949).

Such isolated extremes are of little practical importance; much more useful are values with specified return periods, obtained by applying extreme-value statistical techniques to long series of annual extremes (COURT, 1953). Coldest and hottest temperatures with 100-year return periods, i.e., 1% probability of occurrence in any given year (Fig.35), are no colder than −35°C over about half the country and no hotter than 45°C over a different half.

Mean monthly temperatures in January (Fig.36A) and July (Fig.36B) show generally zonal isotherms over the eastern two-thirds of the United States, away from the confusion of the mountainous western third. Deviations from strict east–west orientation reflect the prevailing air movements.

In January, the southward bulge over the northern plains is the result of periodic invasions of Arctic air from the Mackenzie Basin of western Canada. The trend to the east-northeast over the eastern half of the country reflects the poleward component of the westerlies, in which this Arctic air is mixed, by a series of wave cyclones, with tropical air from the Gulf of Mexico. Very notable is the absence of any marine effect on January temperatures along the eastern coastline, except in New England; winds are consistently off-shore.

In July the 25°C isotherm bulges northward over the Great Plains as the result of the persistent northward flow of tropical air. This air sweeps northwestward across the Gulf and, after it crosses the coastline, is heated from below more rapidly than it cools by ascent over the gently sloping terrain. Hence, average temperatures are warmer in central Texas than along the coast. Along much of the Gulf and Atlantic coasts maximum temperatures, 1931–52, were noticeably warmer in a "tongue" or band some 50–100 km inland than along the coast, or farther inland (BERNSTEIN and HOSLER, 1959).

Interdiurnal variability of daily maximum and minimum temperatures, and also pressures, was mapped for the conterminous United States by LANDSBERG (1966) from data for 75 stations, 1957–61. For neither maximum nor minimum does the variability exceed

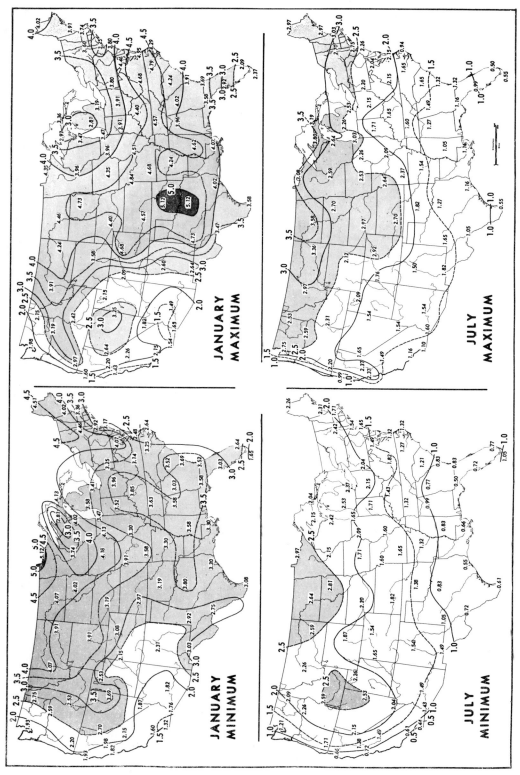

Fig.37. Mean interdiurnal variability of minimum and maximum temperatures (°C) in January and July, based on data at 75 stations, 1957–1961. (Prepared by Stanley Marsh from data of LANDSBERG, 1966.)

5°C in any month. "By and large, the differences between minima are somewhat less than those between the daily maxima" (LANDSBERG, 1968). In January, the day-to-day variation is greatest (5°C) for maxima over the Great Plains, for minima around the Great Lakes. In July, both vary less than in any other month, only 3°C along the central Canadian border (Fig.37).

Extreme horizontal gradients of temperature occur in some places along the Pacific coast: on some days the maximum air temperature may be only 20°C on the coast, thanks to the cold ocean, and 40°C only 20 km inland, behind a range of hills. Contrasts almost as great can also occur in winter: a "frost pocket" behind the coastal hills at Malibu, just northwest of Los Angeles, can have a minimum temperature of −10°C while +5°C is the minimum on the beach, 5 km distant (WACO, 1968).

Freezing temperatures have occurred over all the United States, possibly excepting some off-shore islands. The mean number of days (1921–50) between the last occurrence of 0°C in spring and the first occurrence in autumn, often considered to be the growing season, varies (Fig.38) from less than 60 in the northwestern mountains to more than 330 in southern California, Texas and Florida—the truly subtropical portions of the country. Similar is the period of continuous or intermittent snow cover, from the mean date of its first establishment in autumn to that of its disappearance in spring (Fig.20).

Even in northern United States, winter is not consistently cold. The greatest number of consecutive days with temperatures colder than 0°C during 10 winters (1950–51 to 1959–60) was only 50 in northern Minnesota and less than 20 over most of the country (Fig.39). The Great Lakes rarely freeze completely, and then only in late winter, causing warmer winters and shorter durations of cold temperatures to their east and south than to the west (TATTELMAN, 1968).

Freeze–thaw regions for the United States and Canada have been delineated by WILLIAMS (1964) in terms of the number of times that air temperature went from above +1°C to below −2°C (freeze) or conversely (thaw), and the durations of temperatures above and below these thresholds (actually 34° and 28°F). In the 22 regions of similar freeze–thaw behavior over the United States (5 more applied only to Canada), the total yearly frequency increased from none in the subtropical region to more than 70 alternations along much of the Canadian border and to 140 at a high-elevation station in the west.

## Climates

Three of the world's five major climatic types occur over extensive portions of the conterminous United States, and a fourth, the cold or polar, is found in the higher mountains; only the equatorial rainy type is absent. The western deserts are dry, the Pacific coast and the southeastern quarter of the country are warm, and the northern third is cool. Further subdivisions have been the subject of many fruitless discussions of various proposed classifications: Köppen (ACKERMAN, 1941; KESSELI, 1942; BAILEY, 1948; VILLMOW, 1952; PATTON, 1962; JAMES, 1966), Gorczinski (GORCZINSKI, 1942). Thornthwaite (THORNTHWAITE, 1948), and others.

Depending on the exact criteria adopted, and the stations and record period used, the humid–arid boundary has been placed anywhere from the western edge of the Ozark

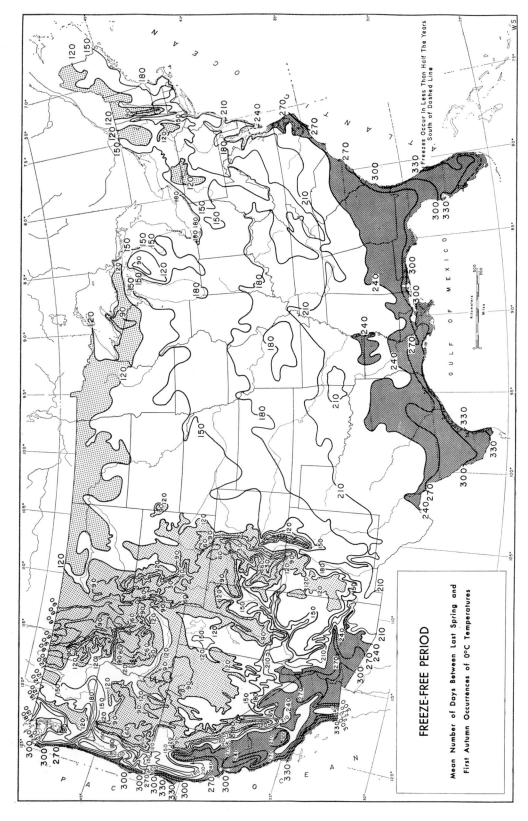

Fig.38. Mean number of days between last spring and first autumn occurrences of 0°C temperatures at 2,565 stations, 1921–1950. (From U.S. Weather Bureau National Atlas map.)

247

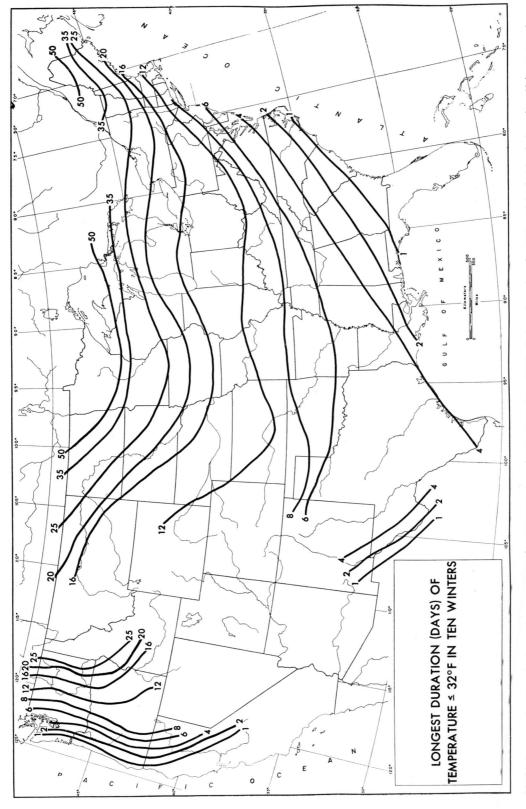

Fig.39. Longest duration, in days, of temperatures below 0°C in ten winters, 1950–51 to 1959–60, based on data for 108 stations in North America, 59 in conterminous United States. (After Tattelman, 1968.)

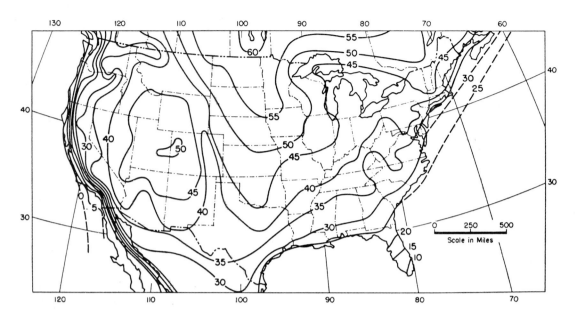

Fig.40. Isolines of equal continentality. (From TREWARTHA, 1961.)

Plateau (95°W) almost to the Rocky Mountains (103°W). Using PATTON's (1962) revisions of Köppen's criterion, KRAMER (1963) found it to be essentially along the 100th meridian from 30°N to 47°N. Similarly, the gradation from warm to cool temperate occurs generally along the Ohio River, but whether a few dozen kilometers north or south of it has been argued heatedly.

Continentality according to CONRAD's (1946) formula varies from 0% along the California coast (D'OOGE, 1955) to 60% in northern Minnesota and North Dakota, is lowered markedly around the Great Lakes (KOPEC, 1965), and in New England decreases from 50% in northern Maine to 30% at Nantucket Island (FOBES, 1954). These aspects appear on TREWARTHA's (1961) map of continentality (Fig.40) for an unspecified number of stations and years of record.

The region of greatest continentality is, naturally, the region of greatest annual range because it has the coldest winter weather: −15°C January mean (Fig.36A), and 50 days continually below freezing once in ten winters (Fig.36B). It has been identified as "the coldest area in the United States" because it has the lowest mean monthly minimum and mean daily temperatures, and the greatest frequency of temperatures below −20°C and −30°C (DE PERCIN, 1954).

Lying directly in the path of most southeastward invasions of Arctic air, this region, in northwest Minnesota and northeast North Dakota, has the worst combination of cold temperatures and strong winds. Heat loss from a human body is estimated, in kilocalories per hour per square meter of body surface, by "wind chill", an empirical formula (COURT, 1948) widely used by military agencies and some newspapers and radio stations (FAL-CONER, 1968). January wind chill varies from less than 500 in the southwest, southern, and southeast extremities of the country to more than 1,300 in northwest Minnesota and northeast North Dakota.

At less extremes of cold, moisture in the air or on the ground is important in human comfort or discomfort. "Cold-wet" conditions have been defined by the research group concerned with the design and issue of military clothing (MEIGS and DE PERCIN, 1957) as having temperatures between $-5°$ and $+10°C$ with more than 0.6 cloudiness, fog, precipitation, or snow on the ground; between 10° and 15°C with fog or precipitation, and 15° and 20°C with fog or precipitation and winds over 2 m/sec. The annual percentage of hours with such conditions is shown in Fig.41.

In warm air, water vapor rather than liquid water governs human comfort. For half a century relative comfort has been measured by Yaglou's empirical "effective temperature" or approximations thereof. Unfortunately, the original definition was in terms of the temperature at which motionless, saturated air would induce the same sensation of comfort (in a sedentary person wearing ordinary clothing) as that induced by the actual air temperature, humidity, and movement. Thus, effective temperature is generally cooler than actual air temperature: at 20°C and 50% relative humidity, the ET is 17°C, and in very dry air (RH 1%) at 30°C the ET is 20°C, the same as in saturated air at 20°C.

Of various efforts to approximate the ET, the most widely used was that of THOM (1956, 1957, 1959), first called the "discomfort index" but quickly renamed, after widespread objection from cities for which high values were reported, the "temperature–humidity index". It is 15 plus 0.4 times the sum of the drybulb and wetbulb temperatures, in Fahrenheit; although it is called an index, not a temperature, its units are those of the temperature scale used. In saturated air, the THI equals air temperature at 75°F (24°C), in accordance with the definition of the effective temperature it is intended to approximate, but is less than air temperature under warmer conditions.

Possibly because of these limitations, its use has declined, and the only available map of THI distribution (Fig.42) is for July, 1951–55. It shows maximum values along the western Gulf coast and in the lower Mississippi Valley because of high humidity in the onshore flow of tropical air, and in the southwestern deserts, because of very warm temperatures. Both Phoenix, Arizona, and Houston, Texas, have THI values around 80 but few people would rate their summer climates as exactly equal in discomfort.

Other efforts to approximate the effective temperature have been made by HENDRICK (1959), LALLY and WATSON (1960), and TERJUNG (1966), who offers maps of his "comfort index" and "wind effect" for January and July and of 17 "physiological climates" in July, and elsewhere (TERJUNG, 1967) of "annual cumulative stress", "proportional cumulative stress", and "annual physioclimatic regime". Terjung's system for cold conditions includes wind chill.

"Temperateness" of climate has been defined by BAILEY (1960, 1964) in terms of the dispersion of the hourly temperatures of a year around the presumed optimum value of 14°C. His index $M$, which ranges from 0 (extreme intemperate) to 100 (supertemperate) is computed from $109 - 30 \log \sigma$, where $\sigma$ is the root mean square difference between the hourly temperatures and 14°C. BAILEY (1968) also provides estimates of $\sigma$ from the annual range. Over the conterminous United States temperateness is least in northern Minnesota and greatest along the California coast, but with large areas of temperate climates in west Texas and the southeastern states (BAILEY, 1964).

Despite these efforts to use radiation, humidity, and wind along with temperature to obtain a more rational and complete evaluation of climatic stress on man and his plants, air temperature alone remains the most widely used criterion. Accumulated hourly or daily temperatures above or below a fixed base define heating or cooling degree days, widely tabulated and disseminated by the official weather service.

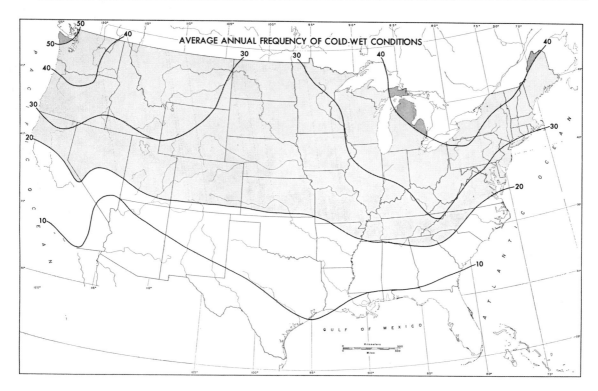

Fig.41. Mean annual frequency (average percent per month) of cold-wet conditions, 1948–1953. (After MEIGS and DE PERCIN, 1957.)

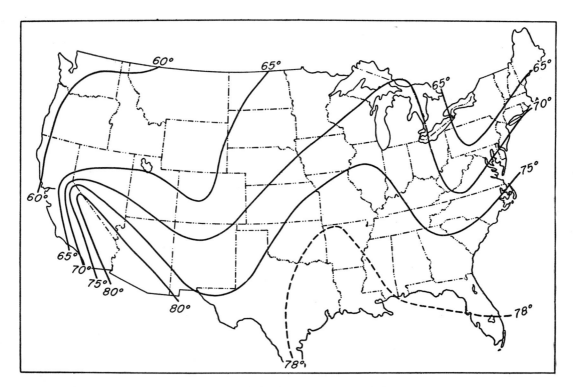

Fig.42. Average temperature-humidity index in July, 1951–1955. (After THOM, 1956.)

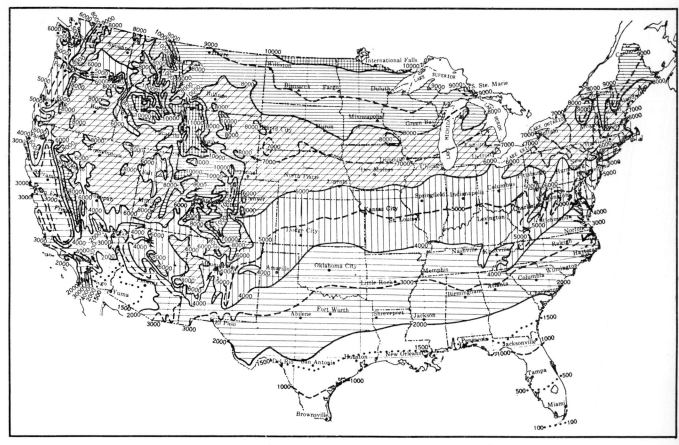

Fig.43. Normal annual total heating degree days, base 65°F (18.3°C), based on normal temperatures, 1931–1960. (From Environmental Data Service.)

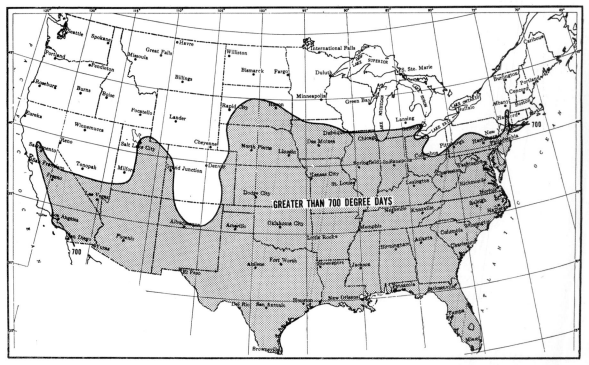

Fig.44. Region with more than 700 cooling degree days, base 65°F (18.3°C) annually, based on normal temperatures, 1931–1960. (From Environmental Data Service.)

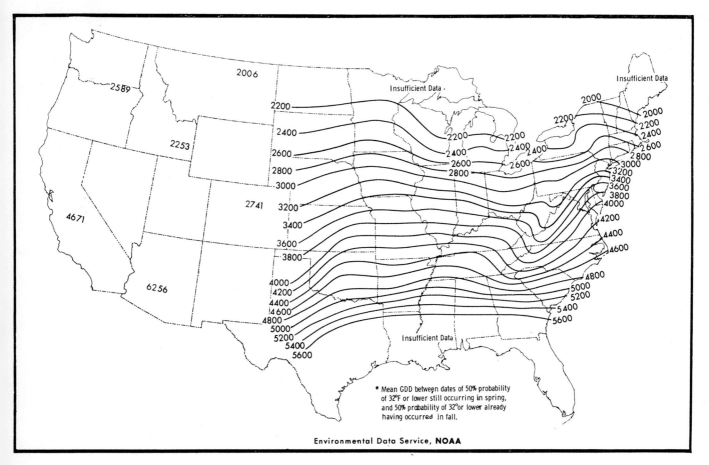

Fig.45. Normal annual growing degree days, based on normal temperatures, 1931–1960. (From Environmental Data Service.)

Heating degree days (Fig.43) are the sums of the negative departures of daily temperatures from 65°F (18.3°C); a positive departure, when a day's mean temperature is warmer than 65°F, counts as zero (THOM, 1952, 1954; BALDWIN, 1971). Fuel consumption for home and industrial heating is almost linearly proportional to the degree days, and the accumulated figures are widely used by fuel merchants in ordering and delivering fuels.

Cooling degree days are calculated similarly for mean daily temperatures above a base, which may be 85°F (29.4°C), 68°F (20°C), or some other threshold. Fig.44 shows the region in which the June–September total cooling degree days, base 65°F (18.3°C) exceeds 700, the accepted requirement for air-conditioning government automobiles.

Growing degree days are calculated similarly for the classic "growing season" between the average dates of last and first freezes (RAHN, 1971). For each day, the growing degree day value is 1.8 times the difference from 10°C of its adjusted mean temperature. This adjusted mean is the average of the minimum and maximum temperatures, except that a minimum $< 10$°C or a maximum $> 30$°C is considered to have that limiting value. Accumulation of these daily values, beginning with some chosen date, provides the total GDD to a given date, or to the end of the season. Mean GDD between median dates of first and last freezes, based on 47 stations for 20 years (1949–68), decrease northward (Fig.45) from 5,600 to 2,200.

**Statistical relations**

Statistical analyses and tests are revealing many interesting relations in the climate of the conterminous United States, as longer series of more accurate observations are being analyzed on bigger and faster computers by more advanced techniques. The studies are of time and space "singularities", lunar influences, serial correlations and space correlations, and probabilistic models for the time behavior of various elements.

Best-known of all the apparent "singularities" in United States climate is the "January thaw" (WAHL, 1952), a warming of several degrees around 20–23 January, primarily in New England. Although its existence has not been demonstrated conclusively, a tendency has been found "for singularities to occur in the first half of the winter during the minor half of the double sunspot cycle and to occur in the second half during the major half" (NEWMAN, 1965). On the opposite side of the continent, 3-day moving averages of daily maximum temperatures at Gonzales Observatory, Victoria, B.C., 1885–1953, show a warming trend at the start of January, then a definite cooling, a minor warming near mid-month, an equally short cooling, and "a major warming from the 24th to the 26th," which was prominent only in those years, 1899–1939, when the mean zonal index of the westerlies between 35°N and 55°N was below its median value (REBMAN, 1954).

The "January warm spell" has been traced by FREDERICK (1966) from around January 9 in the northern Rocky Mountains (Fig.46) eastward to the Atlantic coast about a week later. At Phoenix, Arizona, 1895–1956, minimum temperatures between January 9 and 17 were warmer than during the preceding or following weeks (KANGIESER, 1957). The

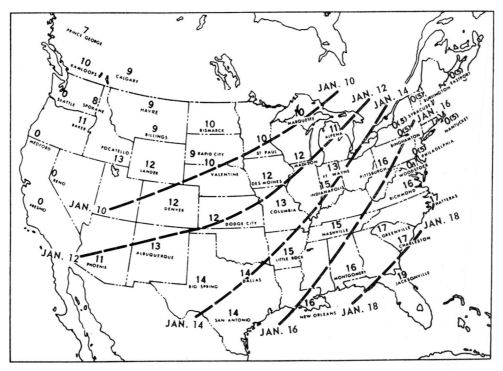

Fig.46. Southeastward progression of mean day of January warm spell. (After FREDERICK, 1966.)

January warm spell in New England was explained by Duquet (1963) as a tendency for cyclonic storms from the Gulf of Mexico to move northeastward along the Appalachian mountains around January 20.

Another "singularity" comes some two months earlier: pronounced cooling at Washington, D.C., between November 21 and 25 (Namias, 1968) is preceded by an apparent first maximum in raininess in southern California in mid-November (De Violini, 1969). A month still earlier, an abrupt change in mean 5-day pressure patterns from 11–15 to 16–20 October causes a sharp increase in both the number of "cold days" and the frequency of snowfall over the northern Plains and the Rocky Mountains (Wahl, 1954). Another phenomenon which could be termed a singularity is the simultaneous increase of precipitation in Arizona and decrease in Oregon in late June (Lowry, 1956).

"No tendency for rainfall anomalies near specific calendar dates," termed "calendaricities", was found in an elaborate statistical study by Brier et al. (1963). But it was based on the total rainfall on each day, 1900–62, at 100 U.S. stations, covering the entire country and thus making no allowance for any eastward or other progression. In these same data, however, "the average amount of rainfall was about 10 % higher a few days after the full moon than a few days before," and maximum rainfalls at 1,544 stations showed a similar agreement with the lunar synodic cycle (Brier and Bradley, 1964).

Further investigation (Brier, 1965) revealed that the total rainfall at 67 stations, 1951–60, was greatest around 3 a.m. and 5 p.m., and that in the 63-year record of daily precipitation at the 1,544 stations "the maximum precipitation occurs on those days of the synodic month when upper lunar transit occurs shortly before 3 a.m. and shortly before 5 p.m." In these data also "the average amount of precipitation over the United States was 20 % higher two days after syzygy than it was two days before."

Thunderstorms of the United States east of the Rocky Mountains also appear to be most frequent two days after full moon (Lethbridge, 1970). The cube root of the number of stations, out of 108 east of 102°W, reporting thunderstorms or distant lightning for each day, 1942–65, was subtracted from the daily mean as smoothed by 11-day moving averages. These daily departures were then investigated by superposed epochs centered on days of lunar ecliptic crossing, of maximum north declination, and of full moon. The first two gave inconclusive results, but a maximum in total thunderstorm activity two days after full moon was indicated by various analyses. The smoothing operations and various data manipulations prevent any statistical validation of this phenomenon, which requires further independent investigation.

If rainfall and thunderstorms are most common after full moon, sunshine should be deficient then. The percent of possible sunshine for each day, 1905–62, at each of 10 stations from Honolulu to Boston was less than average during the first and third weeks of the lunar month and more than average during the second and fourth weeks (Lund, 1965, 1966). No lunar influences on temperature, wind or other elements have been reported recently; Lethbridge (1970) lists nine reports of almost a century ago of lunar effects on pressure as well as rain and thunderstorms.

"Singularities" in the spatial distributions of temperature, precipitation, and other elements may arise from human influences. The well-known "urban heat island" has been found in many cities, and "four major midwest cities without orographic effects" receive 5 to 8 % more annual precipitation and have 5 to 10 % more rainy days than surrounding rural areas (Changnon, 1969). Such increases are variously attributed to increased heat and evaporation, to the friction of buildings and trees, and to pollutants acting as condensation or freezing nuclei.

An even more startling spatial singularity is claimed for LaPorte, in northwest Indiana, which in the past three decades apparently had 31% more annual precipitation, 34% more days with moderate and heavy rain, 38% more thunderstorm days, and 246% more hail days than surrounding stations. These increases are attributed to heat, moisture, and especially pollutants from Chicago, 80 km to the west, and from the steel mills of Gary, Indiana, 32 km west. The reality of this anomaly is supported by a parallel increase in the annual flow of the upper Kankakee River; LaPorte is in the northwest portion of its 1,310 km² drainage basin (HIDORE, 1971). The anomaly is challenged by a claim that its precipitation record is "statistically invalid and physically unacceptable" (HOLZMAN and THOM, 1970), and by botanic evidence that LaPorte is naturally wetter in years when the lake breeze front from Lake Michigan visits it frequently (HARMAN and EL-TON, 1971).

Similar controversy surrounds the finding that annual precipitation during 1947–66 was about 30% greater than during 1929–46 at several locations downwind from large pulp and paper mills in northwest Washington (HOBBS et al., 1970). At issue (ELLIOTT and RAMSEY, 1970) is whether all local areas of increase can be traced to pollution sources, and vice versa.

Intentional modification of weather, primarily by increasing precipitation through introduction of silver iodide crystals to act as ice-forming nuclei in super-cooled clouds, has been a rapidly-increasing activity for two decades. Results have been sought through countless statistical analyses of precipitation, streamflow, hail, and other elements. Most studies have been so local, so specialized, and often so proprietary, that few have contributed materially to the description and understanding of climate. Many have relied upon correlations between the daily, or even hourly, rainfalls in "control" and "target" areas.

Such correlations decrease rather regularly with distance, but not symmetrically. Annual rainfall at Omaha, Nebraska, for 48 to 69 years ending in 1940, showed correlations 0.60 with that at places within 100 km and 0.40 at distances up to 300 km (FOSTER, 1944). Similar relations were found by OLTMAN and TRACY (1951) for annual rainfalls, 1920–48, around four other Great Plains cities: Great Falls, Montana; Bismark, North Dakota; Cheyenne, Wyoming; and Kansas City, Missouri. In all cases, the isocorrs were somewhat elliptical, extending farthest east. Unpublished studies by A. L. Joos show that such correlations become negative at distances of about 1,000 km.

Correlations between climatic elements at various places form a matrix, which can be compressed into eigenvectors. In the matrix of mean monthly precipitation, 1953–63, at 60 Nevada stations, STIDD (1967) found three principal eigenvectors, which he identified with the three major precipitation types (p.213) of HOUGHTON (1969) by mapping their weights at each station.

Correlations between various surface weather elements and upper atmosphere heights, winds, temperatures, and moisture contents are used extensively in developing "objective" models for numerical weather prediction. One of the few such relations of direct climatic interest is between 5-day mean temperatures at a point and similar means ending four days earlier at various other places (KLEIN, 1965). At Dallas, Texas, 5-day mean temperatures in winter, 1948–57, had a serial correlation of 0.47, but the maximum inter-

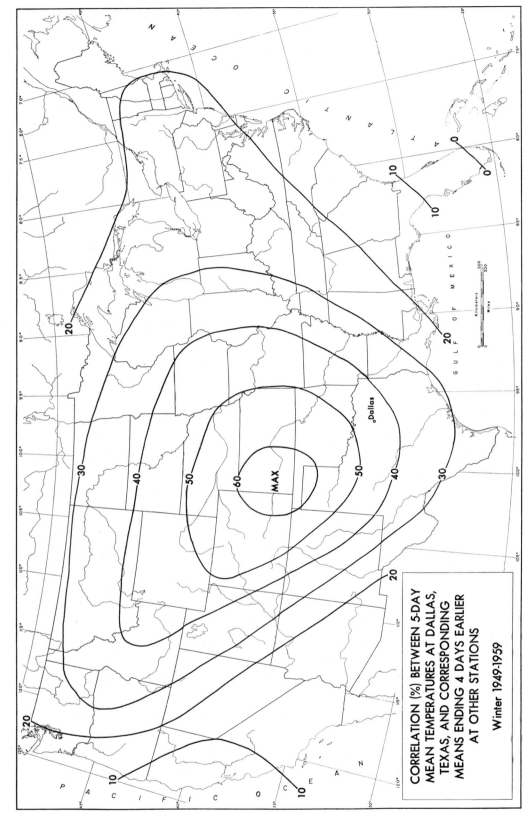

Fig.47. Correlation between 5-day mean temperature at Dallas, Texas, and similar means four days earlier at 30 other cities, based on data for 10 winters (mid-December 1949 to mid-March 1959). (After KLEIN, 1965.)

station lag correlation of 0.60 was about 650 km to the northwest, in southwest Kansas (Fig.47).

For all 30 stations, except for a few west of the Rocky Mountains in summer, the inter-station lag correlations "are nearly all positive and generally higher in winter than in other seasons"; the centers of maximum correlation were to the north or west, farther away in autumn and spring than in winter or summer. Klein concluded that local temperature persistence is greatest in winter, but in autumn and spring is less important than advection in controlling local temperature.

However, Klein's correlations do not imply necessarily that both cold and warm air moves southeastward. Presumably similar patterns exist between 5-day mean temperatures at one point and those three days later at the other places. Weather conditions over most of the conterminous United States tend to change in roughly 7-day cycles, as cyclonic storms and anticyclonic ridges progress eastward. Cold air sweeps south and southeast behind cold fronts, to be replaced in a few days by warm air from the south. Such alternations cause an appreciable serial correlation in daily temperatures; of 14,610 days, 1911–50, at Denver, Colorado, 2,880 were more than 5°C above the mean, and the sequence could not be fitted by a simple binomial process (LEATHERWOOD, 1955). Relative frequencies of wet days, 1951–60, at 33 stations (Fig.48) were used by HERSHFIELD (1970a, b) to construct a nomogram with which to estimate the probability that a wet day would be followed by another wet one; thresholds of 0.01, 0.10, 0.25, and 0.50 inch (0.25, 2.5, 6.4, and 12.7 mm) were used. Such conditional probabilities are needed in fitting wet–dry day sequences by Markov chains of second order, which FEYERHERM and BARK (1967) found provide better fits than do first order chains to sequences at 6 stations in the Mississippi Valley. They applied their procedure to data from most of the individual states in a series of reports, from which farmers and others can derive their own rainfall forecasts.

Seasonal rainfalls, however, exhibit no serial dependence on which analogous long-range forecasts could be based. Serial correlations between the rainfall of one winter (November–April) and the next, one summer and the next, and from winter to summer, at eight Arizona stations with 46 to 87 years of record were not statistically significant (McDONALD, 1960).

Month-to-month persistence of mean temperatures over entire states, 1873–1957, is suggested for the mid-nation in summer, the west from April to May, and the east from December to January (DICKSON, 1967). Average temperatures for each month for each of the 44 states or state groups (e.g., New England) were categorized as below, near, or above normal. Of the 528 contingency tables (or transition matrices) for month-to-month change, only 88 showed serial dependence significant at the 5% level. No significant year-to-year relations were found.

Mean monthly precipitation and temperature for the entire United States, based on areally weighted averages of some 5,000 stations, 1893–1953, showed nonsignificant one-month lag correlations, except for precipitation (0.37) from September to October and temperatures (around +0.45) in the three summer months (ENGER, 1955).

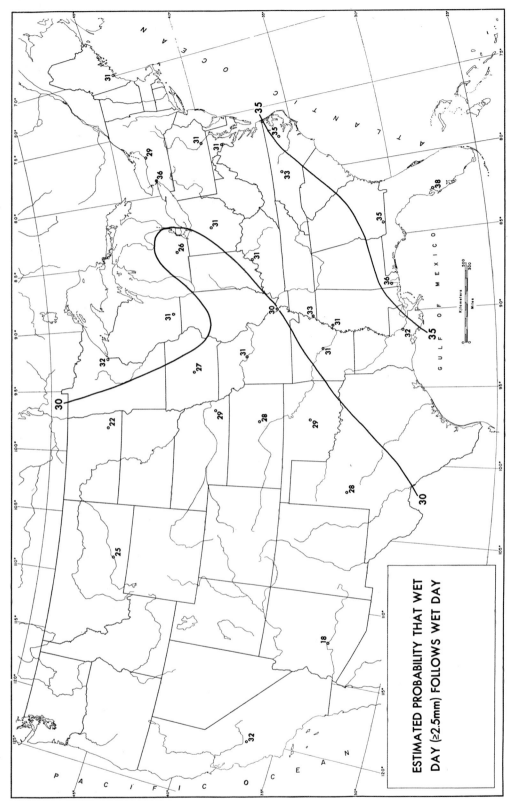

ESTIMATED PROBABILITY THAT WET
DAY (≥2.5mm) FOLLOWS WET DAY

Fig.48. Estimated conditional probability that a wet day (2.5 mm or more) will be followed by a wet day, 1951–1960. (After Hershfield, 1970a.)

**Conclusion**

The climate of the conterminous United States has been described, in the preceding sections, primarily in terms of the three dominant airflows into the country and of the consequences of their inevitable conflicts. The southeastern quarter of the country, largely dominated by air flowing northward from the Gulf of Mexico, is essentially subtropical and warm temperate. The northeastern quarter, affected principally by air from the Great Lakes and the Atlantic, is cool temperate. Over the mountainous western portion of the country, Pacific Ocean air masses are constantly being broken up and modified by the rugged topography. In the middle of the country, the frequent conflict between the various airflows creates the greatest variability in temperature, precipitation and air movements; tornadoes and hailstorms are common in summer, violent blizzards in winter.

Detailed descriptions of the many aspects of the U.S. climate abound. Many of them have been cited in the text and are listed in the references. Many others, however, are not mentioned, some because they are too general, some because they are too detailed, and some, including many excellent studies, because they are too old. The climate of the United States is discussed in encyclopedias, in textbooks on regional geography and climatology, and in other reference books.

Maps of various climatic elements, including the originals of many reproduced here, appear in several compilations. Most lavish is the *National Atlas of the United States*, in which classic climatic subjects are shown on 24 of the 335 pages of maps. Many of these are color revisions of the maps in the 79-page, 40 × 50 cm *Climatic Atlas of the United States* published in 1968 by the Environmental Data Service of the Environmental Science Services Administration (now part of the National Oceanic and Atmospheric Administration). That remains the most complete atlas of basic climatic data for the United States, although it lacks specialized maps of the sort collected and published by VISHER (1954).

Basic climatic data for the United States are collected and published in great profusion by the Environmental Data Service, as well as other government agencies, state agencies, and private groups. Even so, certain aspects of climate are not summarized, such as barometric pressure at station elevation, and vapor pressure. For the tables offered here, mean vapor pressure was computed from mean dew-point.

Selections from this welter of data, maps and discussions have been offered here in hopes of presenting a coherent, complete yet concise account of the climate of the conterminous United States.

**Appendix—Climatic tables** (pp.267–343)

Data in the climatic tables were extracted by Mr. William Slusser, student assistant, from various publications of the Environmental Data Service of the U.S. Weather Bureau (now of the National Oceanic and Atmospheric Administration), chiefly the *Local Climatological Data, Annual*, for each station. Stations were chosen to give as even a coverage as possible, with no significant site changes during the 1931–1960 reference period, and wherever possible with complete data on sunshine, cloudiness, etc.

Mean station pressures were taken from *World Weather Records* and other sources, since such data are not tabulated routinely in the United States; in many cases, the pressure data apply to an elevation differing from the official station elevation given in the title of the table. In this case the different elevation is added between brackets. For temperature and precipitation, means for the standard 1931–1960 period were available, but for other elements values are generally for shorter periods, as indicated in the tables. Values for mean vapor pressure are computed from some other quantity (dew point or relative humidity). Average frequency of days with rain, thunderstorms, etc., are rounded to whole numbers; a value greater than 0 but less than 0.5 is indicated by an asterisk (*). For several stations without pyrheliometers, sunshine data for nearby stations have been used, as indicated on the map (Fig.1). The tables are grouped by states, in alphabetical order, and alphabetically within each state.

# References

ACKERMAN, E. A., 1941. The Köppen classification of climates in North America. *Geogr. Rev.*, 31: 105–111.

BAILEY, H. P., 1948. Proposal for a modification of Koeppen's definitions of the dry climates. *Yearb. Assoc. Pacific Coast Geogr.*, 10: 33–38.

BAILEY, H. P., 1960. A method of determining the warmth and temperateness of climate. *Geogr. Ann.*, 42: 1–16.

BAILEY, H. P., 1964. Toward a unified concept of the temperate climate. *Geogr. Rev.*, 54: 516–545.

BAILEY, H. P., 1968. Hourly temperatures and annual range. *Yearb. Assoc. Pacific Coast Geogr.*, 30: 25–40.

BALDWIN, J. L., 1971. Heating degree days. *Wkly Weather Crop Bull.*, 58 (40): 12.

BEEBE, R. G., 1967. Changes in visibility restrictions over a 20-year period. *Bull. Am. Meteorol. Soc.*, 48: 348.

BENNETT, I., 1959. Glaze, its meteorology and climatology, geographical distribution, and economic effects. *Tech. Rept. Env. Protect. Res. Div., Quartermaster Res. Eng. Center, Natick, Mass.*, EP-105: 217 pp.

BENNETT, I., 1965. Monthly maps of mean daily insolation for the United States. *Solar Energy*, 9:145–158.

BENTON, G. S. and ESTOQUE, M. A., 1954. Water vapor transfer over the North American continent. *J. Meteorol.*, 11: 462–477.

BERNSTEIN, A. B. and HOSLER, C. R., 1959. Temperature patterns along the Atlantic and Gulf coasts. *Mon. Weather Rev.*, 87: 395–400.

BONNER, W. D., 1968. Climatology of the low level jet. *Mon. Weather Rev.*, 99: 833–850.

BORCHERT, J. R., 1950. The climate of the central North American grassland. *Ann. Assoc. Am. Geogr.*, 40: 1–39.

BRIER, G. W., 1965. Diurnal and semidiurnal atmospheric tides in relation to precipitation variations. *Mon. Weather Rev.*, 93: 93–100.

BRIER, G. W. and BRADLEY, D. A., 1964. The lunar synodical period and precipitation in the United States. *J. Atmosph. Sci.*, 21: 386–395.

BRIER, G. W., SHAPIRO, R. and MACDONALD, N. J., 1963. A search for rainfall calendaricities. *J. Atmosph. Sci.*, 20: 529–532.

BROWN, M. J., KRAUSS, R. K. and SMITH, R. M., 1968. Dust deposition and weather. *Weatherwise*, 21: 66–69.

BRUNNSCHWEILER, D. H., 1952. Geographic distribution of air masses in North America. *Vierteljahrsschr., Naturforsch. Ges. Zürich*, 97: 42–49.

BRUNNSCHWEILER, D. H., 1955. Toward a classification of climates on an aerosomatic basis. *Ann. Assoc. Am. Geogr.*, 45: 172.

BRYSON, R. A., 1966. Air masses, streamlines, and the boreal forest. *Geogr. Bull.*, 8: 228–269.

BRYSON, R. A. and LOWRY, W. P., 1955. Synoptic climatology of the Arizona summer precipitation singularity. *Bull. Am. Meteorol. Soc.*, 36: 329–339.

BUE, C. D., 1963. Principal lakes of the United States. *U.S. Geol. Surv., Circ.*, 476: iii + 22 pp.

BUSBY, M. W., 1963. Yearly variations in runoff for the conterminous United States, 1931–1960. *U.S. Geol. Surv., Water-Supply Pap.*, 1669-S: iii + 49 pp.

CHANG, JEN-HU and OKIMOTO, G., 1970. Global water balance according to the Penman approach. *Geogr. Anal.*, 1: 55–67.

CHANGNON JR., S. A., 1968. Precipitation climatology of Lake Michigan basin. *Ill. State Water Surv. Bull.*, 52: iii + 46 pp.

CHANGNON JR., S. A., 1969. Recent studies of urban effects on precipitation in the United States. *Bull. Am. Meteorol. Soc.*, 50: 411–421.

CHANGNON JR., S. A. and JONES, D. M. A., 1972. Review of the influence of the Great Lakes on weather. *Water Resour. Res.*, 8: 360–371.

CHANGNON JR., S. A. and STOUT, G. E., 1967. Crop-hail intensities in central and northwest United States. *J. Appl. Meteorol.*, 6: 542–548.

CONDON, C. R., 1971. North Atlantic tropical cyclones, 1970. *NOAA-EDS, Climatol. Data Natl. Summ.*, 21(13): 60–69.

CONRAD, V., 1946. Usual formulas of continentality and their limits of validity. *Trans., Am. Geophys. Union*, 27: 663–664.

COURT, A., 1948. Wind chill. *Bull. Am. Meteorol. Soc.*, 29: 487–493.

COURT, A., 1949. How hot is Death Valley? *Geogr. Rev.*, 39: 214–220.

COURT, A., 1953. Temperature extremes in the United States. *Geogr. Rev.*, 43: 39–49.

COURT, A., 1970. Tornado incidence maps. *ESSA Res. Lab., Natl. Severe Storms Lab., Tech. Memo.*, ERLTM-NSSL 49: 76 pp.

COURT, A. and GERSTON, R. D., 1966. Fog frequency in the United States. *Geogr. Rev.*, 56: 543–550.

CRAMER, O. P. and LYNOTT, R. E., 1970. Mesoscale analysis of a heat wave in western Oregon. *J. Appl. Meteorol.*, 9: 740–759.

CROWE, P. R., 1951. Rainfall regime in the southeastern United States. In: L. D. STAMP and S. W. WOOL-DRIDGE (Editors), *London Essays in Geography. Rodwell Jones Memorial Volume*. Longmans, Green and Co., London. pp.71–90.

CRY, G. W., 1967. Effects of tropical cyclone rainfall on the distribution of precipitation over the eastern and southern United States. *U.S.Dept. Commer., ESSA, Prof. Pap.* 1: iv + 67 pp.

DE PERCIN, F., 1954. Coldest area in the United States. *Weatherwise*, 7: 152–155.

DE VIOLINI, R., 1969. A November rainfall singularity. *Pac. Missile Range, Misc. Rept.* 69-2: 23 pp.

DICKSON, R. R., 1967. The climatological relationship between temperatures of successive months in the United States. *J. Appl. Meteorol.*, 6: 31–38.

DODD, A. V., 1965. Dew point distribution in the contiguous United States. *Mon. Weather Rev.*, 93: 113–122.

D'OOGE, C. L., 1955. Continentality in the western United States. *Bull. Am. Meteorol. Soc.*, 36: 175–177.

DUQUET, R. T., 1963. The January warm spell and associated large-scale circulation changes. *Mon. Weather Rev.*, 91: 47–60.

EAGLEMAN, J. R., 1968. *Rainfall Variability in the United States*. (Preprint of paper presented at conference and workshop on applied climatology, American Meteorological Society, Asheville, North Carolina.)

EICHENLAUB, V. L., 1970. Lake effect snowfall to the lee of the Great Lakes: its role in Michigan. *Bull. Am. Meteorol. Soc.*, 51: 403–412.

ELDRIDGE, R. G., 1966. Climatic visibilities of the United States. *J. Appl. Meteorol.*, 5: 277–282.

ELLIOTT, W. P. and RAMSEY, F. L., 1970. Comments on Hobbs et al. *J. Atmosph. Sci.*, 27: 1215–1216.

ENGER, I., 1955. Month-to-month persistence. *Bull. Am. Meteorol. Soc.*, 35: 36–37.

FALCONER, R., 1968. Windchill, a useful wintertime weather variable. *Weatherwise*, 21: 227–229, 255.

FEYERHERM, A. M. and BARK, L. D., 1967. Goodness of fit of a Markov chain model for sequences of wet and dry days. *J. Appl. Meteorol.*, 6: 770–773.

FLETCHER, R. J., 1971. Extreme departures from mean monthly precipitation in southeastern United States, 1905–1969. *Proc. Can. Assoc. Geogr.*, 1971: 191–198.

FLOHN, H., 1970. Comments on water budget investigations, especially in tropical and subtropical mountain regions. In: *World Water Balance. Proc. Reading Symp., July 1970*. Int. Assoc. of Sci. Hydrol., Bruxelles, pp.251–262.

FLOWERS, E. C., MCCORMICK, R. A. and KURFIS, K. R., 1969. Atmospheric turbidity over the United States, 1961–1966. *J. Appl. Meteorol.*, 8: 955–962.

FOBES, C. B., 1954. Continentality in New England. *Bull. Am. Meteorol. Soc.*, 35: 197–207.

FOSTER, E. E., 1944. A climatic discontinuity in the areal correlation of annual precipitation in the Middle West. *Bull. Am. Meteorol. Soc.*, 25: 299–306.

FOX, R. L., 1961. Sunshine-cloudiness relationships in the United States. *Mon. Weather Rev.*, 89: 543–548.

FREDERICK, R. H., 1966. Geographical distribution of a 30-year mean January warm spell. *Weather*, 21: 9–13.

GLENN, C. L., 1961. The chinook. *Weatherwise*, 14: 175–182.

GORCZYNSKI, W., 1942. Climatic types of California, according to the decimal scheme of world climates. *Bull. Am. Meteorol. Soc.*, 23: 161–165.

GRILLO, J. N. and SPAR, J., 1971. Rain-snow mesoclimatology of the New York Metropolitan Area. *J. Appl. Meteorol.*, 10: 56–61.

HARMAN, J. R. and ELTON, W. M., 1971. The LaPorte, Indiana, precipitation anomaly. *Ann. Assoc. Am. Geogr.*, 61: 468–480.

HARRIS, M. F., 1969. *Effects of Tropical Cyclones upon Southern California*. M.A. Thesis, San Fernando Valley State College, Northridge, Calif., vii + 89 pp.

HENDRICK, R. L., 1959. An outdoor weather-comfort index for the summer season in Hartford, Connecticut. *Bull. Am. Meteorol. Soc.*, 40: 620–623.

HERSHFIELD, D. M., 1962. A note on the variability of annual precipitation. *J. Appl. Meteorol.*, 1: 575–578.

HERSHFIELD, D. M., 1970a. A comparison of conditional and unconditional probabilities for wet- and dry-day sequences. *J. Appl. Meteorol.*, 9: 825–827.

HERSHFIELD, D. M., 1970b. Frequency and intensity of wet and dry seasons and their interrelations. *Water Resour. Bull.*, 6: 87–93.

HIDORE, J. J., 1971. The effects of accidental weather modification on the flow of the Kankakee river. *Bull. Am. Meteorol. Soc.*, 52: 99–103.

HISER, H. W., 1956. Type distributions of precipitation at selected stations in Illinois. *Trans., Am. Geophys. Union*, 37: 421–424.

HOBBS, P. V., RADKE, L. F. and SHUMWAY, S. E., 1970. Cloud condensation nuclei from industrial sources and their apparent influence on precipitation in Washington State. *J. Atmosph. Sci.*, 27: 81–89.

HOFMANN, W. and RANTZ, S. E., 1968. What is drought? *J. Soil Water Conserv.*, 23: 105–106.

HOLZMAN, B. G. and THOM, H. C. S., 1970. The LaPorte precipitation anomaly. *Bull. Am. Meteorol. Soc.*, 51: 335–337.

HOPE, J. R. and NEWMAN, C. J., 1971. Computer methods applied to Atlantic area tropical storm and hurricane climatology. *Mariners Weather Log*, 15: 272–278.

HORN, L. H. and BRYSON, R. A., 1960. Harmonic analysis of the annual march of precipitation over the United States. *Ann. Assoc. Am. Geogr.*, 50: 157–171.

HOUGHTON, J. G., 1969. *Characteristics of Rainfall in the Great Basin*. University of Nevada, Desert Research Institute, Reno, Nev., xiv + 205 pp.

JAMES, J. W., 1966. A modified Köppen classification of California's climate according to recent data. *Calif. Geogr.*, 7: 1–12.

JETTON, E. V., OEY, H. S. and WOODS, C. E., 1971. Stratospheric circulation and summertime rainfall over southern New Mexico and southwestern Texas. *Weather*, 26: 186–191.

JORGENSEN, D. L., 1967. Climatological probabilities of precipitation for the conterminous United States. *ESSA Tech. Rept., Weather Bur., Silver Spring*, WB-5: iii + 60 pp.

JULIAN, L. T. and JULIAN, P. R., 1969. Boulder's winds. *Weatherwise*, 22: 108–112, 126.

KANGIESER, P. C., 1957. A possible singularity in the January minimum temperature at Phoenix, Arizona. *Mon. Weather Rev.*, 85: 42–44.

KESSELI, J. E., 1942. The climates of California according to the Köppen classification. *Geogr. Rev.*, 32: 476–480.

KLEIN, W. H., 1965. Application of synoptic climatology and short-range numerical prediction to five-day forecasting. *U.S. Weather Bur., Res. Pap.*, 46: vi + 109 pp.

KOHLER, M. A., NORDENSON, T. J. and FOX, W. E., 1955. Evaporation from pans and lakes. *U.S. Weather Bur., Res. Pap.*, 38: 21 pp.

KOPEC, R. J., 1965. Continentality around the Great Lakes. *Bull. Am. Meteorol. Soc.*, 46: 54–57.

KRAMER, F. L., 1963. The Koeppen dry/humid boundary: preliminary test of the Patton formula. *Prof. Geogr.*, 15: 13–14.

LALLY, V. E. and WATSON, B. F., 1960. Humiture revisited. *Weatherwise*, 13: 254–256.

LAMB, H. H., 1969. Climatic fluctuations. In: H. E. LANDSBERG (Editor), *World Survey of Climatology*, Vol. 2. Elsevier, Amsterdam, pp.173–249.

LANDSBERG, H. E., 1960a. Some patterns of rainfall in the north-central U.S.A. *Arch. Meteorol., Geophys. Bioklimatol.*, B-10: 165–174.

LANDSBERG, H. E., 1960b. Do tropical storms play a role in the water balance of the Northern Hemisphere? *J. Geophys. Res.*, 65: 1305–1307.

LANDSBERG, H. E., 1966. Interdiurnal variability of pressure and temperature in the conterminous United States. *U.S. Weather Bur., Wash., Tech. Pap.*, 56: 53 pp.

LANDSBERG, H. E., 1968. Atmospheric variability and climatic determinism. *Yearb., Assoc. Pac. Coast Geogr.*, 30: 13–24. (Also published in: A. COURT (Ed.), *Eclectic Climatology*. Oregon State University Press, Corvallis, Ore., 1968.)

LEATHERWOOD, R. K., 1955. Persistence of warm weather at Denver, Colorado. *Mon. Weather Rev.*, 83: 143–146.

LETHBRIDGE, M. D., 1970. Relationship between thunderstorm frequency and lunar phase and declination. *J. Geophys. Res.*, 75: 5149–5154.

LONG, M. J., 1966. A preliminary climatology of thunderstorm penetrations of the tropopause in the United States. *J. Appl. Meteorol.*, 5: 467–473.

LOWRY, W. P., 1956. Notes on the ending of Oregon's rainy season. *Bull. Am. Meteorol. Soc.*, 37: 426–428.

LUDLUM, D., 1954. Record low temperatures. *Weatherwise*, 7: 26, 45.

LUDLUM, D., 1963. 134°F. *Weatherwise*, 16: 116–117.

LUND, I. A., 1965. Indications of a lunar synodical period in United States observations of sunshine. *J. Atmosph. Sci.*, 22: 24–39.

LUND, I. A., 1966. Further indications of a lunar synodical period in observations of sunshine. *J. Atmosph. Sci.*, 23: 633–634.

MARKHAM, C. C., 1970. Seasonality of precipitation in the United States. *Ann. Assoc. Am. Geogr.*, 60: 593–597.

MARLATT, W. and RIEHL, H., 1963. Precipitation regimes over the upper Colorado River. *J. Geophys. Res.*, 68: 6447–6458.

McDONALD, J. E., 1960. Annual and seasonal persistence of precipitation in Arizona. *Bull. Am. Meteorol. Soc.*, 41: 106.

MEIGS, P. and DE PERCIN, F., 1957. Frequency of cold-wet climatic conditions in the United States. *Mon. Weather Rev.*, 85: 45–52.

MILLER, M. E., CANFIELD, N. L., RITTER, T. A. and WEAVER, C. R., 1972. Visibility changes in Ohio, Kentucky, and Tennessee from 1962 to 1969. *Mon. Weather Rev.*, 100: 67–71.

MORTON, F. I., 1967. Evaporation from large deep lakes. *Water Resour. Res.*, 3: 181–200.

MULLER, R. A., 1966. Snowbelts of the Great Lakes. *Weatherwise*, 19: 248–255.

MURRAY, C. R., 1965. Estimated use of water in the United States. *U.S. Geol. Surv., Circ.*, 556: vi + 53 pp.

NAMIAS, J., 1955. Some meteorological aspects of drought. *Mon. Weather Rev.*, 83: 199–205.

NAMIAS, J., 1968. A late November singularity. *Yearb., Assoc. Pac. Coast Geogr.*, 30: 55–62.

NEWMAN, E., 1965. Statistical investigation of anomalies in the winter temperature record of Boston, Mass. *J. Appl. Meteorol.*, 4: 706–713.

OLTMAN, R. E. and TRACY, H. J., 1951. Trends in climate and in precipitation-runoff relation in Missouri River basin. *U.S. Geol. Surv., Circ.*, 98: 113 pp.

PALMER, W. C., 1965. Meteorological drought. *U.S. Weather Bur., Res. Pap.*, 45: vi + 68 pp.

PALMER, W. C., 1971. Crop moisture index. Persistence of July drought. *Wkly Weather Crop Bull.*, 58(31): 12–13.

PATTON, C. P., 1962. A note on the classification of dry climates in the Köppen system. *Calif. Geogr.*, 3: 105–112.

PAULHUS, J. L. H. and MILLER, J. F., 1964. Average annual precipitation from daily amounts of 0.50 inch or greater. *Mon. Weather Rev.*, 92: 181–186.

PAUTZ, M. E. (Editor), 1969. Severe local storm occurrences, 1955–1967. *U.S. Weather Bur., ESSA Tech. Memo.*, WBTM FCST 12: iv + 77 pp.

PEACE JR., R. L., 1969. Heavy-fog regions in the conterminous United States. *Mon. Weather Rev.*, 97: 116–123.

PYKE, C. B., 1971. *Some Meteorological Aspects of the Seasonal Distribution of Precipitation in the Western United States and Baja California*. Ph.D. Dissertation, University of California, Los Angeles, Calif., 240 pp. (Also published as *Water Resour. Res. Cent., Univ. Calif., Contrib.*, 139: 205 pp.)

RAHN, J. J., 1971. Growing degree days and the 1971 growing season. Spring freeze probabilities and length of the growing season. *Wkly Weather Crop Bull.*, 58(13): 11.

RASMUSSEN, J. L., 1970. Atmospheric water balance and hydrology of the upper Colorado River basin. *Water Resour. Res.*, 6: 62–76.

RASMUSSON, E. M., 1967. Atmospheric water vapor transport and the water balance of North America. Characteristics of the water vapor flux field. *Mon. Weather Rev.*, 95: 403–426.

RASMUSSON, E. M., 1971. A study of the hydrology of eastern North America using atmospheric vapor flux data. *Mon. Weather Rev.*, 99: 119–135.

RASMUSSON, E. M., 1971. Diurnal variation of summertime thunderstorm activity over the United States. *Environ. Tech. Appl. Center, Air Weather Serv., U.S. Air Force, Tech. Note*, 71-4: 12 pp.

REBMAN, E. J., 1954. January temperature profile, Victoria, B.C. A west-coast singularity. *Weather*, 9: 131–134.

REITAN, C. H., 1960a. Mean monthly values of precipitable water over the United States, 1946–1956. *Mon. Weather Rev.*, 88: 25–35.

REITAN, C. H., 1960b. Distribution of precipitable water vapor over the continental United States. *Bull. Am. Meteorol. Soc.*, 41: 79–87.

RICHARDS, T. L. and IRBE, J. C., 1969. Estimates of monthly evaporation losses from the Great Lakes (1950–1968) based on the mass transfer technique. *Proc. Conf. Great Lakes Res., 12th, Ann Arbor*, pp.469–487.

ROBINSON, T. W. and HUNT, C. B., 1961. Some extremes of climate in Death Valley, California. Geological Survey Research in 1961. *U.S. Geol. Surv., Prof. Pap.*, 424-B:192–194.

ROONEY JR., J. F., 1967. The urban snow hazard in the United States: an appraisal of disruption. *Geogr. Rev.*, 57: 538–559.

RUDD, R. D., 1961. Summer frontal precipitation in the United States area of *Daf* climate. *J. Geophys. Res.*, 66: 125–130.

SERGIUS, L. A., ELLIS, G. R. and OGDEN, R. M., 1962. The Santa Ana winds of southern California. *Weatherwise*, 15: 102–105, 121.

SLUSSER, W. F., 1965. Wind rose maps of the United States. *Weatherwise*, 18: 260–263.

SLUSSER, W. F., 1968. *The Median vs. the Mean in Geographical Research*. M.A. Thesis, San Fernando Valley State College, Northridge, Calif., ix + 98 pp.

SPAR, J. (Editor), 1968. *Proceedings of the Conference on Drought in the Northeastern United States. Technical Report 68-3*. Department of Meteorology and Oceanography, New York University, 233 pp.

STIDD, C. K., 1967. The use of eigenvectors for climatic estimates. *J. Appl. Meteorol.*, 6: 255–264.

STOUT, G. E. and CHANGNON JR., S. A., 1968. Climatography of hail in the central United States. *Crop Hail Insur. Actuarial Assoc., Res. Rept.*, 38, ii + 49 pp.

SUGG, A. L., 1968. Beneficial aspects of the tropical cyclone. *J. Appl. Meteorol.*, 7: 39–45.

TATTELMAN, P. I., 1968. Duration of cold temperatures over North America. *U.S. Air Force Camb. Res. Lab., Environ. Res. Pap.*, 286: 23 pp.

TERJUNG, W. H., 1966. Physiologic climates of the conterminous United States: a bioclimatic classification based on man. *Ann. Assoc. Am. Geogr.*, 56: 141–179.

TERJUNG, W. H., 1967. Annual physioclimatic stresses and regimes in the United States. *Geogr. Rev.*, 57: 225–240.

THOM, E. C., 1956. Measuring the need for air conditioning. *Air Conditioning, Heating, Ventilating*, 53(8): 65–70.

THOM, E. C., 1957. A new concept for cooling degree-days. *Air Conditioning, Heating, Ventilating*, 54(6): 73–80.

THOM, E. C., 1958. Cooling degree-days. *Air Conditioning, Heating, Ventilating*, 55(7): 65–72.

THOM, E. C., 1959. The discomfort index. *Weatherwise*, 12: 57–60.

THOM, H. C. S., 1952. Seasonal degree-day statistics for the United States. *Mon. Weather Rev.*, 80: 143–147.

THOM, H. C. S., 1954. The rational relationship between heating degree days and temperature. *Mon. Weather Rev.*, 82: 1–6.

THOMAS, H. E. et al., 1962. Drought in the southwest, 1942–1956. *U.S. Geol. Surv., Profess. Pap.*, 372: 358 pp.

THORNTHWAITE, C. W., 1948. An approach toward a rational classification of climate. *Geogr. Rev.*, 38: 55–94.

TREWARTHA, G. T., 1961. *The Earth's Problem Climates*. University of Wisconsin Press, Madison, Wisc., vi + 334 pp.

TULLER, S. E., 1968. World distribution of mean monthly and annual precipitable water. *Mon. Weather Rev.*, 96: 785–797.

VESTAL, C. K., 1961. The precipitation day statistic. *Mon. Weather Rev.*, 89: 31–37.

VILLMOW, J. R., 1952. The position of the Köppen Da/Db boundary in eastern United States. *Ann. Assoc. Am. Geogr.*, 42: 94–97.

VISHER, S. S., 1954. *Climatic Atlas of the United States.* Harvard University Press, Cambridge, Mass., xii + 403 pp.

WACO, D. E., 1968. Frost pockets in the Santa Monica mountains of southern California. *Weather*, 23: 456–461.

WAHL, E. W., 1952. The January thaw in New England (an example of a weather singularity). *Bull. Am. Meteorol. Soc.*, 33: 380–386.

WAHL, E. W., 1954. A weather singularity over the U.S. in October. *Bull. Am. Meteorol. Soc.*, 35: 351–356.

WAITE, P. J., 1968. Lightning over Iowa. *Iowa Farm Sci.*, 22(11): 8–10.

WILLIAMS, L., 1964. Regionalization of freeze-thaw activity. *Ann. Assoc. Am. Geogr.*, 54: 597–611.

ZEGEL, F. H., 1967. Lightning deaths in the United States: a seven-year survey from 1959 to 1965. *Weatherwise*, 20: 169–173, 179.

## TABLE II

CLIMATIC TABLE FOR BIRMINGHAM, ALA.

Latitude 33°34′N, longitude 86°45′W, elevation 186 m

| Month | Mean sta. press. (mbar) | Temperature (°C) | | | | Mean vap. press. (mbar) | Precipitation (mm) | | Mean snowfall (cm) |
|---|---|---|---|---|---|---|---|---|---|
| | | daily mean | mean daily range | extremes | | | mean | max. in 24 h | |
| | | | | max. | min. | | | | |
| Jan. | 999.0 | 7.5 | 11.3 | 27 | −13 | 8 | 128 | 148 | 1 |
| Feb. | 995.8 | 8.7 | 11.8 | 27 | −19 | 8 | 134 | 117 | 1 |
| Mar. | 994.7 | 11.9 | 12.8 | 30 | −10 | 10 | 152 | 150 | tr. |
| Apr. | 994.0 | 16.7 | 13.5 | 32 | −3 | 13 | 114 | 95 | 0 |
| May | 993.8 | 21.3 | 13.4 | 34 | 2 | 18 | 87 | 74 | 0 |
| June | 993.5 | 25.5 | 12.6 | 39 | 8 | 23 | 102 | 98 | 0 |
| July | 995.0 | 26.7 | 11.6 | 41 | 11 | 25 | 131 | 80 | 0 |
| Aug. | 994.4 | 26.4 | 11.9 | 39 | 11 | 25 | 123 | 130 | 0 |
| Sept. | 994.8 | 23.5 | 12.6 | 38 | 4 | 21 | 85 | 100 | 0 |
| Oct. | 996.3 | 17.6 | 14.2 | 34 | −3 | 14 | 75 | 83 | tr. |
| Nov. | 997.8 | 11.0 | 13.6 | 28 | −15 | 9 | 90 | 124 | * |
| Dec. | 998.4 | 7.7 | 11.6 | 27 | −11 | 8 | 128 | 84 | 1 |
| Annual | 995.6 | 17.1 | 12.6 | 41 | −19 | 15 | 1,347 | 150 | 3 |
| Rec.(yrs.) | 30 | 30 | 30 | 17 | 17 | 10 | 30 | 17 | 17 |

| Month | Number of days with | | | Mean cloudiness (tenths) | Mean sunshine (h) | Wind | |
|---|---|---|---|---|---|---|---|
| | precip. (>0.25 mm) | thunderstorm | fog | | | most freq. direct. | mean speed (m/sec) |
| Jan. | 12 | 2 | 1 | 7.1 | 38 | S | 4 |
| Feb. | 11 | 3 | 1 | 6.6 | 47 | N | 4 |
| Mar. | 11 | 4 | * | 6.4 | 52 | S | 5 |
| Apr. | 9 | 5 | * | 5.7 | 62 | S | 4 |
| May | 9 | 7 | * | 5.9 | 64 | S | 3 |
| June | 10 | 8 | 1 | 5.8 | 64 | SSW | 3 |
| July | 12 | 11 | * | 6.1 | 60 | SSE | 3 |
| Aug. | 10 | 9 | * | 5.6 | 62 | NE | 3 |
| Sept. | 7 | 3 | 1 | 5.3 | 61 | ENE | 3 |
| Oct. | 6 | 1 | 1 | 4.6 | 64 | ENE | 3 |
| Nov. | 9 | 2 | 1 | 5.4 | 56 | N | 4 |
| Dec. | 11 | 1 | 1 | 6.4 | 44 | N | 4 |
| Annual | 117 | 56 | 7 | 5.9 | 56 | S | 4 |
| Rec.(yrs.) | 17 | 17 | 17 | 17 | 17 | 12 | 17 |

TABLE III

Latitude 33°26′N, longitude 112°01′W, elevation 340 m

| Month | Mean sta. press. (mbar) | Temperature (°C) | | | | Mean vap. press. (mbar) | Precipitation (mm) | | Mean snowfall (cm) |
|---|---|---|---|---|---|---|---|---|---|
| | | daily mean | mean daily range | extremes | | | mean | max. in 24 h | |
| | | | | max. | min. | | | | |
| Jan. | 977.9 | 10.4 | 14.8 | 29 | −8 | 5 | 19 | 33 | tr. |
| Feb. | 976.5 | 12.5 | 15.1 | 31 | −6 | 5 | 22 | 27 | tr. |
| Mar. | 974.2 | 15.8 | 16.2 | 33 | −2 | 5 | 17 | 34 | 0 |
| Apr. | 972.2 | 20.4 | 17.0 | 40 | 0 | 7 | 8 | 35 | tr. |
| May | 970.4 | 25.0 | 17.7 | 45 | 6 | 7 | 3 | 24 | 0 |
| June | 969.3 | 29.8 | 17.8 | 47 | 10 | 8 | 2 | 24 | 0 |
| July | 970.7 | 32.9 | 14.8 | 48 | 16 | 16 | 20 | 50 | 0 |
| Aug. | 971.0 | 31.7 | 14.0 | 46 | 16 | 17 | 28 | 78 | 0 |
| Sept. | 970.7 | 29.1 | 15.5 | 48 | 9 | 11 | 19 | 54 | 0 |
| Oct. | 973.3 | 22.3 | 16.2 | 40 | 2 | 8 | 12 | 37 | 0 |
| Nov. | 976.7 | 15.1 | 16.3 | 33 | −4 | 6 | 12 | 27 | 0 |
| Dec. | 977.9 | 11.4 | 15.1 | 31 | −6 | 5 | 22 | 40 | 0 |
| Annual | 973.4 | 21.4 | 15.8 | 48 | −8 | 8 | 184 | 78 | tr. |
| Rec.(yrs.) | 30 | 30 | 30 | 23 | 23 | 10 | 30 | 23 | 23 |

| Month | Number of days with | | | Mean cloud-iness (tenths) | Mean sun-shine (h) | Wind | | Solar radiation (ly/ month) |
|---|---|---|---|---|---|---|---|---|
| | precip. (>0.25 mm) | thunder-storm | fog | | | most freq. direct. | mean speed (m/sec) | |
| Jan. | 4 | * | 1 | 5.1 | 77 | E | 2 | 301 |
| Feb. | 4 | * | * | 4.5 | 79 | E | 2 | 406 |
| Mar. | 3 | * | * | 4.3 | 83 | E | 3 | 531 |
| Apr. | 2 | 1 | 0 | 3.8 | 88 | E | 3 | 640 |
| May | 1 | 1 | 0 | 2.9 | 93 | E | 3 | 718 |
| June | 1 | 1 | 0 | 2.0 | 94 | E | 3 | 730 |
| July | 4 | 6 | 0 | 3.8 | 84 | W | 3 | 659 |
| Aug. | 5 | 7 | 0 | 3.5 | 84 | E | 3 | 606 |
| Sept. | 3 | 3 | 0 | 1.9 | 89 | E | 2 | 555 |
| Oct. | 3 | 1 | 0 | 2.8 | 88 | E | 2 | 447 |
| Nov. | 2 | * | * | 3.1 | 84 | E | 2 | 339 |
| Dec. | 4 | 1 | 1 | 3.5 | 77 | E | 2 | 281 |
| Annual | 34 | 22 | 2 | 3.4 | 86 | E | 2 | 518 |
| Rec.(yrs.) | 21 | 21 | 23 | 15 | 65 | 15 | 15 | 14 |

TABLE IV

CLIMATIC TABLE FOR WINSLOW, ARIZ.
Latitude 35°01′N, longitude 110°44′W, elevation 1,487 m

| Month | Temperature (°C) | | | | Mean vap. press. (mbar) | Precipitation (mm) | | Mean snowfall (cm) |
|-------|------------------|---|---|---|-------------------------|--------------------|---|--------------------|
| | daily mean | mean daily range | extremes | | | mean | max. in 24 h | |
| | | | max. | min. | | | | |
| Jan. | −0.3 | 16.3 | 23 | −28 | 4 | 11 | 14 | 6 |
| Feb. | 3.3 | 17.2 | 26 | −22 | 4 | 12 | 18 | 4 |
| Mar. | 7.8 | 19.0 | 27 | −14 | 4 | 10 | 12 | 3 |
| Apr. | 13.4 | 19.3 | 33 | −8 | 5 | 11 | 12 | 1 |
| May | 18.5 | 19.8 | 38 | −3 | 5 | 8 | 16 | tr. |
| June | 23.8 | 20.3 | 39 | 2 | 6 | 7 | 54 | 0 |
| July | 27.1 | 17.8 | 41 | 8 | 12 | 26 | 37 | 0 |
| Aug. | 25.7 | 16.8 | 39 | 8 | 10 | 36 | 35 | 0 |
| Sept. | 21.8 | 18.4 | 37 | 2 | 6 | 23 | 30 | tr. |
| Oct. | 14.4 | 18.9 | 32 | −7 | 5 | 17 | 26 | tr. |
| Nov. | 5.4 | 18.5 | 26 | −18 | 5 | 9 | 15 | 2 |
| Dec. | 0.6 | 15.8 | 23 | −24 | 4 | 13 | 18 | 5 |
| Annual | 13.4 | 18.2 | 41 | −28 | 6 | 183 | 54 | 21 |
| Rec.(yrs.) | 30 | 30 | 29 | 29 | 10 | 30 | 16 | 29 |

| Month | Number of days with | | | Mean cloud-iness (tenths) | Wind | |
|-------|---------------------|---|---|---------------------------|------|---|
| | precip. (> 0.25 mm) | thunder-storm | fog | | most freq. direct. | mean speed (m/sec) |
| Jan. | 4 | 0 | 1 | 5.6 | SE | 3 |
| Feb. | 4 | * | * | 4.7 | SE | 4 |
| Mar. | 4 | * | * | 4.5 | WSW | 5 |
| Apr. | 3 | 2 | 0 | 4.3 | SW | 5 |
| May | 3 | 2 | 0 | 3.8 | SW | 5 |
| June | 2 | 3 | 0 | 2.6 | SW | 5 |
| July | 7 | 11 | 0 | 4.9 | WSW | 4 |
| Aug. | 9 | 11 | 0 | 4.4 | SW | 4 |
| Sept. | 5 | 5 | 0 | 2.6 | SW | 4 |
| Oct. | 4 | 2 | * | 2.9 | SE | 3 |
| Nov. | 3 | * | 1 | 3.4 | SE | 3 |
| Dec. | 4 | * | 2 | 4.5 | SE | 3 |
| Annual | 53 | 36 | 4 | 4.0 | SW | 4 |
| Rec.(yrs.) | 29 | 12 | 12 | 12 | 22 | 17 |

TABLE V

CLIMATIC TABLE FOR LITTLE ROCK, ARK.
Latitude 34°44′N, longitude 92°14′W, elevation 78 m (109 m)

| Month | Mean sta. press. (mbar) | Temperature (°C) | | | | Mean vap. press. (mbar) | Precipitation (mm) | | Mean snowfall (cm) |
|-------|------|------|------|------|------|------|------|------|------|
| | | daily mean | mean daily range | extremes | | | mean | max. in 24 h | |
| | | | | max. | min. | | | | |
| Jan. | 1007.6 | 4.8 | 11.2 | 28 | −14 | 7 | 133 | 92 | 6 |
| Feb. | 1006.2 | 6.9 | 11.4 | 28 | −21 | 7 | 110 | 131 | 3 |
| Mar. | 1003.5 | 11.0 | 12.1 | 30 | −12 | 8 | 122 | 86 | 1 |
| Apr. | 1002.7 | 16.9 | 12.4 | 33 | 0 | 12 | 125 | 114 | tr. |
| May | 1001.8 | 21.4 | 12.2 | 36 | 4 | 18 | 134 | 196 | 0 |
| June | 1001.7 | 26.1 | 12.1 | 40 | 11 | 23 | 92 | 117 | 0 |
| July | 1002.7 | 27.7 | 12.0 | 42 | 13 | 25 | 85 | 87 | 0 |
| Aug. | 1002.7 | 27.4 | 12.4 | 42 | 14 | 24 | 72 | 75 | 0 |
| Sept. | 1003.7 | 23.5 | 13.4 | 41 | 3 | 18 | 82 | 94 | 0 |
| Oct. | 1005.4 | 17.3 | 14.3 | 34 | −1 | 13 | 73 | 103 | 0 |
| Nov. | 1007.1 | 9.7 | 13.2 | 30 | −8 | 8 | 105 | 129 | * |
| Dec. | 1007.6 | 5.5 | 11.3 | 27 | −13 | 7 | 104 | 81 | 2 |
| Annual | 1004.4 | 16.5 | 12.3 | 42 | −21 | 14 | 1,237 | 196 | 12 |
| Rec.(yrs.) | 30 | 30 | 30 | 19 | 19 | 10 | 30 | 19 | 19 |

| Month | Number of days with | | | Mean cloudiness (tenths) | Mean sunshine (h) | Wind | | Solar radiation (ly/ month) |
|-------|------|------|------|------|------|------|------|------|
| | precip. (>0.25 mm) | thunderstorm | fog | | | most freq. direct. | mean speed (m/sec) | |
| Jan. | 10 | 2 | 2 | 6.9 | 42 | S | 4 | 199 |
| Feb. | 10 | 3 | 2 | 6.3 | 51 | SW | 4 | 264 |
| Mar. | 10 | 5 | 1 | 6.3 | 54 | WNW | 5 | 366 |
| Apr. | 11 | 7 | 1 | 6.0 | 60 | S | 4 | 449 |
| May | 10 | 8 | 1 | 6.1 | 64 | S | 4 | 531 |
| June | 8 | 7 | * | 5.3 | 70 | SSW | 4 | 556 |
| July | 8 | 9 | * | 5.5 | 69 | SW | 3 | 547 |
| Aug. | 7 | 6 | 1 | 4.9 | 72 | SW | 3 | 509 |
| Sept. | 7 | 5 | 1 | 4.6 | 69 | NNE | 3 | 427 |
| Oct. | 6 | 2 | 2 | 4.4 | 70 | SW | 3 | 341 |
| Nov. | 8 | 4 | 1 | 5.2 | 57 | SW | 4 | 238 |
| Dec. | 9 | 2 | 2 | 6.3 | 47 | SW | 4 | 186 |
| Annual | 104 | 60 | 14 | 5.7 | 62 | SW | 4 | 384 |
| Rec.(yrs.) | 19 | 19 | 19 | 19 | 19 | 12 | 19 | 10 |

## TABLE VI

CLIMATIC TABLE FOR BLUE CANYON, CALIF.
Latitude 39°17′N, longitude 120°42′W, elevation 161 m

| Month | Temperature (°C) | | | | Mean vap. press. (mbar) | Precipitation (mm) | | Mean snowfall (cm) |
|---|---|---|---|---|---|---|---|---|
| | daily mean | mean daily range | extremes | | | mean | max. in 24 h | |
| | | | max. | min. | | | | |
| Jan. | 2.3 | 7.3 | 21 | −15 | 5 | 297 | 186 | 138 |
| Feb. | 2.6 | 7.5 | 23 | −14 | 5 | 259 | 211 | 111 |
| Mar. | 3.9 | 8.0 | 22 | −13 | 5 | 215 | 110 | 137 |
| Apr. | 7.3 | 9.0 | 26 | −8 | 7 | 122 | 106 | 75 |
| May | 10.9 | 9.6 | 30 | −6 | 8 | 79 | 131 | 23 |
| June | 15.1 | 10.1 | 33 | −2 | 9 | 24 | 32 | 2 |
| July | 20.3 | 10.6 | 33 | 4 | 9 | 2 | 25 | 0 |
| Aug. | 19.8 | 11.0 | 33 | 3 | 10 | 2 | 18 | 0 |
| Sept. | 17.4 | 10.6 | 34 | −2 | 8 | 16 | 78 | 1 |
| Oct. | 12.1 | 9.4 | 28 | −6 | 8 | 87 | 118 | 5 |
| Nov. | 7.3 | 8.5 | 26 | −11 | 7 | 171 | 220 | 43 |
| Dec. | 4.3 | 7.5 | 24 | −11 | 6 | 260 | 236 | 99 |
| Annual | 10.3 | 9.1 | 34 | −15 | 7 | 1,534 | 236 | 634 |
| Rec.(yrs.) | 30 | 30 | 17 | 17 | 10 | 30 | 21 | 20 |

| Month | Number of days with | | | Mean cloud-iness (tenths) | Wind | |
|---|---|---|---|---|---|---|
| | precip. (>0.25 mm) | thunder-storm | fog | | most freq. direct. | mean speed (m/sec) |
| Jan. | 13 | * | 11 | 7.2 | SSE | 5 |
| Feb. | 12 | * | 10 | 6.3 | SSE | 5 |
| Mar. | 13 | 1 | 11 | 6.5 | SSW | 5 |
| Apr. | 9 | 1 | 8 | 5.9 | ENE | 4 |
| May | 8 | 3 | 6 | 5.1 | ENE | 4 |
| June | 3 | 2 | 2 | 3.0 | SSW | 4 |
| July | 1 | 2 | 0 | 1.5 | SSW | 3 |
| Aug. | 1 | 1 | * | 1.7 | SW | 3 |
| Sept. | 2 | 1 | 1 | 2.4 | WSW | 4 |
| Oct. | 6 | 1 | 5 | 3.8 | ENE | 4 |
| Nov. | 9 | * | 6 | 5.3 | ENE | 4 |
| Dec. | 12 | 0 | 10 | 6.3 | ENE | 4 |
| Annual | 88 | 13 | 68 | 4.6 | ENE | 4 |
| Rec.(yrs.) | 21 | 20 | 19 | 12 | 6 | 6 |

TABLE VII

CLIMATIC TABLE FOR EUREKA, CALIF.
Latitude 40°48′ N, longitude 124°10′W, elevation 13 m (18 m)

| Month | Mean sta. press. (mbar) | Temperature (°C) | | | | Mean vap. press. (mbar) | Precipitation (mm) | | Mean snowfall (cm) |
|---|---|---|---|---|---|---|---|---|---|
| | | daily mean | mean daily range | extremes | | | mean | max. in 24 h | |
| | | | | max. | min. | | | | |
| Jan. | 1017.7 | 8.6 | 7.0 | 24 | −4 | 8 | 170 | 112 | 1 |
| Feb. | 1017.0 | 9.0 | 6.7 | 29 | −3 | 8 | 140 | 124 | tr. |
| Mar. | 1017.2 | 9.3 | 6.5 | 26 | −2 | 8 | 133 | 78 | tr. |
| Apr. | 1017.0 | 10.2 | 5.9 | 26 | 0 | 11 | 68 | 60 | tr. |
| May | 1016.3 | 11.7 | 5.4 | 29 | 3 | 11 | 55 | 57 | 0 |
| June | 1015.6 | 13.1 | 5.1 | 29 | 5 | 12 | 19 | 44 | 0 |
| July | 1015.4 | 13.5 | 4.7 | 24 | 7 | 14 | 3 | 30 | 0 |
| Aug. | 1015.0 | 13.7 | 4.7 | 23 | 7 | 14 | 3 | 23 | 0 |
| Sept. | 1014.1 | 13.6 | 5.8 | 29 | 5 | 14 | 16 | 30 | 0 |
| Oct. | 1015.9 | 12.4 | 6.6 | 28 | 1 | 11 | 81 | 148 | 0 |
| Nov. | 1018.1 | 10.7 | 7.0 | 25 | −2 | 10 | 117 | 116 | 0 |
| Dec. | 1017.5 | 9.4 | 6.9 | 21 | −6 | 8 | 170 | 106 | tr. |
| Annual | 1016.4 | 11.3 | 6.1 | 29 | −6 | 11 | 975 | 148 | 1 |
| Rec.(yrs.) | 30 | 30 | 30 | 50 | 50 | 10 | 30 | 50 | 50 |

| Month | Number of days with | | | Mean cloud-iness (tenths) | Mean sun-shine (h) | Wind | |
|---|---|---|---|---|---|---|---|
| | precip. (>0.25 mm) | thunder-storm | fog | | | most freq. direct.[1] | mean speed (m/sec) |
| Jan. | 17 | 1 | 4 | 7.4 | 40 | SE | 3 |
| Feb. | 14 | 1 | 3 | 7.3 | 44 | SE | 3 |
| Mar. | 15 | * | 2 | 7.3 | 49 | N | 3 |
| Apr. | 12 | * | 2 | 7.2 | 53 | N | 4 |
| May | 9 | * | 1 | 6.9 | 54 | N | 4 |
| June | 5 | * | 2 | 6.4 | 56 | N | 3 |
| July | 2 | * | 3 | 6.6 | 51 | N | 3 |
| Aug. | 2 | * | 5 | 6.9 | 46 | NW | 3 |
| Sept. | 5 | * | 8 | 6.0 | 51 | N | 2 |
| Oct. | 9 | * | 9 | 6.5 | 48 | N | 3 |
| Nov. | 12 | 1 | 7 | 7.1 | 42 | SE | 3 |
| Dec. | 15 | 1 | 4 | 7.5 | 39 | SE | 3 |
| Annual | 118 | 5 | 48 | 6.9 | 49 | N | 3 |
| Rec.(yrs.) | 50 | 50 | 50 | 18 | 30 | 30 | 30 |

[1] To 8 points.

272

## TABLE VIII

CLIMATIC TABLE FOR FRESNO, CALIF.
Latitude 36°46′N, longitude 119°42′W, elevation 100 m

| Month | Temperature (°C) | | | | Mean vap. press. (mbar) | Precipitation (mm) | | Mean snowfall (cm) |
|-------|------------------|---|---------|---|-------|------|----------|--------|
| | daily mean | mean daily range | extremes | | | mean | max. in 24 h | |
| | | | max. | min. | | | | |
| Jan. | 7.5 | 10.8 | 23 | −8 | 7 | 52 | 62 | tr. |
| Feb. | 9.9 | 12.5 | 27 | −4 | 7 | 56 | 49 | 0 |
| Mar. | 12.4 | 14.3 | 32 | −3 | 8 | 50 | 41 | tr. |
| Apr. | 16.0 | 15.8 | 36 | 0 | 9 | 29 | 27 | 0 |
| May | 19.7 | 17.3 | 39 | 3 | 10 | 8 | 24 | 0 |
| June | 23.3 | 18.4 | 43 | 7 | 11 | 2 | 8 | 0 |
| July | 27.0 | 20.0 | 44 | 10 | 12 | tr. | 1 | 0 |
| Aug. | 25.8 | 20.2 | 43 | 11 | 12 | * | 2 | 0 |
| Sept. | 23.3 | 19.5 | 44 | 3 | 12 | 3 | 23 | 0 |
| Oct. | 17.9 | 17.8 | 38 | −1 | 10 | 11 | 21 | 0 |
| Nov. | 11.8 | 15.8 | 32 | −3 | 8 | 24 | 34 | 0 |
| Dec. | 8.0 | 11.0 | 24 | −5 | 7 | 50 | 45 | 0 |
| Annual | 16.9 | 16.1 | 44 | −8 | 9 | 285 | 62 | tr. |
| Rec.(yrs.) | 30 | 30 | 21 | 21 | 10 | 30 | 21 | 21 |

| Month | Number of days with | | | Mean cloud-iness (tenths) | Mean sun-shine (h) | Wind | | Solar radiation (ly/ month) |
|-------|---------------------|---------------|-----|-------|-------|------|------|------|
| | precip. (>0.25 mm) | thunder-storm | fog | | | most freq. direct. | mean speed (m/sec) | |
| Jan. | 8 | * | 11 | 7.0 | 51 | SE | 2 | 182 |
| Feb. | 7 | * | 6 | 6.1 | 64 | NW | 2 | 285 |
| Mar. | 7 | 1 | 2 | 5.4 | 76 | NW | 3 | 425 |
| Apr. | 5 | 1 | * | 4.7 | 82 | NW | 3 | 549 |
| May | 2 | 1 | * | 3.5 | 88 | NW | 4 | 642 |
| June | 1 | * | 0 | 1.7 | 96 | NW | 4 | 701 |
| July | * | * | 0 | 1.1 | 98 | NW | 3 | 684 |
| Aug. | * | * | 0 | 1.1 | 97 | NW | 3 | 621 |
| Sept. | 1 | * | * | 1.5 | 96 | NW | 3 | 506 |
| Oct. | 2 | 1 | 1 | 2.9 | 88 | NW | 2 | 374 |
| Nov. | 4 | * | 6 | 4.9 | 72 | NW | 2 | 248 |
| Dec. | 8 | * | 11 | 6.8 | 48 | SE | 2 | 158 |
| Annual | 46 | 5 | 38 | 3.9 | 82 | NW | 3 | 448 |
| Rec.(yrs.) | 21 | 21 | 21 | 21 | 21 | 12 | 21 | 34 |

TABLE IX

CLIMATIC TABLE FOR LOS ANGELES, CALIF.
Latitude 34°03′N, longitude 118°14′W, elevation 95 m (103 m)

| Month | Mean sta. press. (mbar) | Temperature (°C) | | | | Mean vap. press. (mbar) | Precipitation (mm) | | Mean snowfall (cm) |
|---|---|---|---|---|---|---|---|---|---|
| | | daily mean | mean daily range | extremes | | | mean | max. in 24 h | |
| | | | | max. | min. | | | | |
| Jan. | 1009.7 | 13.2 | 10.2 | 30 | −2 | 9 | 78 | 155 | tr. |
| Feb. | 1009.0 | 13.9 | 9.9 | 32 | 1 | 9 | 85 | 102 | tr. |
| Mar. | 1007.6 | 15.2 | 10.2 | 32 | 3 | 10 | 57 | 87 | 0 |
| Apr. | 1006.5 | 16.6 | 9.8 | 36 | 6 | 12 | 30 | 52 | 0 |
| May | 1005.6 | 18.2 | 9.7 | 37 | 8 | 12 | 4 | 27 | 0 |
| June | 1004.4 | 20.0 | 10.1 | 40 | 10 | 14 | 2 | 4 | 0 |
| July | 1004.5 | 22.8 | 11.5 | 39 | 12 | 16 | tr. | * | 0 |
| Aug. | 1004.5 | 22.8 | 11.3 | 38 | 12 | 17 | 1 | 10 | 0 |
| Sept. | 1003.7 | 22.2 | 11.7 | 43 | 11 | 14 | 6 | 9 | 0 |
| Oct. | 1006.0 | 19.7 | 11.1 | 40 | 8 | 11 | 10 | 16 | 0 |
| Nov. | 1008.6 | 17.1 | 11.8 | 35 | 4 | 10 | 27 | 59 | 0 |
| Dec. | 1009.6 | 14.6 | 10.4 | 32 | 0 | 9 | 73 | 76 | tr. |
| Annual | 1006.7 | 18.0 | 10.7 | 43 | −2 | 12 | 373 | 155 | tr. |
| Rec(yrs.) | 30 | 30 | 30 | 20 | 20 | 10 | 30 | 20 | 20 |

| Month | Number of days with | | | Mean cloud-iness (tenths) | Mean sun-shine (h) | Wind | | Solar radiation (ly/month) |
|---|---|---|---|---|---|---|---|---|
| | precip. (>0.25 mm) | thunder-storm | fog | | | most freq. direct. | mean speed (m/sec) | |
| Jan. | 6 | 1 | 2 | 4.6 | 71 | NE | 3 | 248 |
| Feb. | 5 | 1 | 2 | 4.8 | 72 | W | 3 | 325 |
| Mar. | 6 | 1 | 1 | 4.8 | 74 | W | 3 | 438 |
| Apr. | 4 | 1 | 1 | 5.2 | 67 | W | 3 | 503 |
| May | 2 | * | 1 | 4.7 | 68 | W | 3 | 555 |
| June | 1 | * | 1 | 4.1 | 69 | W | 3 | 578 |
| July | * | * | 1 | 2.8 | 81 | W | 2 | 658 |
| Aug. | 1 | * | 1 | 2.7 | 82 | W | 2 | 592 |
| Sept. | 1 | * | 1 | 2.8 | 81 | W | 2 | 539 |
| Oct. | 2 | * | 3 | 3.8 | 74 | W | 3 | 391 |
| Nov. | 4 | 1 | 2 | 3.5 | 80 | W | 3 | 303 |
| Dec. | 5 | 1 | 2 | 4.4 | 72 | NE | 3 | 257 |
| Annual | 37 | 6 | 17 | 4.0 | 74 | W | 3 | 449 |
| Rec.(yrs.) | 20 | 20 | 20 | 20 | 20 | 20 | 20 | 11 |

TABLE X

CLIMATIC TABLE FOR SAN DIEGO, CALIF.
Latitude 32°44′N, longitude 117°10′W, elevation 4 m (26 m)

| Month | Mean sta. press. (mbar) | Temperature (°C) | | | | Mean vap. press. (mbar) | Precipitation (mm) | | Mean snowfall (cm) |
|---|---|---|---|---|---|---|---|---|---|
| | | daily mean | mean daily range | extremes | | | mean | max. in 24 h | |
| | | | | max. | min. | | | | |
| Jan. | 1015.3 | 13.1 | 10.1 | 31 | −2 | 8 | 51 | 67 | tr. |
| Feb. | 1014.7 | 13.7 | 9.7 | 31 | 2 | 9 | 55 | 43 | 0 |
| Mar. | 1013.4 | 14.7 | 9.2 | 31 | 5 | 10 | 40 | 61 | 0 |
| Apr. | 1012.1 | 16.1 | 8.0 | 33 | 5 | 12 | 20 | 31 | 0 |
| May | 1011.2 | 17.5 | 7.2 | 36 | 9 | 12 | 4 | 11 | 0 |
| June | 1010.0 | 18.2 | 6.6 | 36 | 10 | 14 | 1 | 7 | 0 |
| July | 1009.9 | 20.9 | 6.9 | 34 | 13 | 15 | tr. | 2 | 0 |
| Aug. | 1009.6 | 21.5 | 6.9 | 37 | 14 | 17 | 2 | 21 | 0 |
| Sept. | 1008.7 | 20.8 | 8.0 | 40 | 11 | 14 | 4 | 16 | 0 |
| Oct. | 1011.1 | 18.7 | 8.7 | 37 | 7 | 11 | 12 | 30 | 0 |
| Nov. | 1013.7 | 16.3 | 11.0 | 36 | 4 | 10 | 23 | 62 | 0 |
| Dec. | 1014.7 | 14.2 | 10.5 | 29 | 2 | 8 | 52 | 78 | 0 |
| Annual | 1012.0 | 17.2 | 8.6 | 40 | −2 | 12 | 264 | 78 | tr. |
| Rec.(yrs.) | 30 | 30 | 30 | 20 | 20 | 10 | 30 | 20 | 20 |

| Month | Number of days with | | | Mean cloud-iness (tenths) | Mean sun-shine (h) | Wind | |
|---|---|---|---|---|---|---|---|
| | precip. (>0.25 mm) | thunder-storm | fog | | | most freq. direct. | mean speed (m/sec) |
| Jan. | 7 | * | 4 | 5.0 | 68 | NE | 3 |
| Feb. | 7 | * | 3 | 5.0 | 70 | NE | 3 |
| Mar. | 7 | * | 2 | 4.9 | 70 | WNW | 3 |
| Apr. | 5 | * | 2 | 5.6 | 61 | WNW | 3 |
| May | 2 | * | 1 | 5.4 | 59 | WNW | 3 |
| June | 1 | * | 1 | 5.1 | 58 | SSW | 3 |
| July | * | * | 1 | 4.3 | 67 | WNW | 3 |
| Aug. | * | * | 1 | 4.2 | 68 | WNW | 3 |
| Sept. | 1 | * | 4 | 3.7 | 69 | NW | 3 |
| Oct. | 3 | * | 4 | 4.2 | 65 | WSW | 3 |
| Nov. | 4 | * | 5 | 3.7 | 76 | NE | 2 |
| Dec. | 6 | * | 4 | 4.6 | 70 | NE | 2 |
| Annual | 43 | 3 | 30 | 4.6 | 67 | WNW | 3 |
| Rec.(yrs.) | 20 | 20 | 20 | 20 | 20 | 12 | 20 |

TABLE XI

CLIMATIC TABLE FOR SAN FRANCISCO, CALIF.
Latitude 37°47′N, longitude 122°25′W, elevation 16 m (47 m)

| Month | Mean sta. press. (mbar) | Temperature (°C) | | | | | Mean vap. press. (mbar) | Precipitation (mm) | | Mean snowfall (cm) |
|---|---|---|---|---|---|---|---|---|---|---|
| | | daily mean | mean daily range | extremes | | | | mean | max. in 24 h | |
| | | | | max. | min. | | | | | |
| Jan. | 1015.9 | 10.4 | 5.7 | 24 | −1 | | 9 | 116 | 89 | tr. |
| Feb. | 1014.8 | 11.7 | 6.3 | 24 | 2 | | 9 | 93 | 59 | tr. |
| Mar. | 1014.0 | 12.6 | 6.7 | 28 | 3 | | 9 | 74 | 93 | tr. |
| Apr. | 1013.0 | 13.2 | 6.9 | 30 | 6 | | 11 | 37 | 60 | 0 |
| May | 1011.8 | 14.1 | 6.7 | 33 | 7 | | 11 | 16 | 33 | 0 |
| June | 1010.3 | 15.1 | 6.6 | 35 | 8 | | 13 | 4 | 15 | 0 |
| July | 1010.3 | 14.9 | 6.1 | 33 | 8 | | 14 | tr. | 2 | 0 |
| Aug. | 1010.2 | 15.2 | 6.1 | 31 | 9 | | 14 | 1 | 6 | 0 |
| Sept. | 1009.7 | 16.7 | 7.7 | 36 | 9 | | 14 | 6 | 52 | 0 |
| Oct. | 1012.1 | 16.3 | 7.7 | 32 | 7 | | 11 | 23 | 39 | 0 |
| Nov. | 1015.2 | 14.1 | 7.1 | 29 | 6 | | 11 | 51 | 62 | 0 |
| Dec. | 1015.5 | 11.4 | 5.6 | 24 | 3 | | 9 | 108 | 80 | tr. |
| Annual | 1012.8 | 13.8 | 6.6 | 36 | −1 | | 11 | 529 | 93 | tr. |
| Rec.(yrs.) | 30 | 30 | 30 | 24 | 24 | | 10 | 30 | 24 | 24 |

| Month | Number of days with | | | Mean cloud-iness[1] (tenths) | Mean sun-shine (h) | Wind | |
|---|---|---|---|---|---|---|---|
| | precip. (>0.25 mm) | thunder-storm | fog[1] | | | most freq. direct.[2] | mean speed (m/sec) |
| Jan. | 11 | * | 4 | 6.2 | 54 | N | 3 |
| Feb. | 11 | * | 3 | 5.8 | 59 | W | 3 |
| Mar. | 10 | * | 1 | 5.6 | 66 | W | 4 |
| Apr. | 6 | * | * | 5.3 | 70 | W | 4 |
| May | 4 | * | * | 4.6 | 71 | W | 5 |
| June | 2 | * | * | 3.7 | 74 | W | 5 |
| July | 1 | * | * | 3.2 | 64 | W | 5 |
| Aug. | 1 | * | * | 3.4 | 64 | W | 5 |
| Sept. | 1 | * | 1 | 3.2 | 72 | W | 4 |
| Oct. | 4 | * | 2 | 4.0 | 70 | W | 3 |
| Nov. | 7 | * | 3 | 5.1 | 64 | W | 3 |
| Dec. | 10 | * | 5 | 6.0 | 53 | N | 3 |
| Annual | 67 | 2 | 21 | 4.7 | 66 | W | 4 |
| Rec.(yrs.) | 24 | 24 | 23 | 19 | 24 | 24 | 24 |

[1] Airport data.
[2] To 8 points.

## TABLE XII

CLIMATIC TABLE FOR DENVER, COLO.
Latitude 39°46′N, longitude 104°53′W, elevation 1,610 m (1,613 m)

| Month | Mean sta. press. (mbar) | Temperature (°C) | | | | | Mean vap. press. (mbar) | Precipitation (mm) | | Mean snowfall (cm) |
|---|---|---|---|---|---|---|---|---|---|---|
| | | daily mean | mean daily range | extremes | | | | mean | max. in 24 h | |
| | | | | max. | min. | | | | | |
| Jan. | 835.1 | −1.1 | 14.6 | 22 | −31 | 2 | 14 | 20 | 23 | |
| Feb. | 834.6 | 0.3 | 14.6 | 24 | −34 | 2 | 18 | 26 | 20 | |
| Mar. | 833.6 | 3.3 | 14.5 | 27 | −24 | 3 | 31 | 38 | 31 | |
| Apr. | 834.8 | 8.6 | 14.6 | 29 | −14 | 4 | 54 | 61 | 26 | |
| May | 835.9 | 13.7 | 14.3 | 36 | − 6 | 7 | 69 | 84 | 4 | |
| June | 836.7 | 19.4 | 15.6 | 40 | − 1 | 8 | 37 | 52 | tr. | |
| July | 840.1 | 23.0 | 15.6 | 40 | 6 | 9 | 39 | 42 | 0 | |
| Aug. | 839.9 | 22.2 | 15.3 | 38 | 5 | 8 | 33 | 87 | 0 | |
| Sept. | 839.2 | 17.5 | 16.1 | 36 | − 4 | 7 | 29 | 62 | 4 | |
| Oct. | 838.6 | 11.3 | 15.8 | 31 | −13 | 5 | 26 | 43 | 8 | |
| Nov. | 837.6 | 4.0 | 15.1 | 26 | −22 | 3 | 18 | 25 | 20 | |
| Dec. | 836.0 | 0.6 | 14.6 | 23 | −24 | 2 | 12 | 18 | 15 | |
| Annual | 836.8 | 10.2 | 15.0 | 40 | −34 | 5 | 380 | 87 | 151 | |
| Rec.(yrs.) | 30 | 30 | 30 | 26 | 26 | 10 | 30 | 26 | 26 | |

| Month | Number of days with | | | Mean cloud-iness (tenths) | Mean sun-shine (h) | Wind | |
|---|---|---|---|---|---|---|---|
| | precip. (>0.25 mm) | thunder-storm | fog | | | most freq. direct. | mean speed (m/sec) |
| Jan. | 6 | 0 | 1 | 5.3 | 70 | S | 4 |
| Feb. | 6 | * | 2 | 5.8 | 70 | S | 5 |
| Mar. | 8 | * | 1 | 5.9 | 69 | S | 5 |
| Apr. | 9 | 1 | 1 | 6.4 | 62 | S | 5 |
| May | 11 | 6 | 1 | 6.3 | 64 | S | 5 |
| June | 9 | 10 | 1 | 4.8 | 72 | S | 4 |
| July | 9 | 12 | * | 4.9 | 70 | S | 4 |
| Aug. | 8 | 8 | 1 | 4.8 | 73 | S | 4 |
| Sept. | 6 | 3 | 1 | 4.0 | 77 | S | 4 |
| Oct. | 6 | 1 | 1 | 4.4 | 72 | S | 4 |
| Nov. | 5 | * | 1 | 5.1 | 66 | S | 4 |
| Dec. | 4 | 0 | 1 | 5.0 | 68 | S | 4 |
| Annual | 86 | 42 | 10 | 5.2 | 69 | S | 4 |
| Rec.(yrs.) | 26 | 26 | 20 | 12 | 11 | 12 | 12 |

TABLE XIII

CLIMATIC TABLE FOR WASHINGTON, D.C.
Latitude 38°51′N, longitude 77°03′W, elevation 4 m (34 m)

| Month | Mean sta. press. (mbar) | Temperature (°C) | | | | Mean vap. press. (mbar) | Precipitation (mm) | | Mean snowfall (cm) |
|---|---|---|---|---|---|---|---|---|---|
| | | daily mean | mean daily range | extremes | | | mean | max. in 24 h | |
| | | | | max. | min. | | | | |
| Jan. | 1015.5 | 2.7 | 8.2 | 26 | −15 | 5 | 77 | 44 | 11 |
| Feb. | 1014.1 | 3.2 | 9.3 | 28 | −15 | 5 | 63 | 39 | 10 |
| Mar. | 1012.2 | 7.1 | 10.0 | 32 | −12 | 6 | 82 | 87 | 7 |
| Apr. | 1012.0 | 13.2 | 11.2 | 35 | −4 | 9 | 80 | 45 | tr. |
| May | 1011.3 | 18.8 | 10.8 | 34 | 1 | 14 | 105 | 110 | 0 |
| June | 1010.8 | 23.4 | 10.3 | 38 | 8 | 18 | 82 | 93 | 0 |
| July | 1011.6 | 25.7 | 9.8 | 39 | 13 | 22 | 105 | 75 | 0 |
| Aug. | 1012.2 | 24.7 | 9.5 | 38 | 12 | 21 | 124 | 162 | 0 |
| Sept. | 1014.3 | 20.9 | 10.0 | 38 | 4 | 17 | 97 | 92 | 0 |
| Oct. | 1014.9 | 15.0 | 10.4 | 34 | −1 | 12 | 78 | 126 | tr. |
| Nov. | 1015.2 | 8.7 | 9.8 | 29 | −9 | 7 | 72 | 64 | 2 |
| Dec. | 1016.0 | 3.4 | 8.4 | 24 | −17 | 5 | 71 | 47 | 8 |
| Annual | 1013.3 | 13.9 | 9.8 | 39 | −17 | 12 | 1,036 | 162 | 38 |
| Rec.(yrs.) | 30 | 30 | 30 | 19 | 19 | 10 | 30 | 17 | 17 |

| Month | Number of days with | | | Mean cloud-iness (tenths) | Mean sun-shine (h) | Wind | | Solar radiation (ly/ month) |
|---|---|---|---|---|---|---|---|---|
| | precip. (>0.25 mm) | thunder-storm | fog | | | most freq. direct. | mean speed (m/sec) | |
| Jan. | 11 | 0 | 2 | 6.8 | 45 | NW | 5 | 163 |
| Feb. | 8 | * | 2 | 6.4 | 52 | NW | 5 | 233 |
| Mar. | 12 | 1 | 1 | 6.1 | 55 | NW | 5 | 350 |
| Apr. | 10 | 3 | 1 | 6.5 | 55 | S | 5 | 408 |
| May | 12 | 6 | * | 6.4 | 56 | S | 4 | 514 |
| June | 9 | 5 | * | 5.6 | 65 | S | 4 | 565 |
| July | 10 | 7 | * | 5.9 | 64 | SSW | 4 | 502 |
| Aug. | 10 | 5 | * | 5.6 | 61 | S | 4 | 467 |
| Sept. | 8 | 3 | 1 | 5.4 | 62 | S | 4 | 370 |
| Oct. | 8 | 1 | 2 | 5.3 | 58 | SSW | 4 | 300 |
| Nov. | 8 | 1 | 2 | 5.9 | 55 | SSW | 4 | 194 |
| Dec. | 9 | 0 | 2 | 6.2 | 49 | NW | 4 | 125 |
| Annual | 115 | 32 | 13 | 6.0 | 56 | S | 4 | 394 |
| Rec.(yrs.) | 19 | 12 | 12 | 12 | 12 | 10 | 12 | 48 |

TABLE XIV

CLIMATIC TABLE FOR APALACHICOLA, FLA.
Latitude 29°44′N, longitude 84°59′W, elevation 4 m

| Month | Temperature (°C) | | | | Mean vap. press. (mbar) | Precipitation (mm) | | Mean snowfall (cm) |
|---|---|---|---|---|---|---|---|---|
| | daily mean | mean daily range | extremes | | | mean | max. in 24 h | |
| | | | max. | min. | | | | |
| Jan. | 12.8 | 8.0 | 26 | −8 | 12 | 80 | 96 | tr. |
| Feb. | 13.8 | 8.1 | 27 | −6 | 13 | 99 | 95 | tr. |
| Mar. | 16.1 | 7.8 | 28 | −3 | 14 | 115 | 208 | 0 |
| Apr. | 19.7 | 7.6 | 32 | 3 | 17 | 109 | 192 | 0 |
| May | 23.8 | 7.7 | 36 | 10 | 22 | 73 | 180 | 0 |
| June | 26.8 | 7.2 | 38 | 17 | 25 | 135 | 136 | 0 |
| July | 27.5 | 6.8 | 39 | 19 | 27 | 201 | 149 | 0 |
| Aug. | 27.5 | 6.9 | 37 | 19 | 27 | 197 | 144 | 0 |
| Sept. | 26.1 | 6.6 | 36 | 11 | 26 | 217 | 297 | 0 |
| Oct. | 21.8 | 8.2 | 34 | 4 | 19 | 62 | 143 | 0 |
| Nov. | 16.3 | 8.7 | 31 | −4 | 13 | 66 | 148 | 0 |
| Dec. | 13.2 | 8.2 | 28 | −5 | 12 | 75 | 105 | tr. |
| Annual | 20.4 | 7.7 | 39 | −8 | 19 | 1,429 | 297 | tr. |
| Rec.(yrs.) | 30 | 30 | 31 | 31 | 10 | 30 | 31 | 31 |

| Month | Number of days with | | | Mean cloud-iness (tenths) | Mean sun-shine (h) | Wind | | Solar radiation (ly/ month) |
|---|---|---|---|---|---|---|---|---|
| | precip. (>0.25 mm) | thunder-storm | fog | | | most freq. direct. | mean speed (m/sec) | |
| Jan. | 8 | 1 | 6 | 5.3 | 59 | N | 4 | 293 |
| Feb. | 8 | 2 | 4 | 5.3 | 62 | N | 4 | 362 |
| Mar. | 8 | 4 | 5 | 5.4 | 62 | SE | 4 | 441 |
| Apr. | 6 | 4 | 2 | 4.7 | 71 | SE | 4 | 546 |
| May | 6 | 5 | 1 | 4.3 | 77 | SE | 4 | 609 |
| June | 10 | 10 | * | 5.2 | 70 | SW | 3 | 581 |
| July | 16 | 17 | 0 | 6.0 | 64 | W | 3 | 539 |
| Aug. | 14 | 16 | 0 | 5.7 | 63 | SW | 3 | 507 |
| Sept. | 12 | 10 | * | 5.5 | 63 | NE | 4 | 455 |
| Oct. | 6 | 2 | 1 | 3.9 | 74 | NE | 4 | 458 |
| Nov. | 6 | 1 | 2 | 4.4 | 67 | N | 4 | 357 |
| Dec. | 8 | 1 | 4 | 5.7 | 54 | N | 4 | 286 |
| Annual | 108 | 73 | 25 | 5.1 | 65 | N | 4 | 453 |
| Rec.(yrs.) | 31 | 31 | 31 | 27 | 27 | 22 | 27 | 11 |

TABLE XV

CLIMATIC TABLE FOR JACKSONVILLE, FLA.
Latitude 30°25′N, longitude 81°39′W, elevation 7 m (9 m)

| Month | Mean sta. press. (mbar) | Temperature (°C) | | | | Mean vap. press. (mbar) | Precipitation (mm) | | Mean snowfall (cm) |
|---|---|---|---|---|---|---|---|---|---|
| | | daily mean | mean daily range | extremes | | | mean | max. in 24 h | |
| | | | | max. | min. | | | | |
| Jan. | 1019.5 | 13.3 | 12.1 | 29 | −7 | 11 | 62 | 66 | tr. |
| Feb. | 1018.2 | 14.2 | 12.2 | 31 | −7 | 13 | 74 | 63 | * |
| Mar. | 1016.4 | 16.8 | 12.3 | 32 | −4 | 14 | 89 | 82 | tr. |
| Apr. | 1016.1 | 20.4 | 12.1 | 34 | 2 | 17 | 90 | 124 | 0 |
| May | 1015.0 | 24.3 | 11.8 | 37 | 9 | 22 | 88 | 129 | 0 |
| June | 1014.7 | 27.1 | 10.8 | 39 | 14 | 25 | 161 | 101 | 0 |
| July | 1015.8 | 28.1 | 10.5 | 41 | 18 | 27 | 195 | 94 | 0 |
| Aug. | 1015.1 | 27.9 | 10.1 | 39 | 18 | 28 | 174 | 99 | 0 |
| Sept. | 1014.6 | 26.3 | 9.2 | 38 | 13 | 26 | 192 | 258 | 0 |
| Oct. | 1015.7 | 21.7 | 10.2 | 36 | 3 | 20 | 131 | 169 | 0 |
| Nov. | 1018.1 | 16.5 | 11.7 | 31 | −5 | 13 | 43 | 107 | 0 |
| Dec. | 1019.3 | 13.4 | 11.8 | 29 | −8 | 12 | 56 | 64 | tr. |
| Annual | 1016.5 | 20.8 | 11.2 | 41 | −8 | 19 | 1,355 | 258 | * |
| Rec.(yrs.) | 30 | 30 | 30 | 19 | 19 | 10 | 30 | 19 | 19 |

| Month | Number of days with | | | Mean cloud-iness (tenths) | Mean sun-shine (h) | Wind | |
|---|---|---|---|---|---|---|---|
| | precip. (>0.25 mm) | thunder-storm | fog | | | most freq. direct. | mean speed (m/sec) |
| Jan. | 7 | 1 | 5 | 5.6 | 58 | NW | 4 |
| Feb. | 8 | 1 | 4 | 5.6 | 63 | WNW | 4 |
| Mar. | 9 | 3 | 3 | 5.9 | 62 | NW | 4 |
| Apr. | 7 | 4 | 3 | 5.5 | 68 | SE | 4 |
| May | 8 | 7 | 2 | 5.3 | 69 | WSW | 4 |
| June | 11 | 10 | 1 | 6.2 | 59 | SW | 4 |
| July | 15 | 16 | 1 | 6.3 | 59 | SW | 3 |
| Aug. | 14 | 12 | 1 | 6.1 | 58 | SW | 3 |
| Sept. | 15 | 7 | 1 | 6.9 | 45 | NE | 4 |
| Oct. | 9 | 2 | 3 | 5.7 | 53 | NE | 4 |
| Nov. | 6 | 1 | 6 | 5.1 | 58 | NW | 4 |
| Dec. | 8 | * | 5 | 6.0 | 52 | NW | 4 |
| Annual | 117 | 64 | 35 | 5.9 | 59 | NW | 4 |
| Rec.(yrs.) | 19 | 19 | 16 | 11 | 10 | 11 | 11 |

TABLE XVI

CLIMATIC TABLE FOR MIAMI, FLA.
Latitude 25°48′N, longitude 80°16′W, elevation 2 m (sea level)

| Month | Mean sta. press. (mbar) | Temperature (°C) | | | | Mean vap. press. (mbar) | Precipitation (mm) | | Mean snowfall (cm) |
|---|---|---|---|---|---|---|---|---|---|
| | | daily mean | mean daily range | extremes | | | mean | max. in 24 h | |
| | | | | max. | min. | | | | |
| Jan. | 1019.8 | 19.4 | 10.0 | 31 | 1 | 16 | 52 | 64 | 0 |
| Feb. | 1018.8 | 19.9 | 10.1 | 32 | 0 | 18 | 48 | 52 | 0 |
| Mar. | 1017.5 | 21.4 | 10.4 | 32 | 4 | 19 | 58 | 180 | 0 |
| Apr. | 1016.9 | 23.4 | 9.3 | 34 | 5 | 20 | 99 | 132 | 0 |
| May | 1015.8 | 25.3 | 8.7 | 34 | 12 | 24 | 164 | 214 | 0 |
| June | 1016.1 | 27.1 | 8.1 | 37 | 18 | 27 | 187 | 189 | 0 |
| July | 1017.4 | 27.7 | 7.8 | 36 | 21 | 28 | 171 | 116 | 0 |
| Aug. | 1016.0 | 27.9 | 8.2 | 37 | 20 | 29 | 177 | 80 | 0 |
| Sept. | 1014.3 | 27.4 | 7.5 | 35 | 21 | 28 | 241 | 193 | 0 |
| Oct. | 1014.6 | 25.4 | 7.7 | 33 | 11 | 25 | 209 | 253 | 0 |
| Nov. | 1017.2 | 22.4 | 8.7 | 32 | 4 | 19 | 72 | 201 | 0 |
| Dec. | 1019.1 | 20.1 | 10.0 | 30 | 2 | 17 | 42 | 45 | 0 |
| Annual | 1017.0 | 23.9 | 8.9 | 37 | 1 | 23 | 1,520 | 253 | 0 |
| Rec.(yrs.) | 30 | 30 | 30 | 18 | 18 | 10 | 30 | 18 | 18 |

| Month | Number of days with | | | Mean cloud-iness (tenths) | Mean sun-shine[1] (h) | Wind | | Solar radiation (ly/ month) |
|---|---|---|---|---|---|---|---|---|
| | precip. (>0.25 mm) | thunder-storm | fog | | | most freq. direct. | mean speed (m/sec) | |
| Jan. | 6 | 1 | 1 | 4.5 | 70 | SE | 4 | 344 |
| Feb. | 5 | 1 | 1 | 4.9 | 75 | SE | 4 | 412 |
| Mar. | 6 | 2 | 1 | 5.0 | 73 | SE | 4 | 494 |
| Apr. | 7 | 4 | * | 5.5 | 73 | ESE | 5 | 548 |
| May | 10 | 7 | 1 | 5.5 | 68 | ESE | 4 | 555 |
| June | 13 | 12 | 0 | 6.4 | 62 | SE | 4 | 533 |
| July | 17 | 16 | 0 | 6.4 | 61 | SE | 3 | 539 |
| Aug. | 16 | 16 | * | 6.4 | 64 | SE | 3 | 505 |
| Sept. | 18 | 12 | 0 | 6.5 | 58 | ESE | 4 | 478 |
| Oct. | 15 | 7 | * | 6.0 | 58 | ENE | 4 | 427 |
| Nov. | 8 | 1 | 1 | 5.0 | 68 | N | 4 | 397 |
| Dec. | 7 | 1 | 1 | 5.2 | 65 | N | 4 | 344 |
| Annual | 128 | 80 | 6 | 5.6 | 66 | SE | 4 | 463 |
| Rec.(yrs.) | 18 | 11 | 12 | 12 | 18 | 12 | 11 | 12 |

[1] City data.

TABLE XVII

CLIMATIC TABLE FOR ATLANTA, GA.
Latitude 33°39′N, longitude 84°25′W, elevation 308 m

| Month | Temperature (°C) | | | | Mean vap. press. (mbar) | Precipitation (mm) | | Mean snowfall (cm) |
|---|---|---|---|---|---|---|---|---|
| | daily mean | mean daily range | extremes | | | mean | max. in 24 h | |
| | | | max. | min. | | | | |
| Jan. | 7.1 | 10.3 | 26 | −13 | 8 | 113 | 83 | 2 |
| Feb. | 8.1 | 11.1 | 26 | −15 | 8 | 115 | 82 | 1 |
| Mar. | 11.1 | 12.2 | 28 | −12 | 9 | 136 | 122 | 1 |
| Apr. | 16.1 | 12.3 | 31 | −1 | 12 | 114 | 108 | 0 |
| May | 20.9 | 12.2 | 35 | 5 | 17 | 80 | 130 | 0 |
| June | 24.9 | 11.7 | 38 | 8 | 22 | 97 | 87 | 0 |
| July | 26.0 | 10.7 | 39 | 15 | 23 | 120 | 138 | 0 |
| Aug. | 25.8 | 11.1 | 39 | 15 | 23 | 91 | 128 | 0 |
| Sept. | 22.8 | 11.1 | 37 | 7 | 19 | 83 | 139 | 0 |
| Oct. | 17.1 | 12.2 | 35 | −2 | 13 | 62 | 83 | 0 |
| Nov. | 10.6 | 12.2 | 28 | −16 | 8 | 75 | 104 | tr. |
| Dec. | 6.7 | 10.3 | 23 | −12 | 7 | 111 | 92 | 1 |
| Annual | 16.4 | 11.4 | 39 | −16 | 14 | 1,197 | 139 | 5 |
| Rec.(yrs.) | 30 | 30 | 12 | 12 | 10 | 30 | 26 | 26 |

| Month | Number of days with | | | Mean cloud-iness (tenths) | Mean sun-shine (h) | Wind | |
|---|---|---|---|---|---|---|---|
| | precip. (>0.25 mm) | thunder-storm | fog | | | most freq. direct. | mean speed (m/sec) |
| Jan. | 11 | 1 | 5 | 6.4 | 47 | NW | 5 |
| Feb. | 10 | 2 | 3 | 6.2 | 51 | NW | 5 |
| Mar. | 12 | 4 | 3 | 6.1 | 56 | NW | 5 |
| Apr. | 10 | 5 | 1 | 5.5 | 65 | NW | 5 |
| May | 9 | 6 | 1 | 5.4 | 69 | SW | 4 |
| June | 10 | 9 | 1 | 5.8 | 68 | NW | 4 |
| July | 12 | 10 | 1 | 6.3 | 61 | SW | 3 |
| Aug. | 10 | 8 | 2 | 5.7 | 66 | NW | 3 |
| Sept. | 7 | 3 | 1 | 5.3 | 64 | ENE | 4 |
| Oct. | 7 | 1 | 2 | 4.5 | 66 | NW | 4 |
| Nov. | 8 | 1 | 2 | 5.1 | 58 | NW | 4 |
| Dec. | 11 | 1 | 4 | 6.2 | 48 | NW | 5 |
| Annual | 117 | 51 | 26 | 5.7 | 60 | NW | 4 |
| Rec.(yrs.) | 26 | 26 | 26 | 26 | 26 | 11 | 22 |

TABLE XVIII

CLIMATIC TABLE FOR BOISE, IDAHO
Latitude 43°34′N, longitude 116°13′W, elevation 866 m (835 m)

| Month | Mean sta. press. (mbar) | Temperature (°C) daily mean | mean daily range | extremes max. | min. | Mean vap. press. (mbar) | Precipitation (mm) mean | max. in 24 h | Mean snowfall (cm) |
|---|---|---|---|---|---|---|---|---|---|
| Jan. | 922.9 | −1.9 | 8.3 | 17 | −27 | 4 | 34 | 38 | 19 |
| Feb. | 921.6 | 1.1 | 9.4 | 19 | −23 | 4 | 34 | 25 | 12 |
| Mar. | 919.5 | 5.1 | 11.4 | 24 | −12 | 4 | 34 | 28 | 6 |
| Apr. | 918.5 | 9.9 | 13.8 | 33 | −7 | 6 | 29 | 27 | 1 |
| May | 917.8 | 14.3 | 14.7 | 35 | −3 | 7 | 33 | 38 | tr. |
| June | 917.8 | 18.2 | 15.2 | 43 | 1 | 9 | 23 | 57 | tr. |
| July | 918.3 | 23.7 | 17.4 | 44 | 5 | 9 | 5 | 24 | 0 |
| Aug. | 918.2 | 22.3 | 17.3 | 41 | 3 | 9 | 4 | 20 | 0 |
| Sept. | 919.4 | 17.3 | 16.2 | 39 | −3 | 7 | 10 | 22 | 0 |
| Oct. | 921.4 | 11.4 | 14.1 | 31 | −7 | 6 | 21 | 19 | tr. |
| Nov. | 924.1 | 3.9 | 10.4 | 23 | −19 | 5 | 30 | 18 | 4 |
| Dec. | 923.5 | 0.1 | 8.0 | 17 | −18 | 4 | 34 | 29 | 12 |
| Annual | 920.2 | 10.4 | 13.1 | 44 | −27 | 6 | 290 | 57 | 54 |
| Rec.(yrs.) | 30 | 30 | 30 | 21 | 21 | 10 | 30 | 21 | 21 |

| Month | Number of days with precip. (>0.25 mm) | thunder-storm | fog | Mean cloud-iness (tenths) | Mean sun-shine (h) | Wind most freq. direct. | mean speed (m/sec) | Solar radiation (ly/ month) |
|---|---|---|---|---|---|---|---|---|
| Jan. | 12 | 0 | 5 | 7.5 | 40 | SE | 4 | 146 |
| Feb. | 11 | * | 3 | 7.2 | 49 | SE | 4 | 229 |
| Mar. | 10 | 1 | 1 | 6.8 | 59 | SE | 5 | 341 |
| Apr. | 8 | 1 | * | 6.2 | 67 | SE | 5 | 485 |
| May | 9 | 4 | * | 5.9 | 68 | NW | 4 | 586 |
| June | 7 | 3 | * | 4.7 | 76 | NW | 4 | 640 |
| July | 2 | 3 | 0 | 2.7 | 89 | NW | 4 | 679 |
| Aug. | 2 | 2 | 0 | 3.0 | 86 | NW | 4 | 575 |
| Sept. | 3 | 2 | * | 3.4 | 81 | SE | 4 | 467 |
| Oct. | 7 | 1 | * | 5.0 | 67 | SE | 4 | 309 |
| Nov. | 10 | * | 3 | 6.6 | 46 | SE | 4 | 179 |
| Dec. | 11 | 0 | 6 | 7.5 | 39 | SE | 4 | 124 |
| Annual | 90 | 16 | 19 | 5.5 | 66 | SE | 4 | 397 |
| Rec.(yrs.) | 21 | 21 | 21 | 21 | 21 | 21 | 22 | 12 |

TABLE XIX

CLIMATIC TABLE FOR CAIRO, ILL.
Latitude 37°00′N, longitude 89°10′W, elevation 96 m (109 m)

| Month | Mean sta. press. (mbar) | Temperature (°C) | | | | | Mean vap. press. (mbar) | Precipitation (mm) | | Mean snowfall (cm) |
|---|---|---|---|---|---|---|---|---|---|---|
| | | daily mean | mean daily range | extremes | | | | mean | max. in 24 h | |
| | | | | max. | min. | | | | | |
| Jan. | 1007.6 | 3.0 | 8.0 | 24 | −16 | 6 | 113 | 155 | 7 |
| Feb. | 1006.2 | 4.8 | 8.7 | 23 | −21 | 6 | 93 | 122 | 5 |
| Mar. | 1003.5 | 9.0 | 9.8 | 28 | −14 | 7 | 122 | 94 | 6 |
| Apr. | 1002.6 | 15.2 | 10.6 | 31 | −1 | 11 | 103 | 96 | tr. |
| May | 1002.1 | 20.6 | 10.6 | 37 | 4 | 16 | 112 | 140 | 0 |
| June | 1001.6 | 25.5 | 10.7 | 40 | 11 | 21 | 105 | 109 | 0 |
| July | 1002.7 | 27.3 | 10.3 | 40 | 12 | 23 | 81 | 71 | 0 |
| Aug. | 1002.8 | 26.6 | 10.3 | 39 | 12 | 22 | 79 | 192 | 0 |
| Sept. | 1004.0 | 22.5 | 11.1 | 39 | 6 | 17 | 76 | 86 | 0 |
| Oct. | 1005.4 | 16.6 | 11.1 | 33 | −2 | 12 | 73 | 75 | 0 |
| Nov. | 1006.7 | 8.8 | 9.3 | 28 | −15 | 7 | 98 | 110 | 2 |
| Dec. | 1007.4 | 4.2 | 7.7 | 23 | −15 | 6 | 93 | 63 | 4 |
| Annual | 1004.4 | 15.3 | 9.9 | 40 | −21 | 13 | 1,148 | 192 | 24 |
| Rec.(yrs.) | 30 | 30 | 30 | 18 | 18 | 10 | 30 | 18 | 18 |

| Month | Number of days with | | | Mean cloud-iness (tenths) | Mean sun-shine (h) | Wind | |
|---|---|---|---|---|---|---|---|
| | precip. (>0.25 mm) | thunder-storm | fog | | | most freq. direct. | mean speed (m/sec) |
| Jan. | 11 | 1 | 1 | 7.0 | 42 | SW | 4 |
| Feb. | 10 | 2 | 1 | 6.3 | 52 | NE | 4 |
| Mar. | 12 | 3 | * | 6.2 | 59 | SW | 5 |
| Apr. | 12 | 6 | * | 6.1 | 64 | SW | 5 |
| May | 11 | 7 | * | 6.4 | 68 | SW | 4 |
| June | 10 | 10 | * | 5.8 | 75 | SW | 3 |
| July | 9 | 9 | * | 5.6 | 78 | SW | 3 |
| Aug. | 7 | 6 | * | 5.1 | 80 | NE | 3 |
| Sept. | 8 | 4 | 1 | 4.8 | 74 | NE | 3 |
| Oct. | 7 | 3 | 1 | 4.5 | 73 | S | 3 |
| Nov. | 9 | 2 | 1 | 5.5 | 59 | S | 4 |
| Dec. | 9 | 1 | 1 | 6.5 | 48 | S | 4 |
| Annual | 115 | 54 | 6 | 5.8 | 66 | SW | 4 |
| Rec.(yrs.) | 18 | 18 | 18 | 18 | 18 | 18 | 18 |

TABLE XX

CLIMATIC TABLE FOR CHICAGO, ILL.
Latitude 41°47′N, longitude 87°47′W, elevation 185 m (251 m)

| Month | Mean sta. press. (mbar) | Temperature (°C) | | | | Mean vap. press. (mbar) | Precipitation (mm) | | Mean snowfall (cm) |
|---|---|---|---|---|---|---|---|---|---|
| | | daily mean | mean daily range | extremes max. | min. | | mean | max. in 24 h | |
| Jan. | 990.2 | −3.3 | 8.0 | 19 | −26 | 4 | 47 | 73 | 19 |
| Feb. | 989.8 | −2.3 | 8.3 | 21 | −26 | 4 | 41 | 39 | 18 |
| Mar. | 988.0 | 2.4 | 8.6 | 28 | −22 | 5 | 70 | 64 | 16 |
| Apr. | 987.5 | 9.5 | 10.6 | 31 | −7 | 8 | 77 | 104 | 1 |
| May | 987.3 | 15.6 | 11.6 | 34 | −1 | 12 | 95 | 74 | tr. |
| June | 986.8 | 21.5 | 11.5 | 40 | 2 | 17 | 103 | 116 | 0 |
| July | 988.5 | 24.3 | 11.1 | 39 | 9 | 20 | 86 | 159 | 0 |
| Aug. | 988.9 | 23.6 | 10.9 | 38 | 8 | 19 | 80 | 79 | 0 |
| Sept. | 990.0 | 19.1 | 11.3 | 38 | 2 | 14 | 69 | 65 | 0 |
| Oct. | 990.3 | 13.0 | 10.7 | 33 | −7 | 10 | 71 | 143 | 1 |
| Nov. | 989.4 | 4.4 | 8.8 | 27 | −19 | 6 | 56 | 47 | 9 |
| Dec. | 990.4 | −1.6 | 7.6 | 18 | −26 | 4 | 48 | 60 | 26 |
| Annual | 988.9 | 10.5 | 9.9 | 40 | −26 | 10 | 843 | 159 | 90 |
| Rec.(yrs.) | 30 | 30 | 30 | 18 | 18 | 10 | 30 | 18 | 18 |

| Month | Number of days with | | | Mean cloud-iness (tenths) | Mean sun-shine (h) | Wind | |
|---|---|---|---|---|---|---|---|
| | precip. (>0.25 mm) | thunder-storm | fog | | | most freq. direct. | mean speed (m/sec) |
| Jan. | 10 | * | 3 | 7.1 | 43 | W | 5 |
| Feb. | 10 | 1 | 2 | 6.8 | 48 | W | 5 |
| Mar. | 12 | 3 | 1 | 6.8 | 52 | W | 5 |
| Apr. | 13 | 4 | 1 | 6.8 | 53 | W | 5 |
| May | 12 | 5 | 1 | 6.4 | 59 | SSW | 5 |
| June | 11 | 7 | 1 | 6.2 | 66 | SSW | 4 |
| July | 9 | 6 | 1 | 5.2 | 69 | SW | 3 |
| Aug. | 8 | 5 | 1 | 5.3 | 68 | SW | 3 |
| Sept. | 8 | 3 | 1 | 5.1 | 66 | S | 4 |
| Oct. | 7 | 2 | 1 | 5.2 | 64 | S | 4 |
| Nov. | 10 | 1 | 1 | 7.0 | 44 | W | 5 |
| Dec. | 10 | * | 3 | 6.9 | 44 | W | 5 |
| Annual | 120 | 37 | 17 | 6.2 | 58 | SSW | 4 |
| Rec.(yrs.) | 18 | 18 | 18 | 18 | 18 | 16 | 18 |

TABLE XXI

CLIMATIC TABLE FOR MOLINE, ILL.
Latitude 41°27′N, longitude 90°31′W, elevation 177 m

| Month | Temperature (°C) | | | | Mean vap. press. (mbar) | Precipitation (mm) | | Mean snowfall (cm) |
|---|---|---|---|---|---|---|---|---|
| | daily mean | mean daily range | extremes | | | mean | max. in 24 h | |
| | | | max. | min. | | | | |
| Jan. | −4.9 | 9.5 | 18 | −29 | 3 | 41 | 53 | 17 |
| Feb. | −3.2 | 9.6 | 20 | −31 | 4 | 34 | 48 | 13 |
| Mar. | 2.2 | 10.3 | 28 | −28 | 5 | 61 | 60 | 15 |
| Apr. | 10.0 | 11.9 | 33 | −6 | 8 | 81 | 71 | 1 |
| May | 16.2 | 12.2 | 40 | −1 | 12 | 97 | 66 | tr. |
| June | 21.9 | 12.1 | 40 | 4 | 18 | 111 | 87 | 0 |
| July | 24.2 | 12.7 | 41 | 8 | 21 | 83 | 88 | 0 |
| Aug. | 23.1 | 12.4 | 41 | 4 | 20 | 90 | 117 | 0 |
| Sept. | 18.5 | 13.1 | 38 | −4 | 14 | 83 | 121 | tr. |
| Oct. | 12.4 | 13.2 | 33 | −9 | 9 | 62 | 124 | tr. |
| Nov. | 3.7 | 10.2 | 27 | −19 | 6 | 50 | 42 | 4 |
| Dec. | −2.6 | 9.0 | 19 | −27 | 5 | 42 | 86 | 16 |
| Annual | 10.1 | 11.3 | 41 | −31 | 10 | 835 | 124 | 66 |
| Rec.(yrs.) | 30 | 30 | 28 | 28 | 10 | 30 | 31 | 29 |

| Month | Number of days with | | | Mean cloud-iness (tenths) | Mean sun-shine (h) | Wind | |
|---|---|---|---|---|---|---|---|
| | precip. (>0.25 mm) | thunder-storm | fog | | | most freq. direct. | mean speed (m/sec) |
| Jan. | 9 | * | 3 | 6.7 | 45 | WNW | 5 |
| Feb. | 8 | 1 | 2 | 6.4 | 46 | WNW | 5 |
| Mar. | 11 | 2 | 2 | 6.6 | 51 | NW | 5 |
| Apr. | 11 | 4 | 1 | 6.5 | 54 | NW | 5 |
| May | 12 | 7 | 1 | 6.5 | 57 | S | 5 |
| June | 11 | 8 | 1 | 6.2 | 61 | S | 4 |
| July | 9 | 8 | 1 | 5.3 | 73 | S | 3 |
| Aug. | 9 | 7 | 2 | 5.3 | 70 | E | 3 |
| Sept. | 8 | 4 | 2 | 4.9 | 67 | S | 4 |
| Oct. | 7 | 2 | 3 | 4.9 | 62 | S | 4 |
| Nov. | 7 | 1 | 1 | 6.6 | 44 | WNW | 5 |
| Dec. | 8 | * | 2 | 6.7 | 43 | W | 5 |
| Annual | 110 | 44 | 21 | 6.1 | 58 | W | 4 |
| Rec.(yrs.) | 27 | 26 | 26 | 18 | 17 | 9 | 17 |

TABLE XXII

CLIMATIC TABLE FOR EVANSVILLE, IND.
Latitude 38°03′N, longitude 87°32′W, elevation 116 m

| Month | Temperature (°C) | | | | Mean vap. press. (mbar) | Precipitation (mm) | | Mean snowfall (cm) |
|---|---|---|---|---|---|---|---|---|
| | daily mean | mean daily range | extremes | | | mean | max. in 24 h | |
| | | | max. | min. | | | | |
| Jan. | 1.2 | 9.5 | 24 | −26 | 5 | 101 | 81 | 8 |
| Feb. | 2.6 | 10.2 | 23 | −31 | 6 | 81 | 80 | 6 |
| Mar. | 6.8 | 11.2 | 28 | −23 | 7 | 109 | 131 | 7 |
| Apr. | 13.2 | 12.1 | 32 | −4 | 11 | 101 | 100 | * |
| May | 18.5 | 12.4 | 34 | 1 | 15 | 106 | 109 | 0 |
| June | 23.8 | 12.5 | 40 | 5 | 20 | 95 | 80 | 0 |
| July | 25.7 | 12.3 | 41 | 8 | 23 | 84 | 79 | 0 |
| Aug. | 24.9 | 12.5 | 39 | 8 | 22 | 78 | 66 | 0 |
| Sept. | 20.9 | 13.7 | 39 | −1 | 16 | 73 | 88 | 0 |
| Oct. | 14.8 | 14.0 | 34 | −6 | 11 | 65 | 59 | tr. |
| Nov. | 7.1 | 11.5 | 28 | −19 | 7 | 80 | 68 | 2 |
| Dec. | 2.3 | 9.5 | 24 | −22 | 6 | 78 | 52 | 6 |
| Annual | 13.5 | 11.8 | 41 | −31 | 12 | 1,051 | 131 | 29 |
| Rec.(yrs.) | 30 | 30 | 20 | 20 | 10 | 30 | 20 | 20 |

| Month | Number of days with | | | Mean cloud-iness (tenths) | Mean sun-shine (h) | Wind | |
|---|---|---|---|---|---|---|---|
| | precip. (>0.25 mm) | thunder-storm | fog | | | most freq. direct. | mean speed (m/sec) |
| Jan. | 11 | 1 | 3 | 7.4 | 38 | NW | 4 |
| Feb. | 9 | 2 | 1 | 6.7 | 47 | NW | 5 |
| Mar. | 12 | 3 | 1 | 6.6 | 56 | WNW | 5 |
| Apr. | 12 | 5 | 1 | 6.5 | 62 | SSW | 5 |
| May | 12 | 7 | 1 | 6.4 | 66 | SSW | 4 |
| June | 10 | 9 | * | 5.9 | 73 | SW | 3 |
| July | 10 | 8 | 1 | 5.6 | 77 | SW | 3 |
| Aug. | 7 | 6 | 1 | 5.0 | 78 | SW | 3 |
| Sept. | 7 | 3 | 1 | 4.7 | 74 | SSW | 3 |
| Oct. | 7 | 2 | 2 | 4.8 | 67 | NW | 3 |
| Nov. | 9 | 2 | 1 | 6.1 | 51 | SSW | 4 |
| Dec. | 10 | 1 | 3 | 6.9 | 41 | SSW | 4 |
| Annual | 116 | 49 | 16 | 6.1 | 63 | SSW | 4 |
| Rec.(yrs.) | 20 | 20 | 20 | 20 | 20 | 11 | 20 |

TABLE XXIII

CLIMATIC TABLE FOR INDIANAPOLIS, IND.
Latitude 39°44′N, longitude 86°16′W, elevation 241 m

| Month | Temperature (°C) | | | | Mean vap. press. (mbar) | Precipitation (mm) | | Mean snowfall (cm) |
|---|---|---|---|---|---|---|---|---|
| | daily mean | mean daily range | extremes | | | mean | max. in 24 h | |
| | | | max. | min. | | | | |
| Jan. | −1.6 | 9.0 | 22 | −25 | 4 | 77 | 88 | 10 |
| Feb. | −0.5 | 9.2 | 22 | −28 | 5 | 58 | 59 | 9 |
| Mar. | 3.8 | 10.2 | 27 | −21 | 6 | 87 | 63 | 9 |
| Apr. | 10.4 | 11.6 | 31 | −9 | 9 | 95 | 59 | 1 |
| May | 16.3 | 11.8 | 34 | −2 | 14 | 101 | 67 | tr. |
| June | 21.7 | 11.9 | 39 | 4 | 18 | 117 | 94 | 0 |
| July | 24.0 | 12.1 | 40 | 7 | 21 | 89 | 95 | 0 |
| Aug. | 23.2 | 12.2 | 38 | 6 | 20 | 77 | 69 | 0 |
| Sept. | 19.2 | 12.8 | 38 | −2 | 15 | 82 | 62 | 0 |
| Oct. | 13.0 | 12.7 | 32 | −8 | 11 | 67 | 99 | tr. |
| Nov. | 4.9 | 10.1 | 27 | −19 | 6 | 78 | 77 | 5 |
| Dec. | −0.5 | 8.8 | 21 | −26 | 5 | 68 | 46 | 11 |
| Annual | 11.2 | 11.0 | 40 | −28 | 11 | 996 | 99 | 45 |
| Rec.(yrs.) | 30 | 30 | 23 | 23 | 10 | 30 | 18 | 29 |

| Month | Number of days with | | | Mean cloudiness (tenths) | Mean sunshine (h) | Wind | | Solar radiation (ly/ month) |
|---|---|---|---|---|---|---|---|---|
| | precip. (>0.25 mm) | thunderstorm | fog | | | most freq. direct. | mean speed (m/sec) | |
| Jan. | 12 | 1 | 4 | 7.5 | 37 | NW | 5 | 153 |
| Feb. | 10 | 1 | 3 | 6.9 | 51 | WNW | 6 | 218 |
| Mar. | 12 | 3 | 2 | 6.9 | 54 | WNW | 6 | 308 |
| Apr. | 12 | 5 | 1 | 7.0 | 55 | SW | 6 | 403 |
| May | 13 | 7 | 1 | 6.7 | 61 | SW | 5 | 543 |
| June | 11 | 8 | 1 | 6.2 | 69 | SW | 4 | 552 |
| July | 9 | 8 | 1 | 5.7 | 73 | SW | 4 | 540 |
| Aug. | 8 | 7 | 2 | 5.3 | 74 | SW | 3 | 489 |
| Sept. | 7 | 4 | 1 | 5.1 | 70 | SW | 4 | 408 |
| Oct. | 8 | 2 | 1 | 4.9 | 68 | SW | 4 | 297 |
| Nov. | 10 | 1 | 1 | 6.6 | 46 | SW | 5 | 178 |
| Dec. | 10 | * | 3 | 7.0 | 44 | SW | 5 | 140 |
| Annual | 122 | 47 | 21 | 6.3 | 60 | SW | 5 | 352 |
| Rec.(yrs.) | 21 | 18 | 18 | 18 | 18 | 11 | 12 | 12 |

TABLE XXIV

CLIMATIC TABLE FOR CONCORDIA, KANS.
Latitude 39°33′N, longitude 97°39′W, elevation 448 m (424 m)

| Month | Mean sta. press. (mbar) | Temperature (°C) | | | | Mean vap. press. (mbar) | Precipitation (mm) | | Mean snowfall (cm) |
|---|---|---|---|---|---|---|---|---|---|
| | | daily mean | mean daily range | extremes | | | mean | max. in 24 h | |
| | | | | max. | min. | | | | |
| Jan. | 968.7 | −2.2 | 10.6 | 26 | −32 | 4 | 15 | 44 | 11 |
| Feb. | 968.0 | 0.5 | 11.1 | 27 | −32 | 4 | 22 | 53 | 13 |
| Mar. | 965.4 | 5.3 | 11.8 | 36 | −24 | 5 | 34 | 52 | 14 |
| Apr. | 964.6 | 12.2 | 12.1 | 38 | −10 | 8 | 57 | 78 | 3 |
| May | 964.0 | 17.3 | 11.6 | 39 | −4 | 13 | 92 | 164 | tr. |
| June | 962.9 | 23.3 | 11.8 | 43 | 6 | 19 | 107 | 126 | 0 |
| July | 965.1 | 26.7 | 12.5 | 46 | 8 | 21 | 86 | 145 | 0 |
| Aug. | 965.2 | 25.8 | 12.4 | 47 | 5 | 20 | 81 | 121 | 0 |
| Sept. | 966.4 | 20.8 | 12.6 | 44 | −2 | 14 | 62 | 117 | 0 |
| Oct. | 967.3 | 14.3 | 12.8 | 39 | −9 | 10 | 44 | 111 | 1 |
| Nov. | 968.1 | 5.6 | 11.5 | 28 | −26 | 6 | 29 | 52 | 4 |
| Dec. | 968.3 | −0.1 | 10.2 | 24 | −26 | 4 | 15 | 53 | 8 |
| Annual | 966.2 | 12.5 | 11.7 | 47 | −32 | 11 | 644 | 164 | 54 |
| Rec.(yrs.) | 30 | 30 | 30 | 76 | 76 | 10 | 30 | 76 | 76 |

| Month | Number of days with | | | Mean cloud-iness (tenths) | Mean sun-shine (h) | Wind | |
|---|---|---|---|---|---|---|---|
| | precip. (>0.25 mm) | thunder-storm | fog | | | most freq. direct. | mean speed (m/sec) |
| Jan. | 5 | * | 2 | 5.5 | 60 | N | 3 |
| Feb. | 6 | * | 2 | 5.8 | 60 | N | 4 |
| Mar. | 7 | 1 | 1 | 6.0 | 62 | S | 4 |
| Apr. | 9 | 4 | 1 | 5.9 | 63 | S | 4 |
| May | 11 | 7 | * | 5.9 | 66 | S | 4 |
| June | 10 | 9 | * | 5.0 | 73 | S | 4 |
| July | 8 | 8 | * | 4.5 | 78 | S | 3 |
| Aug. | 9 | 8 | * | 4.2 | 77 | S | 3 |
| Sept. | 7 | 5 | * | 3.9 | 72 | S | 3 |
| Oct. | 6 | 2 | 1 | 4.0 | 70 | S | 3 |
| Nov. | 4 | 1 | 1 | 4.7 | 64 | S | 3 |
| Dec. | 5 | * | 2 | 5.2 | 58 | S | 3 |
| Annual | 87 | 45 | 10 | 5.1 | 68 | S | 4 |
| Rec.(yrs.) | 76 | 76 | 18 | 17 | 53 | 75 | 76 |

TABLE XXV

CLIMATIC TABLE FOR NEW ORLEANS, LA.
Latitude 29°57′N, longitude 90°04′W, elevation 3 m (sea level)

| Month | Mean sta. press. (mbar) | Temperature (°C) | | extremes | | Mean vap. press. (mbar) | Precipitation (mm) | | Mean snowfall (cm) |
|---|---|---|---|---|---|---|---|---|---|
| | | daily mean | mean daily range | max. | min. | | mean | max. in 24 h | |
| Jan. | 1020.7 | 13.3 | 8.5 | 28 | −8 | 11 | 121 | 119 | tr. |
| Feb. | 1019.3 | 14.7 | 9.0 | 29 | −7 | 12 | 106 | 155 | * |
| Mar. | 1017.2 | 17.2 | 9.1 | 32 | −2 | 14 | 167 | 278 | tr. |
| Apr. | 1016.3 | 21.0 | 8.8 | 33 | 5 | 17 | 138 | 356 | 0 |
| May | 1015.2 | 24.5 | 8.6 | 36 | 10 | 23 | 138 | 231 | 0 |
| June | 1015.1 | 27.7 | 8.3 | 39 | 17 | 26 | 141 | 169 | 0 |
| July | 1016.1 | 28.4 | 8.1 | 38 | 19 | 27 | 180 | 139 | 0 |
| Aug. | 1015.6 | 28.6 | 8.0 | 38 | 20 | 28 | 163 | 135 | 0 |
| Sept. | 1015.2 | 26.8 | 7.7 | 37 | 12 | 25 | 148 | 273 | 0 |
| Oct. | 1017.2 | 22.7 | 8.2 | 34 | 4 | 19 | 93 | 347 | 0 |
| Nov. | 1019.9 | 16.9 | 8.5 | 32 | −1 | 13 | 102 | 197 | 0 |
| Dec. | 1020.6 | 13.9 | 8.5 | 29 | −7 | 12 | 116 | 123 | tr. |
| Annual | 1017.4 | 21.3 | 8.4 | 39 | −8 | 19 | 1,613 | 356 | * |
| Rec.(yrs.) | 30 | 30 | 30 | 46 | 46 | | 30 | 46 | 46 |

| Month | Number of days with | | | Mean cloud-iness (tenths) | Mean sun-shine (h) | Mean wind speed (m/sec) |
|---|---|---|---|---|---|---|
| | precip. (>0.25 mm) | thunder-storm | fog | | | |
| Jan. | 10 | 2 | 4 | 6.0 | 49 | 3 |
| Feb. | 9 | 2 | 2 | 6.0 | 51 | 3 |
| Mar. | 9 | 4 | 2 | 5.8 | 57 | 3 |
| Apr. | 7 | 5 | 1 | 5.3 | 64 | 3 |
| May | 9 | 6 | * | 5.1 | 68 | 3 |
| June | 12 | 11 | * | 5.3 | 67 | 3 |
| July | 15 | 16 | * | 6.1 | 61 | 3 |
| Aug. | 14 | 14 | * | 5.8 | 63 | 3 |
| Sept. | 10 | 7 | * | 5.1 | 65 | 3 |
| Oct. | 7 | 2 | 1 | 4.0 | 72 | 3 |
| Nov. | 7 | 2 | 2 | 4.7 | 62 | 3 |
| Dec. | 10 | 2 | 3 | 5.9 | 48 | 3 |
| Annual | 119 | 73 | 15 | 5.4 | 61 | 3 |
| Rec.(yrs.) | 45 | 45 | 45 | 45 | 45 | 45 |

TABLE XXVI

CLIMATIC TABLE FOR CARIBOU, ME.
Latitude 46°52′N, longitude 68°01′W, elevation 190 m

| Month | Temperature (°C) | | | | Mean vap. press. (mbar) | Precipitation (mm) | | Mean snowfall (cm) |
|---|---|---|---|---|---|---|---|---|
| | daily mean | mean daily range | extremes | | | mean | max. in 24 h | |
| | | | max. | min. | | | | |
| Jan. | −11.9 | 10.4 | 11 | −36 | 2 | 54 | 29 | 56 |
| Feb. | −10.8 | 11.0 | 8 | −41 | 2 | 51 | 34 | 63 |
| Mar. | −5.1 | 10.0 | 14 | −29 | 4 | 60 | 42 | 49 |
| Apr. | 2.4 | 9.6 | 27 | −17 | 5 | 67 | 54 | 16 |
| May | 9.9 | 12.0 | 33 | −7 | 8 | 77 | 57 | 1 |
| June | 15.0 | 11.5 | 36 | −1 | 12 | 103 | 60 | tr. |
| July | 18.1 | 11.8 | 35 | 4 | 15 | 103 | 74 | 0 |
| Aug. | 17.0 | 12.0 | 35 | 1 | 15 | 93 | 105 | 0 |
| Sept. | 12.1 | 11.5 | 33 | −5 | 11 | 90 | 158 | tr. |
| Oct. | 6.1 | 10.0 | 26 | −10 | 8 | 85 | 82 | 5 |
| Nov. | −1.0 | 7.4 | 20 | −19 | 6 | 77 | 44 | 29 |
| Dec. | −9.2 | 8.9 | 14 | −31 | 2 | 62 | 53 | 49 |
| Annual | 3.6 | 10.5 | 36 | −41 | 8 | 922 | 158 | 268 |
| Rec.(yrs.) | 30 | 30 | 21 | 21 | 10 | 30 | 21 | 21 |

| Month | Number of days with | | | Mean cloud-iness (tenths) | Wind | | Solar radiation (ly/ month) |
|---|---|---|---|---|---|---|---|
| | precip. (>0.25 mm) | thunder-storm | fog | | most freq. direct. | mean speed (m/sec) | |
| Jan. | 14 | * | 2 | 7.2 | NW | 6 | 137 |
| Feb. | 14 | 0 | 2 | 7.0 | NW | 5 | 232 |
| Mar. | 13 | * | 2 | 6.9 | NW | 6 | 366 |
| Apr. | 13 | 1 | 2 | 7.2 | NW | 5 | 399 |
| May | 13 | 2 | 1 | 7.4 | NW | 5 | 468 |
| June | 15 | 4 | 1 | 7.4 | WSW | 5 | 522 |
| July | 14 | 6 | 2 | 7.1 | WSW | 4 | 501 |
| Aug. | 13 | 4 | 2 | 6.7 | WSW | 4 | 441 |
| Sept. | 12 | 1 | 3 | 6.7 | WSW | 5 | 332 |
| Oct. | 12 | 1 | 2 | 6.8 | NNW | 5 | 209 |
| Nov. | 14 | * | 4 | 8.1 | WSW | 5 | 110 |
| Dec. | 14 | 0 | 3 | 7.4 | WSW | 5 | 108 |
| Annual | 161 | 19 | 26 | 7.2 | WSW | 5 | 319 |
| Rec.(yrs.) | 21 | 20 | 20 | 15 | 12 | 12 | 11 |

TABLE XXVII

CLIMATIC TABLE FOR PORTLAND, ME.
Latitude 43°39′N, longitude 70°19′W, elevation 19 m

| Month | Temperature (°C) | | | | Mean vap. press. (mbar) | Precipitation (mm) | | Mean snowfall (cm) |
|---|---|---|---|---|---|---|---|---|
| | daily mean | mean daily range | extremes | | | mean | max. in 24 h | |
| | | | max. | min. | | | | |
| Jan. | −5.7 | 11.2 | 18 | −29 | 2 | 111 | 48 | 50 |
| Feb. | −5.1 | 11.9 | 18 | −39 | 3 | 97 | 70 | 48 |
| Mar. | −0.3 | 10.4 | 30 | −29 | 4 | 110 | 88 | 35 |
| Apr. | 5.8 | 11.2 | 29 | −22 | 7 | 95 | 62 | 5 |
| May | 11.7 | 12.5 | 33 | −5 | 10 | 87 | 59 | 1 |
| June | 16.7 | 12.2 | 36 | 1 | 14 | 81 | 55 | 0 |
| July | 20.1 | 12.7 | 37 | 5 | 17 | 73 | 57 | 0 |
| Aug. | 19.3 | 12.9 | 38 | 3 | 17 | 61 | 106 | 0 |
| Sept. | 14.8 | 12.8 | 35 | −5 | 14 | 89 | 190 | tr. |
| Oct. | 9.2 | 12.5 | 31 | −8 | 9 | 81 | 79 | tr. |
| Nov. | 3.4 | 10.6 | 23 | −14 | 6 | 106 | 86 | 8 |
| Dec. | −3.4 | 10.6 | 16 | −26 | 4 | 98 | 65 | 31 |
| Annual | 7.2 | 11.8 | 38 | −39 | 9 | 1,089 | 190 | 178 |
| Rec.(yrs.) | 30 | 30 | 20 | 20 | 10 | 30 | 20 | 20 |

| Month | Number of days with | | | Mean cloud-iness (tenths) | Mean sun-shine (h) | Wind | | Solar radiation (ly/ month) |
|---|---|---|---|---|---|---|---|---|
| | precip. (>0.25 mm) | thunder-storm | fog | | | most freq. direct. | mean speed (m/sec) | |
| Jan. | 12 | 0 | 2 | 6.3 | 54 | N | 4 | 159 |
| Feb. | 11 | * | 2 | 5.9 | 60 | N | 4 | 237 |
| Mar. | 11 | 1 | 4 | 6.0 | 58 | W | 5 | 355 |
| Apr. | 12 | 1 | 4 | 6.5 | 56 | S | 5 | 409 |
| May | 13 | 2 | 7 | 6.6 | 56 | S | 4 | 512 |
| June | 12 | 5 | 5 | 6.2 | 62 | S | 4 | 546 |
| July | 9 | 4 | 7 | 5.8 | 68 | S | 3 | 558 |
| Aug. | 9 | 3 | 5 | 5.5 | 66 | S | 3 | 575 |
| Sept. | 8 | 2 | 6 | 5.4 | 63 | S | 4 | 385 |
| Oct. | 9 | 1 | 5 | 5.4 | 59 | WSW | 4 | 272 |
| Nov. | 11 | * | 4 | 6.5 | 48 | W | 4 | 150 |
| Dec. | 10 | * | 2 | 5.9 | 56 | N | 4 | 153 |
| Annual | 127 | 19 | 53 | 6.0 | 59 | S | 4 | 359 |
| Rec.(yrs.) | 20 | 20 | 20 | 17 | 20 | 12 | 20 | 9 |

TABLE XXVIII

CLIMATIC TABLE FOR BOSTON, MASS.
Latitude 42°13′N, longitude 71°07′W, elevation 192 m (195 m)

| Month | Mean sta. press. (mbar) | Temperature (°C) | | | | Mean vap. press. (mbar) | Precipitation (mm) | | Mean snowfall (cm) |
|---|---|---|---|---|---|---|---|---|---|
| | | daily mean | mean daily range | extremes | | | mean | max. in 24 h | |
| | | | | max. | min. | | | | |
| Jan. | 992.0 | −2.8 | 8.3 | 20 | −27 | 4 | 114 | 79 | 39 |
| Feb. | 990.4 | −2.6 | 9.0 | 19 | −29 | 4 | 95 | 123 | 40 |
| Mar. | 989.9 | 1.6 | 8.9 | 29 | −21 | 5 | 115 | 73 | 29 |
| Apr. | 990.7 | 7.6 | 10.2 | 32 | −14 | 7 | 102 | 68 | 8 |
| May | 991.3 | 13.7 | 11.3 | 34 | −3 | 11 | 88 | 95 | * |
| June | 990.8 | 18.4 | 10.7 | 37 | 2 | 15 | 95 | 100 | 0 |
| July | 992.1 | 21.6 | 10.3 | 37 | 8 | 19 | 83 | 113 | 0 |
| Aug. | 992.3 | 20.8 | 10.0 | 38 | 5 | 18 | 103 | 252 | 0 |
| Sept. | 994.7 | 16.9 | 9.7 | 37 | −2 | 15 | 100 | 133 | 0 |
| Oct. | 994.4 | 11.5 | 9.6 | 31 | −6 | 10 | 95 | 153 | 1 |
| Nov. | 993.2 | 5.6 | 8.7 | 27 | −15 | 6 | 115 | 129 | 7 |
| Dec. | 992.5 | −1.1 | 8.3 | 18 | −28 | 4 | 101 | 77 | 26 |
| Annual | 992.1 | 9.3 | 9.6 | 38 | −29 | 10 | 1,206 | 252 | 150 |
| Rec.(yrs.) | 30 | 30 | 30 | 75 | 75 | 10 | 30 | 75 | 75 |

| Month | Number of days with | | | Mean cloud-iness[1] (tenths) | Mean sun-shine (h) | Wind | | Solar radiation (ly/ month) |
|---|---|---|---|---|---|---|---|---|
| | precip. (>0.25 mm) | thunder-storm[1] | fog[1] | | | most freq. direct. | mean speed (m/sec) | |
| Jan. | 13 | * | 2 | 6.3 | 44 | W | 8 | 156 |
| Feb. | 11 | * | 2 | 6.1 | 49 | W | 8 | 229 |
| Mar. | 13 | * | 2 | 6.2 | 48 | W | 8 | 320 |
| Apr. | 12 | 1 | 2 | 6.6 | 49 | NW | 7 | 391 |
| May | 13 | 3 | 3 | 6.5 | 51 | S | 7 | 472 |
| June | 12 | 4 | 2 | 6.3 | 54 | S | 6 | 512 |
| July | 11 | 5 | 3 | 6.1 | 57 | SW | 6 | 500 |
| Aug. | 10 | 4 | 2 | 5.7 | 57 | SW | 6 | 449 |
| Sept. | 10 | 2 | 2 | 5.4 | 55 | SW | 6 | 366 |
| Oct. | 10 | 1 | 2 | 5.5 | 54 | NW | 7 | 275 |
| Nov. | 11 | * | 2 | 6.4 | 47 | W | 7 | 167 |
| Dec. | 11 | * | 1 | 6.1 | 46 | W | 8 | 141 |
| Annual | 137 | 20 | 24 | 6.1 | 51 | W | 7 | 332 |
| Rec.(yrs.) | 75 | 28 | 28 | 28 | 75 | 52 | 52 | 29 |

[1] Airport Station data.

TABLE XXIX

CLIMATIC TABLE FOR DETROIT, MICH.
Latitude 42°24′N, longitude 83°00′W, elevation 189 m

| Month | Temperature (°C) | | | | Mean vap. press. (mbar) | Precipitation (mm) | | Mean snowfall (cm) |
|---|---|---|---|---|---|---|---|---|
| | daily mean | mean daily range | extremes | | | mean | max. in 24 h | |
| | | | max. | min. | | | | |
| Jan. | −2.8 | 6.8 | 19 | −23 | 4 | 52 | 41 | 21 |
| Feb. | −2.7 | 7.5 | 20 | −27 | 4 | 53 | 62 | 20 |
| Mar. | 1.6 | 8.3 | 28 | −18 | 5 | 61 | 47 | 14 |
| Apr. | 8.4 | 10.3 | 31 | −10 | 7 | 76 | 75 | 3 |
| May | 14.7 | 11.2 | 34 | −1 | 11 | 90 | 64 | tr. |
| June | 20.7 | 11.0 | 40 | 3 | 16 | 72 | 67 | 0 |
| July | 23.3 | 11.2 | 41 | 9 | 18 | 72 | 71 | 0 |
| Aug. | 22.4 | 10.7 | 38 | 6 | 18 | 73 | 93 | 0 |
| Sept. | 18.1 | 10.7 | 38 | 0 | 14 | 62 | 65 | 0 |
| Oct. | 12.1 | 10.1 | 32 | −4 | 9 | 67 | 94 | tr. |
| Nov. | 4.7 | 7.5 | 27 | −15 | 6 | 56 | 55 | 8 |
| Dec. | −1.2 | 6.4 | 18 | −21 | 4 | 53 | 39 | 17 |
| Annual | 9.9 | 9.3 | 41 | −27 | 10 | 787 | 94 | 83 |
| Rec.(yrs.) | 30 | 30 | 27 | 27 | 10 | 30 | 27 | 27 |

| Month | Number of days with | | | Mean cloud-iness (tenths) | Mean sun-shine (h) | Wind | |
|---|---|---|---|---|---|---|---|
| | precip. (>0.25 mm) | thunder-storm | fog | | | most freq. direct. | mean speed (m/sec) |
| Jan. | 13 | * | 2 | 7.9 | 31 | NW | 5 |
| Feb. | 12 | 1 | 1 | 7.3 | 43 | NW | 5 |
| Mar. | 13 | 2 | 1 | 7.0 | 50 | NW | 5 |
| Apr. | 12 | 3 | 1 | 6.8 | 52 | W | 5 |
| May | 12 | 5 | 1 | 6.4 | 58 | S | 4 |
| June | 11 | 6 | * | 6.1 | 64 | S | 4 |
| July | 9 | 6 | * | 5.2 | 71 | S | 4 |
| Aug. | 8 | 5 | 1 | 5.3 | 67 | N | 4 |
| Sept. | 9 | 3 | 1 | 5.4 | 61 | S | 4 |
| Oct. | 9 | 1 | 1 | 5.6 | 56 | S | 4 |
| Nov. | 12 | 1 | 1 | 7.5 | 34 | SW | 5 |
| Dec. | 13 | * | 1 | 7.7 | 32 | SW | 5 |
| Annual | 133 | 33 | 11 | 6.5 | 54 | N | 5 |
| Rec.(yrs.) | 27 | 27 | 27 | 27 | 27 | 11 | 27 |

TABLE XXX

CLIMATIC TABLE FOR LANSING, MICH.
Latitude 42°44′N, longitude 84°29′W, elevation 261 m

| Month | Temperature (°C) | | | | Mean vap. press. (mbar) | Precipitation (mm) | | Mean snowfall (cm) |
|---|---|---|---|---|---|---|---|---|
| | daily mean | mean daily range | extremes | | | mean | max. in 24 h | |
| | | | max. | min. | | | | |
| Jan. | −4.3 | 7.6 | 17 | −27 | 4 | 50 | 44 | 28 |
| Feb. | −4.3 | 8.6 | 19 | −32 | 4 | 50 | 110 | 26 |
| Mar. | 0.2 | 9.2 | 28 | −23 | 5 | 61 | 69 | 21 |
| Apr. | 7.6 | 11.3 | 31 | −13 | 7 | 73 | 67 | 6 |
| May | 13.9 | 12.3 | 34 | −5 | 11 | 95 | 73 | 1 |
| June | 19.7 | 12.3 | 37 | 1 | 16 | 85 | 139 | 0 |
| July | 22.1 | 13.1 | 39 | 6 | 18 | 66 | 81 | 0 |
| Aug. | 21.2 | 12.7 | 39 | 3 | 18 | 77 | 70 | 0 |
| Sept. | 16.7 | 12.2 | 36 | −2 | 14 | 66 | 80 | tr. |
| Oct. | 10.7 | 11.3 | 31 | −7 | 9 | 64 | 73 | 1 |
| Nov. | 3.3 | 8.2 | 24 | −16 | 6 | 56 | 42 | 10 |
| Dec. | −2.5 | 7.2 | 18 | −28 | 4 | 51 | 43 | 25 |
| Annual | 8.7 | 10.5 | 39 | −32 | 10 | 794 | 139 | 118 |
| Rec.(yrs.) | 30 | 30 | 53 | 53 | 10 | 30 | 53 | 53 |

| Month | Number of days with | | | Mean cloud-iness (tenths) | Mean sun-shine (h) | Wind | | Solar radiation (ly/ month) |
|---|---|---|---|---|---|---|---|---|
| | precip. (>0.25 mm) | thunder-storm | fog | | | most freq. direct. | mean speed (m/sec) | |
| Jan. | 14 | * | 1 | 7.6 | 34 | SW | 4 | 128 |
| Feb. | 13 | 1 | 1 | 7.0 | 44 | SW | 4 | 213 |
| Mar. | 13 | 2 | 1 | 6.4 | 52 | NW | 4 | 307 |
| Apr. | 12 | 3 | * | 6.1 | 56 | SW | 4 | 361 |
| May | 13 | 3 | 1 | 5.8 | 62 | SW | 4 | 493 |
| June | 11 | 7 | 1 | 5.4 | 67 | S | 3 | 557 |
| July | 9 | 7 | 1 | 4.5 | 72 | SW | 3 | 538 |
| Aug. | 9 | 7 | 1 | 4.9 | 67 | SW | 3 | 468 |
| Sept. | 11 | 5 | 2 | 5.1 | 58 | S | 3 | 375 |
| Oct. | 11 | 2 | 1 | 5.5 | 52 | SW | 3 | 257 |
| Nov. | 13 | 1 | 1 | 7.2 | 32 | SW | 4 | 137 |
| Dec. | 14 | * | 1 | 7.6 | 27 | SW | 4 | 112 |
| Annual | 143 | 38 | 12 | 6.1 | 52 | SW | 4 | 329 |
| Rec.(yrs.) | 43 | 43 | 43 | 39 | 43 | 22 | 22 | 12 |

TABLE XXXI

CLIMATIC TABLE FOR SAULT STE. MARIE, MICH.
Latitude 46°28′N, longitude 84°22′W, elevation 220 m

| Month | Temperature (°C) | | | | Mean vap. press. (mbar) | Precipitation (mm) | | Mean snowfall (cm) |
|---|---|---|---|---|---|---|---|---|
| | daily mean | mean daily range | extremes | | | mean | max. in 24 h | |
| | | | max. | min. | | | | |
| Jan. | −9.3 | 8.7 | 6 | −31 | 2 | 53 | 35 | 53 |
| Feb. | −9.3 | 9.8 | 7 | −32 | 3 | 38 | 24 | 39 |
| Mar. | −4.8 | 9.4 | 24 | −31 | 3 | 46 | 31 | 39 |
| Apr. | 3.1 | 9.9 | 28 | −17 | 6 | 55 | 68 | 12 |
| May | 9.5 | 12.0 | 29 | −6 | 8 | 70 | 62 | 1 |
| June | 14.7 | 12.7 | 33 | −1 | 13 | 84 | 67 | tr. |
| July | 17.8 | 12.8 | 33 | 2 | 16 | 63 | 60 | 0 |
| Aug. | 17.5 | 11.7 | 37 | 0 | 16 | 73 | 56 | 0 |
| Sept. | 12.9 | 10.2 | 34 | −4 | 12 | 97 | 46 | 1 |
| Oct. | 7.4 | 9.1 | 27 | −7 | 8 | 72 | 52 | 6 |
| Nov. | 0.2 | 7.1 | 19 | −21 | 5 | 85 | 38 | 36 |
| Dec. | −6.4 | 7.7 | 11 | −29 | 3 | 58 | 22 | 64 |
| Annual | 4.4 | 10.1 | 37 | −32 | 8 | 794 | 68 | 251 |
| Rec.(yrs.) | 30 | 30 | 19 | 19 | 10 | 30 | 19 | 19 |

| Month | Number of days with | | | Mean cloud-iness (tenths) | Mean sun-shine (h) | Wind | | Solar radiation (ly/ month) |
|---|---|---|---|---|---|---|---|---|
| | precip. (>0.25 mm) | thunder-storm | fog | | | most freq. direct. | mean speed (m/sec) | |
| Jan. | 18 | * | 2 | 7.9 | 32 | E | 5 | 135 |
| Feb. | 15 | * | 2 | 7.4 | 42 | E | 5 | 226 |
| Mar. | 13 | 1 | 3 | 6.8 | 51 | WNW | 5 | 355 |
| Apr. | 12 | 1 | 3 | 6.6 | 53 | WNW | 5 | 423 |
| May | 11 | 3 | 3 | 6.6 | 52 | WNW | 5 | 526 |
| June | 12 | 7 | 4 | 6.3 | 55 | WNW | 4 | 566 |
| July | 10 | 6 | 5 | 5.7 | 63 | WNW | 4 | 573 |
| Aug. | 10 | 5 | 6 | 5.9 | 58 | WNW | 4 | 475 |
| Sept. | 13 | 4 | 7 | 6.9 | 43 | WNW | 4 | 323 |
| Oct. | 11 | 2 | 6 | 6.8 | 44 | E | 4 | 218 |
| Nov. | 18 | 1 | 2 | 8.5 | 22 | E | 5 | 108 |
| Dec. | 19 | 0 | 3 | 8.3 | 25 | E | 5 | 99 |
| Annual | 162 | 30 | 46 | 7.0 | 45 | WNW | 4 | 336 |
| Rec.(yrs.) | 19 | 19 | 19 | 19 | 19 | 11 | 19 | 19 |

TABLE XXXII

CLIMATIC TABLE FOR DULUTH, MINN.
Latitude 46°50′N, longitude 92°11′W, elevation 435 m (345 m)

| Month | Mean sta. press. (mbar) | Temperature (°C) | | | | Mean vap. press. (mbar) | Precipitation (mm) | | Mean snowfall (cm) |
|---|---|---|---|---|---|---|---|---|---|
| | | daily mean | mean daily range | extremes | | | mean | max. in 24 h | |
| | | | | max. | min. | | | | |
| Jan. | 974.6 | −12.9 | 10.3 | 11 | −37 | 2 | 29 | 15 | 42 |
| Feb. | 974.8 | −11.5 | 11.3 | 12 | −34 | 2 | 24 | 18 | 33 |
| Mar. | 973.6 | −5.7 | 10.4 | 26 | −32 | 3 | 41 | 45 | 34 |
| Apr. | 973.6 | 3.1 | 10.7 | 31 | −21 | 5 | 60 | 58 | 17 |
| May | 973.1 | 9.8 | 12.2 | 31 | −7 | 8 | 84 | 57 | 3 |
| June | 971.9 | 15.2 | 12.2 | 33 | −1 | 13 | 108 | 103 | tr. |
| July | 973.7 | 18.9 | 12.3 | 36 | 4 | 16 | 90 | 69 | 0 |
| Aug. | 974.1 | 17.9 | 11.5 | 36 | 3 | 16 | 97 | 74 | 0 |
| Sept. | 974.2 | 12.6 | 11.1 | 32 | −6 | 11 | 73 | 69 | tr. |
| Oct. | 974.2 | 6.7 | 10.5 | 30 | −13 | 8 | 55 | 71 | 3 |
| Nov. | 973.2 | −2.9 | 8.1 | 20 | −27 | 4 | 45 | 57 | 26 |
| Dec. | 974.1 | −10.0 | 9.2 | 10 | −36 | 2 | 29 | 54 | 37 |
| Annual | 973.7 | 3.4 | 10.8 | 36 | −37 | 8 | 735 | 103 | 195 |
| Rec.(yrs.) | 30 | 30 | 30 | 19 | 19 | 10 | 30 | 11 | 17 |

| Month | Number of days with | | | Mean cloud-iness (tenths) | Mean sun-shine (h) | Wind | |
|---|---|---|---|---|---|---|---|
| | precip. (>0.25 mm) | thunder-storm | fog | | | most freq. direct. | mean speed (m/sec) |
| Jan. | 11 | 0 | 3 | 6.9 | 51 | NW | 6 |
| Feb. | 10 | 0 | 2 | 6.3 | 56 | NW | 6 |
| Mar. | 11 | 1 | 3 | 6.5 | 61 | WNW | 6 |
| Apr. | 10 | 1 | 3 | 6.8 | 56 | ENE | 7 |
| May | 13 | 3 | 5 | 6.7 | 57 | E | 6 |
| June | 13 | 7 | 7 | 6.7 | 61 | E | 5 |
| July | 11 | 8 | 6 | 5.8 | 69 | WNW | 5 |
| Aug. | 11 | 8 | 7 | 6.1 | 61 | E | 5 |
| Sept. | 11 | 4 | 5 | 6.7 | 51 | WNW | 5 |
| Oct. | 9 | 2 | 5 | 6.4 | 53 | WNW | 6 |
| Nov. | 11 | * | 3 | 7.5 | 37 | WNW | 6 |
| Dec. | 11 | * | 3 | 7.2 | 45 | NW | 6 |
| Annual | 132 | 34 | 52 | 6.6 | 56 | NW | 6 |
| Rec.(yrs.) | 19 | 18 | 12 | 12 | 10 | 11 | 11 |

TABLE XXXIII

CLIMATIC TABLE FOR MINNEAPOLIS-ST. PAUL, MINN.
Latitude 44°53′N, longitude 93°13′W, elevation 254 m (255 m)

| Month | Mean sta. press. (mbar) | Temperature (°C) | | | | Mean vap. press. (mbar) | Precipitation (mm) | | Mean snowfall (cm) |
|---|---|---|---|---|---|---|---|---|---|
| | | daily mean | mean daily range | extremes | | | mean | max. in 24 h | |
| | | | | max. | min. | | | | |
| Jan. | 986.3 | −10.9 | 11.0 | 14 | −35 | 2 | 18 | 20 | 16 |
| Feb. | 986.4 | −8.9 | 11.6 | 15 | −33 | 3 | 20 | 23 | 19 |
| Mar. | 984.3 | −2.4 | 10.9 | 26 | −33 | 4 | 39 | 36 | 25 |
| Apr. | 983.3 | 7.1 | 12.6 | 33 | −13 | 6 | 47 | 40 | 5 |
| May | 982.6 | 14.2 | 13.3 | 35 | −3 | 9 | 81 | 67 | 1 |
| June | 981.6 | 19.6 | 12.7 | 38 | 1 | 15 | 102 | 64 | 0 |
| July | 983.4 | 22.8 | 13.1 | 40 | 9 | 18 | 83 | 105 | 0 |
| Aug. | 983.9 | 21.4 | 12.6 | 39 | 4 | 18 | 81 | 82 | 0 |
| Sept. | 984.4 | 15.8 | 13.0 | 37 | −3 | 12 | 62 | 90 | * |
| Oct. | 984.7 | 9.1 | 12.8 | 32 | −8 | 8 | 40 | 47 | 1 |
| Nov. | 984.3 | −0.8 | 10.2 | 24 | −23 | 5 | 36 | 74 | 17 |
| Dec. | 985.7 | −7.9 | 10.3 | 17 | −30 | 3 | 22 | 28 | 18 |
| Annual | 984.2 | 6.6 | 12.0 | 40 | −35 | 9 | 631 | 105 | 102 |
| Rec.(yrs.) | 30 | 30 | 30 | 22 | 22 | 10 | 30 | 22 | 22 |

| Month | Number of days with | | | Mean cloud-iness (tenths) | Mean sun-shine (h) | Wind | | Solar radiation[1] (ly/ month) |
|---|---|---|---|---|---|---|---|---|
| | precip. (>0.25 mm) | thunder-storm | fog | | | most freq. direct. | mean speed (m/sec) | |
| Jan. | 8 | 0 | 2 | 6.3 | 50 | NW | 5 | 159 |
| Feb. | 7 | * | 1 | 6.0 | 57 | NW | 5 | 253 |
| Mar. | 11 | 1 | 1 | 6.6 | 54 | NW | 5 | 359 |
| Apr. | 9 | 2 | * | 6.4 | 56 | NW | 6 | 426 |
| May | 11 | 5 | 1 | 6.3 | 58 | SE | 5 | 487 |
| June | 13 | 7 | 1 | 6.0 | 61 | SE | 5 | 543 |
| July | 10 | 8 | 1 | 5.0 | 70 | SE | 4 | 555 |
| Aug. | 10 | 6 | 1 | 5.1 | 66 | SE | 4 | 491 |
| Sept. | 9 | 3 | 1 | 5.2 | 62 | S | 5 | 361 |
| Oct. | 8 | 2 | 1 | 5.2 | 58 | SE | 5 | 246 |
| Nov. | 8 | 1 | 1 | 6.9 | 40 | NW | 5 | 146 |
| Dec. | 8 | * | 1 | 6.8 | 42 | NW | 5 | 124 |
| Annual | 112 | 35 | 12 | 6.0 | 58 | NW | 5 | 346 |
| Rec.(yrs.) | 22 | 22 | 22 | 22 | 22 | 11 | 22 | 10 |

[1] St. Cloud data.

TABLE XXXIV

CLIMATIC TABLE FOR VICKSBURG, MISS.
Latitude 32°21′N, longitude 90°53′W, elevation 71 m (75 m)

| Month | Mean sta. press. (mbar) | Temperature (°C) | | | | Mean vap. press. (mbar) | Precipitation (mm) | | Mean snowfall (cm) |
|---|---|---|---|---|---|---|---|---|---|
| | | daily mean | mean daily range | extremes | | | mean | max. in 24 h | |
| | | | | max. | min. | | | | |
| Jan. | 1011.9 | 9.4 | 9.4 | 27 | −14 | 9 | 130 | 87 | 3 |
| Feb. | 1010.4 | 11.0 | 9.8 | 28 | −14 | 10 | 135 | 123 | 2 |
| Mar. | 1007.9 | 14.2 | 10.6 | 31 | −8 | 12 | 146 | 253 | tr. |
| Apr. | 1007.3 | 18.7 | 10.5 | 32 | 2 | 15 | 125 | 222 | 0 |
| May | 1006.3 | 23.0 | 10.4 | 35 | 6 | 20 | 105 | 98 | 0 |
| June | 1006.1 | 26.5 | 10.0 | 37 | 13 | 25 | 88 | 126 | 0 |
| July | 1007.5 | 27.7 | 9.6 | 38 | 18 | 26 | 99 | 115 | 0 |
| Aug. | 1007.1 | 27.6 | 10.0 | 38 | 15 | 26 | 76 | 170 | 0 |
| Sept. | 1007.4 | 24.8 | 10.4 | 37 | 5 | 22 | 64 | 77 | 0 |
| Oct. | 1009.2 | 19.7 | 11.0 | 34 | −1 | 15 | 52 | 119 | 0 |
| Nov. | 1011.5 | 13.4 | 10.7 | 29 | −7 | 10 | 113 | 135 | tr. |
| Dec. | 1011.2 | 10.2 | 9.3 | 28 | −9 | 9 | 125 | 96 | tr. |
| Annual | 1008.7 | 18.8 | 10.1 | 38 | −14 | 17 | 1,258 | 253 | 5 |
| Rec.(yrs.) | 30 | 30 | 30 | 23 | 23 | 10 | 30 | 23 | 23 |

| Month | Number of days with | | | Mean cloud-iness (tenths) | Mean sun-shine (h) | Wind | |
|---|---|---|---|---|---|---|---|
| | precip. (>0.25 mm) | thunder-storm | fog | | | most freq. direct.[1] | mean speed (m/sec) |
| Jan. | 10 | 3 | 2 | 6.6 | 44 | N | 4 |
| Feb. | 10 | 3 | 2 | 6.7 | 45 | N | 4 |
| Mar. | 10 | 6 | 1 | 6.3 | 53 | S | 4 |
| Apr. | 9 | 6 | 1 | 6.0 | 59 | S | 4 |
| May | 8 | 7 | * | 5.8 | 65 | S | 4 |
| June | 9 | 9 | * | 5.4 | 68 | S | 3 |
| July | 10 | 11 | * | 5.9 | 66 | S | 3 |
| Aug. | 7 | 8 | * | 4.9 | 71 | S | 3 |
| Sept. | 7 | 4 | 1 | 4.9 | 66 | N | 3 |
| Oct. | 6 | 2 | 1 | 4.2 | 68 | N | 3 |
| Nov. | 8 | 2 | 1 | 5.3 | 58 | N | 4 |
| Dec. | 10 | 2 | 2 | 5.9 | 47 | N | 4 |
| Annual | 104 | 63 | 11 | 5.7 | 59 | N | 4 |
| Rec.(yrs.) | 23 | 23 | 23 | 21 | 23 | 22 | |

[1] To 8 points.

TABLE XXXV

CLIMATIC TABLE FOR KANSAS CITY, MO.
Latitude 39°07′N, longitude 94°35′W, elevation 226 m

| Month | Temperature (°C) | | | | Mean vap. press. (mbar) | Precipitation (mm) | | Mean snowfall (cm) |
|---|---|---|---|---|---|---|---|---|
| | daily mean | mean daily range | extremes | | | mean | max. in 24 h | |
| | | | max. | min. | | | | |
| Jan. | −0.7 | 9.2 | 24 | −25 | 4 | 36 | 49 | 13 |
| Feb. | 1.6 | 10.1 | 24 | −25 | 5 | 32 | 42 | 10 |
| Mar. | 6.0 | 10.9 | 32 | −19 | 6 | 63 | 71 | 11 |
| Apr. | 12.9 | 11.6 | 33 | −9 | 9 | 90 | 151 | 2 |
| May | 18.4 | 11.2 | 39 | 0 | 14 | 112 | 67 | tr. |
| June | 24.1 | 11.0 | 42 | 7 | 18 | 116 | 86 | 0 |
| July | 27.2 | 11.1 | 44 | 12 | 22 | 81 | 106 | 0 |
| Aug. | 26.3 | 11.2 | 45 | 9 | 22 | 96 | 112 | 0 |
| Sept. | 21.6 | 12.1 | 43 | 1 | 16 | 83 | 95 | 0 |
| Oct. | 15.4 | 12.2 | 37 | −4 | 11 | 73 | 76 | tr. |
| Nov. | 6.7 | 10.5 | 28 | −15 | 6 | 46 | 53 | 3 |
| Dec. | 1.6 | 8.8 | 23 | −20 | 5 | 39 | 70 | 11 |
| Annual | 13.4 | 10.8 | 45 | −25 | 12 | 867 | 151 | 50 |
| Rec.(yrs.) | 30 | 30 | 27 | 27 | 10 | 30 | 27 | 27 |

| Month | Number of days with | | | Mean cloud-iness (tenths) | Mean sun-shine (h) | Wind | | Solar radiation[1] (ly/month) |
|---|---|---|---|---|---|---|---|---|
| | precip. (>0.25 mm) | thunder-storm | fog | | | most freq. direct. | mean speed (m/sec) | |
| Jan. | 7 | * | 3 | 6.3 | 50 | SSW | 5 | 182 |
| Feb. | 7 | 1 | 2 | 6.2 | 54 | SSW | 5 | 251 |
| Mar. | 9 | 3 | 1 | 6.4 | 56 | ENE | 5 | 335 |
| Apr. | 11 | 5 | 1 | 6.3 | 58 | S | 5 | 435 |
| May | 12 | 8 | 1 | 6.1 | 62 | S | 5 | 462 |
| June | 11 | 9 | * | 5.8 | 69 | S | 5 | 578 |
| July | 8 | 7 | * | 4.7 | 76 | S | 4 | 562 |
| Aug. | 8 | 8 | * | 4.6 | 73 | S | 4 | 529 |
| Sept. | 8 | 5 | 1 | 4.4 | 71 | S | 4 | 443 |
| Oct. | 7 | 3 | 1 | 4.6 | 68 | S | 4 | 324 |
| Nov. | 6 | 1 | 1 | 5.4 | 59 | SSW | 5 | 219 |
| Dec. | 6 | * | 2 | 6.2 | 51 | SSW | 5 | 174 |
| Annual | 100 | 50 | 13 | 5.6 | 64 | SSW | 5 | 375 |
| Rec.(yrs.) | 27 | 27 | 27 | 27 | 27 | 12 | 21 | 12 |

[1] Columbia data.

TABLE XXXVI

CLIMATIC TABLE FOR ST. LOUIS, MO.
Latitude 38°38′N, longitude 90°12′W, elevation 142 m (173 m)

| Month | Mean sta. press. (mbar) | Temperature (°C) | | | | | Mean vap. press. (mbar) | Precipitation (mm) | | Mean snowfall (cm) |
|---|---|---|---|---|---|---|---|---|---|---|
| | | daily mean | mean daily range | extremes | | | | mean | max. in 24 h | |
| | | | | max. | min. | | | | | |
| Jan. | 999.1 | −0.1 | 9.3 | 24 | −24 | 5 | 50 | 99 | 10 |
| Feb. | 998.1 | 1.8 | 9.8 | 24 | −22 | 6 | 52 | 73 | 8 |
| Mar. | 995.5 | 6.2 | 10.8 | 31 | −15 | 6 | 78 | 70 | 9 |
| Apr. | 994.6 | 13.0 | 11.7 | 33 | −7 | 9 | 94 | 160 | * |
| May | 994.2 | 18.7 | 11.6 | 37 | 2 | 14 | 95 | 86 | 0 |
| June | 993.8 | 24.2 | 11.7 | 40 | 9 | 19 | 109 | 222 | 0 |
| July | 995.1 | 26.4 | 11.8 | 44 | 13 | 22 | 84 | 176 | 0 |
| Aug. | 995.3 | 25.4 | 11.7 | 41 | 12 | 22 | 77 | 223 | 0 |
| Sept. | 996.6 | 21.1 | 12.6 | 39 | 2 | 16 | 70 | 106 | 0 |
| Oct. | 997.6 | 14.9 | 12.5 | 34 | −3 | 11 | 73 | 101 | tr. |
| Nov. | 997.8 | 6.7 | 10.7 | 28 | −14 | 6 | 65 | 54 | 3 |
| Dec. | 998.9 | 1.6 | 9.2 | 24 | −17 | 5 | 50 | 75 | 7 |
| Annual | 996.3 | 13.3 | 11.1 | 44 | −24 | 12 | 897 | 223 | 37 |
| Rec.(yrs.) | 30 | 30 | 30 | 23 | 23 | 10 | 30 | 23 | 23 |

| Month | Number of days with | | | Mean cloud-iness (tenths) | Mean sun-shine (h) | Wind | |
|---|---|---|---|---|---|---|---|
| | precip. (>0.25 mm) | thunder-storm | fog | | | most freq. direct. | mean speed (m/sec) |
| Jan. | 9 | 1 | 2 | 6.6 | 45 | NW | 6 |
| Feb. | 9 | 1 | 1 | 6.2 | 50 | NW | 6 |
| Mar. | 11 | 3 | 1 | 5.9 | 56 | S | 6 |
| Apr. | 12 | 5 | 1 | 6.0 | 59 | S | 6 |
| May | 12 | 7 | * | 5.5 | 65 | S | 5 |
| June | 11 | 10 | * | 5.3 | 66 | S | 5 |
| July | 8 | 8 | * | 4.5 | 72 | SW | 4 |
| Aug. | 8 | 7 | * | 4.6 | 70 | S | 4 |
| Sept. | 7 | 4 | * | 4.1 | 70 | S | 5 |
| Oct. | 7 | 3 | 1 | 4.2 | 69 | S | 5 |
| Nov. | 8 | 2 | 1 | 5.4 | 56 | S | 6 |
| Dec. | 8 | * | 1 | 6.5 | 45 | S | 6 |
| Annual | 110 | 51 | 8 | 5.4 | 60 | S | 5 |
| Rec.(yrs.) | 23 | 22 | 22 | 22 | 23 | 31 | 44 |

TABLE XXXVII

CLIMATIC TABLE FOR BILLINGS, MONT.
Latitude 45°48′N, longitude 108°32′W, elevation 1,087 m

| Month | Temperature (°C) | | | | Mean vap. press. (mbar) | Precipitation (mm) | | Mean snowfall (cm) |
|---|---|---|---|---|---|---|---|---|
| | daily mean | mean daily range | extremes | | | mean | max. in 24 h | |
| | | | max. | min. | | | | |
| Jan. | −4.9 | 11.2 | 20 | −34 | 2 | 14 | 15 | 19 |
| Feb. | −3.5 | 11.5 | 21 | −39 | 3 | 15 | 15 | 25 |
| Mar. | 0.7 | 11.3 | 25 | −28 | 4 | 27 | 25 | 26 |
| Apr. | 7.5 | 13.2 | 33 | −21 | 5 | 33 | 63 | 18 |
| May | 13.2 | 13.7 | 36 | −10 | 7 | 48 | 72 | 3 |
| June | 17.6 | 13.8 | 39 | 0 | 9 | 65 | 71 | * |
| July | 22.9 | 17.1 | 41 | 6 | 11 | 23 | 48 | 0 |
| Aug. | 21.6 | 17.0 | 40 | 4 | 10 | 23 | 20 | 0 |
| Sept. | 15.8 | 15.7 | 38 | −3 | 7 | 30 | 46 | 2 |
| Oct. | 10.0 | 14.1 | 30 | −16 | 5 | 28 | 50 | 8 |
| Nov. | 2.3 | 11.0 | 22 | −30 | 4 | 16 | 35 | 16 |
| Dec. | −1.7 | 10.8 | 21 | −27 | 2 | 15 | 18 | 20 |
| Annual | 8.4 | 13.3 | 41 | −39 | 6 | 337 | 72 | 137 |
| Rec.(yrs.) | 30 | 30 | 26 | 26 | 10 | 30 | 26 | 26 |

| Month | Number of days with | | | Mean cloud-iness (tenths) | Mean sun-shine (h) | Wind | |
|---|---|---|---|---|---|---|---|
| | precip. (>0.25 mm) | thunder-storm | fog | | | most freq. direct. | mean speed (m/sec) |
| Jan. | 7 | 0 | 2 | 6.9 | 49 | SW | 6 |
| Feb. | 8 | * | 3 | 6.9 | 53 | SW | 5 |
| Mar. | 9 | * | 2 | 7.2 | 56 | SW | 5 |
| Apr. | 9 | 1 | 2 | 7.0 | 58 | SW | 5 |
| May | 10 | 4 | 1 | 6.4 | 61 | SW | 5 |
| June | 11 | 8 | 1 | 6.0 | 64 | SW | 5 |
| July | 7 | 8 | 1 | 4.0 | 78 | SW | 4 |
| Aug. | 6 | 7 | * | 4.1 | 76 | SW | 4 |
| Sept. | 7 | 2 | 1 | 5.2 | 68 | SW | 5 |
| Oct. | 6 | * | 2 | 5.5 | 63 | SW | 5 |
| Nov. | 6 | 0 | 2 | 6.7 | 48 | SW | 6 |
| Dec. | 6 | * | 1 | 6.7 | 48 | WSW | 6 |
| Annual | 92 | 30 | 18 | 6.1 | 62 | SW | 5 |
| Rec.(yrs.) | 26 | 21 | 13 | 21 | 21 | 12 | 21 |

TABLE XXXVIII

CLIMATIC TABLE FOR HAVRE, MONT.
Latitude 48°34′N, longitude 109°40′W, elevation 758 m (792 m)

| Month | Mean sta. press. (mbar) | Temperature (°C) | | | | Mean vap. press. (mbar) | Precipitation (mm) | | Mean snowfall (cm) |
|---|---|---|---|---|---|---|---|---|---|
| | | daily mean | mean daily range | extremes | | | mean | max. in 24 h | |
| | | | | max. | min. | | | | |
| Jan. | 926.7 | −10.1 | 12.6 | 18 | −49 | 2 | 12 | 17 | 19 |
| Feb. | 926.5 | −8.7 | 13.5 | 22 | −44 | 3 | 11 | 13 | 14 |
| Mar. | 925.6 | −2.8 | 13.3 | 25 | −36 | 4 | 15 | 23 | 15 |
| Apr. | 925.4 | 5.9 | 15.4 | 35 | −22 | 5 | 25 | 46 | 9 |
| May | 925.1 | 12.3 | 16.1 | 37 | −10 | 7 | 39 | 42 | 1 |
| June | 924.6 | 16.1 | 15.3 | 42 | −2 | 9 | 69 | 75 | tr. |
| July | 926.1 | 21.1 | 18.2 | 42 | 3 | 11 | 32 | 48 | 0 |
| Aug. | 926.2 | 19.4 | 18.3 | 41 | −3 | 11 | 28 | 57 | 0 |
| Sept. | 926.5 | 13.4 | 17.1 | 37 | −8 | 7 | 26 | 69 | 2 |
| Oct. | 926.6 | 7.9 | 15.6 | 33 | −22 | 5 | 20 | 35 | 4 |
| Nov. | 926.9 | −1.1 | 13.1 | 23 | −33 | 4 | 13 | 19 | 13 |
| Dec. | 925.7 | −6.2 | 12.6 | 22 | −38 | 3 | 12 | 25 | 18 |
| Annual | 926.0 | 5.6 | 15.1 | 42 | −49 | 6 | 302 | 75 | 95 |
| Rec.(yrs.) | 30 | 30 | 30 | 37 | 37 | 10 | 30 | 37 | 37 |

| Month | Number of days with | | | Mean cloud-iness (tenths) | Mean sun-shine (h) | Wind | | Solar radiation[2] (ly/month) |
|---|---|---|---|---|---|---|---|---|
| | precip. (>0.25 mm) | thunder-storm | fog | | | most freq. direct[1]. | mean speed (m/sec) | |
| Jan. | 8 | 0 | 1 | 6.3 | 49 | SW | 4 | 135 |
| Feb. | 7 | 0 | 1 | 6.0 | 58 | SW | 4 | 224 |
| Mar. | 7 | * | 1 | 5.9 | 61 | SW | 4 | 358 |
| Apr. | 7 | 1 | * | 5.7 | 63 | E | 4 | 429 |
| May | 10 | 3 | * | 5.6 | 63 | E | 4 | 517 |
| June | 12 | 7 | * | 5.3 | 65 | E | 4 | 584 |
| July | 8 | 7 | * | 3.6 | 78 | E | 3 | 626 |
| Aug. | 7 | 6 | * | 3.8 | 75 | E | 3 | 527 |
| Sept. | 7 | 2 | * | 4.8 | 65 | SW | 3 | 398 |
| Oct. | 5 | * | * | 5.5 | 58 | SW | 4 | 260 |
| Nov. | 6 | 0 | 1 | 6.2 | 48 | SW | 4 | 153 |
| Dec. | 6 | 0 | 1 | 6.4 | 46 | SW | 4 | 111 |
| Annual | 90 | 25 | 5 | 5.4 | 62 | SW | 4 | 360 |
| Rec.(yrs.) | 57 | 57 | 56 | 57 | 56 | 37 | 57 | 11 |

[1] To 8 points.
[2] Great Falls data.

TABLE XXXIX

CLIMATIC TABLE FOR OMAHA, NEBR.
Latitude 41°18′N, longitude 95°54′W, elevation 298 m (337 m)

| Month | Mean sta. press. (mbar) | Temperature (°C) | | | | Mean vap. press. (mbar) | Precipitation (mm) | | Mean snowfall (cm) |
|---|---|---|---|---|---|---|---|---|---|
| | | daily mean | mean daily range | extremes | | | mean | max. in 24 h | |
| | | | | max. | min. | | | | |
| Jan. | 978.8 | −5.4 | 10.5 | 21 | −29 | 3 | 21 | 34 | 22 |
| Feb. | 978.3 | −3.1 | 10.6 | 21 | −28 | 4 | 24 | 57 | 19 |
| Mar. | 975.6 | 2.7 | 10.6 | 31 | −27 | 5 | 37 | 37 | 19 |
| Apr. | 974.6 | 10.9 | 12.0 | 33 | −12 | 8 | 65 | 65 | 3 |
| May | 974.0 | 17.2 | 12.1 | 37 | −2 | 12 | 88 | 76 | * |
| June | 973.2 | 22.8 | 11.9 | 41 | 4 | 18 | 115 | 88 | 0 |
| July | 975.4 | 25.8 | 12.5 | 46 | 11 | 20 | 86 | 86 | 0 |
| Aug. | 975.3 | 24.6 | 11.9 | 43 | 6 | 20 | 101 | 86 | 0 |
| Sept. | 976.3 | 19.4 | 12.9 | 40 | −1 | 13 | 67 | 92 | 0 |
| Oct. | 977.1 | 13.2 | 13.2 | 36 | −7 | 9 | 44 | 60 | 1 |
| Nov. | 977.6 | 3.8 | 11.2 | 27 | −19 | 5 | 32 | 64 | 6 |
| Dec. | 978.2 | −2.1 | 10.1 | 22 | −24 | 4 | 20 | 45 | 13 |
| Annual | 976.2 | 10.8 | 11.6 | 46 | −29 | 10 | 700 | 92 | 83 |
| Rec.(yrs.) | 30 | 30 | 30 | 25 | 25 | 10 | 30 | 25 | 25 |

| Month | Number of days with | | | Mean cloud-iness (tenths) | Mean sun-shine (h) | Wind | | Solar radiation (ly/ month) |
|---|---|---|---|---|---|---|---|---|
| | precip. (>0.25 mm) | thunder-storm | fog | | | most freq. direct. | mean speed (m/sec) | |
| Jan. | 7 | * | 2 | 6.1 | 53 | NNW | 5 | 204 |
| Feb. | 7 | * | 2 | 6.3 | 54 | N | 5 | 275 |
| Mar. | 8 | 1 | 1 | 6.6 | 53 | N | 6 | 353 |
| Apr. | 9 | 4 | 1 | 6.3 | 59 | NNW | 6 | 462 |
| May | 11 | 8 | 1 | 6.3 | 61 | SSE | 5 | 509 |
| June | 11 | 10 | 1 | 5.6 | 65 | SSE | 5 | 570 |
| July | 9 | 8 | 1 | 4.7 | 76 | SSE | 4 | 577 |
| Aug. | 10 | 9 | 1 | 4.6 | 70 | SSE | 4 | 519 |
| Sept. | 7 | 5 | 1 | 4.4 | 69 | SSE | 5 | 398 |
| Oct. | 6 | 3 | 1 | 4.5 | 68 | SSE | 5 | 304 |
| Nov. | 5 | 1 | 1 | 5.7 | 53 | SSE | 5 | 201 |
| Dec. | 5 | * | 2 | 6.2 | 48 | SSE | 5 | 170 |
| Annual | 95 | 49 | 15 | 5.6 | 62 | SSE | 5 | 379 |
| Rec.(yrs.) | 25 | 25 | 25 | 25 | 25 | 12 | 25 | 6 |

TABLE XL

CLIMATIC TABLE FOR ELKO, NEV.
Latitude 40°50′N, longitude 115°47′W, elevation 1,547 m

| Month | Temperature (°C) | | | | Mean vap. press. (mbar) | Precipitation (mm) | | Mean snowfall (cm) |
|-------|------------|--------------------|--------|--------|----------------|------|------------------|----------------|
| | daily mean | mean daily range | extremes | | | mean | max. in 24 h | |
| | | | max. | min. | | | | |
| Jan. | −5.2 | 14.0 | 14 | −42 | 3 | 29 | 32 | 27 |
| Feb. | −2.2 | 13.8 | 19 | −38 | 4 | 23 | 23 | 18 |
| Mar. | 1.9 | 14.7 | 23 | −23 | 4 | 21 | 21 | 12 |
| Apr. | 6.8 | 17.6 | 28 | −19 | 5 | 21 | 28 | 4 |
| May | 11.1 | 19.1 | 33 | −9 | 5 | 24 | 23 | 1 |
| June | 15.6 | 21.0 | 38 | −5 | 6 | 18 | 24 | tr. |
| July | 20.9 | 23.6 | 40 | −1 | 7 | 10 | 26 | 0 |
| Aug. | 19.4 | 24.4 | 39 | −4 | 7 | 8 | 26 | 0 |
| Sept. | 14.4 | 24.1 | 37 | −13 | 5 | 9 | 30 | * |
| Oct. | 8.3 | 21.0 | 30 | −13 | 5 | 19 | 33 | 1 |
| Nov. | 1.2 | 16.8 | 23 | −24 | 4 | 23 | 33 | 10 |
| Dec. | −3.1 | 14.1 | 18 | −39 | 3 | 26 | 41 | 19 |
| Annual | 7.4 | 18.7 | 40 | −42 | 5 | 231 | 41 | 92 |
| Rec.(yrs.) | 30 | 30 | 30 | 30 | 10 | 30 | 30 | 29 |

| Month | Number of days with | | | Mean cloud-iness (tenths) | Mean sun-shine (h) | Wind | | Solar radiation[1] (ly/ month) |
|-------|--------------------|----------------|-----|----------------|----------------|----------------|----------------|----------------|
| | precip. (>0.25 mm) | thunder-storm | fog | | | most freq. direct. | mean speed (m/sec) | |
| Jan. | 9 | * | 2 | 7.2 | 64 | SW | 2 | 242 |
| Feb. | 9 | * | 1 | 6.5 | 64 | SW | 3 | 330 |
| Mar. | 8 | * | * | 6.7 | 69 | SW | 3 | 462 |
| Apr. | 7 | 1 | 0 | 6.2 | 66 | WSW | 3 | 558 |
| May | 8 | 4 | * | 5.9 | 69 | SW | 3 | 621 |
| June | 5 | 3 | * | 3.9 | 80 | SW | 3 | 704 |
| July | 3 | 6 | * | 3.2 | 81 | SW | 3 | 659 |
| Aug. | 3 | 4 | 0 | 2.7 | 82 | SW | 3 | 614 |
| Sept. | 3 | 2 | * | 2.9 | 82 | SW | 2 | 520 |
| Oct. | 5 | 1 | * | 4.1 | 75 | SW | 2 | 394 |
| Nov. | 6 | * | * | 5.3 | 68 | SW | 2 | 286 |
| Dec. | 8 | * | 1 | 6.5 | 65 | SW | 2 | 224 |
| Annual | 75 | 21 | 5 | 5.1 | 73 | SW | 3 | 468 |
| Rec.(yrs.) | 30 | 12 | 12 | 12 | 25† | 19 | 12 | 10 |

[1] Data from Ely.

TABLE XLI

CLIMATIC TABLE FOR RENO, NEV.
Latitude 39°30'N, longitude 119'47°W, elevation 1,342 m

| Month | Temperature (°C) | | | | Mean vap. press. (mbar) | Precipitation (mm) | | Mean snowfall (cm) |
|-------|------------------|--|--|--|--|--------------------|--|--|
| | daily mean | mean daily range | extremes | | | mean | max. in 24 h | |
| | | | max. | min. | | | | |
| Jan. | −0.1 | 16.3 | 20 | −27 | 4 | 30 | 60 | 18 |
| Feb. | 2.3 | 16.3 | 21 | −23 | 5 | 26 | 37 | 13 |
| Mar. | 5.0 | 18.2 | 26 | −19 | 5 | 17 | 31 | 13 |
| Apr. | 8.6 | 20.0 | 31 | −11 | 5 | 14 | 42 | 3 |
| May | 11.9 | 21.2 | 34 | −7 | 6 | 13 | 30 | * |
| June | 15.6 | 23.6 | 38 | −4 | 7 | 9 | 17 | tr. |
| July | 20.1 | 25.9 | 40 | 1 | 8 | 7 | 20 | 0 |
| Aug. | 19.2 | 26.5 | 39 | −1 | 7 | 4 | 5 | 0 |
| Sept. | 15.7 | 25.5 | 38 | −6 | 6 | 6 | 20 | tr. |
| Oct. | 10.1 | 23.1 | 32 | −11 | 6 | 13 | 32 | 1 |
| Nov. | 4.1 | 20.5 | 25 | −17 | 5 | 14 | 31 | 4 |
| Dec. | 0.8 | 17.7 | 21 | −21 | 5 | 27 | 55 | 9 |
| Annual | 9.4 | 21.2 | 40 | −27 | 6 | 180 | 60 | 61 |
| Rec.(yrs.) | 30 | 30 | 19 | 19 | 10 | 30 | 19 | 18 |

| Month | Number of days with | | | Mean cloud-iness (tenths) | Mean sun-shine (h) | Wind | |
|-------|---------------------|--|--|--|--|------|--|
| | precip. (>0.25 mm) | thunder-storm | fog | | | most freq. direct. | mean speed (m/sec) |
| Jan. | 6 | 0 | 2 | 6.2 | 64 | S | 3 |
| Feb. | 6 | * | 1 | 5.9 | 68 | S | 3 |
| Mar. | 6 | 0 | * | 5.7 | 71 | WNW | 3 |
| Apr. | 4 | * | * | 5.4 | 76 | WNW | 4 |
| May | 5 | 2 | * | 4.8 | 78 | WNW | 3 |
| June | 3 | 2 | 0 | 3.3 | 83 | WNW | 3 |
| July | 3 | 4 | * | 2.1 | 91 | WNW | 3 |
| Aug. | 2 | 2 | * | 1.8 | 93 | WNW | 3 |
| Sept. | 2 | 1 | * | 2.3 | 91 | WNW | 2 |
| Oct. | 3 | 1 | * | 4.0 | 80 | WNW | 2 |
| Nov. | 4 | 0 | 1 | 5.2 | 71 | S | 2 |
| Dec. | 6 | 0 | 3 | 6.2 | 61 | WNW | 2 |
| Annual | 47 | 12 | 9 | 4.4 | 78 | WNW | 3 |
| Rec.(yrs.) | 18 | 18 | 18 | 18 | 18 | 12 | 18 |

TABLE XLII

CLIMATIC TABLE FOR MOUNT WASHINGTON, N.H.
Latitude 44°16′N, longitude 71°18′W, elevation 1,909 m

| Month | Temperature (°C) | | | | Mean vap. press. (mbar) | Precipitation (mm) | | Mean snowfall (cm) |
|---|---|---|---|---|---|---|---|---|
| | daily mean | mean daily range | extremes | | | mean | max. in 24 h | |
| | | | max. | min. | | | | |
| Jan. | −14.3 | 9.4 | 7 | −44 | 2 | 138 | 86 | 75 |
| Feb. | −14.7 | 9.3 | 6 | −43 | 2 | 132 | 108 | 82 |
| Mar. | −11.3 | 8.1 | 9 | −39 | 3 | 146 | 97 | 82 |
| Apr. | −5.0 | 7.4 | 16 | −29 | 5 | 150 | 96 | 55 |
| May | 1.7 | 7.0 | 17 | −18 | 7 | 148 | 86 | 25 |
| June | 7.2 | 7.0 | 22 | −13 | 10 | 165 | 97 | 3 |
| July | 9.5 | 6.4 | 22 | −2 | 14 | 170 | 80 | * |
| Aug. | 8.7 | 6.2 | 22 | −4 | 12 | 169 | 132 | * |
| Sept. | 5.0 | 6.4 | 19 | −12 | 8 | 178 | 122 | 4 |
| Oct. | −0.6 | 6.8 | 15 | −21 | 7 | 157 | 179 | 24 |
| Nov. | −6.5 | 7.2 | 11 | −29 | 3 | 168 | 87 | 60 |
| Dec. | −12.9 | 8.6 | 7 | −43 | 3 | 160 | 86 | 85 |
| Annual | −2.8 | 7.5 | 22 | −44 | 6 | 1,881 | 179 | 495 |
| Rec.(yrs.) | 30 | 30 | 76 | 76 | 10 | 30 | 28 | 28 |

| Month | Number of days with | | | Mean cloud-iness (tenths) | Mean sun-shine (h) | Wind | |
|---|---|---|---|---|---|---|---|
| | precip. (>0.25 mm) | thunder-storm | fog | | | most freq. direct. | mean speed (m/sec) |
| Jan. | 18 | * | 24 | 7.6 | 33 | W | 21 |
| Feb. | 17 | * | 25 | 7.6 | 33 | W | 21 |
| Mar. | 19 | * | 27 | 7.6 | 34 | W | 20 |
| Apr. | 18 | 1 | 24 | 7.6 | 35 | W | 17 |
| May | 18 | 2 | 24 | 7.6 | 35 | W | 14 |
| June | 16 | 4 | 26 | 8.0 | 31 | W | 13 |
| July | 17 | 4 | 26 | 8.0 | 31 | W | 11 |
| Aug. | 15 | 3 | 26 | 7.6 | 33 | W | 11 |
| Sept. | 15 | 1 | 25 | 7.2 | 37 | W | 13 |
| Oct. | 14 | 1 | 25 | 6.6 | 41 | W | 15 |
| Nov. | 19 | * | 27 | 7.8 | 30 | W | 18 |
| Dec. | 20 | * | 26 | 7.7 | 29 | W | 21 |
| Annual | 206 | 16 | 303 | 7.6 | 34 | W | 16 |
| Rec.(yrs.) | 28 | 28 | 28 | 22 | 22 | 22 | 28 |

TABLE XLIII

CLIMATIC TABLE FOR ATLANTIC CITY, N.J.
Latitude 39°22′N, longitude 74°25′W, elevation 3 m

| Month | Temperature (°C) | | | | Mean vap. press. (mbar) | Precipitation (mm) | | Mean snowfall (cm) |
|-------|------------------|--|--|--|--|--------------------|--|--------------------|
|       | daily mean | mean daily range | extremes | | | mean | max. in 24 h | |
|       | | | max. | min. | | | | |
| Jan. | 1.6 | 9.1 | 20 | −20 | 5 | 90 | 76 | 11 |
| Feb. | 1.5 | 9.6 | 25 | −23 | 6 | 80 | 68 | 12 |
| Mar. | 5.1 | 9.6 | 29 | −13 | 7 | 99 | 80 | 6 |
| Apr. | 10.6 | 10.3 | 32 | −9 | 9 | 87 | 80 | 1 |
| May | 16.3 | 10.8 | 35 | 1 | 14 | 89 | 82 | 0 |
| June | 21.1 | 10.3 | 37 | 7 | 18 | 72 | 107 | 0 |
| July | 23.9 | 9.7 | 39 | 11 | 23 | 94 | 137 | 0 |
| Aug. | 23.2 | 9.5 | 40 | 9 | 22 | 124 | 228 | 0 |
| Sept. | 19.6 | 9.8 | 34 | 3 | 18 | 84 | 232 | 0 |
| Oct. | 14.0 | 10.4 | 33 | −2 | 13 | 81 | 234 | tr. |
| Nov. | 8.2 | 9.8 | 27 | −12 | 8 | 93 | 131 | 1 |
| Dec. | 2.6 | 9.5 | 20 | −22 | 6 | 82 | 89 | 7 |
| Annual | 12.3 | 9.9 | 40 | −23 | 12 | 1,075 | 234 | 38 |
| Rec.(yrs.) | 30 | 30 | 85 | 85 | 10 | 30 | 85 | 75 |

| Month | Number of days with | | | Mean cloud-iness (tenths) | Mean sun-shine (h) | Wind | |
|-------|---------------------|--|--|--|--|------|--|
|       | precip. (>0.25 mm) | thunder-storm | fog | | | most freq. direct. | mean speed (m/sec) |
| Jan. | 12 | * | 2 | 6.2 | 51 | NW | 7 |
| Feb. | 10 | * | 2 | 5.6 | 57 | NW | 7 |
| Mar. | 12 | 1 | 2 | 5.7 | 58 | NW | 8 |
| Apr. | 11 | 2 | 2 | 5.8 | 59 | NW | 7 |
| May | 11 | 3 | 3 | 5.6 | 62 | SW | 7 |
| June | 10 | 5 | 2 | 5.5 | 65 | S | 6 |
| July | 10 | 5 | 2 | 5.3 | 67 | SW | 6 |
| Aug. | 10 | 4 | 1 | 5.2 | 66 | SW | 6 |
| Sept. | 8 | 2 | 1 | 5.1 | 65 | SW | 6 |
| Oct. | 9 | 1 | 1 | 4.8 | 64 | NW | 7 |
| Nov. | 9 | * | 1 | 5.4 | 58 | NW | 7 |
| Dec. | 10 | * | 2 | 6.0 | 52 | NW | 7 |
| Annual | 122 | 23 | 21 | 5.5 | 60 | NW | 7 |
| Rec.(yrs.) | 85 | 84 | 53 | 60 | 62 | 75 | 37 |

TABLE XLIV

CLIMATIC TABLE FOR ALBUQUERQUE, N.M.
Latitude 35°03′N, longitude 106°37′W, elevation 1,620 m (1,515 m)

| Month | Mean sta. press. (mbar) | Temperature (°C) | | | | Mean vap. press. (mbar) | Precipitation (mm) | | Mean snowfall (cm) |
|---|---|---|---|---|---|---|---|---|---|
| | | daily mean | mean daily range | extremes | | | mean | max. in 24 h | |
| | | | | max. | min. | | | | |
| Jan. | 849.1 | 1.7 | 12.7 | 19 | −17 | 4 | 10 | 14 | 5 |
| Feb. | 847.8 | 4.4 | 13.7 | 22 | −21 | 4 | 10 | 12 | 4 |
| Mar. | 846.0 | 7.9 | 15.1 | 27 | −13 | 4 | 12 | 18 | 5 |
| Apr. | 845.9 | 13.2 | 16.1 | 31 | −7 | 4 | 12 | 26 | 1 |
| May | 846.4 | 18.4 | 15.8 | 37 | 1 | 6 | 19 | 27 | tr. |
| June | 847.3 | 23.8 | 16.4 | 38 | 7 | 7 | 14 | 42 | 0 |
| July | 850.1 | 25.8 | 15.2 | 40 | 13 | 12 | 30 | 29 | 0 |
| Aug. | 850.2 | 24.8 | 14.8 | 38 | 12 | 12 | 34 | 29 | 0 |
| Sept. | 849.9 | 21.4 | 14.3 | 37 | 4 | 9 | 24 | 49 | tr. |
| Oct. | 850.0 | 14.7 | 14.7 | 31 | −3 | 7 | 19 | 33 | tr. |
| Nov. | 850.2 | 6.7 | 14.5 | 23 | −12 | 4 | 10 | 19 | 4 |
| Dec. | 849.9 | 2.8 | 12.6 | 22 | −16 | 4 | 12 | 34 | 8 |
| Annual | 848.5 | 13.8 | 14.7 | 40 | −21 | 6 | 206 | 49 | 27 |
| Rec.(yrs.) | 30 | 30 | 30 | 21 | 21 | 10 | 30 | 21 | 21 |

| Month | Number of days with | | | Mean cloud-iness (tenths) | Mean sun-shine (h) | Wind | | Solar radiation (ly/ month) |
|---|---|---|---|---|---|---|---|---|
| | precip. (>0.25 mm) | thunder-storm | fog | | | most freq. direct. | mean speed (m/sec) | |
| Jan. | 4 | * | 1 | 5.0 | 69 | N | 4 | 305 |
| Feb. | 4 | * | 1 | 4.9 | 72 | N | 4 | 388 |
| Mar. | 4 | 1 | 1 | 4.9 | 73 | SE | 5 | 509 |
| Apr. | 4 | 2 | * | 4.8 | 75 | SE | 5 | 623 |
| May | 4 | 4 | 0 | 4.3 | 79 | S | 5 | 694 |
| June | 4 | 5 | 0 | 3.3 | 83 | SE | 4 | 732 |
| July | 9 | 12 | * | 4.4 | 76 | SE | 4 | 682 |
| Aug. | 10 | 12 | * | 4.3 | 76 | SE | 4 | 626 |
| Sept. | 5 | 4 | 0 | 3.1 | 82 | SE | 4 | 548 |
| Oct. | 5 | 3 | * | 3.5 | 78 | SE | 4 | 443 |
| Nov. | 2 | 1 | 1 | 3.6 | 79 | N | 3 | 332 |
| Dec. | 4 | * | 1 | 4.5 | 70 | N | 3 | 280 |
| Annual | 58 | 45 | 5 | 4.2 | 76 | SE | 4 | 514 |
| Rec.(yrs.) | 21 | 21 | 21 | 21 | 21 | 12 | 21 | 15 |

TABLE XLV

CLIMATIC TABLE FOR BINGHAMTON, N.Y.
Latitude 42°06′N, longitude 75°55′W, elevation 261 m

| Month | Temperature (°C) | | | | Mean vap. press. (mbar) | Precipitation (mm) | | Mean snowfall (cm) |
|---|---|---|---|---|---|---|---|---|
| | daily mean | mean daily range | extremes | | | mean | max. in 24 h | |
| | | | max. | min. | | | | |
| Jan. | −3.0 | 8.5 | 21 | −33 | 4 | 62 | 49 | 32 |
| Feb. | −3.1 | 9.6 | 23 | −32 | 4 | 60 | 53 | 33 |
| Mar. | 1.3 | 9.6 | 28 | −24 | 5 | 78 | 50 | 25 |
| Apr. | 8.3 | 11.2 | 33 | −14 | 8 | 78 | 50 | 7 |
| May | 14.6 | 12.5 | 34 | −5 | 11 | 92 | 77 | tr. |
| June | 19.6 | 12.4 | 38 | −1 | 16 | 78 | 67 | 0 |
| July | 22.1 | 12.6 | 39 | 4 | 18 | 97 | 108 | 0 |
| Aug. | 21.0 | 12.5 | 38 | 1 | 18 | 91 | 84 | 0 |
| Sept. | 16.9 | 12.2 | 38 | −4 | 13 | 82 | 116 | tr. |
| Oct. | 11.1 | 11.8 | 33 | −8 | 9 | 77 | 104 | 1 |
| Nov. | 4.8 | 8.8 | 27 | −12 | 6 | 71 | 71 | 10 |
| Dec. | −1.5 | 8.1 | 20 | −30 | 4 | 65 | 60 | 24 |
| Annual | 9.3 | 10.8 | 39 | −33 | 10 | 931 | 116 | 132 |
| Rec.(yrs.) | 30 | 30 | 70 | 70 | 10 | 30 | 70 | 70 |

| Month | Number of days with | | | Mean cloud-iness (tenths) | Mean sun-shine (h) | Wind | |
|---|---|---|---|---|---|---|---|
| | precip. (>0.25 mm) | thunder-storm | fog | | | most freq. direct. | mean speed (m/sec) |
| Jan. | 15 | * | 4 | 8.1 | 35 | WSW | 5 |
| Feb. | 13 | * | 4 | 7.9 | 40 | SSE | 5 |
| Mar. | 14 | 2 | 5 | 7.7 | 45 | NW | 5 |
| Apr. | 14 | 3 | 5 | 7.5 | 49 | WNW | 5 |
| May | 13 | 4 | 5 | 7.2 | 56 | NNW | 5 |
| June | 12 | 7 | 3 | 6.7 | 64 | NNW | 4 |
| July | 12 | 7 | 5 | 6.4 | 66 | WSW | 4 |
| Aug. | 11 | 6 | 4 | 6.6 | 61 | SSW | 4 |
| Sept. | 10 | 3 | 4 | 6.3 | 59 | SSW | 4 |
| Oct. | 11 | 1 | 3 | 6.4 | 54 | WSW | 4 |
| Nov. | 13 | * | 4 | 7.9 | 31 | NNW | 5 |
| Dec. | 14 | * | 4 | 8.2 | 28 | WSW | 5 |
| Annual | 152 | 33 | 50 | 7.2 | 51 | WSW | 5 |
| Rec.(yrs.) | 64 | 12 | 12 | 12 | 12 | 12 | 12 |

TABLE XLVI

CLIMATIC TABLE FOR BUFFALO, N.Y.
Latitude 42°56′N, longitude 78°44′W, elevation 211 m

| Month | Temperature (°C) | | | | Mean vap. press. (mbar) | Precipitation (mm) | | Mean snowfall (cm) |
|-------|------------|---------------|----------|----------|------|------|--------------|------|
| | daily mean | mean daily range | extremes | | | mean | max. in 24 h | |
| | | | max. | min. | | | | |
| Jan. | −4.7 | 7.0 | 22 | −24 | 4 | 72 | 61 | 68 |
| Feb. | −4.9 | 7.7 | 18 | −21 | 4 | 69 | 59 | 55 |
| Mar. | −0.8 | 7.9 | 27 | −19 | 5 | 82 | 54 | 38 |
| Apr. | 6.1 | 10.0 | 31 | −11 | 7 | 76 | 37 | 11 |
| May | 12.4 | 11.3 | 32 | −3 | 11 | 75 | 52 | tr. |
| June | 18.2 | 11.5 | 35 | 2 | 15 | 65 | 44 | 0 |
| July | 21.0 | 11.5 | 34 | 6 | 18 | 65 | 86 | 0 |
| Aug. | 20.2 | 11.4 | 37 | 6 | 17 | 77 | 99 | 0 |
| Sept. | 16.3 | 11.3 | 37 | 0 | 13 | 80 | 68 | tr. |
| Oct. | 10.4 | 10.4 | 31 | −4 | 9 | 76 | 79 | 1 |
| Nov. | 3.7 | 7.7 | 24 | −14 | 6 | 91 | 47 | 29 |
| Dec. | −2.7 | 6.8 | 19 | −22 | 4 | 76 | 52 | 61 |
| Annual | 7.9 | 9.5 | 37 | −24 | 9 | 904 | 99 | 263 |
| Rec.(yrs.) | 30 | 30 | 17 | 17 | 10 | 30 | 10 | 10 |

| Month | Number of days with | | | Mean cloud-iness (tenths) | Mean sun-shine (h) | Wind | |
|-------|---------------------|---------------|-----|------|------|------|------|
| | precip. (>0.25 mm) | thunder-storm | fog | | | most freq. direct. | mean speed (m/sec) |
| Jan. | 20 | * | 2 | 8.3 | 33 | WSW | 7 |
| Feb. | 17 | * | 2 | 8.1 | 41 | SW | 7 |
| Mar. | 16 | 1 | 2 | 7.4 | 49 | SW | 7 |
| Apr. | 15 | 3 | 2 | 7.1 | 52 | SW | 6 |
| May | 12 | 3 | 3 | 6.7 | 59 | SW | 6 |
| June | 9 | 4 | 1 | 6.0 | 68 | SW | 5 |
| July | 10 | 5 | 1 | 5.6 | 71 | SW | 5 |
| Aug. | 12 | 5 | 1 | 5.6 | 68 | SW | 5 |
| Sept. | 10 | 4 | 1 | 5.9 | 62 | S | 5 |
| Oct. | 10 | 1 | 2 | 6.0 | 55 | S | 5 |
| Nov. | 16 | 1 | 1 | 8.1 | 30 | WSW | 6 |
| Dec. | 20 | * | 2 | 8.2 | 31 | WSW | 7 |
| Annual | 167 | 27 | 20 | 6.9 | 52 | SW | 6 |
| Rec.(yrs.) | 10 | 17 | 17 | 17 | 17 | 11 | 21 |

TABLE XLVII

CLIMATIC TABLE FOR NEW YORK (CENTRAL PARK), N.Y.
Latitude 40°47′N, longitude 73°58′W, elevation 40 m (96 m)

| Month | Mean sta. press. (mbar) | Temperature (°C) | | | | Mean vap. press. (mbar) | Precipitation (mm) | | Mean snowfall (cm) |
|---|---|---|---|---|---|---|---|---|---|
| | | daily mean | mean daily range | extremes | | | mean | max. in 24 h | |
| | | | | max. | min. | | | | |
| Jan. | 1006.1 | 0.7 | 7.0 | 22 | −21 | 5 | 84 | 85 | 19 |
| Feb. | 1004.7 | 0.8 | 7.7 | 24 | −26 | 5 | 72 | 75 | 22 |
| Mar. | 1003.2 | 4.7 | 8.1 | 30 | −16 | 6 | 102 | 108 | 14 |
| Apr. | 1003.8 | 10.8 | 9.2 | 33 | −11 | 8 | 87 | 68 | 3 |
| May | 1003.5 | 16.9 | 10.0 | 36 | 0 | 12 | 93 | 97 | 0 |
| June | 1002.8 | 21.9 | 9.8 | 38 | 7 | 16 | 84 | 120 | 0 |
| July | 1003.8 | 24.9 | 9.5 | 41 | 11 | 20 | 94 | 90 | 0 |
| Aug. | 1004.6 | 23.9 | 9.2 | 40 | 10 | 20 | 113 | 122 | 0 |
| Sept. | 1006.8 | 20.3 | 9.3 | 39 | 4 | 17 | 98 | 211 | 0 |
| Oct. | 1006.8 | 14.6 | 8.9 | 34 | −2 | 12 | 80 | 284 | tr. |
| Nov. | 1006.4 | 8.3 | 7.5 | 29 | −15 | 7 | 86 | 93 | 3 |
| Dec. | 1006.8 | 2.2 | 6.9 | 21 | −25 | 5 | 83 | 82 | 16 |
| Annual | 1004.9 | 12.5 | 8.6 | 41 | −26 | 11 | 1,076 | 284 | 77 |
| Rec.(yrs.) | 30 | 30 | 30 | 92 | 92 | 10 | 30 | 49 | 92 |

| Month | Number of days with | | | Mean cloud-iness (tenths) | Mean sun-shine (h) | Wind | | Solar radiation (ly/ month) |
|---|---|---|---|---|---|---|---|---|
| | precip. (>0.25 mm) | thunder-storm | fog[1] | | | most freq. direct. | mean speed (m/sec) | |
| Jan. | 11 | * | 3 | 6.2 | 49 | NW | 5 | 132 |
| Feb. | 10 | * | 2 | 5.8 | 56 | NW | 5 | 199 |
| Mar. | 12 | 1 | 2 | 5.8 | 57 | NW | 5 | 289 |
| Apr. | 11 | 2 | 2 | 6.0 | 59 | NW | 5 | 369 |
| May | 11 | 4 | 2 | 5.8 | 62 | SW | 4 | 430 |
| June | 10 | 5 | 1 | 5.7 | 65 | SW | 4 | 471 |
| July | 11 | 7 | 1 | 5.6 | 66 | SW | 4 | 459 |
| Aug. | 10 | 6 | * | 5.6 | 64 | SW | 4 | 389 |
| Sept. | 8 | 3 | 1 | 5.2 | 64 | SW | 4 | 331 |
| Oct. | 8 | 1 | 2 | 5.1 | 61 | SW | 4 | 242 |
| Nov. | 9 | * | 2 | 5.9 | 53 | NW | 5 | 148 |
| Dec. | 10 | * | 2 | 6.0 | 50 | NW | 5 | 116 |
| Annual | 121 | 30 | 20 | 5.7 | 59 | NW | 4 | 298 |
| Rec.(yrs.) | 92 | 77 | 56 | 36 | 84 | 41 | 41 | 34 |

[1] City data.

312

## TABLE XLVIII

CLIMATIC TABLE FOR ASHEVILLE, N.C.
Latitude 35°36′N, longitude 82°32′W, elevation 671 m

| Month | Temperature (°C) | | | | Mean vap. press. (mbar) | Precipitation (mm) | | Mean snowfall (cm) |
|---|---|---|---|---|---|---|---|---|
| | daily mean | mean daily range | extremes | | | mean | max. in 24 h | |
| | | | max. | min. | | | | |
| Jan. | 4.3 | 10.3 | 25 | −19 | 6 | 81 | 96 | 6 |
| Feb. | 4.8 | 11.2 | 27 | −18 | 6 | 77 | 57 | 9 |
| Mar. | 7.9 | 11.9 | 29 | −15 | 7 | 95 | 64 | 8 |
| Apr. | 13.3 | 12.9 | 31 | −7 | 10 | 81 | 103 | 1 |
| May | 18.0 | 13.2 | 34 | 1 | 15 | 73 | 44 | tr. |
| June | 22.1 | 12.7 | 37 | 6 | 20 | 89 | 81 | 0 |
| July | 23.6 | 11.8 | 37 | 8 | 22 | 109 | 61 | 0 |
| Aug. | 23.1 | 11.7 | 37 | 7 | 20 | 92 | 172 | 0 |
| Sept. | 19.9 | 12.2 | 35 | 2 | 17 | 71 | 90 | 0 |
| Oct. | 14.2 | 13.1 | 32 | −4 | 12 | 63 | 103 | tr. |
| Nov. | 8.1 | 12.2 | 28 | −17 | 7 | 56 | 64 | 1 |
| Dec. | 4.4 | 10.8 | 27 | −17 | 6 | 74 | 110 | 5 |
| Annual | 13.6 | 12.0 | 37 | −19 | 12 | 961 | 172 | 30 |
| Rec.(yrs.) | 30 | 30 | 30 | 30 | 10 | 30 | 30 | 30 |

| Month | Number of days with | | | Mean cloud-iness (tenths) | Mean sun-shine (h) | Wind | |
|---|---|---|---|---|---|---|---|
| | precip. (>0.25 mm) | thunder-storm | fog | | | most freq. direct. | mean speed (m/sec) |
| Jan. | 11 | 1 | 2 | 6.4 | 47 | NW | 4 |
| Feb. | 11 | 1 | 1 | 6.1 | 52 | NW | 4 |
| Mar. | 13 | 2 | 1 | 6.0 | 57 | NW | 4 |
| Apr. | 10 | 4 | 2 | 5.7 | 63 | NW | 4 |
| May | 11 | 6 | 3 | 5.6 | 66 | NW | 3 |
| June | 12 | 10 | 5 | 5.7 | 67 | NW | 3 |
| July | 15 | 13 | 7 | 6.4 | 60 | NW | 3 |
| Aug. | 12 | 10 | 9 | 6.0 | 60 | NW | 3 |
| Sept. | 9 | 4 | 7 | 5.4 | 63 | NW | 3 |
| Oct. | 8 | 1 | 5 | 4.7 | 64 | NW | 3 |
| Nov. | 8 | * | 2 | 5.2 | 58 | NW | 4 |
| Dec. | 10 | * | 2 | 6.1 | 47 | NW | 4 |
| Annual | 130 | 52 | 46 | 5.8 | 59 | NW | 3 |
| Rec.(yrs.) | 30 | 30 | 30 | 30 | 30 | 30 | 30 |

TABLE XLIX

CLIMATIC TABLE FOR GREENSBORO, N.C.
Latitude 36°05′N, longitude 79°57′W, elevation 273 m

| Month | Temperature (°C) | | | | Mean vap. press. (mbar) | Precipitation (mm) | | Mean snowfall (cm) |
|---|---|---|---|---|---|---|---|---|
| | daily mean | mean daily range | extremes | | | mean | max. in 24 h | |
| | | | max. | min. | | | | |
| Jan. | 4.3 | 11.0 | 26 | −22 | 6 | 86 | 78 | 7 |
| Feb. | 5.0 | 12.0 | 27 | −20 | 7 | 84 | 76 | 5 |
| Mar. | 8.6 | 12.8 | 32 | −15 | 8 | 94 | 78 | 4 |
| Apr. | 14.1 | 13.3 | 34 | −6 | 12 | 87 | 69 | tr. |
| May | 19.4 | 13.3 | 37 | 1 | 16 | 84 | 77 | 0 |
| June | 23.8 | 12.6 | 39 | 6 | 21 | 88 | 105 | 0 |
| July | 25.2 | 11.6 | 39 | 9 | 23 | 122 | 113 | 0 |
| Aug. | 24.6 | 11.4 | 38 | 8 | 23 | 117 | 114 | 0 |
| Sept. | 21.2 | 12.0 | 38 | 2 | 19 | 93 | 190 | 0 |
| Oct. | 15.2 | 13.6 | 35 | −5 | 13 | 69 | 159 | 0 |
| Nov. | 8.8 | 13.2 | 29 | −12 | 8 | 68 | 77 | tr. |
| Dec. | 4.4 | 11.5 | 25 | −17 | 7 | 80 | 91 | 4 |
| Annual | 14.6 | 12.4 | 39 | −22 | 14 | 1,072 | 190 | 20 |
| Rec.(yrs.) | 30 | 30 | 32 | 32 | 10 | 30 | 32 | 32 |

| Month | Number of days with | | | Mean cloud-iness (tenths) | Mean sun-shine (h) | Wind | | Solar radiation (ly/ month) |
|---|---|---|---|---|---|---|---|---|
| | precip. (>0.25 mm) | thunder-storm | fog | | | most freq. direct. | mean speed (m/sec) | |
| Jan. | 10 | * | 5 | 6.3 | 50 | SW | 4 | 208 |
| Feb. | 10 | 1 | 3 | 6.0 | 56 | SW | 4 | 271 |
| Mar. | 11 | 2 | 3 | 5.9 | 59 | SW | 4 | 359 |
| Apr. | 10 | 3 | 2 | 5.6 | 64 | SW | 4 | 477 |
| May | 10 | 7 | 2 | 5.6 | 68 | SW | 4 | 536 |
| June | 11 | 9 | 1 | 5.7 | 69 | SW | 3 | 556 |
| July | 13 | 12 | 2 | 6.1 | 65 | SW | 3 | 542 |
| Aug. | 11 | 9 | 2 | 5.8 | 65 | SW | 3 | 484 |
| Sept. | 8 | 4 | 3 | 5.4 | 65 | NE | 3 | 414 |
| Oct. | 7 | 1 | 2 | 4.5 | 67 | NE | 3 | 337 |
| Nov. | 8 | 1 | 3 | 5.1 | 60 | SW | 3 | 243 |
| Dec. | 9 | * | 4 | 5.9 | 54 | SW | 3 | 199 |
| Annual | 118 | 49 | 32 | 5.7 | 62 | SW | 4 | 386 |
| Rec.(yrs.) | 32 | 32 | 32 | 32 | 32 | 12 | 32 | 10 |

TABLE L

CLIMATIC TABLE FOR BISMARCK, N.D.
Latitude 46°46′N, longitude 100°45′W, elevation 502 m (511 m)

| Month | Mean sta. press. (mbar) | Temperature (°C) | | | | Mean vap. press. (mbar) | Precipitation (mm) | | Mean snowfall (cm) |
|---|---|---|---|---|---|---|---|---|---|
| | | daily mean | mean daily range | extremes | | | mean | max. in 24 h | |
| | | | | max. | min. | | | | |
| Jan. | 956.3 | −12.8 | 12.0 | 12 | −42 | 1 | 11 | 17 | 17 |
| Feb. | 956.5 | −10.8 | 12.0 | 20 | −37 | 2 | 11 | 19 | 15 |
| Mar. | 955.1 | −3.8 | 11.0 | 27 | −35 | 3 | 20 | 33 | 22 |
| Apr. | 954.4 | 6.1 | 13.2 | 33 | −19 | 6 | 31 | 39 | 7 |
| May | 953.6 | 13.0 | 14.2 | 37 | −7 | 8 | 50 | 48 | 2 |
| June | 952.3 | 18.1 | 13.4 | 38 | 1 | 13 | 86 | 83 | tr. |
| July | 954.2 | 22.3 | 15.0 | 42 | 4 | 16 | 56 | 45 | 0 |
| Aug. | 954.1 | 21.0 | 15.5 | 43 | 3 | 15 | 44 | 52 | 0 |
| Sept. | 954.7 | 14.8 | 15.5 | 41 | −9 | 9 | 30 | 51 | 1 |
| Oct. | 955.2 | 7.9 | 14.7 | 35 | −15 | 6 | 22 | 32 | 4 |
| Nov. | 955.8 | −2.0 | 11.4 | 23 | −28 | 4 | 15 | 25 | 14 |
| Dec. | 955.7 | −8.4 | 11.3 | 16 | −38 | 2 | 9 | 15 | 13 |
| Annual | 954.9 | 5.4 | 13.3 | 43 | −42 | 7 | 385 | 83 | 95 |
| Rec.(yrs.) | 30 | 30 | 30 | 21 | 21 | 10 | 30 | 21 | 21 |

| Month | Number of days with | | | Mean cloud-iness (tenths) | Mean sun-shine (h) | Wind | | Solar radiation (ly/ month) |
|---|---|---|---|---|---|---|---|---|
| | precip. (>0.25 mm) | thunder-storm | fog | | | most freq. direct. | mean speed (m/sec) | |
| Jan. | 8 | 0 | 1 | 6.6 | 54 | WNW | 5 | 159 |
| Feb. | 7 | 0 | 1 | 6.7 | 56 | WNW | 5 | 244 |
| Mar. | 8 | * | 2 | 7.0 | 56 | WNW | 5 | 350 |
| Apr. | 7 | * | * | 6.6 | 58 | WNW | 6 | 445 |
| May | 9 | 4 | * | 6.4 | 60 | E | 6 | 532 |
| June | 12 | 8 | 1 | 6.0 | 61 | WNW | 5 | 593 |
| July | 9 | 10 | 1 | 4.6 | 75 | SSE | 4 | 607 |
| Aug. | 9 | 9 | 1 | 4.8 | 72 | E | 5 | 524 |
| Sept. | 7 | 3 | 1 | 5.3 | 65 | WNW | 5 | 386 |
| Oct. | 6 | 1 | 1 | 5.6 | 60 | WNW | 5 | 274 |
| Nov. | 7 | * | 1 | 6.9 | 46 | WNW | 5 | 161 |
| Dec. | 7 | 0 | 1 | 6.9 | 48 | WNW | 4 | 125 |
| Annual | 96 | 35 | 11 | 6.1 | 61 | WNW | 5 | 367 |
| Rec.(yrs.) | 21 | 21 | 21 | 21 | 21 | 11 | 21 | 9 |

TABLE LI

CLIMATIC TABLE FOR WILLISTON, N.D.
Latitude 48°09'N, longitude 103°37'W, elevation 579 m

| Month | Temperature (°C) | | | | Mean vap. press. (mbar) | Precipitation (mm) | | Mean snowfall (cm) |
|---|---|---|---|---|---|---|---|---|
| | daily mean | mean daily range | extremes | | | mean | max. in 24 h | |
| | | | max. | min. | | | | |
| Jan. | −12.3 | 10.7 | 13 | −40 | 1 | 14 | 14 | 14 |
| Feb. | −10.3 | 10.8 | 19 | −46 | 2 | 12 | 18 | 13 |
| Mar. | −3.7 | 10.3 | 26 | −34 | 3 | 18 | 31 | 16 |
| Apr. | 6.0 | 12.3 | 33 | −19 | 6 | 24 | 39 | 8 |
| May | 12.9 | 13.0 | 37 | −9 | 8 | 36 | 75 | 2 |
| June | 17.3 | 12.2 | 42 | −1 | 12 | 84 | 78 | tr. |
| July | 21.8 | 14.0 | 43 | 3 | 14 | 48 | 53 | 0 |
| Aug. | 20.4 | 14.3 | 42 | 2 | 13 | 38 | 46 | 0 |
| Sept. | 14.3 | 14.0 | 38 | −8 | 9 | 28 | 50 | 1 |
| Oct. | 7.8 | 13.0 | 33 | −19 | 6 | 19 | 37 | 6 |
| Nov. | −1.9 | 10.0 | 21 | −29 | 4 | 15 | 24 | 11 |
| Dec. | −7.9 | 10.1 | 17 | −36 | 2 | 13 | 19 | 13 |
| Annual | 5.3 | 12.1 | 43 | −46 | 7 | 349 | 78 | 84 |
| Rec.(yrs.) | 30 | 30 | 44 | 44 | | 30 | 44 | 44 |

| Month | Number of days with | | | Mean cloud-iness (tenths) | Mean sun-shine (h) | Wind | |
|---|---|---|---|---|---|---|---|
| | precip. (>0.25 mm) | thunder-storm | fog | | | most freq. direct. | mean speed (m/sec) |
| Jan. | 8 | 0 | 1 | 6.8 | 51 | W | 3 |
| Feb. | 6 | 0 | 1 | 6.9 | 59 | W | 3 |
| Mar. | 7 | 0 | 1 | 7.1 | 60 | SE | 4 |
| Apr. | 7 | 1 | * | 6.6 | 62 | SE | 4 |
| May | 9 | 3 | * | 6.4 | 66 | SE | 4 |
| June | 12 | 7 | * | 6.6 | 66 | SE | 4 |
| July | 9 | 8 | * | 4.6 | 78 | SE | 3 |
| Aug. | 8 | 6 | * | 5.0 | 75 | SE | 3 |
| Sept. | 7 | 2 | 1 | 5.8 | 65 | SE | 3 |
| Oct. | 5 | * | 1 | 6.0 | 60 | SE | 3 |
| Nov. | 6 | 0 | 1 | 6.9 | 48 | W | 3 |
| Dec. | 7 | 0 | 1 | 7.0 | 48 | W | 3 |
| Annual | 91 | 27 | 7 | 6.3 | 63 | SE | 3 |
| Rec.(yrs.) | 44 | 44 | 44 | 19 | 44 | 33 | 33 |

TABLE LII

CLIMATIC TABLE FOR CINCINNATI, OHIO
Latitude 39°09′N, longitude 84°31′W, elevation 232 m (191 m)

| Month | Mean sta. press. (mbar) | Temperature (°C) | | | | Mean vap. press. (mbar) | Precipitation (mm) | | Mean snowfall (cm) |
|---|---|---|---|---|---|---|---|---|---|
| | | daily mean | mean daily range | extremes | | | mean | max. in 24 h | |
| | | | | max. | min. | | | | |
| Jan. | 996.5 | 0.9 | 8.5 | 25 | −27 | 5 | 93 | 115 | 13 |
| Feb. | 995.6 | 1.7 | 9.3 | 24 | −23 | 5 | 71 | 61 | 10 |
| Mar. | 993.5 | 5.9 | 10.4 | 31 | −15 | 6 | 99 | 96 | 8 |
| Apr. | 992.9 | 12.3 | 11.4 | 32 | −8 | 9 | 92 | 87 | 1 |
| May | 992.8 | 17.9 | 11.9 | 35 | 0 | 14 | 97 | 121 | tr. |
| June | 992.4 | 23.0 | 11.6 | 39 | 4 | 19 | 106 | 88 | 0 |
| July | 993.5 | 24.9 | 11.8 | 43 | 10 | 21 | 91 | 103 | 0 |
| Aug. | 994.0 | 24.3 | 12.0 | 39 | 6 | 20 | 83 | 88 | 0 |
| Sept. | 995.2 | 20.6 | 12.6 | 38 | 0 | 15 | 69 | 69 | 0 |
| Oct. | 996.2 | 14.4 | 12.3 | 33 | −7 | 11 | 57 | 65 | * |
| Nov. | 996.2 | 7.0 | 9.6 | 28 | −17 | 6 | 75 | 76 | 4 |
| Dec. | 996.8 | 1.8 | 8.2 | 22 | −25 | 5 | 70 | 85 | 11 |
| Annual | 994.6 | 12.9 | 10.8 | 43 | −27 | 11 | 1,003 | 121 | 47 |
| Rec.(yrs.) | 30 | 30 | 30 | 44 | 44 | 10 | 30 | 44 | 44 |

| Month | Number of days with | | | Mean cloud-iness[1] (tenths) | Mean sun-shine (h) | Wind | |
|---|---|---|---|---|---|---|---|
| | precip. (>0.25 mm) | thunder-storm | fog[1] | | | most freq. direct. | mean speed (m/sec) |
| Jan. | 13 | 1 | 3 | 8.0 | 41 | SW | 4 |
| Feb. | 11 | 1 | 2 | 7.4 | 45 | SW | 4 |
| Mar. | 13 | 3 | 1 | 7.3 | 52 | SW | 4 |
| Apr. | 12 | 4 | 1 | 7.3 | 56 | SW | 4 |
| May | 12 | 7 | 1 | 6.8 | 62 | SW | 3 |
| June | 13 | 9 | 1 | 6.2 | 69 | SW | 3 |
| July | 10 | 10 | 2 | 5.8 | 72 | SW | 2 |
| Aug. | 9 | 8 | 3 | 5.6 | 68 | SW | 2 |
| Sept. | 9 | 4 | 2 | 5.1 | 68 | SW | 2 |
| Oct. | 9 | 2 | 2 | 5.2 | 60 | SW | 3 |
| Nov. | 10 | 1 | 1 | 6.7 | 46 | SW | 4 |
| Dec. | 11 | * | 2 | 7.4 | 40 | SW | 4 |
| Annual | 132 | 50 | 21 | 6.6 | 57 | SW | 3 |
| Rec.(yrs.) | 45 | 45 | 11 | 10 | 45 | 37 | 39 |

[1] Airport data.

TABLE LIII

CLIMATIC TABLE FOR CLEVELAND, OHIO
Latitude 41°24′N, longitude 81°51′W, elevation 237 m

| Month | Temperature (°C) | | | | Mean vap. press. (mbar) | Precipitation (mm) | | Mean snowfall (cm) |
|---|---|---|---|---|---|---|---|---|
| | daily mean | mean daily range | extremes | | | mean | max. in 24 h | |
| | | | max. | min. | | | | |
| Jan. | −2.4 | 8.1 | 23 | −23 | 4 | 68 | 59 | 25 |
| Feb. | −2.2 | 8.3 | 21 | −22 | 4 | 59 | 59 | 25 |
| Mar. | 1.9 | 9.7 | 28 | −21 | 5 | 80 | 70 | 28 |
| Apr. | 8.1 | 12.1 | 31 | −7 | 8 | 87 | 49 | 7 |
| May | 14.2 | 12.5 | 33 | −2 | 11 | 89 | 95 | tr. |
| June | 19.6 | 12.2 | 38 | 3 | 16 | 87 | 71 | 0 |
| July | 21.7 | 12.0 | 39 | 8 | 18 | 84 | 69 | 0 |
| Aug. | 20.8 | 12.1 | 39 | 7 | 18 | 83 | 78 | 0 |
| Sept. | 16.9 | 12.3 | 38 | 0 | 14 | 74 | 47 | tr. |
| Oct. | 11.0 | 12.0 | 32 | −4 | 9 | 61 | 87 | 1 |
| Nov. | 4.1 | 9.6 | 28 | −14 | 6 | 66 | 57 | 17 |
| Dec. | −1.4 | 8.3 | 21 | −23 | 4 | 59 | 32 | 28 |
| Annual | 9.3 | 10.8 | 39 | −23 | 10 | 897 | 95 | 131 |
| Rec.(yrs.) | 30 | 30 | 20 | 20 | 10 | 30 | 19 | 19 |

| Month | Number of days with | | | Mean cloudiness (tenths) | Mean sunshine (h) | Wind | | Sunshine (ly/month) |
|---|---|---|---|---|---|---|---|---|
| | precip. (>0.25 mm) | thunderstorm | fog | | | most freq. direct. | mean speed (m/sec) | |
| Jan. | 16 | * | 2 | 8.1 | 26 | S | 6 | 129 |
| Feb. | 14 | * | 2 | 7.9 | 33 | S | 6 | 178 |
| Mar. | 16 | 2 | 1 | 7.2 | 45 | W | 6 | 288 |
| Apr. | 15 | 4 | 2 | 7.0 | 51 | S | 6 | 378 |
| May | 14 | 5 | 1 | 6.6 | 57 | S | 5 | 508 |
| June | 11 | 8 | 1 | 5.9 | 64 | S | 4 | 555 |
| July | 10 | 7 | 1 | 5.3 | 68 | S | 4 | 540 |
| Aug. | 9 | 5 | 1 | 5.3 | 66 | S | 4 | 478 |
| Sept. | 9 | 3 | * | 5.5 | 61 | S | 4 | 367 |
| Oct. | 10 | 2 | 1 | 5.6 | 55 | S | 5 | 258 |
| Nov. | 15 | 1 | 1 | 7.7 | 33 | S | 6 | 134 |
| Dec. | 15 | * | 1 | 8.0 | 29 | S | 6 | 113 |
| Annual | 154 | 37 | 14 | 6.7 | 49 | S | 5 | 327 |
| Rec.(yrs.) | 19 | 19 | 19 | 19 | 19 | 11 | 19 | 8 |

## TABLE LIV

CLIMATIC TABLE FOR TULSA, OKLA.
Latitude 36°11′N, longitude 95°54′W, elevation 198 m

| Month | Temperature (°C) | | | | Mean vap. press. (mbar) | Precipitation (mm) | | Mean snowfall (cm) |
|---|---|---|---|---|---|---|---|---|
| | daily mean | mean daily range | extremes | | | mean | max. in 24 h | |
| | | | max. | min. | | | | |
| Jan. | 2.9 | 10.8 | 26 | −22 | 6 | 43 | 57 | 10 |
| Feb. | 5.1 | 11.7 | 28 | −18 | 6 | 45 | 38 | 6 |
| Mar. | 9.2 | 12.7 | 33 | −19 | 7 | 62 | 68 | 4 |
| Apr. | 15.2 | 12.4 | 34 | −6 | 10 | 102 | 82 | * |
| May | 19.9 | 11.2 | 35 | 2 | 16 | 134 | 185 | 0 |
| June | 25.2 | 11.1 | 39 | 9 | 21 | 119 | 127 | 0 |
| July | 27.9 | 12.0 | 44 | 14 | 23 | 75 | 83 | 0 |
| Aug. | 27.8 | 12.3 | 43 | 12 | 22 | 77 | 106 | 0 |
| Sept. | 23.5 | 12.6 | 43 | 2 | 17 | 102 | 162 | 0 |
| Oct. | 17.7 | 12.5 | 37 | −3 | 12 | 84 | 139 | tr. |
| Nov. | 9.5 | 11.7 | 31 | −12 | 7 | 58 | 70 | 1 |
| Dec. | 4.8 | 10.5 | 27 | −16 | 6 | 41 | 44 | 4 |
| Annual | 15.7 | 11.8 | 44 | −22 | 13 | 942 | 185 | 25 |
| Rec.(yrs.) | 30 | 30 | 22 | 22 | 10 | 30 | 22 | 22 |

| Month | Number of days with | | | Mean cloud-iness (tenths) | Mean sun-shine (h) | Wind | | Solar radiation[1] (ly/ month) |
|---|---|---|---|---|---|---|---|---|
| | precip. (>0.25 mm) | thunder-storm | fog | | | most freq. direct. | mean speed (m/sec) | |
| Jan. | 6 | 1 | 1 | 6.4 | 49 | N | 5 | 254 |
| Feb. | 8 | 1 | 2 | 6.0 | 53 | S | 5 | 317 |
| Mar. | 8 | 4 | 1 | 6.0 | 54 | S | 6 | 413 |
| Apr. | 9 | 6 | * | 6.1 | 54 | S | 6 | 490 |
| May | 11 | 10 | * | 6.2 | 56 | S | 5 | 540 |
| June | 9 | 9 | * | 5.5 | 65 | S | 5 | 602 |
| July | 7 | 7 | * | 5.0 | 71 | S | 4 | 601 |
| Aug. | 7 | 7 | * | 4.5 | 74 | S | 4 | 581 |
| Sept. | 6 | 5 | * | 4.2 | 71 | S | 4 | 469 |
| Oct. | 7 | 3 | 1 | 4.5 | 66 | S | 4 | 375 |
| Nov. | 6 | 1 | 1 | 4.7 | 64 | S | 5 | 279 |
| Dec. | 7 | 1 | 1 | 5.8 | 54 | S | 5 | 231 |
| Annual | 91 | 55 | 7 | 5.4 | 62 | S | 5 | 429 |
| Rec.(yrs.) | 22 | 22 | 22 | 18 | 18 | 12 | 12 | 10 |

[1] Oklahoma City data.

TABLE LV

CLIMATIC TABLE FOR MEDFORD, OREG.
Latitude 42°22′N, longitude 122°52′W, elevation 396 m

| Month | Temperature (°C) | | | | Mean vap. press. (mbar) | Precipitation (mm) | | Mean snowfall (cm) |
|---|---|---|---|---|---|---|---|---|
| | daily mean | mean daily range | extremes | | | mean | max. in 24 h | |
| | | | max. | min. | | | | |
| Jan. | 2.7 | 8.3 | 20 | −19 | 5 | 80 | 81 | 10 |
| Feb. | 5.3 | 11.1 | 23 | −14 | 6 | 61 | 75 | 4 |
| Mar. | 7.7 | 13.2 | 30 | −9 | 6 | 45 | 40 | 2 |
| Apr. | 10.9 | 15.5 | 33 | −6 | 7 | 27 | 23 | 1 |
| May | 14.4 | 16.1 | 38 | −2 | 8 | 37 | 42 | tr. |
| June | 17.9 | 16.6 | 41 | −1 | 9 | 26 | 50 | 0 |
| July | 22.2 | 19.2 | 46 | 4 | 11 | 5 | 24 | 0 |
| Aug. | 21.5 | 19.6 | 42 | 5 | 11 | 5 | 29 | 0 |
| Sept. | 18.2 | 19.2 | 42 | −2 | 10 | 15 | 34 | 0 |
| Oct. | 12.2 | 15.7 | 36 | −7 | 8 | 49 | 74 | tr. |
| Nov. | 6.3 | 10.7 | 24 | −9 | 7 | 66 | 76 | 1 |
| Dec. | 3.6 | 7.3 | 18 | −16 | 6 | 86 | 77 | 2 |
| Annual | 11.9 | 14.4 | 46 | −19 | 8 | 502 | 81 | 20 |
| Rec.(yrs.) | 30 | 30 | 31 | 31 | 10 | 30 | 31 | 31 |

| Month | Number of days with | | | Mean cloud-iness (tenths) | Wind | | Solar radiation (ly/ month) |
|---|---|---|---|---|---|---|---|
| | precip. (>0.25 mm) | thunder-storm | fog | | most freq. direct. | mean speed (m/sec) | |
| Jan. | 14 | * | 10 | 8.2 | SSE | 2 | 122 |
| Feb. | 12 | * | 5 | 7.6 | S | 2 | 211 |
| Mar. | 12 | * | 1 | 7.2 | NNW | 2 | 331 |
| Apr. | 9 | 1 | * | 6.6 | NNW | 3 | 480 |
| May | 9 | 2 | * | 5.9 | NW | 3 | 509 |
| June | 6 | 2 | * | 4.8 | NW | 3 | 659 |
| July | 1 | 2 | 0 | 2.0 | WNW | 3 | 699 |
| Aug. | 1 | 1 | * | 2.2 | WNW | 2 | 600 |
| Sept. | 4 | 1 | * | 3.3 | WNW | 2 | 451 |
| Oct. | 8 | * | 4 | 5.5 | S | 2 | 276 |
| Nov. | 11 | * | 9 | 7.3 | SE | 1 | 149 |
| Dec. | 14 | 0 | 14 | 8.6 | SSE | 1 | 94 |
| Annual | 101 | 9 | 45 | 5.8 | WNW | 2 | 382 |
| Rec.(yrs.) | 31 | 31 | 31 | 31 | 11 | 11 | 13 |

TABLE LVI

CLIMATIC TABLE FOR PORTLAND, OREG.
Latitude 45°32′N, longitude 122°40′W, elevation 9 m (47 m)

| Month | Mean sta. press. (mbar) | Temperature (°C) | | | | Mean vap. press. (mbar) | Precipitation (mm) | | Mean snowfall (cm) |
|---|---|---|---|---|---|---|---|---|---|
| | | daily mean | mean daily range | extremes | | | mean | max. in 24 h | |
| | | | | max. | min. | | | | |
| Jan. | 1013.1 | 4.6 | 5.2 | 18 | −14 | 7 | 161 | 117 | 13 |
| Feb. | 1012.4 | 6.6 | 6.3 | 20 | −14 | 7 | 124 | 68 | 5 |
| Mar. | 1011.9 | 8.7 | 7.7 | 28 | −6 | 7 | 121 | 64 | 2 |
| Apr. | 1012.2 | 11.9 | 9.7 | 34 | −1 | 9 | 62 | 50 | * |
| May | 1011.8 | 15.1 | 10.5 | 37 | 2 | 11 | 52 | 46 | tr. |
| June | 1011.4 | 17.4 | 10.2 | 39 | 5 | 12 | 43 | 55 | 0 |
| July | 1011.8 | 20.3 | 11.8 | 42 | 6 | 13 | 10 | 34 | 0 |
| Aug. | 1011.1 | 20.1 | 11.5 | 39 | 7 | 13 | 18 | 33 | 0 |
| Sept. | 1010.8 | 18.1 | 10.8 | 39 | 2 | 12 | 44 | 73 | 0 |
| Oct. | 1011.9 | 13.6 | 8.3 | 32 | −2 | 11 | 99 | 62 | tr. |
| Nov. | 1013.8 | 8.4 | 6.1 | 22 | −9 | 8 | 153 | 113 | * |
| Dec. | 1012.7 | 6.2 | 5.0 | 18 | −16 | 7 | 188 | 127 | 3 |
| Annual | 1012.1 | 12.6 | 8.6 | 42 | −16 | 10 | 1,075 | 127 | 23 |
| Rec.(yrs.) | 30 | 30 | 30 | 58 | 58 | 10 | 30 | 58 | 58 |

| Month | Number of days with | | | Mean cloud-iness (tenths) | Mean sun-shine (h) | Wind | | Solar radiation[1] (ly/ month) |
|---|---|---|---|---|---|---|---|---|
| | precip. (>0.25 mm) | thunder-storm | fog | | | most freq. direct. | mean speed (m/sec) | |
| Jan. | 20 | * | 3 | 8.7 | 21 | ESE | 5 | 97 |
| Feb. | 16 | * | 3 | 8.4 | 32 | ESE | 4 | 159 |
| Mar. | 17 | * | 2 | 8.4 | 34 | ESE | 4 | 269 |
| Apr. | 14 | 1 | 1 | 7.4 | 48 | NW | 3 | 375 |
| May | 12 | 2 | * | 7.2 | 48 | NW | 3 | 491 |
| June | 9 | 1 | * | 7.2 | 45 | NW | 3 | 488 |
| July | 3 | 1 | * | 4.4 | 69 | NW | 3 | 527 |
| Aug. | 4 | 1 | * | 5.2 | 60 | NW | 3 | 454 |
| Sept. | 8 | 1 | 3 | 5.7 | 58 | NW | 3 | 360 |
| Oct. | 12 | * | 7 | 7.2 | 38 | ESE | 3 | 206 |
| Nov. | 17 | * | 6 | 8.0 | 30 | ESE | 4 | 111 |
| Dec. | 19 | * | 4 | 9.0 | 21 | ESE | 4 | 78 |
| Annual | 151 | 7 | 29 | 7.2 | 44 | NW | 3 | 301 |
| Rec.(yrs.) | 58 | 20 | 18 | 12 | 11 | 12 | 12 | 10 |

[1] Astoria data.

TABLE LVII

CLIMATIC TABLE FOR HARRISBURG, PA.
Latitude 40°13′N, longitude 76°51′W, elevation 102 m

| Month | Temperature (°C) | | | | Mean vap. press. (mbar) | Precipitation (mm) | | Mean snowfall (cm) |
|---|---|---|---|---|---|---|---|---|
| | daily mean | mean daily range | extremes | | | mean | max. in 24 h | |
| | | | max. | min. | | | | |
| Jan. | −0.1 | 8.0 | 23 | −20 | 4 | 70 | 49 | 21 |
| Feb. | 0.3 | 8.7 | 24 | −17 | 4 | 59 | 41 | 19 |
| Mar. | 4.6 | 9.7 | 30 | −13 | 6 | 87 | 54 | 18 |
| Apr. | 11.0 | 11.4 | 33 | −6 | 8 | 77 | 55 | 1 |
| May | 17.1 | 11.9 | 36 | 0 | 13 | 99 | 79 | tr. |
| June | 21.8 | 11.6 | 38 | 6 | 16 | 87 | 71 | 0 |
| July | 24.3 | 11.5 | 38 | 9 | 19 | 89 | 99 | 0 |
| Aug. | 23.1 | 11.2 | 38 | 8 | 19 | 93 | 85 | 0 |
| Sept. | 19.1 | 11.3 | 39 | 0 | 15 | 72 | 111 | 0 |
| Oct. | 13.2 | 11.2 | 36 | −4 | 10 | 75 | 66 | * |
| Nov. | 6.6 | 9.3 | 29 | −11 | 7 | 75 | 74 | 5 |
| Dec. | 0.8 | 8.0 | 22 | −22 | 5 | 74 | 51 | 16 |
| Annual | 11.8 | 10.3 | 39 | −22 | 11 | 957 | 111 | 80 |
| Rec.(yrs.) | 30 | 30 | 22 | 22 | 10 | 30 | 22 | 22 |

| Month | Number of days with | | | Mean cloud-iness (tenths) | Mean sun-shine (h) | Wind | |
|---|---|---|---|---|---|---|---|
| | precip. (>0.25 mm) | thunder-storm | fog | | | most freq. direct. | mean speed (m/sec) |
| Jan. | 12 | * | 3 | 6.8 | 46 | WNW | 4 |
| Feb. | 9 | * | 2 | 6.7 | 53 | WNW | 4 |
| Mar. | 12 | 1 | 2 | 6.7 | 58 | WNW | 4 |
| Apr. | 12 | 2 | 1 | 7.0 | 58 | W | 4 |
| May | 13 | 6 | 1 | 6.7 | 59 | W | 3 |
| June | 11 | 7 | 1 | 6.0 | 65 | W | 3 |
| July | 10 | 7 | 1 | 6.0 | 69 | W | 3 |
| Aug. | 10 | 6 | 1 | 5.9 | 67 | W | 3 |
| Sept. | 9 | 3 | 2 | 5.7 | 63 | WSW | 3 |
| Oct. | 9 | 1 | 3 | 5.6 | 57 | W | 3 |
| Nov. | 9 | * | 2 | 6.6 | 48 | WNW | 3 |
| Dec. | 10 | * | 3 | 6.8 | 46 | WNW | 3 |
| Annual | 126 | 33 | 22 | 6.4 | 57 | WNW | 3 |
| Rec.(yrs.) | 22 | 22 | 22 | 10 | 22 | 11 | 22 |

TABLE LVIII

CLIMATIC TABLE FOR CHARLESTON, S.C.
Latitude 32°54′N, longitude 80°02′W, elevation 12 m (sea level)

| Month | Mean sta. press. (mbar) | Temperature (°C) | | | | Mean vap. press. (mbar) | Precipitation (mm) | | Mean snowfall (cm) |
|---|---|---|---|---|---|---|---|---|---|
| | | daily mean | mean daily range | extremes | | | mean | max. in 24 h | |
| | | | | max. | min. | | | | |
| Jan. | 1020.8 | 10.2 | 12.2 | 28 | −8 | 9 | 65 | 57 | tr. |
| Feb. | 1019.3 | 10.8 | 12.3 | 28 | −10 | 10 | 84 | 83 | tr. |
| Mar. | 1017.5 | 13.7 | 12.6 | 32 | −6 | 11 | 100 | 168 | tr. |
| Apr. | 1017.3 | 17.9 | 12.9 | 34 | −2 | 14 | 73 | 104 | 0 |
| May | 1016.4 | 22.2 | 12.3 | 37 | 5 | 19 | 92 | 77 | 0 |
| June | 1015.9 | 25.7 | 11.2 | 39 | 11 | 24 | 127 | 107 | 0 |
| July | 1017.0 | 26.7 | 10.1 | 38 | 14 | 26 | 196 | 148 | 0 |
| Aug. | 1016.5 | 26.5 | 10.2 | 39 | 14 | 26 | 168 | 104 | 0 |
| Sept. | 1017.0 | 24.2 | 10.4 | 37 | 9 | 24 | 148 | 225 | 0 |
| Oct. | 1018.1 | 19.0 | 12.3 | 34 | −3 | 17 | 72 | 147 | 0 |
| Nov. | 1019.8 | 13.3 | 13.3 | 30 | −9 | 11 | 53 | 54 | tr. |
| Dec. | 1020.9 | 10.0 | 12.6 | 27 | −9 | 9 | 72 | 60 | * |
| Annual | 1018.0 | 18.3 | 11.9 | 39 | −10 | 17 | 1,250 | 225 | * |
| Rec.(yrs.) | 30 | 30 | 30 | 18 | 18 | 10 | 30 | 18 | 18 |

| Month | Number of days with | | | Mean cloud-iness (tenths) | Mean sun-shine (h) | Wind | | Solar radiation (ly/ month) |
|---|---|---|---|---|---|---|---|---|
| | precip. (>0.25 mm) | thunder-storm | fog | | | most freq. direct. | mean speed (m/sec) | |
| Jan. | 9 | 1 | 4 | 6.2 | 62 | SW | 4 | 251 |
| Feb. | 9 | 1 | 2 | 6.0 | 57 | NNE | 5 | 310 |
| Mar. | 11 | 2 | 2 | 6.2 | 66 | SSW | 5 | 390 |
| Apr. | 8 | 3 | 3 | 5.5 | 75 | SSW | 5 | 447 |
| May | 14 | 7 | 2 | 5.9 | 74 | SW | 4 | 549 |
| June | 10 | 11 | 2 | 6.2 | 68 | SW | 4 | 555 |
| July | 15 | 15 | 1 | 6.7 | 69 | SW | 4 | 526 |
| Aug. | 12 | 12 | 2 | 6.0 | 70 | SW | 3 | 498 |
| Sept. | 10 | 6 | 2 | 6.5 | 64 | NNE | 4 | 410 |
| Oct. | 6 | 1 | 3 | 5.5 | 71 | NNE | 4 | 350 |
| Nov. | 7 | 1 | 4 | 5.0 | 62 | N | 4 | 285 |
| Dec. | 8 | * | 4 | 5.9 | 69 | SW | 4 | 230 |
| Annual | 119 | 60 | 31 | 6.0 | 67 | SW | 4 | 400 |
| Rec.(yrs.) | 18 | 18 | 11 | 11 | 5 | 11 | 11 | 13 |

TABLE LIX

Latitude 44°23′N, longitude 98°13′W, elevation 391 m (393 m)

| Month | Mean sta. press. (mbar) | Temperature (°C) | | | | Mean vap. press. (mbar) | Precipitation (mm) | | Mean snowfall (cm) |
|---|---|---|---|---|---|---|---|---|---|
| | | daily mean | mean daily range | extremes | | | mean | max. in 24 h | |
| | | | | max. | min. | | | | |
| Jan. | 970.9 | −10.3 | 11.8 | 17 | −37 | 2 | 12 | 40 | 16 |
| Feb. | 970.7 | − 8.3 | 11.8 | 22 | −34 | 3 | 15 | 32 | 21 |
| Mar. | 968.5 | − 1.3 | 11.1 | 32 | −31 | 4 | 28 | 48 | 22 |
| Apr. | 967.6 | 7.8 | 13.6 | 33 | −12 | 6 | 47 | 45 | 6 |
| May | 966.6 | 14.4 | 14.3 | 37 | −7 | 9 | 60 | 89 | 1 |
| June | 965.4 | 20.1 | 13.8 | 41 | 0 | 15 | 80 | 79 | 0 |
| July | 967.3 | 24.2 | 15.4 | 43 | 4 | 18 | 46 | 56 | 0 |
| Aug. | 967.4 | 22.9 | 15.2 | 43 | 2 | 17 | 53 | 105 | 0 |
| Sept. | 968.2 | 16.9 | 15.7 | 39 | −7 | 11 | 39 | 66 | tr. |
| Oct. | 969.0 | 9.9 | 15.3 | 36 | −12 | 8 | 29 | 103 | 1 |
| Nov. | 969.6 | 0.1 | 12.3 | 24 | −28 | 4 | 17 | 31 | 13 |
| Dec. | 969.8 | − 6.5 | 11.0 | 22 | −31 | 3 | 14 | 32 | 13 |
| Annual | 968.4 | 7.5 | 13.5 | 43 | −37 | 8 | 440 | 105 | 93 |
| Rec.(yrs.) | 30 | 30 | 30 | 21 | 21 | 10 | 30 | 21 | 21 |

| Month | Number of days with | | | Mean cloud-iness (tenths) | Mean sun-shine (h) | Wind | |
|---|---|---|---|---|---|---|---|
| | precip. (>0.25 mm) | thunder-storm | fog | | | most freq. direct. | mean speed (m/sec) |
| Jan. | 6 | * | 2 | 6.6 | 53 | SSE | 5 |
| Feb. | 7 | 0 | 2 | 6.5 | 59 | NW | 5 |
| Mar. | 9 | * | 1 | 7.2 | 56 | NW | 6 |
| Apr. | 9 | 2 | * | 6.6 | 61 | SSE | 6 |
| May | 10 | 5 | 1 | 6.2 | 64 | SSE | 6 |
| June | 11 | 10 | 1 | 5.7 | 67 | SSE | 5 |
| July | 8 | 9 | 1 | 4.5 | 78 | SSE | 5 |
| Aug. | 9 | 9 | 1 | 4.9 | 73 | SSE | 5 |
| Sept. | 7 | 4 | 1 | 4.9 | 68 | SSE | 5 |
| Oct. | 6 | 2 | 1 | 5.0 | 63 | SSE | 5 |
| Nov. | 6 | * | 1 | 6.6 | 49 | NW | 6 |
| Dec. | 6 | * | 2 | 6.7 | 48 | SSE | 5 |
| Annual | 94 | 41 | 14 | 6.0 | 63 | SSE | 5 |
| Rec.(yrs.) | 21 | 21 | 21 | 17 | 21 | 11 | 21 |

TABLE LX

CLIMATIC TABLE FOR RAPID CITY, S.D.
Latitude 44°02′N, longitude 103°03′W, elevation 965 m (993 m)

| Month | Mean sta. press. (mbar) | Temperature (°C) | | | | | Mean vap. press. (mbar) | Precipitation (mm) | | Mean snowfall (cm) |
|---|---|---|---|---|---|---|---|---|---|---|
| | | daily mean | mean daily range | extremes | | | | mean | max. in 24 h | |
| | | | | max. | min. | | | | | |
| Jan. | 900.4 | −5.6 | 13.5 | 23 | −33 | 2 | 9 | 32 | 14 |
| Feb. | 900.3 | −4.4 | 13.1 | 23 | −28 | 2 | 12 | 25 | 15 |
| Mar. | 899.3 | −0.5 | 12.7 | 28 | −26 | 4 | 26 | 56 | 22 |
| Apr. | 899.7 | 6.9 | 13.6 | 32 | −13 | 6 | 42 | 76 | 14 |
| May | 900.0 | 13.2 | 13.5 | 34 | −8 | 9 | 68 | 62 | 2 |
| June | 899.7 | 18.3 | 13.5 | 38 | −1 | 12 | 78 | 81 | 1 |
| July | 902.2 | 23.2 | 15.5 | 43 | 4 | 14 | 45 | 64 | 0 |
| Aug. | 901.8 | 22.2 | 15.6 | 41 | 4 | 14 | 31 | 36 | 0 |
| Sept. | 902.1 | 16.4 | 15.6 | 40 | −4 | 9 | 24 | 50 | tr. |
| Oct. | 902.2 | 10.0 | 14.6 | 34 | −10 | 6 | 20 | 36 | 4 |
| Nov. | 901.8 | 1.7 | 13.2 | 25 | −28 | 4 | 10 | 28 | 11 |
| Dec. | 900.6 | −2.7 | 12.8 | 21 | −28 | 2 | 8 | 13 | 9 |
| Annual | 900.8 | 8.2 | 13.9 | 43 | −33 | 7 | 373 | 81 | 92 |
| Rec.(yrs.) | 30 | 30 | 30 | 18 | 18 | 10 | 30 | 18 | 18 |

| Month | Number of days with | | | Mean cloud-iness (tenths) | Mean sun-shine (h) | Wind | | Solar radiation (ly/ month) |
|---|---|---|---|---|---|---|---|---|
| | precip. (>0.25 mm) | thunder-storm | fog | | | most freq. direct. | mean speed (m/sec) | |
| Jan. | 6 | 0 | 2 | 6.4 | 54 | NNW | 4 | 186 |
| Feb. | 7 | 0 | 2 | 6.4 | 59 | NNW | 5 | 274 |
| Mar. | 9 | * | 2 | 6.7 | 60 | NNW | 6 | 397 |
| Apr. | 9 | 1 | 2 | 6.6 | 61 | NNW | 6 | 481 |
| May | 12 | 7 | 1 | 6.5 | 57 | NNW | 5 | 522 |
| June | 13 | 11 | 1 | 5.7 | 60 | NNW | 5 | 587 |
| July | 9 | 12 | * | 4.2 | 72 | NNW | 5 | 588 |
| Aug. | 8 | 11 | * | 4.3 | 71 | NNW | 5 | 538 |
| Sept. | 5 | 3 | * | 4.5 | 67 | NNW | 5 | 423 |
| Oct. | 5 | 1 | 1 | 4.8 | 64 | NNW | 5 | 310 |
| Nov. | 6 | * | 1 | 6.2 | 56 | NNW | 5 | 202 |
| Dec. | 5 | 0 | 1 | 6.2 | 53 | NNW | 5 | 170 |
| Annual | 94 | 46 | 14 | 5.7 | 62 | NNW | 5 | 390 |
| Rec.(yrs.) | 18 | 18 | 18 | 18 | 18 | 10 | 10 | 12 |

TABLE LXI

CLIMATIC TABLE FOR MEMPHIS, TENN.
Latitude 35°03′N, longitude 89°59′W, elevation 80 m (122 m)

| Month | Mean sta. press. (mbar) | Temperature (°C) | | extremes | | Mean vap. press. (mbar) | Precipitation (mm) | | Mean snowfall (cm) |
|---|---|---|---|---|---|---|---|---|---|
| | | daily mean | mean daily range | max. | min. | | mean | max. in 24 h | |
| Jan. | 1006.0 | 5.6 | 9.6 | 26 | −17 | 7 | 154 | 98 | 5 |
| Feb. | 1004.6 | 7.0 | 10.3 | 26 | −24 | 8 | 119 | 74 | 4 |
| Mar. | 1002.0 | 10.9 | 11.0 | 29 | −11 | 8 | 129 | 91 | 2 |
| Apr. | 1001.1 | 16.6 | 11.3 | 33 | −2 | 12 | 118 | 89 | 0 |
| May | 1000.7 | 21.6 | 11.5 | 36 | 3 | 17 | 107 | 125 | 0 |
| June | 1000.3 | 26.1 | 11.5 | 40 | 10 | 22 | 93 | 56 | 0 |
| July | 1001.4 | 27.7 | 11.5 | 41 | 11 | 25 | 90 | 84 | 0 |
| Aug. | 1001.3 | 27.2 | 11.9 | 41 | 9 | 24 | 75 | 81 | 0 |
| Sept. | 1002.2 | 23.6 | 12.5 | 39 | 2 | 19 | 72 | 118 | 0 |
| Oct. | 1003.9 | 17.6 | 13.5 | 35 | −4 | 13 | 69 | 55 | 0 |
| Nov. | 1005.4 | 10.3 | 12.1 | 29 | −13 | 8 | 111 | 83 | tr. |
| Dec. | 1005.9 | 6.4 | 10.0 | 26 | −12 | 7 | 125 | 56 | 1 |
| Annual | 1002.9 | 16.7 | 11.4 | 41 | −24 | 14 | 1,262 | 125 | 12 |
| Rec.(yrs.) | 30 | 30 | 30 | 19 | 19 | 10 | 30 | 10 | 10 |

| Month | Number of days with | | | Mean cloud-iness (tenths) | Mean sun-shine (h) | Wind | |
|---|---|---|---|---|---|---|---|
| | precip. (>0.25 mm) | thunder-storm | fog | | | most freq. direct. | mean speed (m/sec) |
| Jan. | 10 | 3 | 2 | 7.4 | 43 | S | 5 |
| Feb. | 10 | 3 | 2 | 6.6 | 49 | S | 5 |
| Mar. | 11 | 4 | 1 | 6.5 | 52 | S | 5 |
| Apr. | 10 | 7 | 1 | 6.1 | 61 | S | 5 |
| May | 8 | 7 | * | 6.0 | 68 | S | 4 |
| June | 8 | 7 | * | 5.4 | 74 | S | 4 |
| July | 9 | 8 | * | 5.7 | 75 | S | 3 |
| Aug. | 7 | 6 | * | 4.8 | 78 | S | 3 |
| Sept. | 6 | 3 | 1 | 4.6 | 74 | NE | 3 |
| Oct. | 5 | 2 | 1 | 4.6 | 69 | S | 4 |
| Nov. | 8 | 2 | 1 | 5.0 | 62 | S | 4 |
| Dec. | 10 | 1 | 2 | 6.2 | 51 | S | 5 |
| Annual | 102 | 53 | 11 | 5.7 | 63 | S | 4 |
| Rec.(yrs.) | 10 | 10 | 10 | 12 | 10 | 12 | 10 |

TABLE LXII

CLIMATIC TABLE FOR NASHVILLE, TENN.
Latitude 36°07′N, longitude 86°41′W, elevation 176 m (166 m)

| Month | Mean sta. press. (mbar) | Temperature (°C) | | | | | Mean vap. press. (mbar) | Precipitation (mm) | | Mean snowfall (cm) |
|---|---|---|---|---|---|---|---|---|---|---|
| | | daily mean | mean daily range | extremes | | | | mean | max. in 24 h | |
| | | | | max. | min. | | | | | |
| Jan. | 1000.8 | 4.4 | 10.0 | 26 | −26 | 6 | 139 | 112 | | 9 |
| Feb. | 999.6 | 5.6 | 10.5 | 25 | −25 | 7 | 115 | 103 | | 7 |
| Mar. | 997.1 | 9.5 | 11.5 | 28 | −15 | 8 | 132 | 118 | | 5 |
| Apr. | 996.6 | 15.3 | 12.5 | 32 | −4 | 11 | 95 | 65 | | * |
| May | 996.1 | 20.3 | 12.5 | 36 | 2 | 16 | 94 | 91 | | tr. |
| June | 995.9 | 25.2 | 12.3 | 41 | 7 | 21 | 83 | 125 | | 0 |
| July | 996.7 | 26.8 | 11.7 | 42 | 11 | 23 | 94 | 90 | | 0 |
| Aug. | 996.8 | 26.2 | 12.0 | 40 | 8 | 22 | 73 | 70 | | 0 |
| Sept. | 997.9 | 22.7 | 13.0 | 41 | 2 | 17 | 73 | 78 | | 0 |
| Oct. | 999.1 | 16.4 | 13.6 | 34 | −3 | 12 | 59 | 58 | | 0 |
| Nov. | 1000.4 | 9.2 | 11.8 | 29 | −18 | 8 | 83 | 92 | | 2 |
| Dec. | 1000.8 | 5.2 | 10.0 | 24 | −15 | 6 | 106 | 99 | | 3 |
| Annual | 998.2 | 15.6 | 11.8 | 42 | −26 | 13 | 1,146 | 125 | | 26 |
| Rec.(yrs.) | 30 | 30 | 30 | 21 | 21 | 10 | 30 | 21 | | 19 |

| Month | Number of days with | | | Mean cloud-iness (tenths) | Mean sun-shine (h) | Wind | | Sunshine (ly/month) |
|---|---|---|---|---|---|---|---|---|
| | precip. (>0.25 mm) | thunder-storm | fog | | | most freq. direct. | mean speed (m/sec) | |
| Jan. | 12 | 2 | 2 | 7.4 | 36 | S | 4 | 155 |
| Feb. | 11 | 2 | 1 | 6.7 | 45 | S | 4 | 229 |
| Mar. | 12 | 4 | 1 | 6.5 | 52 | S | 4 | 320 |
| Apr. | 11 | 5 | 1 | 6.1 | 60 | S | 4 | 433 |
| May | 11 | 8 | 1 | 6.0 | 61 | S | 3 | 484 |
| June | 9 | 8 | 1 | 5.4 | 68 | S | 3 | 591 |
| July | 10 | 10 | 1 | 5.6 | 63 | S | 3 | 528 |
| Aug. | 9 | 8 | 1 | 5.2 | 65 | S | 3 | 473 |
| Sept. | 7 | 4 | 1 | 4.8 | 65 | S | 3 | 402 |
| Oct. | 7 | 2 | 2 | 4.7 | 62 | S | 3 | 311 |
| Nov. | 9 | 2 | 1 | 5.6 | 54 | S | 4 | 203 |
| Dec. | 11 | 1 | 2 | 6.6 | 41 | S | 4 | 148 |
| Annual | 119 | 56 | 15 | 5.9 | 56 | S | 3 | 356 |
| Rec.(yrs.) | 19 | 19 | 19 | 20 | 19 | 19 | 19 | 20 |

TABLE LXIII

CLIMATIC TABLE FOR ABILENE, TEXAS
Latitude 32°26′N, longitude 99°41′W, elevation 506 m (530 m)

| Month | Mean sta. press. (mbar) | Temperature (°C) | | | | Mean vap. press. (mbar) | Precipitation (mm) | | Mean snowfall (cm) |
|---|---|---|---|---|---|---|---|---|---|
| | | daily mean | mean daily range | extremes | | | mean | max. in 24 h | |
| | | | | max. | min. | | | | |
| Jan. | 956.9 | 7.0 | 13.1 | 32 | −23 | 6 | 22 | 31 | 4 |
| Feb. | 955.5 | 9.1 | 13.5 | 32 | −17 | 7 | 28 | 44 | 2 |
| Mar. | 953.0 | 12.8 | 14.9 | 36 | −14 | 7 | 26 | 50 | 1 |
| Apr. | 952.0 | 17.9 | 14.4 | 37 | −4 | 10 | 58 | 95 | tr. |
| May | 951.5 | 22.1 | 13.0 | 41 | 3 | 15 | 110 | 69 | 0 |
| June | 951.6 | 26.8 | 12.7 | 41 | 11 | 19 | 68 | 93 | 0 |
| July | 953.8 | 28.4 | 12.3 | 42 | 13 | 20 | 58 | 95 | 0 |
| Aug. | 953.5 | 28.3 | 12.3 | 43 | 13 | 18 | 37 | 96 | 0 |
| Sept. | 954.3 | 24.4 | 12.8 | 41 | 2 | 16 | 53 | 78 | 0 |
| Oct. | 955.6 | 19.0 | 13.8 | 38 | −2 | 12 | 72 | 130 | 0 |
| Nov. | 957.1 | 11.7 | 13.7 | 33 | −9 | 8 | 28 | 39 | 1 |
| Dec. | 956.9 | 7.8 | 13.2 | 32 | −13 | 6 | 32 | 58 | 1 |
| Annual | 954.4 | 17.9 | 13.3 | 43 | −23 | 12 | 592 | 130 | 9 |
| Rec.(yrs.) | 30 | 30 | 30 | 21 | 21 | 10 | 30 | 21 | 21 |

| Month | Number of days with | | | Mean cloud-iness (tenths) | Mean sun-shine (h) | Wind | | Solar radiation[1] (ly/ month) |
|---|---|---|---|---|---|---|---|---|
| | precip. (>0.25 mm) | thunder-storm | fog | | | most freq. direct. | mean speed (m/sec) | |
| Jan. | 5 | * | 1 | 5.8 | 64 | S | 6 | 287 |
| Feb. | 6 | 2 | 1 | 5.9 | 68 | S | 6 | 362 |
| Mar. | 4 | 3 | 1 | 5.5 | 72 | S | 7 | 473 |
| Apr. | 7 | 5 | * | 5.2 | 66 | SSE | 7 | 553 |
| May | 8 | 8 | * | 5.4 | 73 | SSE | 6 | 606 |
| June | 7 | 6 | * | 4.2 | 86 | SSE | 6 | 616 |
| July | 4 | 5 | 0 | 4.6 | 83 | SSE | 5 | 674 |
| Aug. | 4 | 5 | 0 | 4.2 | 84 | SSE | 5 | 578 |
| Sept. | 5 | 3 | * | 4.0 | 73 | SSE | 5 | 516 |
| Oct. | 6 | 3 | * | 4.3 | 70 | S | 5 | 400 |
| Nov. | 4 | 1 | 1 | 4.3 | 71 | S | 5 | 317 |
| Dec. | 4 | 1 | 1 | 5.0 | 65 | SSW | 6 | 295 |
| Annual | 64 | 42 | 5 | 4.9 | 71 | S | 6 | 473 |
| Rec.(yrs.) | 21 | 21 | 21 | 21 | 15 | 12 | 16 | 9 |

[1] Midland data.

## TABLE LXIV

Latitude 35°14′N, longitude 101°42′W, elevation 1,099 m

| Month | Temperature (°C) | | | | Mean vap. press. (mbar) | Precipitation (mm) | | Mean snowfall (cm) |
|---|---|---|---|---|---|---|---|---|
| | daily mean | mean daily range | extremes | | | mean | max. in 24 h | |
| | | | max. | min. | | | | |
| Jan. | 2.3 | 14.1 | 27 | −24 | 4 | 17 | 21 | 11 |
| Feb. | 4.3 | 14.6 | 28 | −26 | 5 | 16 | 27 | 6 |
| Mar. | 7.8 | 16.0 | 33 | −19 | 5 | 21 | 25 | 6 |
| Apr. | 13.3 | 15.6 | 34 | −10 | 7 | 34 | 40 | 1 |
| May | 18.3 | 14.6 | 39 | −2 | 10 | 86 | 171 | tr. |
| June | 24.2 | 15.1 | 42 | 6 | 14 | 73 | 156 | 0 |
| July | 26.2 | 14.6 | 40 | 12 | 17 | 59 | 104 | 0 |
| Aug. | 25.6 | 14.5 | 41 | 9 | 16 | 66 | 108 | 0 |
| Sept. | 21.3 | 14.6 | 39 | 1 | 13 | 48 | 87 | tr. |
| Oct. | 15.3 | 14.7 | 35 | −4 | 9 | 45 | 88 | tr. |
| Nov. | 7.5 | 15.2 | 29 | −16 | 5 | 17 | 30 | 4 |
| Dec. | 3.8 | 13.8 | 27 | −15 | 4 | 20 | 79 | 6 |
| Annual | 14.2 | 14.7 | 42 | −26 | 9 | 502 | 171 | 34 |
| Rec.(yrs.) | 30 | 30 | 20 | 20 | 10 | 30 | 20 | 20 |

| Month | Number of days with | | | Mean cloud-iness (tenths) | Mean sun-shine (h) | Wind | |
|---|---|---|---|---|---|---|---|
| | precip. (>0.25 mm) | thunder-storm | fog | | | most freq. direct. | mean speed (m/sec) |
| Jan. | 4 | * | 2 | 5.2 | 67 | SW | 6 |
| Feb. | 4 | * | 4 | 5.4 | 66 | SW | 6 |
| Mar. | 4 | 1 | 4 | 5.1 | 70 | SW | 7 |
| Apr. | 6 | 4 | 2 | 5.2 | 71 | SW | 7 |
| May | 9 | 9 | 2 | 5.3 | 70 | S | 6 |
| June | 8 | 9 | 1 | 4.4 | 76 | S | 6 |
| July | 9 | 11 | 1 | 4.5 | 77 | S | 5 |
| Aug. | 8 | 9 | * | 4.1 | 78 | S | 5 |
| Sept. | 5 | 4 | 1 | 3.6 | 77 | S | 6 |
| Oct. | 5 | 3 | 2 | 4.0 | 73 | SW | 6 |
| Nov. | 3 | * | 2 | 3.7 | 76 | SW | 6 |
| Dec. | 4 | * | 2 | 4.7 | 68 | SW | 6 |
| Annual | 69 | 50 | 23 | 4.6 | 73 | S | 6 |
| Rec.(yrs.) | 20 | 20 | 20 | 20 | 20 | 12 | 20 |

TABLE LXV

CLIMATIC TABLE FOR BROWNSVILLE, TEXAS
Latitude 25°54′N, longitude 97°26′W, elevation 5 m

| Month | Temperature (°C) | | | | Mean vap. press. (mbar) | Precipitation (mm) | | Mean snowfall (cm) |
|---|---|---|---|---|---|---|---|---|
| | daily mean | mean daily range | extremes | | | mean | max. in 24 h | |
| | | | max. | min. | | | | |
| Jan. | 16.3 | 10.2 | 31 | −5 | 14 | 34 | 75 | tr. |
| Feb. | 17.8 | 10.3 | 34 | −6 | 15 | 38 | 126 | tr. |
| Mar. | 19.9 | 9.9 | 37 | 0 | 17 | 26 | 47 | tr. |
| Apr. | 23.3 | 9.3 | 38 | 7 | 21 | 39 | 94 | 0 |
| May | 26.1 | 9.1 | 38 | 12 | 24 | 60 | 102 | 0 |
| June | 28.2 | 8.9 | 38 | 9 | 27 | 75 | 208 | 0 |
| July | 28.9 | 9.5 | 39 | 20 | 28 | 43 | 92 | 0 |
| Aug. | 28.9 | 9.7 | 38 | 19 | 27 | 70 | 112 | 0 |
| Sept. | 27.3 | 9.6 | 40 | 13 | 26 | 127 | 138 | 0 |
| Oct. | 24.4 | 10.3 | 35 | 6 | 20 | 90 | 169 | 0 |
| Nov. | 19.8 | 10.5 | 34 | 1 | 15 | 34 | 92 | 0 |
| Dec. | 17.2 | 10.4 | 31 | −2 | 14 | 44 | 145 | 0 |
| Annual | 23.2 | 9.8 | 40 | −6 | 21 | 680 | 208 | tr. |
| Rec.(yrs.) | 30 | 30 | 21 | 21 | 10 | 30 | 21 | 21 |

| Month | Number of days with | | | Mean cloud-iness (tenths) | Mean sun-shine (h) | Wind | | Solar radiation (ly/ month) |
|---|---|---|---|---|---|---|---|---|
| | precip. (>0.25 mm) | thunder-storm | fog | | | most freq. direct. | mean speed (m/sec) | |
| Jan. | 7 | 1 | 4 | 6.5 | 47 | SSE | 5 | 286 |
| Feb. | 7 | 1 | 5 | 6.5 | 49 | SSE | 6 | 344 |
| Mar. | 5 | 1 | 3 | 6.7 | 48 | SE | 6 | 400 |
| Apr. | 4 | 2 | 3 | 6.7 | 53 | SE | 6 | 467 |
| May | 4 | 3 | 1 | 5.8 | 66 | SE | 6 | 561 |
| June | 5 | 2 | * | 5.3 | 72 | SE | 6 | 606 |
| July | 4 | 2 | 0 | 4.7 | 81 | SE | 5 | 624 |
| Aug. | 7 | 4 | 0 | 4.9 | 77 | SE | 5 | 566 |
| Sept. | 10 | 4 | * | 5.2 | 68 | SE | 4 | 467 |
| Oct. | 7 | 2 | 1 | 4.7 | 68 | SE | 4 | 409 |
| Nov. | 7 | 1 | 2 | 6.0 | 52 | SSE | 5 | 297 |
| Dec. | 6 | * | 5 | 6.5 | 47 | NNW | 5 | 254 |
| Annual | 73 | 23 | 24 | 5.8 | 61 | SE | 5 | 440 |
| Rec.(yrs.) | 18 | 18 | 18 | 18 | 18 | 11 | 18 | 11 |

TABLE LXVI

CLIMATIC TABLE FOR DALLAS, TEXAS
Latitude 32°51′N, longitude 96°51′W, elevation 146 m

| Month | Temperature (°C) | | | | Mean vap. press. (mbar) | Precipitation (mm) | | Mean snowfall (cm) |
|---|---|---|---|---|---|---|---|---|
| | daily mean | mean daily range | extremes | | | mean | max. in 24 h | |
| | | | max. | min. | | | | |
| Jan. | 7.7 | 11.0 | 31 | −17 | 7 | 59 | 131 | 3 |
| Feb. | 9.7 | 11.2 | 31 | −14 | 8 | 65 | 69 | 1 |
| Mar. | 13.4 | 12.1 | 36 | −12 | 9 | 72 | 135 | * |
| Apr. | 18.3 | 11.6 | 36 | −1 | 13 | 102 | 130 | 0 |
| May | 22.7 | 10.9 | 37 | 4 | 19 | 123 | 158 | 0 |
| June | 27.4 | 10.7 | 41 | 12 | 23 | 82 | 104 | 0 |
| July | 29.4 | 10.7 | 44 | 16 | 24 | 49 | 137 | 0 |
| Aug. | 29.4 | 11.1 | 43 | 16 | 23 | 49 | 233 | 0 |
| Sept. | 25.5 | 11.6 | 41 | 2 | 19 | 72 | 83 | 0 |
| Oct. | 19.9 | 12.2 | 37 | −1 | 14 | 69 | 166 | 0 |
| Nov. | 12.7 | 12.0 | 32 | −8 | 9 | 69 | 127 | tr. |
| Dec. | 8.9 | 11.1 | 32 | −11 | 7 | 68 | 85 | 1 |
| Annual | 18.8 | 11.3 | 44 | −17 | 15 | 879 | 233 | 5 |
| Rec.(yrs.) | 30 | 30 | 20 | 20 | 10 | 30 | 20 | 20 |

| Month | Number of days with | | | Mean cloud-iness (tenths) | Mean sun-shine (h) | Wind | | Solar radiation[1] (ly/ month) |
|---|---|---|---|---|---|---|---|---|
| | precip. (>0.25 mm) | thunder-storm | fog | | | most freq. direct. | mean speed (m/sec) | |
| Jan. | 7 | 1 | 1 | 6.3 | 47 | S | 5 | 254 |
| Feb. | 8 | 3 | 1 | 6.1 | 50 | S | 5 | 321 |
| Mar. | 8 | 4 | 1 | 5.8 | 56 | S | 6 | 428 |
| Apr. | 9 | 6 | * | 6.0 | 58 | SSE | 6 | 487 |
| May | 9 | 7 | * | 5.8 | 63 | S | 5 | 565 |
| June | 6 | 5 | * | 4.6 | 75 | S | 6 | 638 |
| July | 5 | 4 | * | 4.3 | 78 | S | 4 | 616 |
| Aug. | 6 | 4 | * | 4.0 | 77 | SSE | 4 | 592 |
| Sept. | 5 | 3 | * | 4.0 | 74 | SE | 4 | 493 |
| Oct. | 6 | 3 | 1 | 4.4 | 66 | SE | 4 | 399 |
| Nov. | 6 | 2 | 1 | 4.7 | 63 | S | 5 | 294 |
| Dec. | 6 | 1 | 2 | 5.6 | 54 | SSE | 5 | 241 |
| Annual | 81 | 43 | 7 | 5.1 | 66 | S | 5 | 444 |
| Rec.(yrs.) | 20 | 20 | 20 | 20 | 20 | 12 | 20 | 12 |

[1] Fort Worth data.

TABLE LXVII

CLIMATIC TABLE FOR EL PASO, TEXAS
Latitude 31°48′N, longitude 106°24′W, elevation 1,194 m (1,152 m)

| Month | Mean sta. press. (mbar) | Temperature (°C) | | | | Mean vap. press. (mbar) | Precipitation (mm) | | Mean snowfall (cm) |
|---|---|---|---|---|---|---|---|---|---|
| | | daily mean | mean daily range | extremes | | | mean | max. in 24 h | |
| | | | | max. | min. | | | | |
| Jan. | 887.6 | 6.6 | 13.8 | 24 | −21 | 4 | 12 | 15 | 4 |
| Feb. | 886.6 | 9.8 | 14.3 | 27 | −13 | 4 | 10 | 22 | 2 |
| Mar. | 884.4 | 12.7 | 15.1 | 31 | −8 | 4 | 9 | 44 | 2 |
| Apr. | 884.0 | 17.4 | 15.4 | 35 | −2 | 5 | 7 | 23 | tr. |
| May | 883.8 | 22.2 | 15.6 | 40 | 3 | 7 | 10 | 31 | 0 |
| June | 883.9 | 26.9 | 15.5 | 43 | 10 | 9 | 18 | 29 | 0 |
| July | 886.2 | 27.4 | 13.9 | 43 | 16 | 14 | 33 | 48 | 0 |
| Aug. | 886.3 | 26.6 | 13.5 | 39 | 15 | 15 | 30 | 51 | 0 |
| Sept. | 886.4 | 23.9 | 14.0 | 39 | 5 | 12 | 29 | 73 | 0 |
| Oct. | 887.5 | 18.6 | 14.9 | 34 | 0 | 8 | 23 | 45 | 0 |
| Nov. | 888.5 | 11.2 | 15.7 | 29 | −9 | 6 | 8 | 30 | 1 |
| Dec. | 888.5 | 7.3 | 13.8 | 24 | −15 | 5 | 12 | 27 | 3 |
| Annual | 886.1 | 17.6 | 14.6 | 43 | −21 | 8 | 201 | 73 | 12 |
| Rec.(yrs.) | 30 | 30 | 30 | 21 | 21 | 10 | 30 | 21 | 21 |

| Month | Number of days with | | | Mean cloud-iness (tenths) | Mean sun-shine (h) | Wind | | Solar radiation (ly/ month) |
|---|---|---|---|---|---|---|---|---|
| | precip. (>0.25 mm) | thunder-storm | fog | | | most freq. direct. | mean speed (m/sec) | |
| Jan. | 3 | * | 1 | 4.7 | 75 | N | 5 | 344 |
| Feb. | 3 | * | * | 4.2 | 79 | N | 5 | 431 |
| Mar. | 2 | 1 | * | 4.5 | 81 | WSW | 6 | 550 |
| Apr. | 2 | 1 | * | 3.8 | 85 | WSW | 6 | 658 |
| May | 2 | 3 | 0 | 3.3 | 88 | W | 6 | 576 |
| June | 4 | 5 | 0 | 3.0 | 88 | W | 5 | 662 |
| July | 8 | 10 | 0 | 4.7 | 77 | S | 5 | 668 |
| Aug. | 7 | 11 | 0 | 4.1 | 80 | S | 4 | 639 |
| Sept. | 4 | 3 | * | 3.0 | 82 | S | 4 | 570 |
| Oct. | 4 | 2 | * | 3.3 | 82 | N | 4 | 465 |
| Nov. | 2 | * | * | 3.2 | 84 | N | 4 | 365 |
| Dec. | 3 | * | * | 4.1 | 77 | N | 4 | 313 |
| Annual | 44 | 36 | 1 | 3.8 | 82 | N | 5 | 519 |
| Rec.(yrs.) | 21 | 21 | 21 | 18 | 18 | 12 | 18 | 13 |

TABLE LXVIII

CLIMATIC TABLE FOR HOUSTON, TEXAS
Latitude 29°46′N, longitude 95°22′W, elevation 12 m (16 m)

| Month | Mean sta. press. (mbar) | Temperature (°C) | | | | Mean vap. press. (mbar) | Precipitation (mm) | | Mean snowfall (cm) |
|---|---|---|---|---|---|---|---|---|---|
| | | daily mean | mean daily range | extremes | | | mean | max. in 24 h | |
| | | | | max. | min. | | | | |
| Jan. | 1018.3 | 12.6 | 8.9 | 28 | −12 | 11 | 94 | 81 | 1 |
| Feb. | 1016.7 | 13.9 | 9.0 | 32 | −10 | 12 | 82 | 99 | * |
| Mar. | 1014.3 | 16.9 | 9.5 | 36 | −6 | 14 | 61 | 93 | 0 |
| Apr. | 1013.2 | 20.7 | 9.2 | 33 | 4 | 16 | 87 | 96 | 0 |
| May | 1012.3 | 24.6 | 9.2 | 35 | 8 | 23 | 113 | 138 | 0 |
| June | 1012.4 | 27.9 | 9.2 | 37 | 14 | 26 | 97 | 210 | 0 |
| July | 1013.7 | 28.8 | 9.0 | 41 | 20 | 28 | 131 | 176 | 0 |
| Aug. | 1013.1 | 28.9 | 9.3 | 39 | 17 | 27 | 90 | 230 | 0 |
| Sept. | 1012.9 | 26.6 | 9.5 | 37 | 7 | 25 | 97 | 132 | 0 |
| Oct. | 1014.8 | 22.4 | 10.3 | 36 | 3 | 18 | 91 | 194 | 0 |
| Nov. | 1017.6 | 16.4 | 9.8 | 32 | −2 | 13 | 103 | 275 | 0 |
| Dec. | 1018.0 | 13.6 | 9.0 | 28 | −7 | 11 | 104 | 86 | tr. |
| Annual | 1014.8 | 21.1 | 9.3 | 41 | −12 | 19 | 1,150 | 275 | 1 |
| Rec.(yrs.) | 30 | 30 | 30 | 22 | 22 | 10 | 30 | 22 | 22 |

| Month | Number of days with | | | Mean cloud-iness (tenths) | Mean sun-shine (h) | Wind | |
|---|---|---|---|---|---|---|---|
| | precip. (>0.25 mm) | thunder-storm | fog | | | most freq. direct. | mean speed (m/sec) |
| Jan. | 10 | 2 | 3 | 6.6 | 46 | N | 5 |
| Feb. | 10 | 2 | 2 | 6.7 | 44 | SE | 5 |
| Mar. | 9 | 2 | 2 | 6.4 | 51 | SE | 5 |
| Apr. | 8 | 5 | 1 | 6.4 | 54 | SE | 5 |
| May | 8 | 6 | 1 | 6.2 | 63 | SE | 5 |
| June | 8 | 7 | * | 5.5 | 70 | S | 4 |
| July | 10 | 11 | 0 | 5.9 | 69 | S | 4 |
| Aug. | 9 | 9 | * | 5.7 | 68 | S | 4 |
| Sept. | 9 | 5 | * | 5.4 | 64 | SE | 4 |
| Oct. | 7 | 3 | 2 | 4.7 | 67 | SE | 4 |
| Nov. | 8 | 2 | 2 | 5.6 | 57 | SE | 5 |
| Dec. | 10 | 2 | 3 | 6.2 | 48 | SE | 5 |
| Annual | 106 | 56 | 16 | 5.9 | 59 | SE | 5 |
| Rec.(yrs.) | 22 | 22 | 22 | 22 | 22 | 22 | 22 |

TABLE LXIX

CLIMATIC TABLE FOR SAN ANTONIO, TEXAS
Latitude 29°32′N, longitude 98°28′W, elevation 241 m

| Month | Temperature (°C) | | | | Mean vap. press. (mbar) | Precipitation (mm) | | Mean snowfall (cm) |
|---|---|---|---|---|---|---|---|---|
| | daily mean | mean daily range | extremes | | | mean | max. in 24 h | |
| | | | max. | min. | | | | |
| Jan. | 11.1 | 11.5 | 31 | −18 | 9 | 44 | 72 | 1 |
| Feb. | 13.0 | 11.9 | 33 | −14 | 11 | 42 | 58 | * |
| Mar. | 16.1 | 12.7 | 36 | −6 | 11 | 42 | 60 | tr. |
| Apr. | 20.1 | 11.8 | 37 | 1 | 15 | 72 | 70 | 0 |
| May | 24.1 | 11.0 | 38 | 7 | 20 | 88 | 109 | 0 |
| June | 27.7 | 10.8 | 39 | 14 | 24 | 75 | 157 | 0 |
| July | 28.9 | 11.2 | 41 | 18 | 24 | 53 | 177 | 0 |
| Aug. | 28.8 | 11.6 | 41 | 17 | 23 | 60 | 141 | 0 |
| Sept. | 25.9 | 11.0 | 39 | 5 | 22 | 89 | 175 | 0 |
| Oct. | 21.4 | 12.2 | 35 | 1 | 16 | 64 | 134 | 0 |
| Nov. | 15.3 | 12.2 | 33 | −5 | 11 | 35 | 49 | tr. |
| Dec. | 12.1 | 13.0 | 32 | −10 | 9 | 44 | 73 | tr. |
| Annual | 20.4 | 11.7 | 41 | −18 | 16 | 708 | 177 | 1 |
| Rec.(yrs.) | 30 | 30 | 18 | 18 | 10 | 30 | 18 | 18 |

| Month | Number of days with | | | Mean cloud-iness (tenths) | Mean sun-shine (h) | Wind | | Solar radiation (ly/ month) |
|---|---|---|---|---|---|---|---|---|
| | precip. (>0.25 mm) | thunder-storm | fog | | | most freq. direct. | mean speed (m/sec) | |
| Jan. | 8 | 1 | 6 | 6.5 | 46 | NE | 4 | 274 |
| Feb. | 9 | 2 | 4 | 6.5 | 49 | NE | 4 | 347 |
| Mar. | 7 | 3 | 3 | 6.3 | 56 | NE | 5 | 416 |
| Apr. | 7 | 4 | 1 | 6.4 | 57 | SE | 5 | 449 |
| May | 7 | 7 | 1 | 6.1 | 60 | SE | 5 | 535 |
| June | 6 | 4 | * | 5.3 | 69 | SE | 5 | 602 |
| July | 4 | 4 | * | 5.0 | 77 | SE | 4 | 630 |
| Aug. | 5 | 4 | 0 | 4.7 | 75 | SE | 4 | 584 |
| Sept. | 7 | 4 | * | 4.8 | 71 | SE | 4 | 489 |
| Oct. | 6 | 2 | 2 | 4.8 | 67 | N | 4 | 397 |
| Nov. | 6 | 2 | 3 | 5.5 | 57 | N | 4 | 292 |
| Dec. | 7 | 1 | 5 | 5.8 | 53 | N | 4 | 248 |
| Annual | 79 | 38 | 25 | 5.6 | 62 | SE | 4 | 439 |
| Rec.(yrs.) | 18 | 18 | 18 | 18 | 18 | 12 | 18 | 11 |

TABLE LXX

CLIMATIC TABLE FOR SALT LAKE CITY, UTAH
Latitude 40°46′N, longitude 111°58′W, elevation 1,286 m (1,329 m; 40°41′N, 111°58′W)

| Month | Mean sta. press. (mbar) | Temperature (°C) | | | | Mean vap. press. (mbar) | Precipitation (mm) | | Mean snowfall (cm) |
|---|---|---|---|---|---|---|---|---|---|
| | | daily mean | mean daily range | extremes | | | mean | max. in 24 h | |
| | | | | max. | min. | | | | |
| Jan. | 869.8 | −2.1 | 9.6 | 16 | −30 | 4 | 34 | 35 | 35 |
| Feb. | 868.1 | 0.6 | 10.1 | 20 | −34 | 4 | 30 | 27 | 24 |
| Mar. | 866.3 | 4.7 | 11.8 | 26 | −15 | 4 | 40 | 46 | 21 |
| Apr. | 865.5 | 9.9 | 13.9 | 29 | −10 | 5 | 45 | 61 | 8 |
| May | 865.1 | 14.7 | 15.1 | 34 | −3 | 5 | 36 | 52 | 1 |
| June | 865.2 | 19.4 | 16.5 | 39 | 2 | 7 | 25 | 48 | tr. |
| July | 867.4 | 24.7 | 17.5 | 42 | 5 | 9 | 15 | 42 | 0 |
| Aug. | 867.3 | 23.6 | 17.0 | 39 | 4 | 9 | 22 | 50 | 0 |
| Sept. | 867.5 | 18.3 | 17.1 | 37 | −1 | 5 | 13 | 23 | tr. |
| Oct. | 868.7 | 11.5 | 15.0 | 31 | −8 | 5 | 29 | 37 | 1 |
| Nov. | 870.7 | 3.4 | 11.5 | 23 | −26 | 4 | 33 | 29 | 16 |
| Dec. | 870.1 | −0.2 | 9.3 | 19 | −29 | 4 | 32 | 26 | 27 |
| Annual | 867.6 | 10.7 | 13.7 | 42 | −34 | 5 | 354 | 61 | 133 |
| Rec.(yrs.) | 30 | 30 | 30 | 32 | 32 | 10 | 30 | 32 | 32 |

| Month | Number of days with | | | Mean cloud-iness (tenths) | Mean sun-shine (h) | Wind | | Solar radiation (ly/ month) |
|---|---|---|---|---|---|---|---|---|
| | precip. (>0.25 mm) | thunder-storm | fog | | | most freq. direct. | mean speed (m/sec) | |
| Jan. | 10 | * | 4 | 7.0 | 47 | SSE | 3 | 174 |
| Feb. | 9 | 1 | 2 | 7.0 | 53 | SE | 4 | 256 |
| Mar. | 10 | 1 | * | 6.5 | 61 | SSE | 4 | 381 |
| Apr. | 9 | 2 | * | 6.1 | 68 | SE | 4 | 490 |
| May | 8 | 5 | 0 | 5.4 | 73 | SE | 4 | 588 |
| June | 5 | 5 | 0 | 4.2 | 79 | SSE | 4 | 654 |
| July | 4 | 7 | 0 | 3.5 | 82 | SSE | 4 | 647 |
| Aug. | 6 | 8 | 0 | 3.4 | 82 | SSE | 4 | 569 |
| Sept. | 5 | 4 | 0 | 3.4 | 84 | SE | 4 | 462 |
| Oct. | 6 | 2 | * | 4.3 | 73 | SE | 4 | 329 |
| Nov. | 7 | * | 1 | 5.6 | 56 | SSE | 3 | 208 |
| Dec. | 9 | * | 3 | 6.8 | 46 | SSE | 3 | 151 |
| Annual | 86 | 35 | 10 | 5.3 | 69 | SSE | 4 | 409 |
| Rec.(yrs.) | 32 | 32 | 32 | 25 | 23 | 29 | 31 | 9 |

TABLE LXXI

CLIMATIC TABLE FOR BURLINGTON, VE.
Latitude 44°28′N, longitude 73°09′W, elevation 101 m

| Month | Temperature (°C) | | | | Mean vap. press. (mbar) | Precipitation (mm) | | Mean snowfall (cm) |
|-------|-------|-------|-------|-------|-------|-------|-------|-------|
| | daily mean | mean daily range | extremes | | | mean | max. in 24 h | |
| | | | max. | min. | | | | |
| Jan. | −7.7 | 10.3 | 17 | −34 | 2 | 50 | 37 | 47 |
| Feb. | −7.0 | 10.8 | 16 | −32 | 2 | 45 | 42 | 48 |
| Mar. | −1.6 | 10.2 | 29 | −29 | 4 | 54 | 33 | 26 |
| Apr. | 6.2 | 11.3 | 29 | −13 | 6 | 67 | 36 | 4 |
| May | 13.2 | 12.9 | 33 | −4 | 8 | 76 | 57 | tr. |
| June | 18.7 | 12.7 | 36 | 2 | 14 | 89 | 65 | 0 |
| July | 21.4 | 12.7 | 37 | 7 | 16 | 98 | 62 | 0 |
| Aug. | 20.1 | 12.7 | 38 | 4 | 16 | 86 | 91 | 0 |
| Sept. | 15.5 | 12.0 | 34 | −3 | 12 | 84 | 63 | tr. |
| Oct. | 9.2 | 11.1 | 29 | −6 | 9 | 75 | 43 | tr. |
| Nov. | 2.7 | 8.5 | 24 | −19 | 6 | 67 | 46 | 14 |
| Dec. | −5.0 | 9.1 | 17 | −30 | 4 | 54 | 66 | 36 |
| Annual | 7.2 | 11.2 | 38 | −34 | 8 | 845 | 91 | 175 |
| Rec.(yrs.) | 30 | 30 | 17 | 17 | 10 | 30 | 17 | 17 |

| Month | Number of days with | | | Mean cloud-iness (tenths) | Mean sun-shine (h) | Wind | |
|-------|-------|-------|-------|-------|-------|-------|-------|
| | precip. (>0.25 mm) | thunder-storm | fog | | | most freq. direct. | mean speed (m/sec) |
| Jan. | 14 | 0 | 1 | 7.5 | 35 | S | 5 |
| Feb. | 13 | 0 | 1 | 7.3 | 44 | S | 4 |
| Mar. | 13 | * | 1 | 7.0 | 49 | S | 4 |
| Apr. | 13 | 1 | 1 | 7.2 | 46 | S | 4 |
| May | 13 | 3 | 1 | 6.9 | 52 | S | 4 |
| June | 11 | 5 | 1 | 6.7 | 58 | S | 4 |
| July | 12 | 6 | * | 6.3 | 63 | S | 4 |
| Aug. | 11 | 5 | 1 | 6.0 | 62 | S | 3 |
| Sept. | 12 | 2 | 2 | 6.2 | 53 | S | 4 |
| Oct. | 11 | 1 | 2 | 6.5 | 48 | S | 4 |
| Nov. | 14 | * | 1 | 8.1 | 29 | S | 4 |
| Dec. | 14 | * | 1 | 7.8 | 31 | S | 5 |
| Annual | 151 | 23 | 13 | 7.0 | 48 | S | 4 |
| Rec.(yrs.) | 17 | 17 | 17 | 17 | 17 | 17 | 17 |

TABLE LXXII

CLIMATIC TABLE FOR RICHMOND, VA.
Latitude 37°30'N, longitude 77°20'W, elevation 49 m

| Month | Temperature (°C) | | | | Mean vap. press. (mbar) | Precipitation (mm) | | Mean snowfall (cm) |
|-------|-------|-------|-------|-------|-------|-------|-------|-------|
| | daily mean | mean daily range | extremes | | | mean | max. in 24 h | |
| | | | max. | min. | | | | |
| Jan. | 3.7 | 10.7 | 27 | −24 | 6 | 88 | 82 | 11 |
| Feb. | 4.4 | 11.9 | 28 | −23 | 6 | 74 | 40 | 6 |
| Mar. | 8.2 | 12.7 | 34 | −12 | 7 | 87 | 52 | 7 |
| Apr. | 13.9 | 13.7 | 36 | −3 | 10 | 80 | 53 | * |
| May | 19.2 | 13.2 | 38 | −1 | 16 | 94 | 58 | 0 |
| June | 23.7 | 12.5 | 40 | 5 | 20 | 95 | 72 | 0 |
| July | 25.6 | 11.5 | 40 | 11 | 23 | 142 | 124 | 0 |
| Aug. | 24.7 | 11.2 | 39 | 8 | 22 | 141 | 223 | 0 |
| Sept. | 21.2 | 11.8 | 39 | 3 | 18 | 93 | 97 | 0 |
| Oct. | 15.1 | 12.7 | 37 | −5 | 13 | 76 | 101 | tr. |
| Nov. | 9.2 | 12.7 | 30 | −12 | 8 | 77 | 103 | 2 |
| Dec. | 4.3 | 11.3 | 26 | −18 | 6 | 75 | 80 | 5 |
| Annual | 14.4 | 12.2 | 40 | −24 | 13 | 1,122 | 223 | 31 |
| Rec.(yrs.) | 30 | 30 | 31 | 31 | 10 | 30 | 23 | 23 |

| Month | Number of days with | | | Mean cloud-iness (tenths) | Mean sun-shine (h) | Wind | |
|-------|-------|-------|-------|-------|-------|-------|-------|
| | precip. (>0.25 mm) | thunder-storm | fog | | | most freq. direct. | mean speed (m/sec) |
| Jan. | 10 | * | 3 | 6.7 | 49 | S | 4 |
| Feb. | 9 | * | 2 | 6.1 | 54 | WSW | 4 |
| Mar. | 11 | 1 | 2 | 6.2 | 57 | W | 4 |
| Apr. | 10 | 2 | 2 | 6.2 | 63 | S | 4 |
| May | 11 | 6 | 2 | 6.3 | 65 | SSW | 3 |
| June | 9 | 7 | 2 | 5.9 | 69 | S | 3 |
| July | 12 | 10 | 2 | 5.9 | 70 | SSW | 3 |
| Aug. | 10 | 7 | 3 | 6.0 | 65 | S | 3 |
| Sept. | 8 | 3 | 4 | 5.8 | 64 | S | 3 |
| Oct. | 8 | 1 | 4 | 5.5 | 58 | NNE | 3 |
| Nov. | 9 | 1 | 2 | 5.7 | 54 | S | 3 |
| Dec. | 9 | * | 3 | 6.0 | 52 | SW | 3 |
| Annual | 116 | 38 | 31 | 6.0 | 60 | S | 3 |
| Rec.(yrs.) | 23 | 23 | 31 | 15 | 10 | 12 | 12 |

TABLE LXXIII

CLIMATIC TABLE FOR SEATTLE, WASH.
Latitude 47°36′N, longitude 122°20′W, elevation 4 m

| Month | Temperature (°C) | | | | Mean vap. press. (mbar) | Precipitation (mm) | | Mean snowfall (cm) |
|---|---|---|---|---|---|---|---|---|
| | daily mean | mean daily | extremes | | | mean | max. in 24 h | |
| | | | max. | min. | | | | |
| Jan. | 5.1 | 4.9 | 19 | −12 | 6 | 132 | 62 | 12 |
| Feb. | 6.4 | 5.8 | 21 | −11 | 7 | 99 | 68 | 4 |
| Mar. | 8.0 | 7.0 | 24 | −6 | 7 | 84 | 59 | 2 |
| Apr. | 11.0 | 8.5 | 31 | −1 | 8 | 50 | 39 | tr. |
| May | 14.1 | 9.3 | 33 | 2 | 10 | 40 | 34 | 0 |
| June | 16.3 | 9.2 | 38 | 7 | 12 | 36 | 27 | 0 |
| July | 18.7 | 10.6 | 38 | 9 | 12 | 16 | 31 | 0 |
| Aug. | 18.3 | 9.9 | 36 | 9 | 12 | 19 | 20 | 0 |
| Sept. | 16.2 | 8.7 | 33 | 6 | 12 | 42 | 49 | 0 |
| Oct. | 12.4 | 6.7 | 26 | −1 | 11 | 83 | 50 | tr. |
| Nov. | 8.3 | 5.5 | 21 | −11 | 8 | 127 | 81 | 2 |
| Dec. | 6.6 | 4.7 | 18 | −6 | 7 | 138 | 84 | 2 |
| Annual | 11.8 | 7.6 | 38 | −12 | 9 | 866 | 84 | 22 |
| Rec.(yrs.) | 30 | 30 | 27 | 27 | 10 | 30 | 27 | 27 |

| Month | Number of days with | | | Mean cloud-iness (tenths) | Mean sun-shine (h) | Wind | | Solar radiation (ly/ month) |
|---|---|---|---|---|---|---|---|---|
| | precip. (>0.25 mm) | thunder-storm | fog*1 | | | most freq. direct. | mean speed (m/sec) | |
| Jan. | 19 | * | 6 | 8.0 | 27 | S | 3 | 71 |
| Feb. | 15 | * | 4 | 7.7 | 34 | S | 3 | 128 |
| Mar. | 16 | * | 3 | 7.4 | 42 | S | 4 | 244 |
| Apr. | 13 | * | 1 | 6.9 | 48 | S | 4 | 357 |
| May | 11 | 1 | 1 | 6.4 | 52 | S | 3 | 437 |
| June | 9 | 1 | 1 | 6.4 | 48 | SW | 3 | 475 |
| July | 5 | 1 | 3 | 4.9 | 62 | NNW | 3 | 504 |
| Aug. | 6 | 1 | 4 | 5.3 | 55 | NNW | 3 | 444 |
| Sept. | 8 | 1 | 8 | 5.6 | 52 | SSE | 3 | 315 |
| Oct. | 14 | 1 | 9 | 7.2 | 36 | SSE | 3 | 174 |
| Nov. | 17 | * | 7 | 8.0 | 28 | SSE | 3 | 94 |
| Dec. | 19 | * | 7 | 8.1 | 24 | SSE | 3 | 60 |
| Annual | 151 | 6 | 54 | 6.8 | 45 | S | 3 | 275 |
| Rec.(yrs.) | 27 | 24*2 | 19 | 24 | 27 | 11 | 10 | 11 |

*1 Airport data.
*2 1934–1959.

TABLE LXXIV

CLIMATIC TABLE FOR TATOOSH ISLAND, WASH.
Latitude 48°23'N, longitude 124°44'W, elevation 31 m (sea level)

| Month | Mean sta. press. (mbar) | Temperature (°C) | | | | Mean vap. press. (mbar) | Precipitation (mm) | | Mean snowfall (cm) |
|---|---|---|---|---|---|---|---|---|---|
| | | daily mean | mean daily range | extremes | | | mean | max. in 24 h | |
| | | | | max. | min. | | | | |
| Jan. | 1015.3 | 5.6 | 3.6 | 18 | −10 | 7 | 275 | 93 | 9 |
| Feb. | 1015.5 | 6.2 | 3.9 | 18 | −9 | 8 | 221 | 116 | 4 |
| Mar. | 1015.7 | 6.8 | 4.2 | 21 | −4 | 8 | 212 | 121 | 3 |
| Apr. | 1017.1 | 8.6 | 4.6 | 24 | 1 | 9 | 133 | 94 | tr. |
| May | 1017.7 | 10.6 | 4.6 | 27 | 2 | 11 | 76 | 56 | tr. |
| June | 1017.7 | 12.2 | 4.3 | 29 | 6 | 13 | 72 | 70 | 0 |
| July | 1018.9 | 13.1 | 4.4 | 31 | 7 | 13 | 59 | 94 | 0 |
| Aug. | 1017.9 | 13.3 | 4.6 | 26 | 7 | 14 | 50 | 58 | 0 |
| Sept. | 1016.9 | 12.7 | 4.9 | 27 | 4 | 13 | 90 | 96 | 0 |
| Oct. | 1016.2 | 11.1 | 4.3 | 25 | 1 | 11 | 209 | 150 | tr. |
| Nov. | 1016.7 | 8.4 | 3.7 | 20 | −7 | 9 | 267 | 111 | 1 |
| Dec. | 1014.7 | 6.9 | 3.6 | 16 | −7 | 8 | 309 | 102 | 2 |
| Annual | 1016.7 | 9.6 | 4.2 | 31 | −10 | 10 | 1,973 | 150 | 19 |
| Rec.(yrs.) | 30 | 30 | 30 | 58 | 58 | 10 | 30 | 58 | 58 |

| Month | Number of days with | | | Mean cloud-iness (tenths) | Mean sun-shine (h) | Wind | |
|---|---|---|---|---|---|---|---|
| | precip. (>0.25 mm) | thunder-storm | fog | | | most freq. direct. | mean speed (m/sec) |
| Jan. | 22 | 1 | 1 | 8.0 | 26 | E | 9 |
| Feb. | 18 | * | 1 | 7.4 | 36 | E | 8 |
| Mar. | 20 | * | 1 | 7.4 | 39 | E | 7 |
| Apr. | 17 | * | 2 | 7.2 | 44 | W | 6 |
| May | 14 | * | 3 | 7.2 | 47 | W | 5 |
| June | 12 | * | 5 | 7.2 | 45 | W | 4 |
| July | 10 | * | 11 | 6.8 | 48 | S | 4 |
| Aug. | 10 | * | 16 | 7.0 | 44 | S | 4 |
| Sept. | 11 | * | 10 | 6.5 | 47 | S | 5 |
| Oct. | 17 | 1 | 6 | 7.0 | 38 | E | 7 |
| Nov. | 21 | 1 | 2 | 8.0 | 26 | E | 8 |
| Dec. | 23 | 1 | 1 | 8.0 | 24 | E | 9 |
| Annual | 197 | 5 | 58 | 7.3 | 40 | E | 6 |
| Rec.(yrs.) | 58 | 58 | 58 | 54 | 50 | 21 | 29 |

TABLE LXXV

CLIMATIC TABLE FOR WALLA WALLA, WASH.
Latitude 46°02′N, longitude 118°20′W, elevation 289 m (302 m)

| Month | Mean sta. press. (mbar) | Temperature (°C) | | | | Mean vap. press. (mbar) | Precipitation (mm) | | Mean snowfall (cm) |
|---|---|---|---|---|---|---|---|---|---|
| | | daily mean | mean daily range | extremes | | | mean | max. in 24 h | |
| | | | | max. | min. | | | | |
| Jan. | 983.3 | 0.7 | 6.4 | 21 | −27 | 5 | 48 | 36 | 21 |
| Feb. | 982.0 | 3.6 | 7.3 | 21 | −26 | 5 | 39 | 34 | 12 |
| Mar. | 980.3 | 7.8 | 9.3 | 26 | −11 | 6 | 40 | 30 | 3 |
| Apr. | 979.6 | 12.1 | 11.2 | 34 | −7 | 7 | 36 | 40 | 1 |
| May | 979.0 | 16.1 | 12.3 | 37 | −2 | 8 | 38 | 47 | tr. |
| June | 978.2 | 19.6 | 12.8 | 41 | 5 | 9 | 31 | 51 | 0 |
| July | 978.4 | 24.4 | 14.7 | 44 | 8 | 10 | 5 | 29 | 0 |
| Aug. | 978.3 | 23.2 | 14.2 | 41 | 7 | 9 | 8 | 26 | 0 |
| Sept. | 979.2 | 18.9 | 13.2 | 39 | −3 | 9 | 20 | 35 | 0 |
| Oct. | 981.3 | 12.8 | 10.6 | 31 | −9 | 8 | 39 | 35 | * |
| Nov. | 984.3 | 5.7 | 7.3 | 25 | −17 | 7 | 44 | 36 | 4 |
| Dec. | 983.1 | 3.2 | 6.7 | 23 | −26 | 5 | 47 | 34 | 11 |
| Annual | 980.6 | 12.3 | 10.5 | 44 | −27 | 7 | 395 | 51 | 52 |
| Rec.(yrs.) | 30 | 30 | 30 | 46 | 46 | 10 | 30 | 46 | 46 |

| Month | Number of days with | | | Mean cloud-iness (tenths) | Mean sun-shine (h) | Wind | |
|---|---|---|---|---|---|---|---|
| | precip. (>0.25 mm) | thunder-storm | fog | | | most freq. direct[1]. | mean speed (m/sec) |
| Jan. | 14 | * | 3 | 8.3 | 25 | S | 2 |
| Feb. | 11 | * | 2 | 7.8 | 36 | S | 2 |
| Mar. | 12 | * | * | 7.1 | 51 | S | 3 |
| Apr. | 9 | 1 | 0 | 6.2 | 63 | S | 3 |
| May | 9 | 2 | * | 5.7 | 67 | S | 3 |
| June | 7 | 2 | 0 | 5.1 | 72 | S | 2 |
| July | 3 | 2 | 0 | 2.4 | 86 | S | 2 |
| Aug. | 3 | 2 | 0 | 3.0 | 83 | S | 2 |
| Sept. | 5 | 1 | * | 4.1 | 72 | S | 2 |
| Oct. | 9 | * | * | 5.7 | 59 | S | 2 |
| Nov. | 11 | * | 3 | 7.9 | 33 | S | 2 |
| Dec. | 14 | 0 | 5 | 8.7 | 19 | S | 2 |
| Annual | 107 | 11 | 14 | 6.0 | 60 | S | 2 |
| Rec.(yrs.) | 46 | 46 | 46 | 19 | 45 | 45 | 45 |

[1] To 8 points.

TABLE LXXVI

CLIMATIC TABLE FOR PARKERSBURG, W.VA.
Latitude 39°16′N, longitude 81°34′W, elevation 187 m

| Month | Temperature (°C) | | | | Mean vap. press. (mbar) | Precipitation (mm) | | Mean snowfall (cm) |
|---|---|---|---|---|---|---|---|---|
| | daily mean | mean daily range | extremes | | | mean | max. in 24 h | |
| | | | max. | min. | | | | |
| Jan. | 1.4 | 9.2 | 26 | −27 | 4 | 85 | 75 | 17 |
| Feb. | 1.9 | 10.0 | 25 | −33 | 5 | 72 | 73 | 15 |
| Mar. | 5.9 | 11.0 | 32 | −19 | 6 | 90 | 61 | 11 |
| Apr. | 12.3 | 12.5 | 34 | −9 | 9 | 83 | 86 | 2 |
| May | 17.8 | 12.7 | 36 | −2 | 12 | 94 | 76 | tr. |
| June | 22.6 | 12.1 | 37 | 3 | 17 | 108 | 91 | 0 |
| July | 24.3 | 11.7 | 40 | 8 | 20 | 104 | 122 | 0 |
| Aug. | 23.6 | 11.8 | 41 | 7 | 19 | 96 | 91 | 0 |
| Sept. | 20.1 | 12.5 | 39 | 0 | 15 | 69 | 76 | 0 |
| Oct. | 13.9 | 12.6 | 33 | −7 | 11 | 52 | 86 | * |
| Nov. | 7.1 | 10.6 | 28 | −16 | 6 | 60 | 82 | 5 |
| Dec. | 2.1 | 9.1 | 24 | −23 | 5 | 72 | 68 | 11 |
| Annual | 12.8 | 11.3 | 41 | −33 | 11 | 985 | 122 | 61 |
| Rec.(yrs.) | 30 | 30 | 72 | 72 | 10 | 30 | 72 | 72 |

| Month | Number of days with | | | Mean cloud-iness (tenths) | Mean sun-shine (h) | Mean wind speed (m/sec) |
|---|---|---|---|---|---|---|
| | precip. (>0.25 mm) | thunder-storm | fog | | | |
| Jan. | 16 | * | 1 | 7.3 | 30 | 3 |
| Feb. | 13 | 1 | * | 6.9 | 36 | 3 |
| Mar. | 14 | 2 | 1 | 6.4 | 42 | 3 |
| Apr. | 13 | 3 | * | 6.0 | 49 | 3 |
| May | 12 | 7 | * | 5.6 | 56 | 3 |
| June | 13 | 9 | 1 | 5.3 | 60 | 2 |
| July | 11 | 9 | 1 | 5.0 | 63 | 2 |
| Aug. | 10 | 7 | 1 | 5.1 | 60 | 2 |
| Sept. | 9 | 4 | 2 | 4.8 | 60 | 2 |
| Oct. | 9 | 1 | 2 | 5.2 | 53 | 2 |
| Nov. | 11 | * | 1 | 6.6 | 37 | 3 |
| Dec. | 13 | * | 1 | 7.3 | 29 | 3 |
| Annual | 144 | 43 | 11 | 6.0 | 48 | 3 |
| Rec.(yrs.) | 72 | 72 | 61 | 71 | 63 | 72 |

TABLE LXXVII

CLIMATIC TABLE FOR MILWAUKEE, WISC.
Latitude 42°57′N, longitude 87°54′W, elevation 205 m

| Month | Temperature (°C) | | | | Mean vap. press. (mbar) | Precipitation (mm) | | Mean snowfall (cm) |
|-------|------|------|------|------|------|------|------|------|
| | daily mean | mean daily range | extremes | | | mean | max. in 24 h | |
| | | | max. | min. | | | | |
| Jan. | −6.3 | 8.6 | 17 | −31 | 4 | 46 | 43 | 33 |
| Feb. | −5.3 | 8.7 | 16 | −28 | 4 | 36 | 42 | 20 |
| Mar. | −0.6 | 8.7 | 27 | −23 | 5 | 59 | 65 | 22 |
| Apr. | 6.4 | 10.6 | 29 | −11 | 7 | 64 | 54 | 2 |
| May | 11.9 | 11.7 | 32 | −2 | 11 | 80 | 52 | tr. |
| June | 17.4 | 11.8 | 37 | 1 | 16 | 92 | 80 | 0 |
| July | 20.4 | 11.4 | 38 | 7 | 19 | 75 | 110 | 0 |
| Aug. | 19.9 | 11.1 | 38 | 7 | 18 | 78 | 103 | 0 |
| Sept. | 15.7 | 11.6 | 37 | −2 | 11 | 69 | 134 | tr. |
| Oct. | 10.0 | 11.2 | 30 | −6 | 8 | 53 | 66 | tr. |
| Nov. | 2.1 | 9.2 | 25 | −21 | 6 | 55 | 55 | 8 |
| Dec. | −4.1 | 8.3 | 17 | −24 | 4 | 41 | 49 | 24 |
| Annual | 7.3 | 10.2 | 38 | −31 | 9 | 748 | 134 | 109 |
| Rec.(yrs.) | 30 | 30 | 20 | 20 | 10 | 30 | 20 | 20 |

| Month | Number of days with | | | Mean cloud-iness (tenths) | Mean sun-shine (h) | Wind | | Solar radiation[1] (ly/month) |
|-------|------|------|------|------|------|------|------|------|
| | precip. (>0.25 mm) | thunder-storm | fog | | | most freq. direct. | mean speed (m/sec) | |
| Jan. | 10 | * | 3 | 7.0 | 41 | WNW | 6 | 148 |
| Feb. | 9 | * | 2 | 6.7 | 44 | WNW | 6 | 220 |
| Mar. | 11 | 1 | 2 | 6.7 | 50 | WNW | 6 | 313 |
| Apr. | 11 | 4 | 3 | 6.5 | 53 | NNE | 6 | 394 |
| May | 13 | 5 | 3 | 6.4 | 57 | NNE | 6 | 466 |
| June | 11 | 7 | 3 | 6.1 | 62 | NNE | 5 | 514 |
| July | 9 | 7 | 1 | 5.2 | 71 | SW | 4 | 531 |
| Aug. | 9 | 6 | 2 | 5.2 | 66 | SW | 4 | 452 |
| Sept. | 8 | 4 | 1 | 5.3 | 62 | SW | 5 | 348 |
| Oct. | 8 | 2 | 2 | 5.3 | 58 | SW | 5 | 241 |
| Nov. | 10 | 1 | 2 | 7.1 | 42 | WNW | 6 | 145 |
| Dec. | 10 | * | 2 | 7.0 | 40 | WNW | 6 | 115 |
| Annual | 119 | 37 | 26 | 6.2 | 55 | WNW | 6 | 324 |
| Rec.(yrs.) | 20 | 20 | 20 | 20 | 20 | 11 | 20 | 44 |

[1] Madison data.

## TABLE LXXVIII

CLIMATIC TABLE FOR LANDER, WYO.
Latitude 42°49′N, longitude 108°44′W, elevation 1,696 m

| Month | Temperature (°C) | | | | Mean vap. press. (mbar) | Precipitation (mm) | | Mean snowfall (cm) |
|---|---|---|---|---|---|---|---|---|
| | daily mean | mean daily range | extremes | | | mean | max. in 24 h | |
| | | | max. | min. | | | | |
| Jan. | −7.1 | 12.8 | 16 | −35 | 2 | 12 | 21 | 23 |
| Feb. | −4.4 | 13.6 | 20 | −33 | 2 | 18 | 22 | 31 |
| Mar. | 0.1 | 13.7 | 21 | −27 | 3 | 29 | 32 | 42 |
| Apr. | 6.2 | 14.1 | 28 | −12 | 4 | 62 | 50 | 41 |
| May | 11.6 | 14.6 | 33 | −8 | 5 | 67 | 62 | 14 |
| June | 16.7 | 15.6 | 38 | −4 | 5 | 35 | 90 | 7 |
| July | 21.4 | 16.9 | 38 | 4 | 7 | 20 | 18 | 0 |
| Aug. | 20.4 | 16.5 | 37 | 2 | 7 | 12 | 15 | 0 |
| Sept. | 15.0 | 16.0 | 34 | −4 | 5 | 26 | 50 | 3 |
| Oct. | 8.4 | 14.8 | 29 | −13 | 4 | 31 | 43 | 19 |
| Nov. | −0.6 | 13.3 | 20 | −26 | 3 | 23 | 29 | 34 |
| Dec. | −4.9 | 12.7 | 14 | −29 | 3 | 11 | 25 | 20 |
| Annual | 6.9 | 14.5 | 38 | −35 | 4 | 346 | 90 | 234 |
| Rec.(yrs.) | 30 | 30 | 17 | 17 | 10 | 30 | 17 | 17 |

| Month | Number of days with | | | Mean cloud-iness (tenths) | Mean sun-shine (h) | Wind | | Solar radiation (ly/month) |
|---|---|---|---|---|---|---|---|---|
| | precip. (>0.25 mm) | thunder-storm | fog | | | most freq. direct. | mean speed (m/sec) | |
| Jan. | 5 | 0 | * | 5.7 | 68 | SW | 3 | 230 |
| Feb. | 6 | 0 | * | 5.9 | 69 | SW | 3 | 320 |
| Mar. | 8 | * | * | 5.9 | 73 | SW | 3 | 450 |
| Apr. | 8 | 1 | * | 5.7 | 70 | SW | 4 | 540 |
| May | 9 | 5 | 0 | 6.2 | 66 | SW | 3 | 581 |
| June | 6 | 7 | * | 4.8 | 74 | SW | 3 | 670 |
| July | 6 | 9 | 0 | 4.0 | 77 | SW | 3 | 641 |
| Aug. | 4 | 7 | 0 | 4.3 | 75 | SW | 3 | 573 |
| Sept. | 5 | 4 | 0 | 4.1 | 73 | SW | 3 | 472 |
| Oct. | 4 | * | 0 | 4.5 | 73 | SW | 3 | 353 |
| Nov. | 5 | 0 | 1 | 5.6 | 63 | SW | 3 | 237 |
| Dec. | 4 | 0 | 1 | 5.4 | 68 | SW | 2 | 196 |
| Annual | 69 | 33 | 2 | 5.2 | 71 | SW | 3 | 439 |
| Rec.(yrs.) | 17 | 17 | 17 | 17 | 17 | 17 | 17 | 11 |

*Chapter 4*

# The Climate of Mexico

PEDRO A. MOSIÑO ALEMÁN AND ENRIQUETA GARCÍA

## Introduction

Although the atmosphere extends to great heights, its lower layers, especially if unstirred as they usually are in mountainous, tropical countries, influence greatly the environment of plants and animals that live in them. And at times those layers take a decisive part in the weather processes that occur aloft, even if only as a result of their relative sluggishness. It is because of this capacity of land-locked air to take part through braking action in the momentum exchange with swifter upper air currents, that the relatively small irregularities of the planet Earth, i.e., the mountains, acquire enormous importance. Notwithstanding their short projection into the atmosphere, they may alter the distribution of air properties aloft.

Thus, whereas in middle latitudes high-momentum, high-altitude atmospheric currents steer the lower level disturbances, in Mexico slow-moving low-level airstreams blocked by extensive mountain ranges and complexes greatly affect precipitation distribution, temperature and low-level wind directions. These are likely to affect the distribution of plant and animal resources that make possible the existence of man in the tropics.

The Mexican Republic, extending between 14°30′ and 32°42′N, is a long strip of land crossed by large mountain ranges with extensive plateaus in between, resulting in an average elevation of nearly 1,800 m above sea level for the central part of the country.

Only a thorough understanding of the great influence that the orography of Mexico exerts on air temperature, precipitation distribution and even air circulation, can explain the diversity of climates found in this country. Other factors, of course, modify locally this generalized picture to a greater or smaller extent. As a rule, the climate of Mexico is thought of as either hot, humid tropical or hot, dry desert. This is an incomplete and contradictory picture which springs from consideration of its latitude only, and the modifying influence of altitude is frequently ignored.

In what follows, therefore, we acknowledge the presence of those great currents, largely in the form of jet streams, that often visit our latitudes, accounting for the weather changes that occur over the high plateau in the colder part of the year. But this does not contradict at all the thesis of a pervasive orographic effect, but rather complements it. The great altitude of the Plateau determines that all the atmosphere below its average altitude will be under the control of orography, through damming or deflection of the lower air currents, while the top of the plateau suffers the vagaries of the upper air circulation. (See Chapter 1.)

Thus, over the Plateau, particularly during the colder part of the year, weather phenomena are determined by the physical characteristics of the extensive upper air currents and their planetary scale disturbances that move at great heights within the upper troposphere.

These currents are not able to avoid, however, the more subtle and less well understood effects exerted by the large continental masses and the thermal contrasts between these and the oceans. These finer effects are most clearly visible in the seasonal changes of speed and latitudinal position, as well as local direction, of the upper wind systems, which thereby acquire a different rhythm from that of atmospheric motions at lower levels.

Accordingly, the latitudinal position of the large upper air currents are responsible for the duration and intensity of seasonal phenomena, like the rainy season and the degree of dryness of the cold season over the Mexican Plateau; whereas more transitory circulations have their seat in the lower levels of the atmosphere, and are hence strongly influenced by orography.

In the case of Mexico, whose southern half lies within the tropics and the northern half in the subtropics, it is necessary to visualize the Mexican Plateau as a tropical highland subject to both wind systems above and to lower air currents on its flanks. In the following account both tropical and middle latitude meteorology will be used in treating characteristics of upper wind systems.

The ruggedness of the Mexican land areas, together with other geographic controls, determine the major weather regions within which many climatic varieties are found, because of the predominance of one or the other of these controls. Representative climatic charts of the whole country are drawn, however, without paying too much attention to detail over rough terrain. Many more weather stations would be needed to depict climatic elements with the required detail in mountainous areas. Recourse has been made, therefore, to the regularity imposed by altitude on some elements to interpolate approximately their values over rough terrain, attention being paid to orographic influences.

Most of the climatological data come from two types of meteorological stations: (*1*) "estaciones climatológicas", whose number is over 1,800, where temperature and rainfall measurements are made, and at a few of them, evaporation measurements; (*2*) "observatorios meteorológicos", where observations of temperature, precipitation, pressure and wind are made together with cloud weather types and supplemented with sunshine measurements. They are about 100 in number and correspond roughly with the standard synoptic stations; most of these, and a large number of the climatological stations, belong to the Dirección de Geografía y Meteorología. The rest of them belong to the Secretaría de Recursos Hidráulicos, la Comisión Federal de Electricidad and to Civil Aviation. In the last group, hourly observations of temperature, dew point, wind direction and speed are made, supplemented by cloud-type determinations and other data of aviation interest.

The times of observation are variable: 08h00 and 20h00 in the network of the Secretaría de Recursos Hidráulicos; 07h00, 15h00 and 19h00 in that of the Dirección de Geografía y Meteorología; nevertheless, the average values of those observations are used in the compilation of both monthly and annual means.

In general, unless otherwise specified, the charts have been drawn with data for the period 1931–1960; this is particularly true for the tabular data of the 14 observatories included (pp.391–404). However, data from a few climatological stations with less than 30 years have been used in areas lacking enough observations, this being the case in the unpopulated mountain areas of northern Mexico.

Publications of climatic data for the period concerned used to be very irregular in this country. The *Boletín Anual del Servicio Meteorológico* appeared until 1943 when it was discontinued. It was not until recently (1960) that its publication was resumed. The authors had to compile and average anew most of the data taken from original weather records at the Central Meteorological Observatory in Tacubaya and the Secretaría de Recursos Hidráulicos. The lack of publications of climatic data in a summarized form has made it impossible to compile data for other common elements like thunderstorm days, fog days and cloudiness not discussed here for this reason.

However, this does not mean that no climatic studies were published at all within the period 1931–1960. There are some works which deserve to be mentioned, such as the *Atlas Climatológico de Mexico* (3rd ed., 1939), which contains charts that were revised and brought up to date by Vivó and Gómez (1946) in their *Climatología de Mexico*. They include some upper air data and a classification of climates of Mexico to Köppen's scheme. There was also *Mapa de las Provincias Climatologicas de la Republica Mexicana* by Contreras Arias (1942) in which use was made of Thornthwaite's 1931 climatic classification to describe the climates of this country. However, no interpretations of weather phenomena are given in the first two works mentioned, though the last one contains a brief account of surface weather. In the present monograph, a special effort has been made to explain climatic facts in the light of modern meteorological doctrines. Only a dynamic interpretation of the weather phenomena can lead to a thorough understanding of the factors that make the climate of a country like Mexico, where orographic complexity introduces so many difficulties into its study and description.

## The permanent controls

Attention is given in this section to the geographic and, therefore, fixed factors that constitute the stage on which more variable factors like the semipermanent features of air circulation and the characteristic weather disturbances can act.

The geographic controls, although important in determining the distribution and magnitude of weather phenomena, cannot of themselves produce the climatic characteristics observed. One wonders what the climate of the Mexican Plateau would be, for instance, if Mexico as a whole were located outside of the tropics. One immediately realizes that the characteristic weather phenomena would certainly be quite different, if only because the Mexican highlands would be in that case under the influence of different wind belts.

### Latitude

As is well known, the effect of latitude is felt strongly in countries far from the equator, as it determines the inclination or obliquity of the sun's rays, regulating both the intensity and the potential daily duration of sunshine at a locality.

In the southern half of Mexico, which astronomically speaking is located within the tropics, this effect is less important, as the sun passes directly overhead twice a year, and never goes farther than 47° from the vertical at noon. The result is a fairly even distribution of insolation throughout the year. It is only in the northern part of the country where the variations mentioned above are slightly felt. In Mexico as a whole the potential duration of sunshine varies little with latitude—somewhat more than one hour between the southernmost point (Chiapas) and the Sonoran desert to the northwest.

The influence of latitude, however, shows up more clearly in the annual temperature curve. To the north of the Tropic of Cancer the annual curve shows only one peak in summer, with a minimum in winter; while to the south of it, two maxima are often found, corresponding as a rule to the dual passages of the sun overhead, taking into account the temperature lag with respect to insolation.

The latitudinal effect manifests itself, also, as in other regions of the world, in the geographical location over Mexico of the great atmospheric currents that occupy different positions during the year, although they are well defined during a given season, in particular at lower latitude localities (tropical and subtropical).

Thus, the southern half of Mexico lies in the Northern Hemisphere trade winds, whose steadiness and regularity both in direction and speed are well known. During the period between May and October, the Intertropical Convergence Zone exerts some influence in the genesis of unstable weather along the southern Pacific coast of Mexico.

On the other hand, the northern half of the country, which on the average lies within the subtropical high pressure belt, is far enough to the north to be visited by the westerlies during the colder half of the year when the belt moves southward. Westerly winds often blow over the entire Mexican Plateau during the winter; but their presence in this wide area is generally due to its altitude above sea level rather than its latitudinal position.

**Orography**

The importance of altitude in determining the climate of Mexico can be realized from the following figures: more than 50% of the Mexican land areas is found above the 1,000 m contour line; the coastal plains, including the Yucatán Peninsula with an elevation below 500 m, comprise only 32% of the total area of the country.

It is necessary, then, for climatic studies like the present one to begin with a description of the main orographic systems that form the abrupt relief of the country.

The outstanding feature of Mexico's orography is its huge Plateau, with an average elevation over 1,500 m above mean sea level. It rises southwards from levels below this average at the Mexico–United States border as far as latitude 19°N, where it attains its maximum altitude at the Mesa de Anahuac (more than 2,200 m; Fig.1).

On the western edge of the Plateau high terrain forms a wide and complex cordillera called Sierra Madre Occidental. This is roughly oriented from north-northwest to south-southeast, and drops by successive steps towards the coastal plains on the Pacific side.

On the eastern edge of the Plateau the ground descends precipitously towards the coastal plains of the Gulf of Mexico, there being no continuous border ridge as on the western side; nevertheless, some authors call this region Sierra Madre Oriental (see p.10.).

The southern edge of the Plateau is marked by a belt of volcanoes that extends roughly from west to east between latitude 19° and 20°N, called Sierra Volcánica Transversal or Eje Volcánico, where many snowcapped mountains rise to great heights.

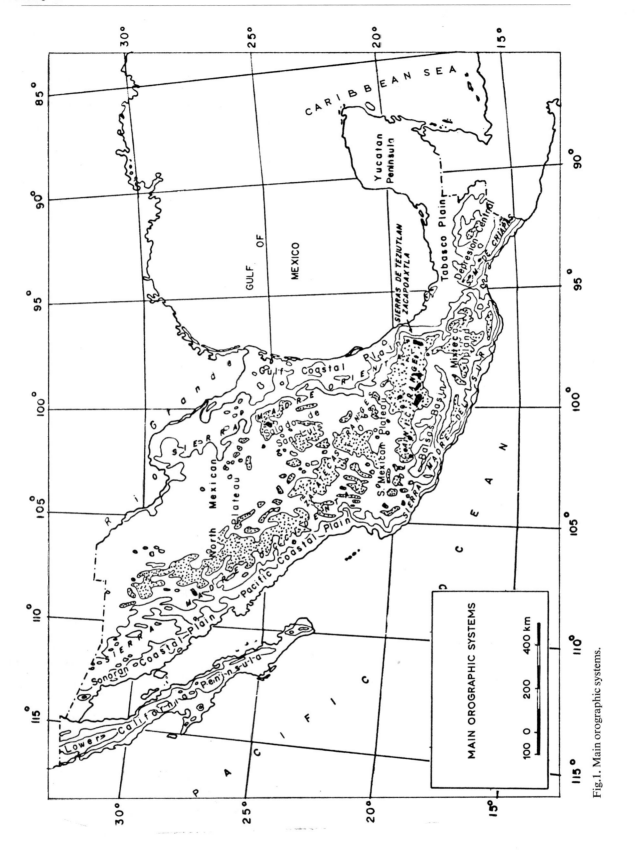

Fig.1. Main orographic systems.

The Mexican Plateau is traditionally divided into two parts. The northwestern portion, called "Mesa del Norte", is a series of basins into which some seasonal endorreic streams from the Sierra Madre Occidental drain; and the southern portion, called "Mesa Central", which is higher than the former, is crossed diagonally from northwest to southeast by a system of ridges that constitutes the Sierras de Zacatecas and Guanajuato. These sierras divide the Mesa Central into two triangular portions: one sloping gently to the northeast, up to the edge of the Plateau (Sierra Madre Oriental), and the other towards the southwest—limited on its southern side by the Sierra Volcánica Transversal, and on its western flank by the Sierra Madre Occidental, which at this latitude is somewhat lower.

On the Sierra Volcánica Transversal there are many small lacustrine basins separated by volcanic ridges; those basins located on the eastern portion are grouped to constitute the Mesa de Anáhuac with an average elevation of about 2,200 m.

Between the Sierra Volcánica Transversal and the Sierra Madre del Sur, which borders the Pacific coast from west to east south of parallel 19°N, extends a wide but rugged basin containing the Balsas River system. This river forces its way through the wall of the Sierra Madre del Sur near 102°W.

Southeastward of the Balsas Basin, between the Sierra Madre del Sur and the mountain chains that border the Gulf of Mexico, there exists a mountain complex called the Mixteca upland with an average elevation comparable to that of the Mesa de Anáhuac, but without the high peaks which dot the latter.

To the east of the Mixteca upland lies the Tehuantepec Pass which connects the southern coastal plains of the Gulf of Mexico with the Gulf of Tehuantepec.

Beyond the Tehuantepec Isthmus, the Chiapas mountain complex, which is an extension towards the northwest of the Guatemala uplands, shows several ridges roughly parallel to the Pacific coast, with a very narrow coastal plain on this side. Towards the Gulf of Mexico, it lowers to a wide plain by successive steps.

The Yucatán Peninsula to the northeast of the Tehuantepec Isthmus is an almost flat platform with small elevation above sea level.

On the northwestern part of Mexico the Baja California Peninsula has a long but narrow mountain chain that parallels the eastern shoreline, leaving wide coastal plains between the foot of the mountains and the irregular western coast of the peninsula.

The continental divide in Mexico follows the crest of the Sierra Madre Occidental on the western edge of the Plateau down to the 24°N parallel, turns southeastward along the Sierras de Zacatecas and Guanajuato; it then swings up to the eastern edge of the Mexican Plateau, down across the Mixteca upland to the Tehuantepec Isthmus, and beyond along the Sierra Madre de Chiapas.

*Orographic effects*

The most significant effects of orography on the atmosphere in Mexico can be summarized as follows: (*1*) damming effects; (*2*) deflecting effects; (*3*) blocking effects; (*4*) altitude above sea level; (*5*) forced ascent; (*6*) adiabatic heating by descent.

It goes without saying that these effects are not independent from one another, but the above classification facilitates their discussion.

(*1*) The damming of stable air occurs wherever there are high and extensive ridges or cordilleras, which if they are long enough, are able to contribute to the development of a steep pressure gradient between the windward and the leeward of the mountain range. The most outstanding example in Mexico is found on the eastern slopes of the Sierra Madre Oriental where polar continental air coming from the United States and beyond during the winter is detained by that range, stopping its passage towards the interior of the Mexican land areas. Often the continental air masses are too shallow to overcome the average altitude of that mountain barrier, whose passes are at about 1,500 m. Something similar occurs in the portion of the southern Gulf of Mexico from the Tehuantepec Pass to the Guatemalan border, where the high mountains of Chiapas rise as barriers that are overcome by polar continental air only once in a while. The same is observed along the western portion of the longitudinal ridge in the Baja California Peninsula and along the Sierra Madre Occidental, where stable air masses from the Pacific are hindered from entering the Mexican Plateau.

(2) Deflecting effects. One consequence of the damming of stable air by an extensive mountain range is the deflection of the winds upon meeting such a barrier, as they escape the control exerted by the Coriolis force, and are left at the mercy of the strong pressure gradients that the mountain range itself helps to produce through damming. Thus, for instance, the trade winds that blow from the northeast at the surface over the Gulf of Mexico, upon meeting the mountain range which borders the Plateau, are unable to cross over it and are instead deflected southwards through the Tehuantepec Pass where they become northerly winds of unusual violence. The escape of this air northward along the coastal plains of the Gulf of Mexico is usually prevented by the high pressures normally present at higher latitudes.

The effect is felt as far north as Tampico on the coast of the Gulf of Mexico, accounting for the presence of strong northerly winds in Veracruz, even when there is not a low pressure area at all in the southern Gulf of Mexico, especially if there is an anticyclone to the north, a frequent case during the winter.

(3) Blocking effects are those where orographic obstacles interfere with the motion of atmospheric disturbances. A remarkable instance of this is the blocking action exerted by the mountain massifs of central Mexico and Guatemala, which often interfere with the motion of easterly depressions from the Caribbean Sea. Upon meeting these obstacles, the depressions frequently split into two portions, which later on take different paths.

(4) Altitude above sea level plays a decisive role in the distribution of climates in lower latitudes. Crossed as it is by long and abrupt mountain ranges, in many instances rising to very high altitudes, Mexico shows temperature stories or thermal levels, which succeed each other from sea level up to the highest mountain peaks, where the ambient temperature always remains below freezing. This is the source of the familiar nomenclature used by Mexicans, who refer to these temperature stories as the *tierra caliente* (from sea level to 1,500 m), *tierra templada* (from 1,500 to 3,000 m), and the *tierra fría* (from 3,000 m upwards). Thus isotherms in southern Mexico follow contour lines very closely, showing clearly the influence of altitude on this element.

But even if this thermal effect were not enough to qualify it as a very important one in the genesis of the different climates of the Mexican Republic, the effect of altitude on air humidity and, therefore, on rainfall distribution, would be sufficient to do so. Admittedly, the oceans are the main sources of water vapor, which they lose in great quantities by evaporation, the winds carrying it to great distances and heights by turbulent diffusion. However, water vapor, like temperature, decreases rapidly with altitude so that if the moisture-laden winds meet a steep mountain barrier, high enough and sufficiently long, they are unable in general to overcome it and the supply of moisture inland is cut off, allowing only the reduced water vapor content of the upper air winds to pass over the mountain ridges.

This is mostly the case in the central and southern plateau area of Mexico, especially during the winter, since a temperature inversion often exists between the lower, moist

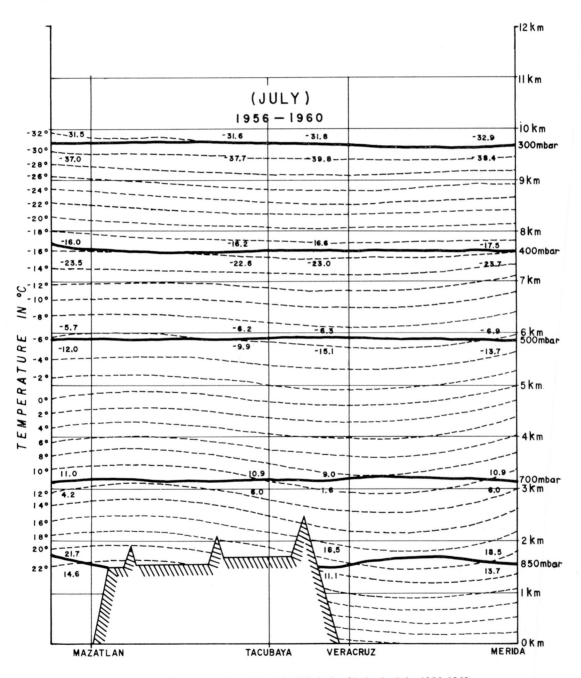

Fig.2. Cross-section of the atmosphere along the 20° circle of latitude, July, 1956–1960.

marine layer and the dry upper air at this time of the year. During the summer, however, upward turbulent diffusion is aided by suitable air instability, so that a less steep hygrolapse sets in, causing the incoming air from the adjacent oceanic areas to produce heavy rains inland over the plateau. An exception to this is found only in the extreme northwestern portion of Mexico where stability is greatest at this season.

Finally, a marked thermal effect exerted by the extensive Mexican Plateau makes itself felt on account of its considerable altitude. We shall refer to this as the elevated-heat-source-effect. It operates as follows: upper air soundings made at coastal stations and surface observations made over the Plateau show that the surface temperature over the latter is often higher than the free air temperature at the same level over the Gulf of Mexico. This means that air masses carried by the wind undergo heating from below in moving from high above the sea surface towards elevated terrain, having suffered hardly any mechanical lifting in passing from one place to the other. Thus, if the hygrolapse is adequate, the potential instability of these air masses is released over central Mexico, causing heavy rains inland, quite independently of the immediate influence of ascent of moist air along the slopes of the cordilleras bordering the Mexican Plateau (Fig.2). This elevated heat source effect accounts, in this manner, for the character of the convective activity frequently found over the Mesa Central, where thunderstorms often occur during the rainy season.

(5) Forced ascent. Moist unstable air from the ocean carried inland by a wind perpendicular to a mountain range is forced to ascend, cooling itself adiabatically. Eventually it is led to release its moisture in the form of heavy rains over the mountain slopes. This is generally the case in mountain ranges extending across a deep current of air, like the Sierra Madre Oriental which is oriented at right angles to the surface trades that blow over the Gulf of Mexico.

(6) Adiabatic heating by descent. The above-mentioned situation gives rise, further, to a foehn effect on the lee side of the mountains due to heating of the air by adiabatic compression following descent behind the mountain range, causing, in turn, a decrease of its relative humidity.

These physical changes in air that is able to cross orographic barriers explain the strong contrasts in rainfall found over the Mexican Republic, where we have for instance, Teapa, in Tabasco state, with 3,900 mm of rainfall a year, against 818 mm for the same period in Cintalapa, Chiapas state, these two places being only 100 km apart.

**Distribution of land and water**

The latitudinally elongated shape of the Mexican land areas, which narrow like a horn southward and continue through the isthmus of Tehuantepec, gives to southern Mexico its tropical maritime thermal regime, and is largely responsible for the steadiness of weather and mildness of climate of the Central Plateau.

In fact, notwithstanding the modifying influence of altitude upon moisture content, the narrower part of Mexico, because of its nearness to the sea, enjoys, at least over the warmer part of the year, the thermostatic effect of the oceans on both sides. On the Atlantic side, shallow, relatively warm seas, like the Gulf of Mexico and the Caribbean Sea, in which an unimpeded air circulation causes intense evaporation from the ocean surface, constitute a good source of moisture and heat. These play an important role in the weather and climate of Mexico. The moist air carried inland by the trades is the raw material for the abundant rainfall of the southern part of the country.

On the Pacific side, the presence of the peninsula of Baja California introduces great complications in the climatic picture, the factor presently under discussion becoming more conspicuous here. First, it is worthwhile to point out the low efficiency of the moderating influence of the shallow body of water which forms the northern half of the Gulf of California, practically confined by the Angel de la Guarda and Tiburón islands. Here temperature oscillations from summer to winter and from day to night are at their greatest; while in contrast close to this region, on the western slopes of the backbone ranges of Baja California, the annual temperature variation becomes considerably smaller. This contrast is clearly related to the presence of the cold California current along the western coastline of the peninsula, and to the seclusion of the inner waters of the Gulf of Cortés by the latter, which extends southward in such a way as to impede the entrance of cold water from the California current.

It is only here of all Mexican waters where the effect of a marine current manifests itself, this being the reason why more attention has not been given to such currents in the present work. However, we believe that in the whole northwestern part of Mexico no other single factor except orography is more significant than this one.

### The characteristic atmospheric phenomena

#### General

As mentioned previously, Mexico is both a tropical and a subtropical country so far as latitude is concerned. However, its complex orography and its altitude above sea level cause it to share the characteristic weather phenomena of low latitude countries with those of the middle latitudes of the world.

Thus, although its southern half is within the trade wind belt and its northern part is located in the high pressure belt of the Northern Hemisphere, the great altitude of its land areas makes it liable to the disturbances of the upper westerlies, which, at least during the colder half of the year, dominate the weather processes and are even evident at ground level over the Plateau. During the summer, as the trade current builds up into higher levels and moves toward higher latitudes, the entire country comes under the influence of the deep and wide easterly current which, coming from the Caribbean Sea and beyond, penetrates the Gulf of Mexico with generally light but moist winds.

The atmosphere over this country does not undergo large weather changes, in comparison with those observed at higher latitudes, except when a tropical cyclone moves into the Gulf of Mexico from the Caribbean or along the southern Pacific coast (especially from August through November).

#### The westerlies

The high pressure belt located on the average near 35°N, along which the oceanic anticyclones of both the North Pacific and the North Atlantic oceans are found, as well as the great deserts of the Northern Hemisphere, extends across northern Mexico during the summer. To the south of this belt the equatorial margins of the oceanic anticyclones give way to a wide and deep easterly current of air, i.e., the trades.

During most of winter and early spring, when the high pressure belt moves southward, northwestern Mexico is under the influence of the westerly current of middle latitudes, causing winds to blow with some force at the surface over the lower California Peninsula. These are the classical westerlies, which manifest themselves over the subtropics and even over the tropics at great altitudes. Not infrequently these high-level winds contain a jet of fast moving air aloft, and then ground-level winds blow with great force across the Sierra Madre Occidental, and even over the barren highlands of the Mesa del Norte, where they lift dust storms and cause an unusual chill factor.

Often, still during the colder part of the year, the upper westerlies are found as far as southern Mexico, including Mexico City and the city of Oaxaca. Here altitude plays an important role as the high terrain penetrates the base of the westerly current aloft.

When the westerlies prevail at the ground over the Mexican Plateau, the disturbances proper to that basic current continuously affect it. Upper troughs travel within that current and bring to the elevated Plateau surface the windshifts and other phenomena typical of such troughs.

On the west coast and west of the continental divide, subtropical cold vortices aloft moving in slowly from the Pacific are common during late winter and early spring. They are generally associated with a small moist tongue ahead of the vortex center, which often causes heavy rains and even snowfall at high places over the Sierra Madre Occidental and over the Mesa del Norte.

**The trades**

During the summer season, as the high pressure belt of the Northern Hemisphere moves northward, the deep trades prevail all over Mexico, blowing along the southern margin of the Bermuda anticyclone. These winds cause a continuous feeding of the Mexican land area with moist tropical air. The isobaric pattern over most of the eastern portion of the country is shown in Fig.3, where the direction of the isobars suggests the existence of air trajectories along which warm moist tropical air travels into Mexico at mid-tropospheric levels.

The invasion of tropical air usually takes the shape of a tongue of moist air following shortly behind a heat wave, which moves from southeast to northwest along the Mexican Plateau. The heat wave splits into two wave fronts upon reaching the Mexico–U.S. border. One of these leads westward toward the Gulf of California, and the other toward the northwestern part of the Gulf of Mexico. The moist tongue is believed to contribute to the general pattern of heat invasion, as it causes the decrease in temperature following the maximum. It appears that this decrease in air temperature is due to cooling of the air by evaporation of raindrops, falling in connection with widespread rains associated with convective activity within the moist tongue.

Along the western coast of Mexico a stationary pressure trough prevails most of the time aloft, clearly connected with the surface heat low over the dry areas of southwestern United States. During the summer months this upper air trough moves off westward, away from the coast, while pressure over the Gulf of Mexico rises with a concurrent retrograde motion of the Bermuda anticyclone. This is not an unusual event over the

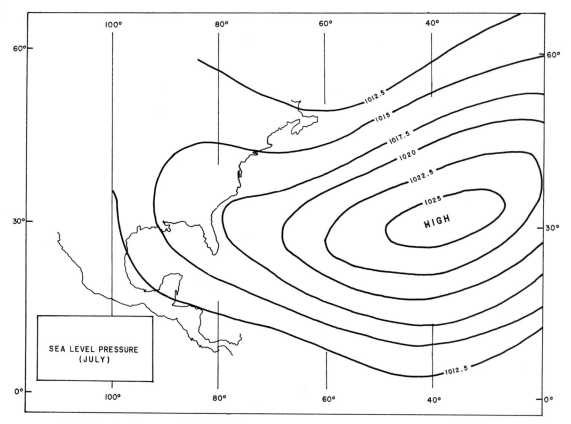

Fig.3. Sea level pressure (mbar) in July.

Mexican land areas and is responsible for most of the rainy types of easterly flow characteristic of the warmer half of the year, as shown by MosiÑo (1964a,b). In this connection it is worth noting that regular rainy periods in central Mexico are very often associated with positive 24-h pressure changes, whereas negative ones, especially if they affect large areas of the surface chart, are accompanied by excellent weather along the coastal plains of the Gulf of Mexico.

The only exception to this rule takes place when a deep tropical storm is approaching the coast, and then moves inland, for example, near or around the Yucatán Peninsula. This makes 24-h pressure changes hard to use without a weather chart at hand and a thorough knowledge of Mexican meteorology. However, 24-h pressure changes can be a very useful tool if proper consideration is given to the pressure field over the wide areas of nearby seas in particular during the winter.

As is well known, the surface trades blow into the Intertropical Convergence Zone, which in some cases reaches as far north as southern Mexico. This is not evidenced by instrumental data which are mostly lacking here, but the approach of the equatorial trough of low pressure to the southern portion of the country may account for most of the heavy rains experienced in that area every year. However, tropical storms that occur during late summer in and around the Gulf of Tehuantepec are responsible for some of the rain in the area.

### Moisture surges and easterly waves

It is well known that minor wave disturbances, called easterly waves (RIEHL, 1945), travel in the trade current over the Caribbean Sea. However, these are not regular features of the windfield over the western Gulf of Mexico or over the Mexican Plateau. Apparently, the easterly current experiences at low levels the blocking effect of the great mountain masses of southern Mexico long before it enters the western Gulf. The lower trade current divides itself into two branches. One branch finds its way out through the Tehuantepec Pass towards the south, while the other extends northward into the U.S. along the eastern slopes of the Sierra Madre Oriental. This low-level flow pattern is crowned by deep, straight easterlies above the general level of the Plateau.

Thus, easterly waves traveling over the Yucatán Peninsula are overtaken by the windfield which tends to disintegrate their structure, leaving only at high levels (above 2,000 m) a moisture surge that is carried by the winds against the mountains of the eastern edge of the Plateau. This splitting is also the fate of wave-like disturbances traveling at lower latitudes, between parallels 14° and 16°N, which meet the mountain masses of the Chiapas state in Mexico and Guatemala in central America. These are the "temporales" of the isthmian part of America.

It could be argued that cloudiness and precipitation during the summer season over most of Mexico occur in series of three or four consecutive rainy days intermingled with drier periods, and that this could be indicative of weather systems organized through wave motion. But if we consider easterly waves as features of the windfield characteristic of the lower levels of the atmosphere in the tropics, with their maximum windshifts at about 1,500 m, as originally described by RIEHL (1945), no such shifts could be observed over the southern Plateau, whose altitude is over 2,000 m. Moreover, pressure changes like those associated with the Caribbean easterly waves are hardly detectable over Mexico.

Thus, instead of a line of minimum pressure across the low-level easterlies, a slight pressure jump is observed in advance of the moisture surge associated with the beginning of each series of rainy days over central Mexico.

### "Northers" of the Gulf of Mexico

During the winter half of the year, invasions of cool, modified continental polar air masses enter the Gulf of Mexico from the north, in the shape of a shallow wedge of dense air with northeastern winds at intermediate levels. This air, unable in most cases to climb the steep slopes of the mountains that border the Mexican Plateau, is instead deflected southwards into the southern Gulf of Mexico. Here it finds a low mountain pass through which it can slip into the Gulf of Tehuantepec, where it blows as a dry, gusty wind.

All over the western Gulf of Mexico these outbreaks of polar air cause very strong winds which are known as "northers" in southern United States and as "el norte" in Mexico, variously described by GUZMÁN (1903), LÓPEZ (1922b) and DOMÍNGUEZ (1950).

The polar air mass moves regularly southwards behind a line which in many instances is only a shear line, and less frequently takes the form of a true front. The polar air often picks up moisture from the relatively warm waters of the Gulf of Mexico and produces heavy rains over the eastern slopes of the Sierra Madre Oriental, as well as over the seaward slopes of the mountains in the Mexican states of Oaxaca and Chiapas. The invasion of relatively cold, dense air over the Gulf of Mexico is clearly associated with an anticyclone centered somewhere in southern United States or, in some cases, over the western plains. The approach of this anticyclonic area causes strong rises of pressure, so that Mexican meteorologists have learned to associate windy and cloudy weather over the Gulf of Mexico with unusual pressure rises. Apparently, horizontal convergence of air motion caused by the damming of the wedge of dense and moist air against the Sierra Madre Oriental overcompensates the subsidence called for by the anticyclonically curved isobars behind the cold front.

These outbursts of bad weather, caused by polar outbreaks in the Gulf of Mexico, are usually preceded by a general lowering of barometric pressure over most of the area, often including the Atlantic plains of Central America, up to where the influence of an elongated trough of low pressure reaches. This trough is commonly found ahead of the cold front along a line that extends southward from the advancing wedge of continental polar air. Concurrently, a general subsidence within the tropical maritime air takes place in advance of the frontal line as a horizontally divergent windfield within the trough sets in. This causes shallow temperature inversions over most of the southern Gulf of Mexico beneath which dense radiation fogs form at night in places near the coast. Southerly winds often prevail at intermediate levels over the area, whereas strong southwest or west winds are often observed at the ground over the plateau in advance of such polar invasions.

Here the rule that relates negative 24-h pressure changes to fine weather over all Mexico east of the continental divide is at its best either at sea-level or over the Plateau.

Positive pressure changes are related to bad weather over the area covered by modified continental polar air; the only exception again is the formation of a young wave cyclone in the middle of the Gulf of Mexico. This frontal wave is usually present only with very low values of the zonal index over North America. Indeed, under low index conditions there are times when unexpected retrograde motions of extratropical lows are observed along the eastern seaboard of the United States, reflecting themselves upon the kinematics of air motion at lower latitudes.

**The polar trough in Mexico**

It is also during low index conditions of zonal flow over North America when deep, cold air masses moving from the Great Basin may enter the country behind an upper air trough, invading the northern Plateau area and the Gulf of Mexico.

On this occasion, the passage of the upper air trough over the Plateau causes light rains and snowfalls at high places ahead of the trough line, whereas behind it a general clearing over high terrain is observed. Nevertheless, over the seaward slopes of the Sierra Madre Oriental a solid overcast and heavy rains are prevalent after the passage of the trough aloft, again with pressure rises at lower levels, due to horizontal convergence of air against the mountains.

On the other hand, high index conditions produce over the central United States only shallow air masses that travel southwards causing the regular "nortes" over the Gulf of Mexico, as mentioned above. Then small-amplitude, fast-moving upper air troughs cross northern Mexico from west to east. In these instances clear weather and strong westerly winds prevail over the plateau, as the bad weather effect of horizontal convergence of air is restricted to the coastal plains of the Gulf of Mexico, due to the shallowness of the continental polar air masses.

## Tropical storms

These are the well-known revolving storms characteristic of the tropics, of which hurricanes or tropical cyclones, as they are called in this country, constitute the more dangerous members. The season of these storms runs generally from late May through November, September having the greatest frequency. Most of the cyclones that appear in Mexican waters stay out at sea but some have been observed to cross the Plateau. Coastal regions are generally struck more frequently from September through November. Because of the presence of deep polar troughs over the Mexican Plateau, the last two months are those of highest frequency once the polar troughs induce early recurvature (see p.225).

The presence of a major upper-air trough over the Mexican Plateau seems to have a strong bearing upon the likelihood of an existing tropical storm entering either the Pacific coast or the Gulf coast. For instance, if the storm is found over the Gulf of Mexico and the upper westerlies extend down to the ground over the Plateau, late during the season the storm is likely to move in a northerly or northeasterly direction, missing the Gulf of Mexico. With the same conditions aloft, if the storm is on the west coast of Mexico, it is likely to recurve, hitting the coast. Otherwise, storm tracks are usually oriented from southeast to northwest along this coast.

Early summer tracks, however, follow a more westerly direction away from the coast on the Pacific side, and into and over the Yucatán Peninsula and the western coast of the Gulf of Mexico on the Atlantic side.

The speed of motion of tropical storms generally ranges from 2 to 8 m/sec. Winds in the more severe tropical storms are believed to have reached the 50 or 60 m/sec mark.

Tropical storms on the Pacific coast of Mexico are characteristically small in diameter and frequently they are crowned by an open wave aloft resembling an easterly wave. The extension of the trough line or crest of this wave into and over the Mexican Plateau causes heavy rains over this part of the country during the passage of the storms along the coast. This is the reason why in some quarters they are thought beneficial, as high winds are usually restricted to the coasts (CONTRERAS ARIAS, 1957).

Recent observations from meteorological satellites seem to indicate that a higher frequency of these storms reaches hurricane force than previously thought.

## Sea-breezes

On the Pacific side, deep sea-breezes, if unopposed under the sheltering action of the mountains, play the same role as a weather producing factor as do the trades on the Atlantic side. There, anabatic and katabatic winds from nearby mountain slopes aid the sea-breezes and land-breezes respectively in the production of convection, both over the mountain slopes during the day, and offshore over the sea at night. This combined effect of land and sea is very powerful, especially during the early rainy season when tropical revolving storms have not yet made their appearance. Also the effect of air cooling by sea-breezes helps to make the heat of the day more bearable for the population along the extended Pacific coastline. In fact, air circulation changes brought by the approach of a tropical disturbance along the coast, which blocks the onset of the daily sea-breeze, often make people complain of extreme heat. This "bochorno" constitutes a warning of the approach of bad weather along the coast.

### Storm tracks

Under this heading, both cyclonic and anticyclonic tracks are considered, because both types of pressure system are possible weather-producing features of the windfield in the hilly country.

Thus, anticyclonic centers, frequently associated with polar invasions with the "northers" of the Gulf of Mexico, come into lower latitudes along the well-known tracks studied by KLEIN (1957).

Among the cyclonic tracks those followed by tropical storms are the most interesting. As mentioned above, tropical revolving storms are regular features of the windfield over Mexican waters, both over the Gulf of Mexico and along the Pacific coast, from May through November. They travel embedded in the deflected easterly current which forms the equatorial margin of the Bermuda–Azores anticyclone.

The western end of this enormous high-pressure area extends well into Mexico during the hurricane season. Because of the peculiar shape of the Mexican land areas, most of the tracks of these storms are thus parallel to the coast, particularly those found over the eastern Pacific Ocean.

## Temperature

### General

The permanent factors that help to determine the climate of Mexico were discussed previously. Prominent among them is the altitude of the different regions of the country. Second to this, latitude, closely related to the insolation and the length of the day, plays an important role in the distribution of the types of annual range of temperature. The distribution of land and water masses controls in no slight measure the time of the year when annual maxima and minima of temperature occur. On the other hand, the incidence of cloud and precipitation also influences the time of occurrence of the annual temperature maxima.

As a first approximation and regarding only the effect of altitude on mean annual temperatures, we can think of the thermal stories mentioned on p.351 as being delimited by nearly horizontal mean isothermal surfaces within the lower troposphere. It is clear that mean annual isotherms, considered as intersection lines of the isothermal surface with high ground, will nearly coincide with contour lines on a topographic map. Bearing this in mind, and considering the departures from this general picture introduced by other factors, charts of mean isotherms are easily interpreted over most of Mexico since the effect of altitude in the temperature distribution predominates over other factors.

According to GARCÍA (1964), the Mexican land areas are divided by mean annual isotherms into the following thermal stories (Fig.4).

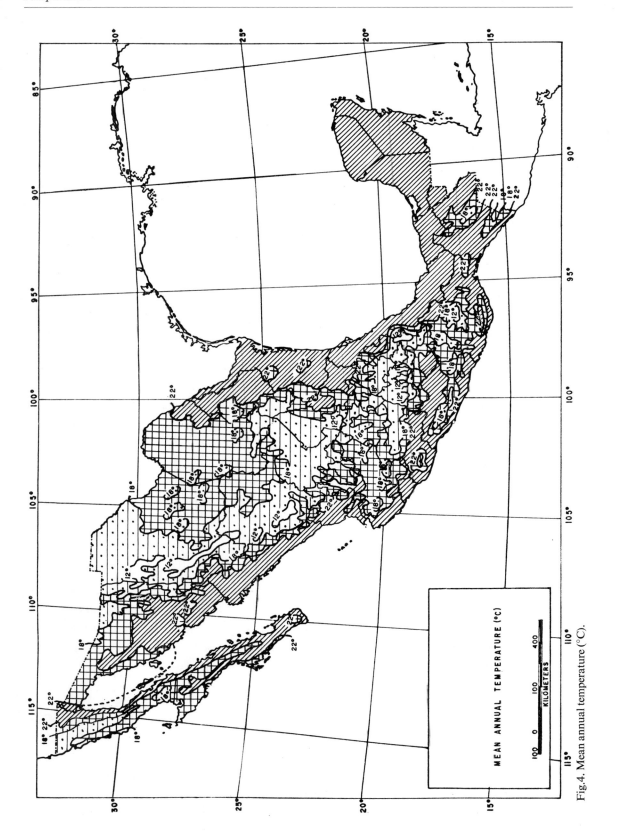

Fig. 4. Mean annual temperature (°C).

The *hot story*, which corresponds with the "tierra caliente" (HERNÁNDEZ, 1923), with a mean annual temperature over 22°C, comprises the coastal plains of both the Pacific and the Gulf of Mexico from sea level to 800 or 1,000 m of altitude in the south and from sea level to 100 or 200 m in the north, the Yucatán Peninsula, the Balsas River Basin, the central depression of Chiapas, as well as an inner strip along the western foot of the mountains of the Baja California Peninsula.

The *warm story*, between the hot and the temperate, with a mean annual temperature between 18° and 22°C, covers the eastern slopes of the Sierra Madre Oriental, the central uplands of Chiapas, the intermediate slopes of the Sierra Madre del Sur, the highest parts of the Balsas Basin, the westward slopes of the Sierra Madre Occidental and the coasts of the Baja California Peninsula.

The *temperate story*, with a mean annual temperature between 12° and 18°C, comprises a great part of the Sierra Madre Occidental, the Sierra Volcánica Transversal, the Mesa Central, the highest parts of the Sierra Madre del Sur, the Mixteca upland and the highest portions of the Chiapas mountains.

The *cool story*, between the temperate and the cold one, with mean annual temperatures between 5° and 12°C, covers part of the highest areas of the volcanic belt and part of the Sierra Madre Occidental.

The *cold and very cold stories*, with mean annual temperatures under 5°C, comprise only the highest snowcapped peaks of the Sierra Volcánica Transversal.

One important departure from the above picture is caused by latitude. The isotherms show a small but definite tendency to descend towards the north from the highlands to sea level, as if they were railroad lines with a very small grade, during the colder half of the year.

Departures from the above picture are noticeable over flat terrain, as in the intermountain areas that lie between the high sierras forming the edges of the Mexican Plateau, as well as over the coastal plains. In the former places the direction of the isotherms is controlled by other factors like radiative cooling which causes shallow temperature inversions in the valley bottoms during the coldest part of the year. Over the coastal plains the run of the isotherms is affected by advective cooling by cold air masses which, highly modified, are still able to influence the temperature over the unobstructed coastal plains on both Pacific and Gulf of Mexico flanks. Nevertheless, both radiation and advective cooling are responsible only in a minor way for the overall temperature distribution of the whole country.

### The distribution of mean annual temperature

Fig.5 shows the normal January isotherms. Temperatures higher than 25°C are found only along the Pacific coastal plain south of 19°N and over the lower parts of the Balsas Basin. Temperatures between 20° and 25°C are found on the Pacific coast south of 22°N over the mountain slopes that surround the Balsas Basin, on the coast of the Gulf of Mexico south of 20°N, and over the central depression of Chiapas. The Pacific coast is more than 5°C warmer than the plains of the Gulf of Mexico. Isotherms between 10° and 15°C cover all the Mexican Plateau, both western and eastern slopes of the Sierra Madre Occidental, the mountains of the Baja California Peninsula, some parts of the Chiapas complex and the Mixteca upland. Isotherms under 5°C are found over the northern

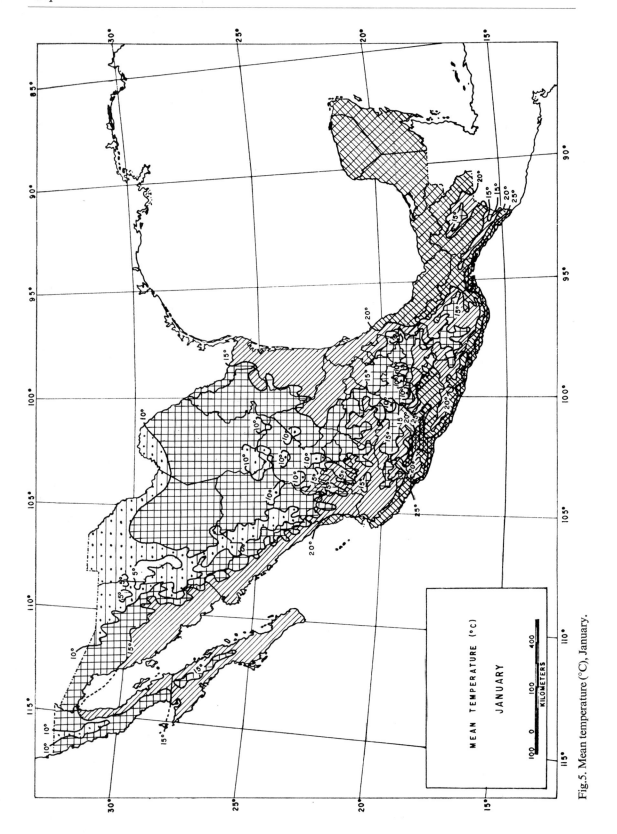

Fig.5. Mean temperature (°C), January.

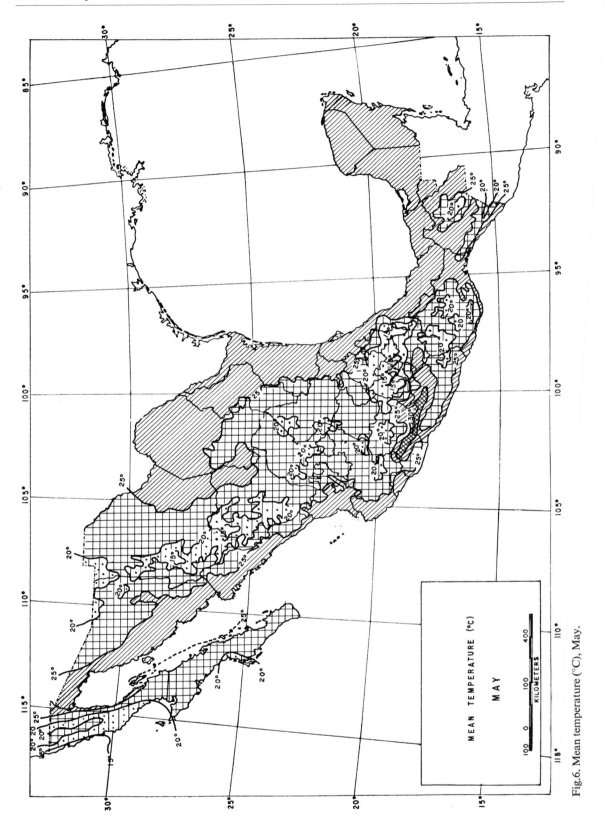

Fig.6. Mean temperature (°C), May.

portions of the Sierra Madre Occidental and over the highest ridges of the Sierra Volcánica Transversal. The coldest portions are the northwestern parts of the former near the U.S.–Mexico border.

During the warmer part of the year, an opposite tendency, namely for the isotherms to decline southward, is also detectable in some parts of the country. This is clearly due to the increase in heating of the air experienced in the northern portions of Mexico from winter to summer in comparison with the cooling of the air by evaporation of raindrops in the southern part of the country. Sea breezes are also able to alter the direction of mean isotherms near the coasts in summer.

On the other hand, the concentration of rainy periods within the summer months accounts for the most remarkable feature of the march of temperature in Mexico, i.e., a wide range in the time of occurrence of the temperature maxima from spring through summer and early fall. A map of isotherms for May is included in addition to that for July, because May is the warmest month over most of the central and southern areas of the country. This map (Fig.6) shows that the areas with temperatures above 25°C cover the northeastern part of the Mesa del Norte, the lower parts of the Mesa Central, the coastal plains of the Pacific as well as those of the Gulf of Mexico, including the Yucatán Peninsula, the inner coasts of the Baja California Peninsula, the Balsas River Basin and the central depression of Chiapas. The hottest part is the Balsas Basin where temperatures rise to over 30°C. However, temperatures between 15° and 20°C are found at the highest points of the Mesa Central, over the Mesa de Anáhuac, the central and northern parts of the Sierra Madre Occidental, the northwestern coasts of the Baja California Peninsula, the Mixteca upland, the Sierra Madre del Sur and the highest parts of the Chiapas complex. Isotherms under 10°C are almost lacking in Mexico at this time of the year.

The map for July (Fig.7) shows that the areas with temperatures above 25°C comprise the Mesa del Norte, the coastal plains of both the Pacific Ocean and the Gulf of Mexico, the inner coastal regions of the Baja California Peninsula, the Balsas Basin and the central depression of Chiapas. The warmest parts are the northernmost portions of the coastal plains of the Gulf of Mexico and the coasts of the Gulf of California north of 24°N where temperatures rise above 30°C. Temperatures between 15° and 20°C are found over the southernmost parts of the Plateau, at the highest points of the Zacatecas and Guanajuato ridges, the Sierra Madre Occidental, the Sierra Madre del Sur, and over the higher parts of the Chiapas complex. Temperatures under 10°C are found only at the highest points of the Sierra Volcánica Transversal.

**The mean annual range of temperature**

A classification of types of annual temperature variation over the country has been made through a consideration of annual ranges and shapes of the temperature curves, paying special attention to the time of occurrence of annual temperature maxima.

Annual ranges, although smaller in comparison with more continental areas of the world, are distributed in a rather regular fashion, with increasing magnitude from south to north over the Mexican land areas. In this description the entire area of Mexico has been subdivided into three zones or belts.

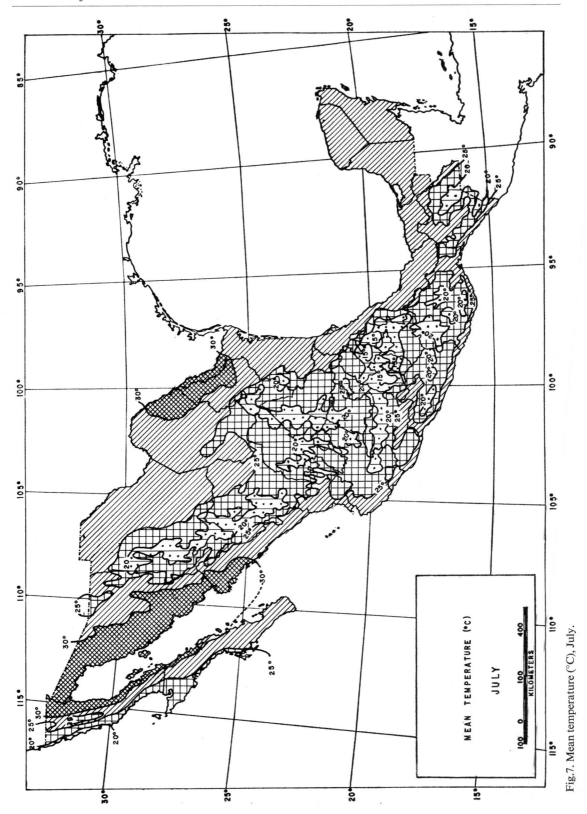

Fig.7. Mean temperature (°C), July.

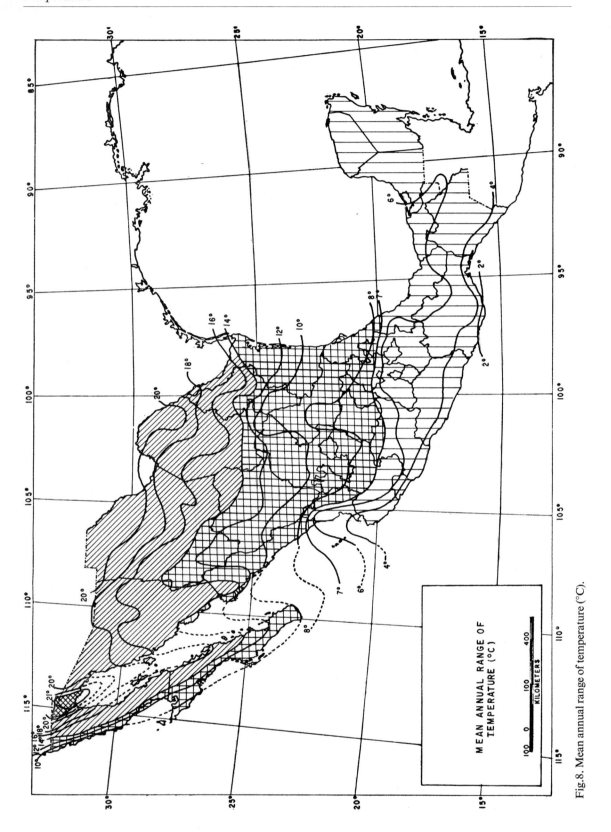

Fig.8. Mean annual range of temperature (°C).

(*1*) A nearly isothermal belt with an annual range under 7°C (GARCÍA et al., 1961), found roughly south of the northern reaches of the Sierra Volcánica Transversal.

(*2*) An intermediate belt with annual ranges between 7° and 14°C.

(*3*) Another belt with ranges higher than the latter value, located along the Mexico–U.S. border, where a mild winter season is experienced (mild by middle latitude standards, since for most southern Mexicans this area represents a region where seasonal contrasts in temperature have induced them to call it "extremosa", meaning "with marked temperature extremes") (Fig.8).

Annual ranges of temperature increase gradually towards the north, as would be expected, except near the Baja California Peninsula, across which annual ranges increase rapidly from the west coast to the head of the Gulf of California, where their values reach a maximum of 21°C in Mexicali, B.C. This shows again the tremendous contrast between the small air temperature variation over the cold waters of the California current and that over the enclosed waters of the upper Gulf of California.

The small ranges of annual temperature west of the backbone sierras of the California Peninsula are apparently due to upwelling of cold water by wind stress on the ocean, known to operate off the western coast of the continent, which gives to the coastal and nearby inland areas of Ensenada and Tijuana the maritime climate that characterizes this corner of the country.

In southern Mexico annual temperature ranges are very small. For instance, at the southernmost tip of the state of Oaxaca, temperature ranges attain a value of only 2°C.

**Times of annual temperature maxima**

Another feature of the annual temperature variation in Mexico is that the annual temperature maxima can occur in all months from April to September (LÓPEZ, 1922a; HERNÁNDEZ, 1923; PAGE, 1930).

From May to September a regular time sequence of arrival of temperature maxima can thus be followed across Mexico. It happens much as if the maximum propagates itself from southeast to northwest along the highlands, spreading out and finally splitting off into two branches before reaching the Mexico–U.S. border, one leading eastward and another westward towards the subtropical maritime areas of Mexico where August, even September, are the months of maximum temperature. In this regard the Yucatán Peninsula shows the characteristics of the southern plateau area, since its temperature maximum occurs mostly in May. Fig.9 shows clearly this sequence of events.

In contrast with most of the middle latitude areas of the Northern Hemisphere where either July or August are the warmest months, the temperature peak in southern Mexico arrives before summer solstice, whereas in some areas of the Baja California Peninsula it is delayed until September. The latter is probably due to the influence of the cold California current on air temperature. The water temperature of this current, due to the high thermal capacity of water, shows an annual wave that lags behind the air temperature curve over the continent. Thus, when air temperatures over land have already diminished during fall and the beginning of winter, the sea water temperatures still remain relatively high; on the other hand, during summer, when air temperatures have begun to increase inland, water temperatures are still fairly low and, in fact, reach a minimum during this season.

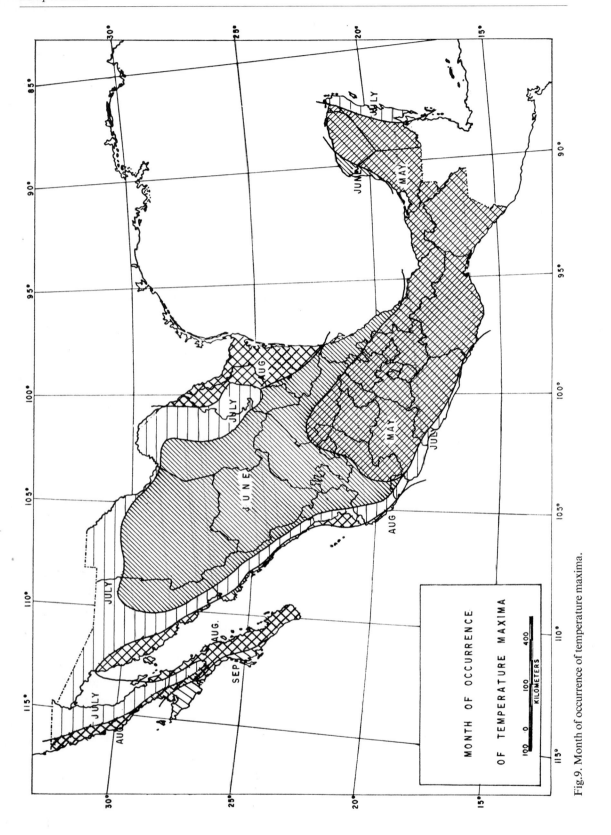

Fig.9. Month of occurrence of temperature maxima.

In northeastern Mexico an August peak is apparent, this being explained by peculiarities of the mid-summer air circulation and the geographic–temporal distribution of rain, to be described in its proper place (cf. section "The precipitation", p.373).

The early arrival of the annual temperature maximum in tropical Mexico has frequently been attributed to the motion of the insolation peak with the northward increase in the sun's declination (LÓPEZ, 1922a; PAGE, 1930). However, this would imply the existence of a second maximum late in the summer season, which should follow an inverse time sequence, as the sun's declination decreases again after the Northern Hemisphere's summer solstice. Apparently, this second temperature peak, which is nevertheless shown in some graphs of the annual march of temperature for certain stations in central Mexico, is impeded from reaching the same level as the first peak by cooling of the air. This is induced either by screening of solar radiation by clouds or, equally well, by evaporation of rain falling through the air.

Thus, in central Mexico the mean temperature curve begins to rise early in February or March; it attains a peak in May just before the beginning of the summer rainy period and falls towards a secondary minimum in July or early August, only to start again increasing towards a second peak in late August or even September. However, few stations recover their warmth after the secondary minimum; most of them remain cool, the temperature fall continuing, at a reduced rate at least, until the rainy season is over. The end of the rainy season is a definite turning point at which temperatures start decreasing rapidly towards their main annual minimum which, over most of the country, occurs generally in December or January.

The presence of the mid-summer moist air tongue mentioned earlier (cf. section "The characteristic atmospheric phenomena", p.354) fits in very well with the above description. The rains associated with the presence of this tongue over the Plateau bring a general relief from heat to places which have successively been invaded by the temperature wave. Temperature drops upon the arrival of the rains, that is, causing the annual temperature maxima at places in southern and central Mexico to occur one or two months earlier than the summer solstice.

As a general temperature rise takes place in the southern part of Mexico, beginning in May or even April and continuing northward and laterally over the coastal plains on both sides of the Plateau, rains begin shortly afterward in southern and eastern Mexico, the edge of the rainy area spreading from southeast to northwest in the same fashion as the edge of the temperature wave did before.

The area where the temperature maximum takes place before July extends far to the north of the Tropic of Cancer, roughly over the same area where summer concentration of rainfall prevails, showing the close connection between the temperature maximum shift and the onset of the summer rainy season.

On the other hand, in the areas where the temperature peak is delayed, as over the Baja California Peninsula, mid-summer rains are very scarce. Indeed, winter rains, though slight, often exceed in amount those of summer, as in the extreme northwestern portion of Mexico, where the rainy season is six months out of phase with that of the rest of the country.

**Regions of annual temperature variation**

Combining the chart which shows the sequence of arrival of the annual temperature maxima and that of the magnitude of the temperature ranges throughout the country, a chart of regions of annual temperature variation has been constructed (Fig.10). There are 11 regions which are designated by a combination of capital letters and numbers (see Table I).

The orderly fashion in which the several regions extend from south to north and hence towards both the Gulf of Mexico and the Pacific coast, according to the month of annual temperature maximum, shows the validity of this classification, as it describes very well the movement of the heat wave mentioned above. The numerical indexes refer to the belts of annual range of temperature and distinguish contiguous regions with the same month of incidence of temperature maxima. The degree of continentality differs, however, taking the annual range of temperature as a good measure of the continental character of an area, regardless of its proximity to the sea. From this point of view, most of southern Mexico shows a marked maritime influence, whereas the northern part is subject to temperature changes that go from moderate to strong throughout the year.

Region $M_1$ comprises most of the southern Plateau area, the Balsas Basin, the Mixteca upland, and the southeastern portion of the country, including the peninsula of Yucatán. It shows a rather small rise of temperature from January to May, the mean temperature of May being taken here as the upper limit, since this is the warmest month of the year in this region.

Region $M_2$ with its temperature peak in the same month as $M_1$, but with an annual range between 7° and 14°C, shows, however, a tendency to present a bimodal type of curve in its annual march of temperature or at least a flattening off of the temperature curve after the onset of the rainy season. It occupies most of the Mesa Central.

Regions $N_1$, $N_2$ and $N_3$, which constitute the area where the temperature peak occurs in June, extend over most of the Mesa del Norte and the Sierra Madre Occidental with southward branches along the eastern and western ends of the Sierra Volcánica Transversal.

The shape of both the M and N areas shows how the temperature wave propagates most rapidly over high ground, avoiding the coastal plains, where it arrives in July or even August.

TABLE I

DESIGNATION OF TEMPERATURE REGIONS

| Belts of annual range of temperature | Month of occurrence of temperature maximum | | | | |
|---|---|---|---|---|---|
| | May | June. | July | Aug. | Sept. |
| Small range (0°–7°C) | $M_1$ | $N_1$ | $O_1$ | | |
| Medium range (7°–14°C) | $M_2$ | $N_2$ | $O_2$ | $P_2$ | $Q_2$ |
| High range (higher than 14°C) | | $N_3$ | $O_3$ | $P_3$ | |

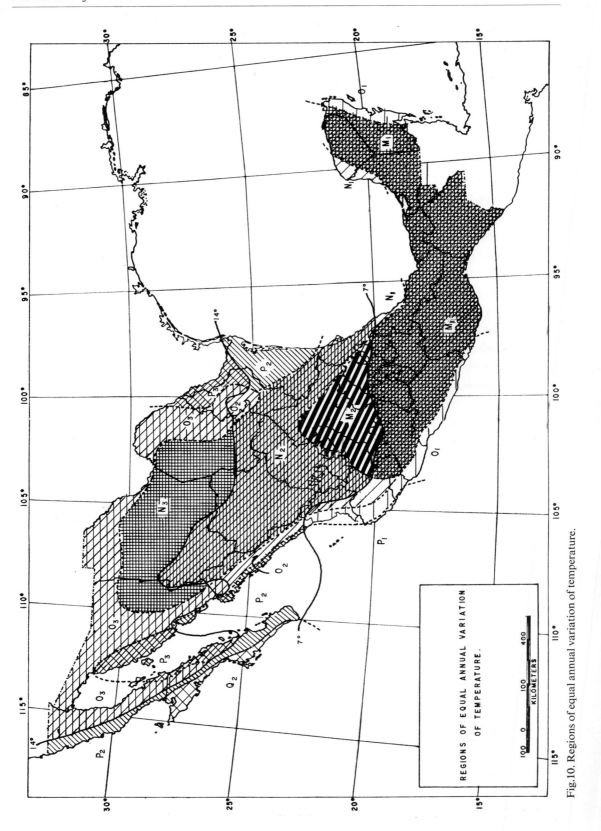

Fig.10. Regions of equal annual variation of temperature.

Regions $O_1$, $O_2$ and $O_3$ which cover areas where the temperature maximum occurs in July, have their broadest extent over the coastal plains of the northern Gulf of California as well as over the northern portion of the Mesa del Norte and the northern part of the coastal plains of the Gulf of Mexico. This fact supports again the suggestion made above, i.e., the tendency of the temperature wave to move northward faster over high terrain.

Regions $P_2$ and $P_3$ belong to areas where the annual temperature peak falls in August, and are found over both coastal plains of the Pacific and Gulf of Mexico north of the Tropic of Cancer, these being the true subtropical maritime areas of Mexico together with region $Q_2$, with a temperature maximum in September, which occupies the south-western coast of the Baja California Peninsula. Regions $P_2$ and $P_3$ show clearly the influence of water masses counteracting the effect of air heating by insolation. $Q_2$ is a region close to a relatively cold current. Regions $P_2$ and $P_3$ of the northern portion of the coastal plains of the Gulf of Mexico show the influence of clear skies in raising the air temperature during the "canicula" (midsummer drought characteristic of this portion of the country) (cf. section "The precipitation").

## The precipitation

### General

In previous sections the importance of orographic systems in determining the meteorological processes characteristic of Mexico has been emphasized. However, in dealing with rain-producing processes it is necessary to qualify further the effect of mountains upon the geographical distribution of rainfall as it is not obvious to what extent this powerful control prevails over other factors. First of all, there is the orthodox opinion that synoptic weather systems play the leading role in creating that distribution; whereas the opposite school of thought postulates the overall preponderance of orography in determining the rainfall patterns over Mexico. Mosiño (1964b) has discussed both points of view elsewhere.

It appears that the truth lies in between these extremes; namely, that there are at least two factors or controls that share the privilege of being responsible for the observed distribution of precipitation in Mexico. (*1*) Orography is, of course, the geographically permanent factor without which much of the observed rainfall of southern Mexico could not possibly be explained. (*2*) However, relief by itself, though a necessary condition, is not a sufficient one. Atmospheric disturbances must operate and when this factor does not work drought conditions prevail throughout the country.

We have repeatedly stated that the synoptic features at *sea level* do not, with a few exceptions, determine the precipitation areas, as the divergence patterns of air motion at this level are highly distorted by the presence of the high mountain systems of Mexico. Here again, we find the all-pervading effect of orography and relief on the *lower layers* of the atmosphere.

Although it may seem that we are attaching too much importance to orography, it is only the shallow synoptic features that are being eliminated as a cause of the main rainfall-producing patterns. After this elimination we are left with a scale of atmospheric disturbances that are in a greater measure independent of relief; i.e., the long planetary waves that determine, even during the warmer half of the year, many of the actual weather processes observed in Mexico.

RAMASWAMY (1956) in India has shown that the direction and curvature of the upper air currents are instrumental in shaping atmospheric events at the ground. According to his arguments, air circulation at high levels in the atmosphere plays an important role in the distribution of rainfall in the tropics, and lower circulation features play a secondary role in the production of large-scale convection.

**Causes of precipitation over the Plateau**

It is in connection with the broad upper-air currents that the elevated-heat-source effect mentioned above operates at its best, since it is during the establishment of an anomalous airflow aloft when that effect can be best detected. There is then usually no conventional synoptic disturbance apparent on the weather map. The only condition for the release of convective instability is an adequate moisture content and direction of upper air currents. Here we quote MOSIÑO (1964b) regarding the characteristic weather phenomena of the Mexican Plateau.

"It is a striking feature of the weather over the Mexican Plateau that the instability of the air often seems to be released very suddenly—sometimes without the slightest warning or clue as to the source of instability, if attention is paid only to the surface weather map. Here the upper-air charts are very useful in providing a hint to the development of convection over the Plateau.

For, even if no signs of instability are revealed by the surface chart over the coastal plains and the seas that surround Mexico, experience has shown that such meteorological events are associated with the presence aloft of broad air currents which bring moist, potentially unstable air over the Plateau. This air, which does not release its potential instability over lower ground, becomes ready to do so when subject to mechanical lifting over the Plateau or to the heating from below produced by the high-level source provided by the elevated ground.

The direction and position of those currents are clearly dependent upon the configurations of the airflow aloft and not upon the surface patterns. Those configurations, being of a much larger scale and of longer period than the surface patterns, are closely related to—if not identical with—the long planetary waves. Thus, it is not surprising that the rhythm of the weather changes over the Mexican Plateau be set by the slow progression of upper-air troughs and ridges rather than by the passage of fronts or other features which attain their highest intensity at the surface. This is the case in winter, while in summer the equatorial branches of the circulation and the interruption of the trades are responsible for the weather changes over the Plateau.

It has been observed that the passage of a westerly trough, for instance, over the Mexican Plateau takes about a week, whereas cold fronts take only a few hours to cross the Gulf of Mexico at the surface during the colder season of the year. In summer the evolution of the easterly waves and troughs over the Plateau is slower than at the surface over the Caribbean. All this points to the different time and space scale to which these features belong. However, we must point out an exception to these rules: the tropical cyclones, which being very definite features at the surface, are still apparent at higher levels without losing their shape and characteristics. These are deep structures imbedded in deep air currents."

Thus, it is not possible in general to explain the rainfall patterns over the Mexican Plateau by recourse to the presence of conventional sea-level features. Not all the rainfall is due to orography, however, much of it being due to the general atmospheric conditions associated with the motions and evolution in time and space of the large planetary waves and tropical cyclones.

The moisture is obviously carried inland by the winds, whereas the triggering action of the instability is provided "in situ" by the elevated-heat-source effect mentioned in early sections. The synoptic features that in middle latitudes are the efficient causes of precipitation, like frontal action, trough lines and squall lines, are ineffective over the Mexican Plateau. [The authors are tempted to include in the above negative list some "shallow" tropical disturbances recently described (RIEHL, 1954) as occurring in tropical countries, like the easterly waves, wind shear lines, etc. However, they do not do so because although it is difficult to find them over the Mexican Plateau, these disturbances may be observed in the future with better means than those used up to the present.]

**Dry and wet seasons in Mexico**

In general, easterly flow aloft over Mexico brings moist air, and so is responsible for most of the rainfall recorded during the rainy season in this country, whereas westerly upper-airflow brings dry air over all continental Mexico. These rules work regardless of the sense of curvature of the streamlines.

Thus, two distinct seasons, i.e., dry and wet, are found in most of Mexico, according to which type of airflow predominates over this country. These two seasons are the well-known "tiempo de aguas" and "tiempo de secas" of the vernacular language. The rainy or "wet" season lasts from May through November and in the rest of the year drought prevails over much of the country. The exceptions include the coastal plains of the southern Gulf of Mexico and the extreme northwestern part of the Baja California Peninsula. In the former, northers play a significant role in the production of rainfall during the winter season; while in the latter, summer dryness is the rule. One characteristic common to most areas of southern and central Mexico is the fact that the larger part of the annual precipitation falls from May through October (Fig.11), this being the season of the greatest moisture content of the air over the Plateau.

In connection with Fig.11 we can see that most of the southern coast of the Pacific has more than 90% of its precipitation in summer, this percentage decreasing rapidly towards the northwest over the Baja California Peninsula, where the prevailing winter rains account for the low values. The high percentages shown along the southern Pacific coast are probably related to cyclonic rains from tropical storms traveling along this coast in summer.

On the other hand, along the coast of the Gulf of Mexico lower summer concentrations show clearly the influence of winter rains due to the "northers". Over the northern Plateau area, east of the Sierra Madre Occidental, a tongue of high summer concentration seems to indicate the presence of the mid-summer tongue of moist air mentioned below, whereas a finger of low values located along the crest of the Sierra Madre Occidental shows the influence of winter precipitation over that area due to the southwesterly flow aloft.

Before discussing the moist tongue of the summer season, we shall describe the occurrence

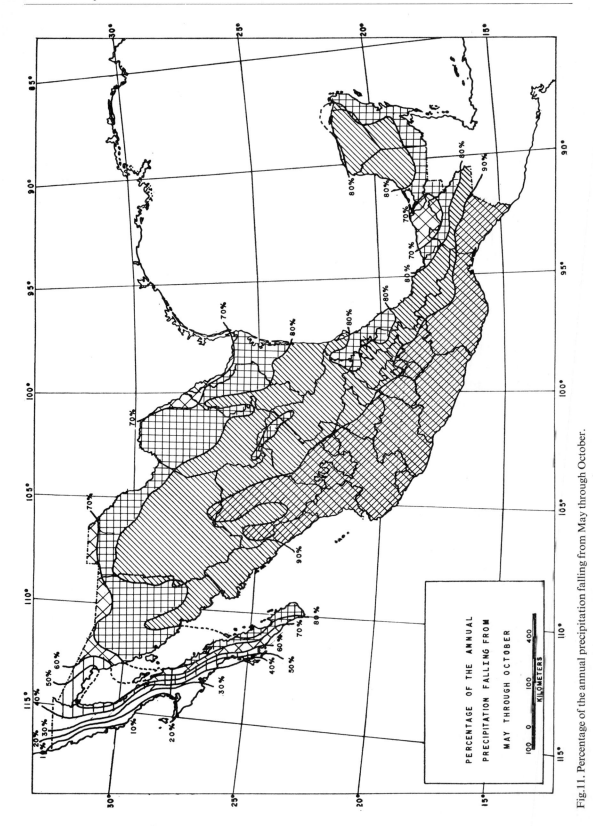

Fig.11. Percentage of the annual precipitation falling from May through October.

of a period of relative dryness which takes place over most of the eastern half of the country. This is known as the August drought, which in recent studies by the authors (Mosiño and García, 1966) has been found characteristic of July as well in some parts of the country.

### The mid-summer drought

This is a dry spell that occurs, as its name implies, in the middle of the rainy season over most of the coastal plains of the Gulf of Mexico and the southern portion of the Pacific coast. There are some indications that this phenomenon is not confined to the Mexican area, but extends well into the tropics as far south as Panama and Colombia (Vivó, 1964). Although the period over which the dry spell extends can be very long (from 2 to 4 months) accounting for dry years in some parts of Mexico, it is not rainless, but is rather a relatively dry period in which a diminished number of rainy days cause a decrease in the monthly amounts of rain; the annual march of precipitation thus shows a distinct bimodal distribution.

Explanations of this dry period are always given in terms of the variation of the sun's declination. The rainfall maxima, however, do not fall on about the same dates every year, as one would expect from such a simple dependence upon an astronomical event as the passage of the sun above the zenith.

It seems rather that, granted that insolation affects air stability, the position of the sun in the sky does not act in such a simple way, but through changes in the air circulation in the tropics.

In Mexico at least it appears that definite variations in the character of the air circulation over the Gulf of Mexico are the immediate cause of the mid-summer relative dry spell. Statistics on frequency of circulation types show a tendency for the summer air circulation to revert occasionally to winter-type regimes of airflow in the Gulf of Mexico. This reversion interrupts the deep trades over the eastern Gulf of Mexico, which give way at the surface to patterns that resemble those of the winter season, if not with respect to temperature, at least with respect to airflow.

The cut-off of the trade current is accomplished by a major upper air trough which lies along the eastern seaboard of the U.S. and extends southwards from the Florida peninsula across Cuba, down to Yucatán and Central America (Mosiño, 1964b).

When this upper-air trough develops, the typical trade disturbances like easterly waves and tropical cyclones do not enter the Gulf of Mexico, but show a tendency to recurve northwards along the eastern part of the trough.

In this situation a decrease of rainfall is experienced in the coastal plains of the Gulf of Mexico and over the Mexican Plateau, giving rise to what is called "la canícula" or "August drought" which in central and northern South America is known as "el veranillo" (Ellis, 1962).

It is possible that the interruption of the trades by polar troughs over the western Atlantic is further related to rain-deficient years in Mexico. For if the polar troughs establish themselves in such a position during late winter and spring, the deep trades cannot enter the Mexican Plateau until the summer season is well advanced. Dry-farming activities suffer from the lack of opportune rains in such years.

Fig.12 shows the time interval that the mid-summer drought comprises. It can be clearly

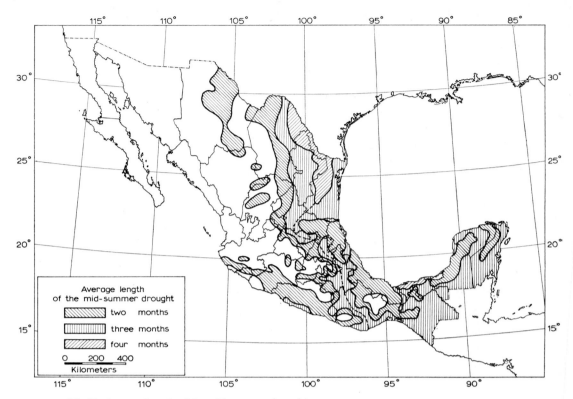

Fig.12. Average length of the mid-summer drought.

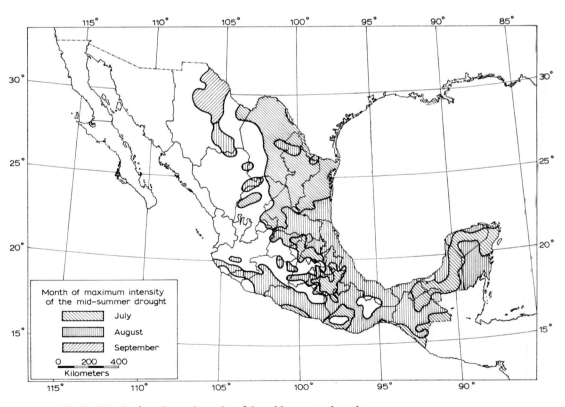

Fig.13. Month of maximum intensity of the mid-summer drought.

378

seen that the area over which this phenomenon extends in the Mexican Republic covers the eastern half of the country, including a belt along the southern Pacific coast. This area extends across the several precipitation zones and shows a variation of the relative dry period from two to four months. Although there is no correlation in the length of time covered by this phenomenon with the annual precipitation amounts, the maximal duration of four months, over the lower Río Grande Valley and along the limits of the states of Nuevo Leon and Tamaulipas, gives a hint of the relationship that the mid-summer drought may have to the belt of irregular rains that thins out, in both winter and summer seasons, in the upper Río Grande Valley.

On the other hand, over the southern plains of the Gulf of Mexico, where heavy rainfall amounts are the rule throughout the year with a definite predominance of summer rainfall, the length of the dry period can also be from two to four months, although here the mid-summer drought may not be as damaging as over the rain-deficient areas cited above.

Fig.13 shows the areal distribution of the month of occurrence of maximum intensity of the mid-summer drought, this being the month of lowest rainfall within such a period. The westward limit of the area affected by the mid-summer drought is shown clearly in Fig.12 and Fig.13. West of this line the stations show only one annual peak, whereas east and south of it the stations have at least two months with lower diminished rainfall amounts.

### The moist tongue over the Plateau in summer

Some charts drawn in the recent past in meteorological forecasting centers in the United States show the presence of a powerful moist air tongue that enters that country from Mexico near El Paso, Texas, during the summer months. According to some descriptions this moist tongue comes round the southern edge of the Bermuda–Azores anticyclone into the Mexican Plateau where mechanical lifting should cause release of its convective instability, carrying farther into the United States huge amounts of moisture which were not entirely depleted by precipitation of convective character over the Mexican land areas (Mosiño, 1964a).

This feature had long been suspected from the times when isentropic charts were drawn at forecasting centers in the United States in the 1930's. At that time there were no radiosonde stations in Mexico and very soon after the establishment of a limited network of this type of upper-air stations in this country isentropic analysis was discontinued; study of the structure of the moist tongue hence became more difficult. Nevertheless, a hint of its existence was provided by Bryson (1957) through his harmonic analysis of precipitation over the southwestern United States and northwestern Mexico. Systematic statistics about its vertical structure are still lacking, however.

From circulation studies it has been concluded that the summer tongue of moist air over Mexico enters from the Caribbean Sea and the Gulf of Mexico. However, Sands (1959) has suggested the possibility that the root of this moist tongue might come from the eastern Pacific and beyond, from the Southern Hemisphere through the deflected southeastern trades which, upon crossing the Equator, would meet the northeastern trades of the Northern Hemisphere and rise to high levels along the Intertropical Convergence Zone. This hypothesis has not been substantiated up to the present because of the scarcity of reliable upper air data over the eastern Pacific Ocean. However, the hypothetical axis of the moist tongue postulated by Sands coincides fairly well with the general direction of precipitation patterns produced recently by R. A. Bryson (written personal communication, 1965) through further harmonic analysis of Mexican rainfall.

**Distribution of mean annual rainfall** (Fig.14)

An inspection of the rainfall charts included in this section shows that the distribution of areas with most precipitation over the Mexican land areas follows roughly a pattern which resembles the letter U (Fig.15).

The right-hand stroke of this U falls along the seaward slopes of the Sierra Madre Oriental and the coastal plains of the southern Gulf of Mexico, including in its tail the area with heavy precipitation of Chiapas and the southern portion of the Yucatán Peninsula.

The left-hand stroke of the U is constituted by the areas of heavy amounts of rain along the Sierra Madre Occidental including those of the states of Nayarit and Jalisco. The horizontal stroke of the letter, which bridges the above-mentioned parts, is formed by the area of heavy amounts of rainfall along the Sierra Volcánica Transversal, and the Sierra Madre del Sur.

Between the vertical strokes, northward of the horizontal bar of this highly deformed U, are found areas with lower precipitation comprising the northern Plateau area to the north and a few small and scattered areas embracing the interior basins of the Balsas River and those of the upper watersheds of the Mixteca upland.

West of the Sierra Madre Occidental the driest section of the country is found to the north of the Gulf of California and the middle part of the Baja California Peninsula, including the Sonoran desert to the east of the Colorado River mouth.

Most of southern Mexico has plenty of precipitation with a noticeable decrease towards the northwestern coastal area of the Yucatán Peninsula.

The rainiest area, with a mean annual rainfall over 1,500 mm, is located south of parallel 22°N; it includes the mountains of the central and southern parts of the country sloping towards the coastal plains of the Gulf of Mexico.

The heavy amounts located along the eastern slopes of the Sierra Madre Oriental are evidently due to the prevailing wind directions which have a strong easterly component from the Gulf of Mexico inland. During the summer, because of the shifting northwards of the North Atlantic anticyclone, the deep trades of the Northern Hemisphere, with a general direction from northeast to southwest at the surface and from east to west aloft, prevail over the area, carrying inland great amounts of moisture evaporated from the warm waters of the Gulf of Mexico.

As they meet the mountains that border the coastal plains of this water body, the trades are forced to ascend, delivering great quantities of rain over the seafacing buttresses of the Plateau. The amount of rain is considerably increased by the presence of tropical cyclones which originate over the Caribbean Sea and beyond in late summer and early fall. These cyclones, which follow tracks roughly parallel to the mountain chains from the Caribbean Sea into the southern Gulf of Mexico, are able to produce upslope winds which bring torrential rains to these areas during disturbed weather periods. Both the deflected trades and tropical storms account for the early fall maximum of rainfall.

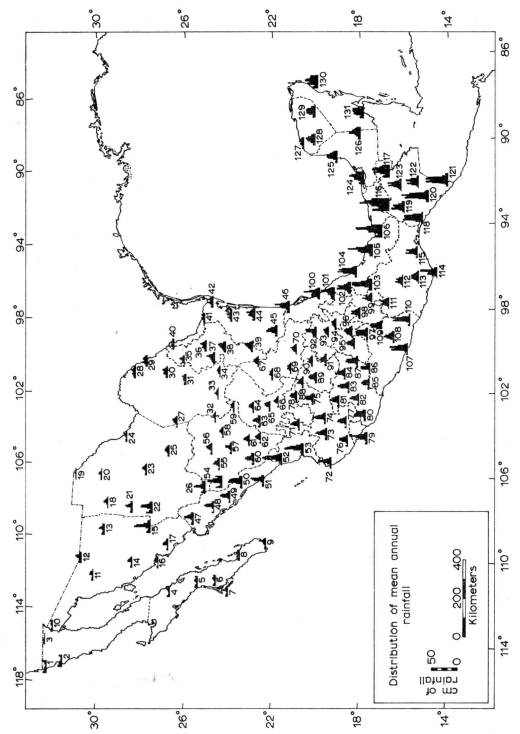

Fig.14. Distribution of mean monthly rainfall (from GARCÍA, 1965).

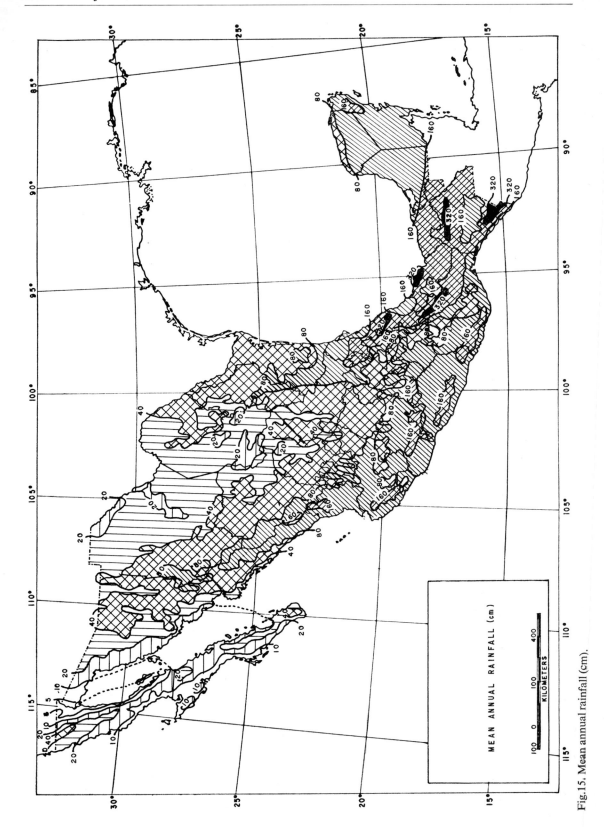

Fig.15. Mean annual rainfall (cm).

In general, over this area precipitation decreases during the colder half of the year due to the shifting southward of the subtropical high pressure belt and the consequent drift of the trade wind belt towards the south. Nevertheless, the "northers" caused by invasions of cool, modified cP air from central U.S. in passing over the Gulf of Mexico pick up moisture, which is released as heavy rain over this region in winter. Here also the rainfall increases locally over the slopes of mountain chains that run perpendicularly to the main direction of humid winds, creating great differences in the amount of rain over slopes of diverse orientation.

There are four main areas with annual rainfall over 3,500 mm (GARCÍA, 1965): the region located immediately north of parallel 20°N, the region located south of parallel 18°N, the northern slopes of the Chiapas complex, and the eastward slopes of the isolated Sierra de los Tuxtlas located on the coastal plain south of parallel 19°N. The greatest amounts of rain are found neither over the coastal belt nor at the top of the mountains, but rather over an area with an altitude between 100 and 600 m above sea level.

It is worth pointing out two small zones with precipitation under 1,500 mm in this region; their presence is probably due to the local configuration of the relief, which in this area acts as a barrier to the moist winds.

As shown in Fig.15, the rain is more abundant, as a rule, on the Gulf of Mexico side than on the Pacific side. Whereas the former shows great areas with more than 2,000 mm of rain a year, the latter only receives more than 2,000 mm in isolated places. However, a zone with a mean annual precipitation over 3,500 mm exists on the Pacific side over the southeastern portion of the Sierra Madre de Chiapas. Here the abundant rain may be variously attributed to orography, to the influence of tropical cyclones which affect this coast, originated both in the Caribbean Sea and the Gulf of Tehuantepec, and perhaps to the Intertropical Convergence Zone, which in summer may reach this latitude.

Along the seaward slopes of the Sierra Madre Occidental and the Sierra Madre del Sur moderate annual amounts of precipitation are regularly experienced, apparently due to their nearness to the tracks of tropical cyclones, which lie roughly parallel to the Pacific coast of Mexico.

An alternative explanation for the rainfall amounts found over the western slopes of the Sierra Madre Occidental might be the presence over the area of the mid-summer moist tongue mentioned earlier which according to SANDS (1959) enters the Mexican territory near the coast of southern Sinaloa and Nayarit states. Satellite photos also suggest deep penetration of Pacific moisture into Mexico. There is evidence that the depth of this moist stream is related to its previous history over the Pacific. On the other hand, one wonders to what extent both explanations are valid, as tropical cyclones are known to be often associated with moist tongues aloft.

The southern part of the Mexican Plateau has from 600 to 1,000 mm a year over the flat areas and more than 1,000 mm over the mountains. In general, the precipitation over this region is both of convective and orographic character, the rainy season being the summer, with July as the wettest month. During the winter, drought conditions prevail over the region, associated with the westerly winds that dominate the entire Plateau at this time of the year, as the subtropical high pressure belt shifts southwards. When the westerlies prevail at the ground level over the Plateau, the disturbances usual to that basic current may occasionally affect it, producing strong winds, temperature drops and some precipitation, which over the highest mountains can be snow.

When cold air masses associated with the "northers" on the Gulf of Mexico are deep, they may have some influence on the Plateau weather causing temperature drops, cloudiness and some precipitation. During late summer and early fall the precipitation increases, especially over the highest parts of the area, apparently because of the presence of tropical cyclones in both the Gulf of Mexico and the Pacific Ocean. The southern slopes of the Sierra Volcánica Transversal have from 1,200 to 1,500 mm of rain a year with a maximum in September.

The interior basins south of 20°N, like the central depression of Chiapas, the Balsas River basin and the small basins of the Mixteca upland, because of their isolation from the seas and their leeward location with respect to the prevailing winds, receive from 600 to 1,000 mm a year, and some places even less than 600 mm.

The northern part of the Plateau is an extensive area with scarce precipitation. The dryness of this area may be attributed to its location within the subtropical high pressure belt and to the presence of the sierras that limit and isolate it from the seas. The driest area, with less than 300 mm a year, extends over the north central part of the region and covers the flat areas of the Plateau from 24°N to the Mexico–U.S. border. Here the summer rains are probably due to convection, dependent for its humidity upon the mid-summer tongue of moist air discussed above. In winter, the scarce precipitation, which falls generally as snow, is associated with middle latitude disturbances (polar troughs) embedded within the westerlies that prevail at this time of the year over the area.

As a matter of fact, during this time of the year stability conditions of the air masses over the eastern Pacific Ocean appear to be adequate for rapid evaporation from the sea surface. This moisture is carried inland into northern Mexico by the westerlies. Most of the moisture is precipitated over the western slopes of the Sierra Madre Occidental, but some is carried forward into the northern Plateau area. The snowfalls experienced here often are triggered by westerly disturbances of a vortical character which stagnate or even retreat westwards, including by their presence over the southwestern U.S. winds from southeasterly and easterly quarters that bring in moist air from the Gulf of Mexico. These situations are often accompanied by a moist tongue ahead of a cold core vortex traveling slowly in an easterly direction. This same winter moist tongue is also responsible for the winter precipitation over the northeastern part of the country, once the upper air vortex becomes involved with frontal action related to a polar outbreak over the area.

Furthermore, the northern Plateau area, as mentioned earlier, is affected by cold air masses coming from the Great Basin region in the United States which enter Mexico behind a polar trough, producing freezing temperatures and an increase of the likelihood of snowfalls at high levels.

The northeastern part of the coastal plain of the Gulf of Mexico receives from 500 mm on the north to 1,200 mm of rain a year on the south; it is considerably drier than the southern part because of the divergent nature of airflow often found over the area.

The driest part of the country is the northwestern corner of the Pacific coastal plain which, located within the subtropical high pressure belt, has a mean altitude under 200 m and is thus subject to widespread subsidence as mentioned in previous sections. In this area, the mouth of the Colorado River has less than 50 mm of rain a year, the scarce summer precipitation being due mainly to convection of moist air brought in by the moist tongue noted above, and that of the wintertime due to cold vortices carried inland by the westerlies.

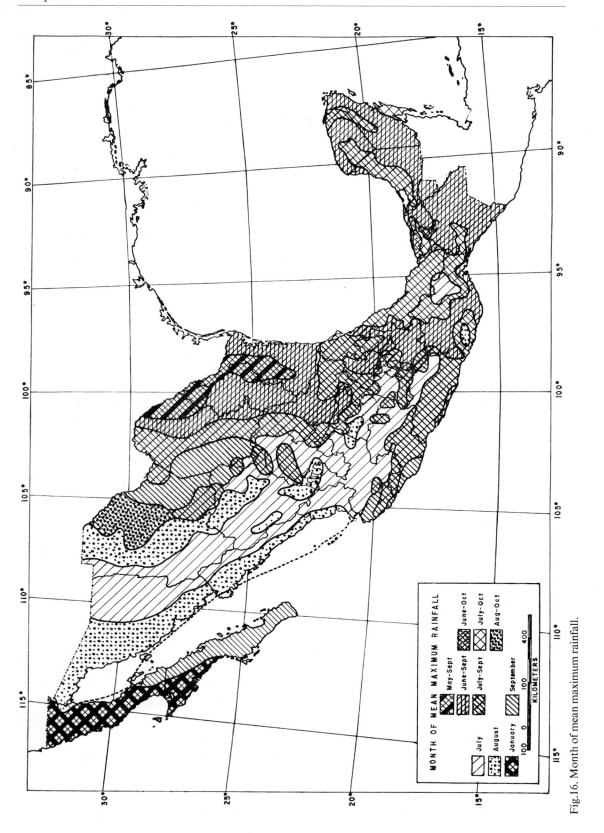

Fig.16. Month of mean maximum rainfall.

The Baja California Peninsula, close to the former, is another region with scarce precipitation. It receives generally less than 300 mm a year, except the highest parts of the mountain chain that runs its whole length where from 400 to 600 mm a year are possible. The rainiest season on the western side of the peninsula north of parallel 26°N is the winter, precipitation being due to the southward shift of the subtropical high pressure belt which lets typical mid-latitude disturbances come into the area. During the summer the cold California current stabilizes the air from below and drought conditions prevail along this coast, where some places completely lack rain for one or two months for many consecutive years.

The almost flat Yucatán Peninsula receives about 1,500 mm in its southern portion. The precipitation decreases towards the north, reaching 500 mm on the northwest coast. The existence of this area with low precipitation, which extends well into the Gulf of Mexico, is apparently related to divergence of air crossing the shoreline, as the prevailing winds are from land to sea in this part of the peninsula.

**Month of mean maximum precipitation**

Since there are two seasons which alternate in producing rain and drought over most of Mexico, it is interesting to show the times at which the mean monthly maximum precipitation occurs. This is shown in Fig.16. The salient feature of this chart is the parallelism of the areas with the same time of occurrence of precipitation maxima to the coasts, apparently related to the tracks of tropical disturbances which parallel the coasts of both the Pacific and the Atlantic oceans.

Over most of the central Plateau and along the highest parts of the Sierra Madre Occidental, July and August are the months of mean maximum precipitation, with a few small areas east of the continental divide and in the isthmian part of the country where either June or September is the month of peak rainfall.

Over the northeastern states and on the eastern portion of the Mesa del Norte, September is the peak month, with a secondary peak either in May, June or July over the areas affected by the mid-summer drought discussed above.

**Interannual variability of precipitation**

WALLÉN (1955) in his thorough study of precipitation of Mexico has discussed several measures of annual variability. We reproduce as Fig.17 his fig.9, where it can be seen that most of the northern Plateau area has a very high coefficient of variability of annual rainfall. A tongue of high values extends from El Paso, Texas, southeastward into Mexico. As a counterpart of this tongue of high values, there are several areas of low variability extending over most of the southern Plateau area and along the highest parts of the Sierra Madre Occidental, with an axis over southern Mexico along the Sierra Volcánica Transversal.

This chart suggests again the position of the mid-summer tongue of moist air repeatedly cited in previous sections. The importance to agriculture of this area of low variability of annual rainfall is stressed by WALLÉN (1955, 1956).

This chart stresses the tremendous risks to which dry-farming is subject in some parts of Mexico, whereas more fortunate areas of the southern Plateau enjoy more stable annual

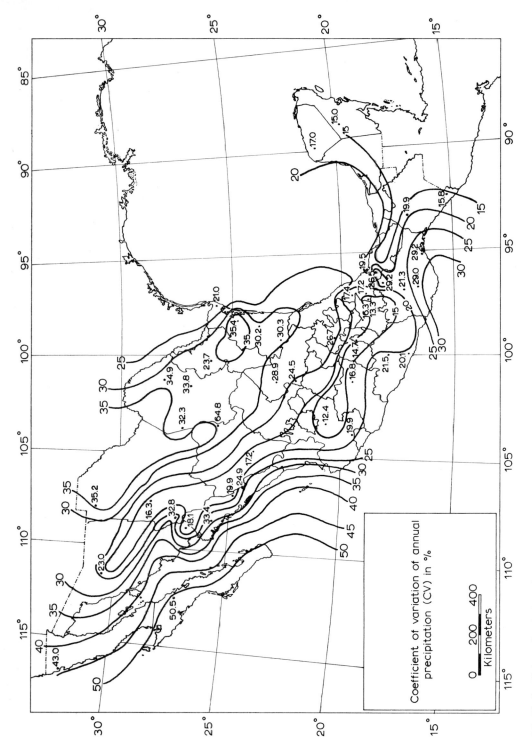

Fig.17. Coefficient of variation of annual precipitation (CV) in % (taken from WALLÉN, 1955).

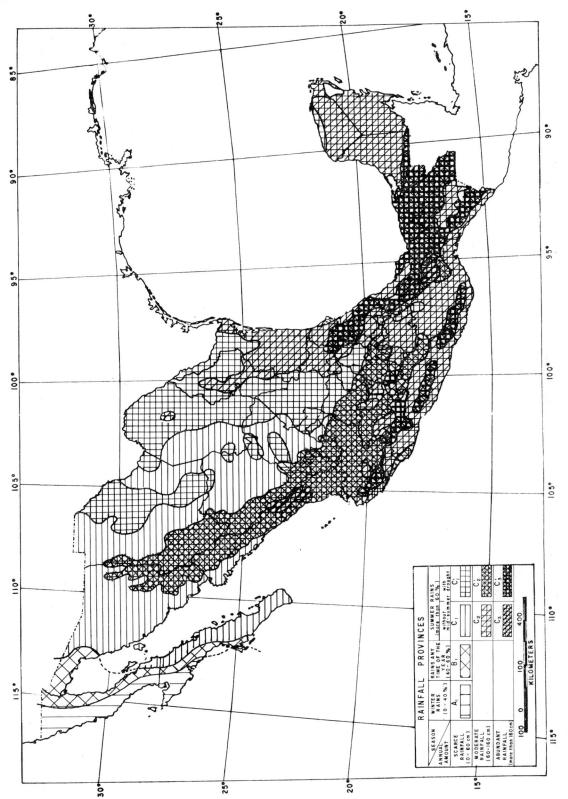

Fig.18. Rainfall provinces.

amounts of rainfall. That explains why this part of Mexico is the most densely populated. It is worth pointing out the coincidence of low variability of rain over the northeastern part of the Yucatán Peninsula, which is an agricultural area surpassed only by the most populated areas of central Mexico, notwithstanding the low amounts of annual rainfall received.

**Rainfall provinces of Mexico**

Several authors (PAGE, 1930; WALLÉN, 1955; and others) have classified the different regions of Mexico into rainfall provinces according to two criteria of classification. One obvious criterion is the annual rainfall amount. The other is the time of the year of maximum precipitation or some other characteristic.

The present authors also take into account the extent of the mid-summer drought, since this phenomenon is a definite feature of the rainfall distribution in about half the Mexican land areas.

As shown in "The mid-summer drought" discussion on p.377 the relatively dry spell occurs in the middle of the rainy season, its intensity being responsible, to some extent, for the variability of rainfall from one year to another, particularly over the areas with moderate annual amounts of rain.

A new chart of rainfall provinces has therefore been constructed, taking into account the extent and duration of the mid-summer drought. Fig.18 is the result of the use of this classification for Mexico.

As a first criterion we used the mean annual rainfall amounts, designating as areas of scarce precipitation those with less than 60 cm a year, as areas with moderate amounts those with 60 to 160 cm, and as areas with abundant rain those with more than 160 cm.

The second criterion is the percentage of rain falling during the period May through October, taking as 100% the mean total annual rainfall at each station. According to the characteristic distribution of rain throughout the year, we have called the areas with less than 40% of rainfall within the period mentioned areas with winter rainfall. Those with percentages from 40 to 60 are considered areas with no distinct rainfall season, and those with more than 60% are regions with summer rainfall.

An outstanding characteristic of the stations with winter rainfall in Mexico is that most of them have less than 60 cm of annual rainfall. Only the areas with a definite summer precipitation show amounts larger than 60 cm.

The areas with large amounts of rain during the warmer half of the year may show the effect of the mid-summer drought discussed on p.377 so that these areas have been subdivided according to whether a definite mid-summer drought is present or not.

The result of this classification is shown in Fig.18 which we have called the rainfall provinces of Mexico. This map is considerably different from those drawn by the authors quoted above. It has eight major rainfall provinces whose description is given in the table attached to the figure.

**References**

BRYSON, R. A., 1957. The annual march of precipitation in Arizona, New Mexico and northwestern Mexico. *Inst. Atm. Phys., Univ. Ariz., Publ.*, 6: 24 pp.

CONTRERAS ARIAS, A., 1942. *Mapa de las Provincias Climatológicas de la República Mexicana.* Secretaría de Agricultura y Fomento, Dirección de Geografía, Meteorología e Hidrología. México, D.F. Publ., 7: 52 pp.

CONTRERAS ARIAS, A., 1957. Los dos aspectos del efecto de la actividad Ciclónica Tropical sobre el Territorio Mexicano. *Seminario Sobre los Huracanes del Caribe, Ciudad Trujillo, D.N., República Dominicana*, 1957, pp.339–509.

DOMÍNGUEZ, E., 1950. *Meteorología Náutica.* Veracruz, Ver.

ELLIS, G. R., 1962. The "Veranillo" in Panamá. *Geofís. Int.*, 2(2): 15.

GARCÍA, E., 1964. *Modificaciones al Sistema de Clasificación Climática de Köppen.* Offset Larios, S.A. México, D.F., viii + 71 pp.

GARCÍA, E., 1965. Distribución de la Precipitación en la República Mexicana. *Publ. Inst. Geogr.*, 1: 175–191.

GARCÍA, E., SOTO, C. and MIRANDA, F., 1961. Larrea y clima. *An. Inst. Biol., México, D.F.*, 31: 133–171.

GUZMÁN, J., 1903. Climatología de la República Mexicana desde el punto de vista higiénico. *Mem. Rev. Soc. Cient. "Antonio Alzate"*, 20: 181–289.

HERNÁNDEZ, J., 1923. The temperature of Mexico. *Mon. Weather Rev., Suppl.*, 23: 37 pp.

KLEIN, W. H., 1957. Principal tracks and mean frequencies of cyclones and anticyclones in the Northern Hemisphere. *U.S. Dept. Commerce, Weather Bur., Wash., D.C., Res. Pap.*, 40: 54 pp.

LÓPEZ, E., 1922a. Climatología de la República Mexicana. *Mem. Rev. Soc. Cient. "Antonio Alzate"*, 40: 109–144.

LÓPEZ, E., 1922b. Estudio sobre "Nortes". *Mem. Rev. Soc. Cient. "Antonio Alzate"*, 41: 91–108.

LUCIO, R. (w.d.). *Los Vientos Boreales en la República Mexicana.* Secretaría de Agricultura y Fomento, Dirección de Geografía, Meteorología e Hidrología, Mexico, D.F., Publ., 7: 10 pp.

MOSIÑO ALEMÁN, P. A., 1964a. Meteorología marítima de la Costa Sur de México. *An. Inst. Geofís.*, 1: 17–24.

MOSIÑO ALEMÁN, P. A., 1964b. Surface weather and upper-airflow patterns in Mexico. *Geofís. Int., México, D.F.*, 4 (3): 117–168.

MOSIÑO ALEMÁN, P. A. and GARCÍA, E., 1966. *Evaluación de la Sequía Intraestival de la República Mexicana.* Unión Geográfica Internacional, Soc. Mex. de Geografía y Estadistica, Mexico.

PAGE, J. L., 1930. Climate of Mexico. *Mon. Weather Rev., Suppl.*, 33: 30 pp.

RAMASWAMY, C., 1956. On the subtropical jet stream and its role in the development of large-scale convection. *Tellus*, 8 (1): 26–60.

RIEHL, H., 1945. Waves in the easterlies and the polar front in the tropics. *Dept. Meteorol., Univ. Chicago, Chicago, Ill., Misc. Rept.*, 17: 79 pp.

RIEHL, H., 1954. *Tropical Meteorology.* McGraw-Hill, New York, London, Toronto, 392 pp.

SANDS, R. D., 1959. *A Study in the Regional Climatology of Mexico with Precipitation as the Correlative Factor.* Doctor's Thesis, Clark University, Worcester, Mass., 134 pp.

SERVICIO METEOROLÓGICO MEXICANO, 1939. *Atlas Climatológico de México* (3rd ed). Secretaría de Agricultura y Fomento, Dirección de Geografía, Meteorología e Hidrología, México, D.F., 77 pp., 155 maps.

THORNTHWAITE, C. W., 1931. The climates of North America according to a new classification. *Geograph. Rev.*, 21:633–655.

VIVÓ, J. A., 1964. Weather and climate of Mexico and central America. In: R. WAUCHOPE (Editor), *Handbook of Middle American Indians.* Univ. of Texas Press, Austin, Texas, 1:187–215.

VIVÓ, J. A. and GÓMEZ, J. C., 1946. *Climatología de México.* Instituto Panamericano de Geografía e Historia y Dirección de Geografía, Meteorología e Hidrología, XI: 73 pp.; 54 maps.

WALLÉN, C. C., 1955. Some characteristics of precipitation in Mexico. *Geogr. Ann.* 37: 51–85.

WALLÉN, C. C., 1956. Fluctuations and variability in Mexican rainfall. *Am .Assoc. Advan. Sci., Publ.*, 43: 141–155.

TABLE II[1]

CLIMATIC TABLE FOR SALINA CRUZ, OAXACA, MEXICO
Latitude 16°12′N, longitude 95°12′W, elevation 56 m

| Month | Mean sta. press. (mbar) | Temperature (°C) | | | | Mean vap. press. (mbar) | Precipitation (mm) | |
|---|---|---|---|---|---|---|---|---|
| | | daily mean | mean daily range | extremes | | | mean | max. in 24 h |
| | | | | max. | min. | | | |
| Jan. | 1009.1 | 25.6 | 8.3 | 35.3 | 16.0 | 20.0 | 4.4 | 22.5 |
| Feb. | 1008.4 | 25.9 | 8.2 | 36.5 | 13.6 | 21.4 | 3.8 | 34.7 |
| Mar. | 1007.3 | 27.0 | 7.9 | 37.2 | 16.4 | 23.4 | 1.6 | 13.1 |
| Apr. | 1006.6 | 28.4 | 7.8 | 38.8 | 18.0 | 25.1 | 0.6 | 7.3 |
| May | 1005.9 | 29.5 | 7.7 | 39.1 | 19.6 | 26.8 | 48.1 | 134.4 |
| June | 1006.3 | 28.3 | 7.5 | 38.7 | 19.0 | 27.9 | 264.3 | 225.9 |
| July | 1007.3 | 28.4 | 7.7 | 38.2 | 19.2 | 26.7 | 207.4 | 242.1 |
| Aug. | 1007.1 | 28.5 | 8.1 | 37.7 | 18.8 | 26.9 | 176.1 | 168.5 |
| Sept. | 1006.3 | 27.6 | 7.6 | 37.4 | 20.2 | 27.0 | 240.4 | 286.7 |
| Oct. | 1007.0 | 27.4 | 7.7 | 36.1 | 18.7 | 23.9 | 87.9 | 264.6 |
| Nov. | 1008.3 | 26.7 | 8.0 | 35.5 | 16.7 | 21.1 | 8.6 | 66.4 |
| Dec. | 1008.7 | 26.0 | 8.2 | 36.0 | 13.0 | 19.9 | 3.6 | 18.7 |
| Annual | 1007.4 | 27.4 | 7.9 | 39.1 | 13.0 | 24.2 | 1,046.8 | 286.7 |

| Month | Number of days with | | | Mean cloud-iness (tenths) | Mean sun-shine (h) | Wind | |
|---|---|---|---|---|---|---|---|
| | precip. (>0.25 mm) | thunder-storm | fog | | | most freq. direct. | mean speed (m/sec) |
| Jan. | <1 | 0 | <1 | 2 | 259.0 | NNE | 6.1 |
| Feb. | <1 | <1 | 1 | 2 | 255.1 | NNE | 5.5 |
| Mar. | <1 | <1 | <1 | 2 | 280.9 | NNE | 4.8 |
| Apr. | <1 | <1 | <1 | 3 | 235.3 | NNE | 4.1 |
| May | 3 | 2 | <1 | 4 | 221.7 | NNE | 3.5 |
| June | 10 | 5 | 1 | 7 | 163.5 | NNE | 2.3 |
| July | 8 | 4 | <1 | 7 | 200.3 | NNE | 2.8 |
| Aug. | 8 | 4 | <1 | 6 | 214.8 | NNE | 2.8 |
| Sept. | 9 | 4 | 1 | 7 | 166.6 | NNE | 2.2 |
| Oct. | 4 | <1 | <1 | 5 | 229.5 | NNE | 4.5 |
| Nov. | <1 | <1 | <1 | 2 | 255.5 | NNE | 6.0 |
| Dec. | <1 | 0 | <1 | 2 | 252.8 | NNE | 6.0 |
| Annual | 47 | 20 | 6 | 4 | 2,735.0 | NNE | 4.2 |

[1] Tables II–XV data for the period 1931–1960. See also pp.346–347.

TABLE III

CLIMATIC TABLE FOR ACAPULCO, GUERRERO, MEXICO
Latitude 16°50′N, longitude 96°56′W, elevation 3 m

| Month | Mean sta. press. (mbar) | Temperature (°C) | | | | Mean vap. press. (mbar) | Precipitation (mm) | |
|---|---|---|---|---|---|---|---|---|
| | | daily mean | mean daily range | extremes | | | mean | max. in 24 h |
| | | | | max. | min. | | | |
| Jan. | 1013.7 | 26.7 | 8.7 | 36.0 | 10.9 | 25.6 | 6.3 | 43.9 |
| Feb. | 1013.0 | 26.5 | 8.5 | 35.8 | 18.0 | 25.6 | 1.3 | 23.0 |
| Mar. | 1012.4 | 26.7 | 8.6 | 37.6 | 18.1 | 25.8 | 0.3 | 7.7 |
| Apr. | 1011.7 | 27.5 | 8.2 | 37.0 | 18.2 | 26.9 | 1.2 | 11.4 |
| May | 1011.3 | 28.5 | 7.4 | 40.5 | 20.0 | 28.8 | 36.0 | 147.5 |
| June | 1011.4 | 28.6 | 7.1 | 37.5 | 21.0 | 29.8 | 281.0 | 228.0 |
| July | 1012.3 | 28.7 | 7.6 | 37.6 | 21.4 | 29.9 | 256.1 | 208.1 |
| Aug. | 1012.1 | 28.8 | 8.0 | 37.0 | 21.0 | 30.0 | 252.4 | 192.5 |
| Sept. | 1010.9 | 28.1 | 7.5 | 36.8 | 20.4 | 29.7 | 349.1 | 235.4 |
| Oct. | 1011.4 | 28.1 | 7.5 | 37.0 | 20.5 | 29.7 | 159.2 | 245.8 |
| Nov. | 1012.7 | 27.7 | 8.2 | 37.0 | 19.4 | 28.3 | 28.1 | 224.1 |
| Dec. | 1012.8 | 26.7 | 8.4 | 32.4 | 18.2 | 26.6 | 7.7 | 33.5 |
| Annual | 1012.1 | 27.7 | 8.0 | 40.5 | 10.9 | 28.1 | 1,378.7 | 245.8 |

| Month | Number of days with | | | Mean cloud-iness (tenths) | Mean sun-shine (h) | Mean wind speed (m/sec)* |
|---|---|---|---|---|---|---|
| | precip. (>0.25 mm) | thunder-storm | fog | | | |
| Jan. | 1 | 0 | 0 | 3 | 255.3 | 2.2 |
| Feb. | 0 | 0 | 0 | 2 | 244.6 | 2.9 |
| Mar. | 0 | 0 | 0 | 2 | 259.6 | 3.3 |
| Apr. | 0 | 0 | 0 | 3 | 211.8 | 3.2 |
| May | 3 | 0 | 0 | 4 | 205.5 | 3.1 |
| June | 13 | 1 | 0 | 7 | 194.3 | 3.1 |
| July | 14 | 1 | 0 | 7 | 213.3 | 3.0 |
| Aug. | 13 | 1 | 0 | 7 | 220.7 | 3.4 |
| Sept. | 16 | 1 | 0 | 8 | 178.3 | 3.9 |
| Oct. | 9 | 0 | 0 | 6 | 216.2 | 2.8 |
| Nov. | 2 | 0 | 1 | 3 | 251.6 | 2.3 |
| Dec. | 1 | 0 | 0 | 3 | 256.2 | 1.8 |
| Annual | 72 | 4 | 1 | 5 | 2,707.4 | 2.9 |

* Data for some years missing.

TABLE IV

CLIMATIC TABLE FOR MORELIA, MICHOACAN, MEXICO
Latitude 19°42′N, longitude 101°07′W, elevation 1,923 m

| Month | Mean sta. press. (mbar) | Temperature (°C) | | | | Mean vap. press. (mbar) | Precipitation (mm) | |
|---|---|---|---|---|---|---|---|---|
| | | daily mean | mean daily range | extremes | | | mean | max. in 24 h |
| | | | | max. | min. | | | |
| Jan. | 811.6 | 14.4 | 13.3 | 26.0 | −0.5 | 9.1 | 11.8 | 21.3 |
| Feb. | 810.9 | 16.1 | 13.8 | 27.6 | 0.8 | 9.4 | 5.4 | 14.8 |
| Mar. | 809.9 | 18.1 | 14.1 | 30.2 | 2.4 | 9.4 | 6.6 | 33.9 |
| Apr. | 810.4 | 19.8 | 13.9 | 30.7 | 7.0 | 9.9 | 16.7 | 66.5 |
| May | 810.2 | 20.8 | 12.5 | 31.5 | 8.1 | 12.2 | 42.5 | 46.1 |
| June | 810.6 | 20.0 | 9.6 | 31.5 | 10.5 | 14.9 | 134.5 | 59.3 |
| July | 811.9 | 18.7 | 8.6 | 28.9 | 10.5 | 15.6 | 170.1 | 55.4 |
| Aug. | 811.7 | 18.6 | 8.8 | 27.7 | 11.2 | 15.3 | 154.0 | 52.2 |
| Sept. | 810.7 | 18.3 | 8.8 | 27.2 | 2.5 | 15.3 | 130.7 | 60.8 |
| Oct. | 811.4 | 17.5 | 9.9 | 27.5 | 4.7 | 13.6 | 55.6 | 45.3 |
| Nov. | 812.0 | 15.9 | 12.2 | 26.5 | 1.5 | 11.7 | 16.3 | 27.0 |
| Dec. | 811.9 | 14.6 | 13.2 | 25.5 | 0.3 | 10.3 | 10.5 | 27.2 |
| Annual | 811.1 | 17.7 | 11.6 | 31.5 | −0.5 | 12.2 | 754.7 | 66.5 |

| Month | Number of days with | | | Mean cloud-iness (tenths) | Mean sun-shine (h) | Wind | |
|---|---|---|---|---|---|---|---|
| | precip. (>0.25 mm) | thunder-storm | fog | | | most freq. direct. | mean speed (m/sec) |
| Jan. | 2 | <1 | <1 | 4 | 208.3 | SW | 1.9 |
| Feb. | 2 | <1 | <1 | 3 | 221.6 | SW | 2.1 |
| Mar. | 2 | <1 | <1 | 3 | 243.0 | SW | 2.5 |
| Apr. | 3 | 1 | 0 | 3 | 219.4 | SW | 2.3 |
| May | 8 | 2 | <1 | 4 | 207.4 | SW | 2.1 |
| June | 18 | 2 | 1 | 7 | 176.5 | SW | 2.0 |
| July | 22 | 3 | 1 | 8 | 177.2 | Var. | 1.6 |
| Aug. | 21 | 3 | 3 | 7 | 192.9 | S | 1.4 |
| Sept. | 18 | 1 | 4 | 7 | 165.4 | NNE | 1.5 |
| Oct. | 10 | 1 | 4 | 5 | 209.1 | NNE | 1.4 |
| Nov. | 3 | <1 | 3 | 3 | 213.8 | SW | 1.4 |
| Dec. | 2 | <1 | 1 | 4 | 205.1 | SW | 1.5 |
| Annual | 111 | 13 | 17 | 5 | 2,439.7 | SW | 1.8 |

TABLE V

CLIMATIC TABLE FOR ISLA DE COZUMEL, Q. ROO, MEXICO
Latitude 20°31′N, longitude 86°57′W, elevation 3 m

| Month | Mean sta. press. (mbar) | Temperature (°C) | | | | Mean vap. press. (mbar) | Precipitation (mm) | |
|---|---|---|---|---|---|---|---|---|
| | | daily mean | mean daily range | extremes | | | mean | max. in 24 h |
| | | | | max. | min. | | | |
| Jan. | 1016.6 | 22.9 | 8.6 | 31.8 | 6.5 | 23.9 | 86.8 | 180.0 |
| Feb. | 1015.8 | 23.3 | 9.3 | 32.1 | 6.2 | 23.7 | 63.4 | 124.0 |
| Mar. | 1014.2 | 24.5 | 9.5 | 35.0 | 7.8 | 25.0 | 48.2 | 92.0 |
| Apr. | 1013.7 | 26.0 | 9.5 | 35.0 | 9.0 | 26.9 | 53.0 | 137.0 |
| May | 1012.8 | 26.9 | 9.1 | 35.0 | 11.4 | 28.7 | 140.8 | 156.0 |
| June | 1013.0 | 27.2 | 7.9 | 35.2 | 18.2 | 30.8 | 200.2 | 182.0 |
| July | 1014.6 | 27.2 | 8.7 | 35.6 | 17.4 | 31.0 | 111.5 | 89.0 |
| Aug. | 1013.3 | 27.2 | 9.2 | 35.6 | 15.6 | 31.2 | 152.0 | 276.0 |
| Sept. | 1011.4 | 26.8 | 8.2 | 35.2 | 16.4 | 31.2 | 243.2 | 176.0 |
| Oct. | 1012.5 | 26.0 | 7.4 | 35.8 | 14.0 | 29.1 | 234.4 | 175.0 |
| Nov. | 1014.7 | 24.6 | 7.7 | 33.4 | 10.4 | 26.4 | 110.8 | 112.0 |
| Dec. | 1016.1 | 23.4 | 8.1 | 32.4 | 9.0 | 24.5 | 108.5 | 91.0 |
| Annual | 1014.1 | 25.5 | 8.6 | 35.8 | 6.2 | 27.7 | 1,552.8 | 276.0 |

| Month | Number of days with | | | Mean cloud-iness (tenths) | Most freq. wind direct. |
|---|---|---|---|---|---|
| | precip. (>0.25 mm) | thunder-storm | fog | | |
| Jan. | 8 | 0 | 0 | 6 | E |
| Feb. | 6 | 0 | 0 | 5 | SE |
| Mar. | 4 | 1 | 0 | 5 | SE |
| Apr. | 4 | 2 | 0 | 6 | SE |
| May | 8 | 3 | 0 | 6 | SE |
| June | 13 | 4 | 0 | 7 | SE |
| July | 12 | 5 | 0 | 7 | SE |
| Aug. | 12 | 6 | 0 | 7 | c. |
| Sept. | 16 | 6 | 0 | 7 | SE |
| Oct. | 14 | 3 | 0 | 7 | NE |
| Nov. | 10 | 1 | 0 | 6 | NNE |
| Dec. | 10 | 1 | 0 | 6 | NE |
| Annual | 117 | 32 | 0 | 6 | SE |

## TABLE VI

CLIMATIC TABLE FOR GUADALAJARA, JALISCO, MEXICO
Latitude 20°41′N, longitude 103°20′W, elevation 1,589 m

| Month | Mean sta. press. (mbar) | Temperature (°C) | | | | Mean vap. press. (mbar) | Precipitation (mm) | |
|---|---|---|---|---|---|---|---|---|
| | | daily mean | mean daily range | extremes | | | mean | max. in 24 h |
| | | | | max. | min. | | | |
| Jan. | 847.4 | 14.7 | 17.1 | 29.2 | −5.5 | 8.3 | 14.8 | 33.1 |
| Feb. | 846.6 | 16.6 | 18.1 | 30.0 | −2.0 | 8.2 | 4.3 | 25.8 |
| Mar. | 845.7 | 18.4 | 19.7 | 35.0 | −1.0 | 8.0 | 4.0 | 25.2 |
| Apr. | 845.7 | 21.1 | 19.5 | 35.0 | 1.6 | 8.3 | 5.2 | 28.9 |
| May | 845.0 | 22.9 | 17.2 | 35.8 | 5.4 | 10.7 | 24.0 | 46.5 |
| June | 845.6 | 22.3 | 12.6 | 38.0 | 9.9 | 15.4 | 172.2 | 69.7 |
| July | 847.0 | 20.5 | 10.4 | 31.1 | 10.3 | 16.8 | 251.4 | 97.0 |
| Aug. | 847.0 | 20.5 | 10.9 | 31.0 | 11.0 | 17.0 | 194.3 | 65.6 |
| Sept. | 846.0 | 19.9 | 10.5 | 31.0 | 6.4 | 16.6 | 158.1 | 55.8 |
| Oct. | 846.8 | 19.1 | 13.5 | 30.8 | 2.4 | 14.1 | 47.8 | 55.6 |
| Nov. | 847.4 | 16.9 | 16.4 | 31.0 | −2.8 | 10.9 | 10.1 | 22.5 |
| Dec. | 847.5 | 15.2 | 16.8 | 28.8 | −3.6 | 9.2 | 8.3 | 47.0 |
| Annual | 846.5 | 19.0 | 15.2 | 38.0 | −5.5 | 12.0 | 894.5 | 97.0 |

| Month | Mean evap. (mm) | Number of days with | | | Mean cloud-iness (tenths) | Mean sun-shine (h) | Mean wind speed (m/sec) |
|---|---|---|---|---|---|---|---|
| | | precip. (>0.25 mm) | thunder-storm | fog | | | |
| Jan. | 131.7 | 2 | 0 | 1 | 5 | 199.2 | 1.5 |
| Feb. | 173.0 | 1 | 0 | 1 | 4 | 215.9 | 1.8 |
| Mar. | 244.7 | 1 | 0 | 0 | 3 | 261.1 | 2.3 |
| Apr. | 274.1 | 1 | 0 | 2 | 4 | 250.5 | 2.3 |
| May | 284.9 | 4 | 1 | 0 | 4 | 262.4 | 2.3 |
| June | 214.1 | 16 | 3 | 1 | 7 | 202.1 | 1.9 |
| July | 167.8 | 23 | 4 | 3 | 8 | 192.0 | 1.6 |
| Aug. | 155.0 | 20 | 3 | 3 | 7 | 207.0 | 1.3 |
| Sept. | 136.8 | 17 | 1 | 4 | 7 | 179.3 | 1.4 |
| Oct. | 137.3 | 7 | 1 | 4 | 6 | 212.1 | 1.4 |
| Nov. | 125.7 | 2 | 0 | 2 | 4 | 216.9 | 1.3 |
| Dec. | 119.8 | 2 | 0 | 1 | 5 | 195.3 | 1.4 |
| Annual | 2,264.9 | 96 | 13 | 22 | 5 | 2,593.8 | 1.7 |

## TABLE VII

CLIMATIC TABLE FOR MERIDA, YUCATÁN, MEXICO
Latitude 20°58′N, longitude 89°38′W, elevation 22 m

| Month | Mean sta. press. (mbar) | Temperature (°C) | | | | Mean vap. press. (mbar) | Precipitation (mm) | |
|---|---|---|---|---|---|---|---|---|
| | | daily mean | mean daily range | extremes | | | mean | max. in 24 h |
| | | | | max. | min. | | | |
| Jan. | 1016.0 | 23.0 | 9.9 | 33.0 | 11.2 | 20.7 | 30.8 | 68.0 |
| Feb. | 1014.8 | 23.8 | 10.9 | 35.0 | 9.2 | 20.5 | 24.1 | 101.7 |
| Mar. | 1012.7 | 25.6 | 11.6 | 37.2 | 11.0 | 21.6 | 16.9 | 41.2 |
| Apr. | 1011.8 | 27.1 | 11.7 | 38.8 | 14.2 | 23.1 | 21.1 | 48.5 |
| May | 1011.1 | 27.8 | 10.9 | 40.2 | 17.0 | 25.1 | 83.0 | 139.3 |
| June | 1011.6 | 27.7 | 9.8 | 37.8 | 19.0 | 27.1 | 134.1 | 117.6 |
| July | 1013.6 | 27.3 | 9.6 | 35.4 | 18.0 | 27.4 | 130.3 | 66.2 |
| Aug. | 1012.7 | 27.4 | 9.3 | 35.0 | 17.2 | 27.7 | 148.3 | 188.7 |
| Sept. | 1010.9 | 27.1 | 8.8 | 35.0 | 18.2 | 28.3 | 180.1 | 76.9 |
| Oct. | 1012.2 | 25.9 | 8.3 | 34.0 | 15.1 | 26.0 | 91.2 | 69.5 |
| Nov. | 1014.7 | 24.2 | 8.8 | 33.6 | 13.2 | 22.6 | 35.4 | 65.9 |
| Dec. | 1015.8 | 23.0 | 9.1 | 35.2 | 10.2 | 21.2 | 33.9 | 67.7 |
| Annual | 1013.2 | 25.8 | 9.9 | 40.2 | 9.2 | 24.3 | 929.2 | 188.7 |

| Month | Mean evap. (mm) | Number of days with | | | Mean cloud-iness (tenths) | Mean sun-shine (h) | Wind | |
|---|---|---|---|---|---|---|---|---|
| | | precip. (>0.25 mm) | thunder-storm | fog | | | most freq. direct. | mean speed (m/sec) |
| Jan. | 128.6 | 4 | <1 | 18 | 5 | 162.3 | ESE | 2.3 |
| Feb. | 131.5 | 4 | <1 | 19 | 4 | 153.2 | ESE | 2.5 |
| Mar. | 165.3 | 3 | <1 | 19 | 4 | 188.5 | ESE | 3.0 |
| Apr. | 172.1 | 3 | <1 | 17 | 4 | 184.5 | ESE | 3.0 |
| May | 187.1 | 7 | 1 | 17 | 5 | 216.4 | ESE | 2.7 |
| June | 160.5 | 12 | 1 | 15 | 6 | 198.8 | ESE | 2.2 |
| July | 162.9 | 15 | 2 | 16 | 6 | 205.0 | E | 1.5 |
| Aug. | 161.7 | 14 | 4 | 15 | 6 | 202.4 | ESE | 1.4 |
| Sept. | 141.0 | 16 | 4 | 14 | 6 | 174.1 | E | 1.3 |
| Oct. | 137.0 | 10 | 1 | 14 | 5 | 171.5 | NE | 1.5 |
| Nov. | 127.9 | 5 | <1 | 15 | 5 | 157.0 | NNE | 1.8 |
| Dec. | 123.5 | 5 | <1 | 16 | 5 | 158.6 | NNE | 1.8 |
| Annual | 1,799.1 | 98 | 13 | 195 | 5 | 2,172.3 | ESE | 2.1 |

TABLE VIII

CLIMATIC TABLE FOR GUANAJUATO, GUANAJUATO, MEXICO
Latitude 21°01′N, longitude 101°15′W, elevation 2,037 m

| Month | Mean sta. press. (mbar) | Temperature (°C) | | | | Mean vap. press. (mbar) | Precipitation (mm) | |
|-------|------|------|------|------|------|------|------|------|
| | | daily mean | mean daily range | extremes | | | mean | max. in 24 h |
| | | | | max. | min. | | | |
| Jan. | 802.0 | 14.2 | 12.7 | 27.4 | −1.3 | 7.2 | 13.6 | 28.2 |
| Feb. | 801.3 | 15.8 | 13.9 | 29.0 | −0.8 | 6.9 | 5.1 | 24.4 |
| Mar. | 800.6 | 18.2 | 14.5 | 31.8 | 1.6 | 6.7 | 5.1 | 25.1 |
| Apr. | 800.7 | 20.2 | 14.4 | 31.6 | 5.5 | 7.1 | 16.3 | 28.3 |
| May | 800.7 | 21.6 | 13.6 | 33.6 | 9.1 | 9.4 | 32.7 | 25.9 |
| June | 801.1 | 20.3 | 11.8 | 33.3 | 10.6 | 12.3 | 126.8 | 66.1 |
| July | 802.5 | 19.0 | 10.8 | 29.8 | 10.5 | 13.1 | 138.1 | 71.9 |
| Aug. | 802.3 | 19.1 | 11.0 | 29.6 | 10.1 | 13.2 | 136.0 | 106.5 |
| Sept. | 801.3 | 18.4 | 10.0 | 29.5 | 8.8 | 12.9 | 124.0 | 70.2 |
| Oct. | 802.0 | 17.7 | 11.6 | 30.0 | 5.5 | 10.9 | 42.6 | 47.0 |
| Nov. | 802.4 | 16.1 | 12.6 | 28.0 | 2.7 | 9.1 | 15.3 | 54.4 |
| Dec. | 802.2 | 14.7 | 12.6 | 26.9 | 0.3 | 7.9 | 11.3 | 34.7 |
| Annual | 801.6 | 17.9 | 12.5 | 33.6 | −1.3 | 9.7 | 667.9 | 106.5 |

| Month | Mean evap. (mm) | Number of days with | | | Mean cloud-iness (tenths) | Mean sun-shine (h) | Wind | |
|-------|------|------|------|------|------|------|------|------|
| | | precip. (>0.25 mm) | thunder-storm | fog | | | most freq. direct. | mean speed (m/sec) |
| Jan. | 138.2 | 3 | 0 | 0 | 5 | 232.2 | c. | 1.4 |
| Feb. | 163.0 | 2 | 0 | 0 | 4 | 244.9 | c. | 1.8 |
| Mar. | 235.1 | 2 | 0 | 0 | 4 | 283.6 | SW | 1.9 |
| Apr. | 255.1 | 3 | 1 | 0 | 4 | 264.7 | c. | 1.9 |
| May | 246.3 | 7 | 2 | 0 | 5 | 250.5 | NE | 2.2 |
| June | 202.8 | 13 | 2 | 1 | 7 | 215.6 | ENE | 2.5 |
| July | 172.9 | 14 | 2 | 2 | 7 | 211.4 | ENE | 2.4 |
| Aug. | 174.1 | 14 | 3 | 1 | 7 | 235.3 | ENE | 2.3 |
| Sept. | 164.3 | 12 | 2 | 2 | 7 | 205.8 | ENE | 2.8 |
| Oct. | 171.0 | 6 | 1 | 1 | 5 | 246.1 | ENE | 2.5 |
| Nov. | 139.5 | 3 | 1 | 0 | 4 | 247.5 | NE | 1.9 |
| Dec. | 127.1 | 2 | 0 | 1 | 5 | 233.3 | c. | 1.4 |
| Annual | 2,189.4 | 81 | 14 | 8 | 5 | 2,870.9 | ENE | 2.1 |

## TABLE IX

CLIMATIC TABLE FOR TAMPICO, TAMAULIPAS, MEXICO
Latitude 22°12′N, longitude 97°51′W, elevation 73 m

| Month | Mean sta. press. (mbar) | Temperature (°C) | | | | Mean vap. press. (mbar) | Precipitation (mm) | |
|---|---|---|---|---|---|---|---|---|
| | | daily mean | mean daily range | extremes | | | mean | max. in 24 h |
| | | | | max. | min. | | | |
| Jan. | 1016.7 | 19.2 | 7.6 | 33.5 | 0.0 | 18.4 | 43.0 | 118.6 |
| Feb. | 1015.2 | 20.4 | 7.6 | 36.2 | 0.7 | 19.1 | 15.7 | 37.0 |
| Mar. | 1012.5 | 22.0 | 7.5 | 42.3 | 6.0 | 21.3 | 11.7 | 40.6 |
| Apr. | 1011.3 | 24.6 | 7.2 | 42.7 | 10.8 | 24.8 | 22.7 | 79.0 |
| May | 1010.4 | 26.8 | 6.7 | 39.3 | 16.0 | 28.1 | 47.1 | 94.0 |
| June | 1010.8 | 28.0 | 6.8 | 39.0 | 19.2 | 30.2 | 121.1 | 93.7 |
| July | 1012.9 | 28.0 | 7.1 | 36.5 | 16.9 | 30.1 | 153.3 | 193.0 |
| Aug. | 1012.1 | 28.2 | 7.1 | 38.5 | 15.6 | 30.1 | 134.8 | 225.7 |
| Sept. | 1011.4 | 27.2 | 7.2 | 38.5 | 10.3 | 28.9 | 280.6 | 184.9 |
| Oct. | 1013.7 | 25.6 | 7.5 | 34.5 | 5.0 | 25.7 | 132.0 | 136.2 |
| Nov. | 1016.3 | 22.0 | 7.7 | 35.0 | 3.5 | 21.0 | 46.3 | 195.6 |
| Dec. | 1016.8 | 19.9 | 7.4 | 30.0 | 6.5 | 19.1 | 26.2 | 32.2 |
| Annual | 1013.3 | 24.3 | 7.3 | 42.7 | 0.0 | 24.7 | 1,034.5 | 225.7 |

| Month | Number of days with | | | Mean cloud-iness (tenths) | Wind | |
|---|---|---|---|---|---|---|
| | precip. (>0.25 mm) | thunder-storm | fog | | most freq. direct. | mean speed (m/sec) |
| Jan. | 6 | <1 | 5 | 7 | N | 1.7 |
| Feb. | 4 | <1 | 5 | 6 | N | 2.0 |
| Mar. | 4 | <1 | 6 | 6 | ESE | 2.1 |
| Apr. | 4 | 1 | 6 | 6 | E | 2.3 |
| May | 5 | 2 | 3 | 5 | E | 2.0 |
| June | 9 | 2 | 1 | 5 | ENE | 1.8 |
| July | 11 | 1 | 2 | 5 | E | 1.3 |
| Aug. | 10 | 2 | 1 | 5 | E | 1.6 |
| Sept. | 15 | 2 | 1 | 6 | NE | 1.4 |
| Oct. | 9 | 1 | 3 | 5 | NE | 1.5 |
| Nov. | 7 | <1 | 6 | 6 | N | 1.6 |
| Dec. | 5 | 0 | 6 | 6 | N | 1.5 |
| Annual | 89 | 11 | 45 | 6 | NNE | 1.7 |

TABLE X

CLIMATIC TABLE FOR ZACATECAS, ZAC., MEXICO
Latitude 22°47′N, longitude 102°34′W, elevation 2,612 m

| Month | Mean sta. press. (mbar) | Temperature (°C) | | | | Mean vap. press. (mbar) | Precipitation (mm) | |
|---|---|---|---|---|---|---|---|---|
| | | daily mean | mean daily range | extremes | | | mean | max. in 24 h |
| | | | | max. | min. | | | |
| Jan. | 748.5 | 9.6 | 7.3 | 22.0 | −8.1 | 4.9 | 7.3 | 26.3 |
| Feb. | 748.0 | 10.7 | 8.2 | 23.3 | −9.2 | 4.9 | 3.1 | 9.5 |
| Mar. | 747.3 | 12.9 | 9.3 | 26.0 | −3.8 | 4.3 | 1.8 | 34.3 |
| Apr. | 747.5 | 15.1 | 9.4 | 26.5 | −1.6 | 4.6 | 3.6 | 22.5 |
| May | 747.6 | 16.6 | 9.4 | 28.7 | 6.0 | 6.8 | 10.8 | 24.1 |
| June | 748.5 | 16.2 | 8.7 | 27.4 | 5.9 | 10.2 | 18.4 | 56.3 |
| July | 749.8 | 14.6 | 7.8 | 24.8 | 7.2 | 10.8 | 64.1 | 76.4 |
| Aug. | 749.6 | 14.8 | 7.5 | 24.8 | 7.5 | 10.9 | 65.6 | 34.3 |
| Sept. | 748.9 | 13.7 | 6.8 | 24.6 | 3.1 | 11.0 | 53.8 | 33.0 |
| Oct. | 749.4 | 13.2 | 7.0 | 23.9 | 1.7 | 9.2 | 22.9 | 30.5 |
| Nov. | 749.2 | 11.5 | 7.4 | 22.8 | −3.2 | 6.9 | 9.8 | 43.2 |
| Dec. | 749.0 | 10.1 | 7.0 | 20.5 | −6.8 | 5.7 | 4.1 | 14.6 |
| Annual | 748.6 | 13.3 | 8.0 | 28.7 | −9.2 | 7.5 | 265.3 | 76.4 |

| Month | Number of days with | | | Mean cloudiness (tenths) | Mean sunshine (h) | Wind | |
|---|---|---|---|---|---|---|---|
| | precip. (>0.25 mm) | thunderstorm | fog | | | most freq. direct. | mean speed (m/sec) |
| Jan. | 2 | <1 | 4 | 4 | 215.1 | SSW | 7.4 |
| Feb. | 1 | <1 | 3 | 3 | 216.4 | SSW | 7.5 |
| Mar. | <1 | <1 | 2 | 3 | 245.3 | SSW | 7.5 |
| Apr. | <1 | <1 | 2 | 3 | 247.8 | SW | 7.0 |
| May | 4 | <1 | 3 | 4 | 273.2 | SW | 5.6 |
| June | 9 | <1 | 7 | 5 | 227.3 | E | 4.9 |
| July | 12 | 1 | 12 | 6 | 215.0 | E | 5.0 |
| Aug. | 12 | <1 | 11 | 6 | 230.3 | E | 4.7 |
| Sept. | 11 | <1 | 16 | 6 | 182.7 | E | 5.3 |
| Oct. | 6 | <1 | 11 | 4 | 210.5 | E | 5.1 |
| Nov. | 2 | <1 | 6 | 4 | 227.2 | E | 5.5 |
| Dec. | 2 | <1 | 5 | 4 | 206.1 | SSW | 6.2 |
| Annual | 63 | 3 | 82 | 4 | 2,696.9 | E | 6.0 |

TABLE XI

CLIMATIC TABLE FOR MAZATLAN, SINALOA, MEXICO
Latitude 23°11′N, longitude 106°25′W, elevation 78 m

| Month | Mean sta. press. (mbar) | Temperature (°C) | | | | Mean vap. press. (mbar) | Precipitation (mm) | |
|---|---|---|---|---|---|---|---|---|
| | | daily mean | mean daily range | extremes | | | mean | max. in 24 h |
| | | | | max. | min. | | | |
| Jan. | 1006.3 | 19.8 | 4.8 | 27.4 | 11.0 | 17.9 | 13.1 | 65.4 |
| Feb. | 1005.8 | 19.8 | 4.8 | 26.3 | 11.2 | 17.9 | 5.8 | 26.6 |
| Mar. | 1004.8 | 20.5 | 5.0 | 27.8 | 11.8 | 18.9 | 3.1 | 30.1 |
| Apr. | 1004.1 | 22.1 | 5.1 | 30.8 | 13.9 | 20.8 | 0.9 | 4.6 |
| May | 1003.1 | 24.6 | 4.7 | 31.6 | 14.8 | 23.9 | 1.2 | 62.2 |
| June | 1002.5 | 27.1 | 3.8 | 32.7 | 21.0 | 27.6 | 33.0 | 79.7 |
| July | 1003.5 | 28.0 | 4.3 | 32.7 | 20.5 | 29.6 | 188.3 | 215.4 |
| Aug. | 1003.2 | 28.1 | 4.3 | 34.0 | 20.0 | 30.3 | 226.9 | 213.7 |
| Sept. | 1001.7 | 27.9 | 4.6 | 32.8 | 20.0 | 30.6 | 244.9 | 152.0 |
| Oct. | 1003.0 | 27.2 | 4.7 | 32.9 | 20.5 | 28.6 | 56.6 | 111.0 |
| Nov. | 1004.7 | 24.3 | 4.9 | 30.8 | 19.0 | 22.9 | 16.4 | 69.2 |
| Dec. | 1005.6 | 21.5 | 4.8 | 28.8 | 12.4 | 19.3 | 15.3 | 43.5 |
| Annual | 1004.0 | 24.2 | 4.7 | 32.4 | 11.0 | 24.0 | 805.4 | 215.4 |

| Month | Number of days with | | | Mean cloud-iness (tenths) | Mean sun-shine (h) | Wind | |
|---|---|---|---|---|---|---|---|
| | precip. (>0,25 mm) | thunder-storm | fog | | | most freq. direct. | mean speed (m/sec) |
| Jan. | 2 | <1 | 1 | 5 | 188.7 | NW | 3.4 |
| Feb. | 1 | 0 | 2 | 4 | 192.1 | NW | 3.4 |
| Mar. | <1 | <1 | 4 | 4 | 243.0 | NW | 3.1 |
| Apr. | <1 | <1 | 4 | 4 | 248.7 | WNW | 2.9 |
| May | <1 | <1 | 6 | 3 | 268.5 | WNW | 2.6 |
| June | 4 | 1 | 1 | 5 | 242.3 | W | 2.6 |
| July | 15 | 5 | <1 | 7 | 210.5 | W | 2.7 |
| Aug. | 15 | 6 | <1 | 7 | 213.1 | WNW | 2.7 |
| Sept. | 13 | 4 | 1 | 6 | 189.5 | WNW | 2.6 |
| Oct. | 4 | 1 | 1 | 4 | 242.4 | WNW | 2.4 |
| Nov. | 1 | <1 | 2 | 4 | 214.1 | WNW | 2.6 |
| Dec. | 2 | <1 | 2 | 4 | 181.0 | NW | 3.1 |
| Annual | 57 | 17 | 24 | 5 | 2,633.9 | WNW | 2.8 |

TABLE XII

CLIMATIC TABLE FOR LA PAZ, BAJA CALIFORNIA, MEXICO
Latitude 24°10′N, longitude 110°07′W, elevation 18 m

| Month | Mean sta. press. (mbar) | Temperature (°C) | | | | Mean vap. press. (mbar.) | Max. precip. in 24 h (mm) |
|---|---|---|---|---|---|---|---|
| | | daily mean | mean daily range | extremes max. | min. | | |
| Jan. | 1014.8 | 18.3 | 8.0 | 30.5 | 2.2 | 13.8 | 66.7 |
| Feb. | 1014.2 | 19.0 | 8.9 | 34.5 | 1.9 | 14.2 | 30.0 |
| Mar. | 1012.8 | 20.7 | 10.4 | 35.0 | 6.0 | 15.5 | 18.0 |
| Apr. | 1011.6 | 22.8 | 11.9 | 38.0 | 8.6 | 17.0 | 21.6 |
| May | 1010.1 | 24.9 | 12.5 | 39.2 | 10.0 | 19.2 | 4.2 |
| June | 1009.1 | 26.8 | 12.1 | 40.2 | 12.8 | 20.9 | 35.0 |
| July | 1009.4 | 29.4 | 9.8 | 40.6 | 16.0 | 25.0 | 66.0 |
| Aug. | 1009.3 | 29.8 | 8.9 | 40.0 | 16.7 | 26.2 | 90.0 |
| Sept. | 1008.1 | 29.0 | 8.4 | 40.0 | 17.5 | 26.0 | 179.0 |
| Oct. | 1010.3 | 26.4 | 9.2 | 39.0 | 14.3 | 23.0 | 75.2 |
| Nov. | 1012.7 | 23.1 | 8.7 | 36.2 | 12.0 | 19.2 | 60.0 |
| Dec. | 1014.0 | 20.0 | 7.9 | 33.2 | 7.8 | 16.7 | 43.5 |
| Annual | 1011.4 | 24.2 | 9.7 | 40.6 | 1.9 | 19.7 | 179.0 |

| Month | Number of days with | | Mean cloud-iness (tenths) | Wind | |
|---|---|---|---|---|---|
| | precip. (>0.25 mm) | thunder-storm | | most freq. direct. | mean speed (m/sec) |
| Jan. | <1 | 0 | 4 | NE | 2.2 |
| Feb. | <1 | 0 | 3 | NE | 1.9 |
| Mar. | <1 | 0 | 3 | NE | 1.9 |
| Apr. | <1 | <1 | 3 | SW | 2.1 |
| May | 0 | 0 | 2 | SW | 2.2 |
| June | <1 | 0 | 2 | SW | 2.6 |
| July | 1 | <1 | 4 | SW | 2.0 |
| Aug. | 3 | <1 | 4 | SW | 1.9 |
| Sept. | 4 | <1 | 4 | SW | 1.7 |
| Oct. | 1 | <1 | 3 | SW | 1.6 |
| Nov. | 1 | 0 | 3 | NE | 2.0 |
| Dec. | 2 | 0 | 4 | NE | 1.9 |
| Annual | 14 | <1 | 3 | SW | 2.0 |

TABLE XIII

CLIMATIC TABLE FOR CIUDAD LERDO, DURANGO, MEXICO
Latitude 25°30′N, longitude 103°32′W, elevation 1,140 m

| Month | Mean sta. press. (mbar) | Temperature (°C) | | | | Mean vap. press. (mbar) | Precipitation (mm) | |
|---|---|---|---|---|---|---|---|---|
| | | daily mean | mean daily range | extremes | | | mean | max. in 24 h |
| | | | | max. | min. | | | |
| Jan. | 889.9 | 13.6 | 16.6 | 31.0 | − 8.7 | 7.7 | 6.4 | 18.0 |
| Feb. | 888.9 | 16.1 | 16.8 | 38.4 | − 8.0 | 8.0 | 5.6 | 33.4 |
| Mar. | 887.0 | 19.2 | 18.2 | 36.5 | − 3.5 | 8.1 | 1.8 | 13.5 |
| Apr. | 886.5 | 22.8 | 18.3 | 38.5 | − 2.0 | 9.5 | 4.4 | 21.0 |
| May | 886.2 | 25.8 | 18.1 | 40.4 | 7.4 | 12.6 | 16.4 | 54.8 |
| June | 886.7 | 27.7 | 15.3 | 40.8 | 11.0 | 17.0 | 26.6 | 49.0 |
| July | 888.5 | 26.8 | 13.8 | 38.0 | 14.5 | 18.9 | 40.2 | 46.0 |
| Aug. | 888.7 | 26.3 | 14.0 | 38.2 | 14.0 | 18.9 | 41.9 | 50.9 |
| Sept. | 888.5 | 24.2 | 13.6 | 38.0 | 7.0 | 18.3 | 57.0 | 90.5 |
| Oct. | 889.3 | 21.2 | 15.5 | 36.2 | 3.0 | 15.0 | 32.6 | 69.5 |
| Nov. | 890.5 | 16.5 | 16.5 | 34.2 | − 5.0 | 10.3 | 6.8 | 24.0 |
| Dec. | 890.6 | 13.6 | 16.1 | 32.0 | − 5.5 | 8.7 | 8.5 | 25.5 |
| Annual | 888.4 | 21.2 | 16.1 | 40.8 | − 8.7 | 12.8 | 248.2 | 90.5 |

| Month | Mean evap. (mm) | Number of days with | | | Mean cloud- iness (tenths) | Mean wind speed (m/sec) |
|---|---|---|---|---|---|---|
| | | precip. (>0.25 mm) | thunder- storm | fog | | |
| Jan. | 117.4 | 2 | 0 | 3 | 4 | 1.0 |
| Feb. | 141.4 | 2 | 0 | 1 | 4 | 1.2 |
| Mar. | 214.0 | 1 | 0 | 1 | 3 | 1.2 |
| Apr. | 248.4 | 1 | 0 | 2 | 3 | 1.2 |
| May | 289.9 | 3 | 1 | 2 | 3 | 1.3 |
| June | 285.2 | 5 | 2 | 0 | 3 | 1.3 |
| July | 262.1 | 6 | 2 | 0 | 4 | 1.2 |
| Aug. | 242.1 | 6 | 2 | 1 | 4 | 1.0 |
| Sept. | 180.1 | 6 | 1 | 1 | 4 | 0.9 |
| Oct. | 148.5 | 4 | 0 | 1 | 3 | 0.7 |
| Nov. | 116.1 | 2 | 0 | 1 | 3 | 0.8 |
| Dec. | 95.5 | 3 | 0 | 2 | 4 | 0.9 |
| Annual | 2,340.7 | 41 | 8 | 15 | 4 | 1.0 |

TABLE XIV

CLIMATIC TABLE FOR MONTERREY, NUEVO LEON, MEXICO
Latitude 25°40′N, longitude 100°18′W, elevation 534 m

| Month | Mean sta. press. (mbar) | Temperature (°C) | | | | Mean vap. press. (mbar) | Precipitation (mm) | |
|---|---|---|---|---|---|---|---|---|
| | | daily mean | mean daily range | extremes | | | mean | max. in 24 h |
| | | | | max. | min. | | | |
| Jan. | 956.5 | 15.4 | 10.9 | 36.6 | −6.8 | 11.9 | 18.9 | 41.7 |
| Feb. | 955.7 | 17.0 | 11.5 | 37.0 | −3.5 | 13.0 | 16.9 | 21.7 |
| Mar. | 953.2 | 20.3 | 12.6 | 40.8 | 0.4 | 14.4 | 12.7 | 24.5 |
| Apr. | 952.9 | 23.4 | 11.8 | 40.7 | 7.0 | 18.2 | 26.3 | 50.1 |
| May | 951.6 | 25.9 | 11.1 | 41.7 | 11.7 | 21.7 | 35.9 | 46.5 |
| June | 952.3 | 27.8 | 11.2 | 42.1 | 13.5 | 24.0 | 62.9 | 117.5 |
| July | 953.6 | 28.1 | 11.5 | 39.7 | 16.0 | 24.1 | 61.4 | 81.4 |
| Aug. | 953.1 | 28.0 | 11.2 | 39.0 | 18.0 | 24.3 | 110.5 | 129.4 |
| Sept. | 953.3 | 25.6 | 9.6 | 38.7 | 12.1 | 23.1 | 156.1 | 159.5 |
| Oct. | 955.0 | 22.5 | 9.8 | 35.5 | 8.4 | 19.8 | 91.2 | 232.4 |
| Nov. | 955.9 | 17.9 | 10.1 | 36.5 | −0.5 | 14.8 | 23.4 | 44.5 |
| Dec. | 956.6 | 15.6 | 10.8 | 36.0 | −2.5 | 12.1 | 17.7 | 31.8 |
| Annual | 954.1 | 22.3 | 11.0 | 42.1 | −6.8 | 18.5 | 633.9 | 232.4 |

| Month | Number of days with | | | Mean cloud-iness (tenths) | Mean sun-shine (h) | Wind | |
|---|---|---|---|---|---|---|---|
| | precip. (>0.25 mm) | thunder-storm | fog | | | most freq. direct. | mean speed (m/sec) |
| Jan. | 4 | 0 | 6 | 5 | 123.8 | NE | 1.3 |
| Feb. | 4 | <1 | 6 | 5 | 119.6 | E | 1.6 |
| Mar. | 3 | <1 | 6 | 5 | 155.9 | NE | 2.0 |
| Apr. | 6 | 1 | 6 | 5 | 136.2 | E | 2.3 |
| May | 5 | 1 | 6 | 5 | 165.7 | E | 2.1 |
| June | 6 | 1 | 4 | 5 | 192.5 | E | 2.4 |
| July | 5 | 1 | 3 | 5 | 213.5 | E | 2.4 |
| Aug. | 7 | 1 | 2 | 5 | 199.2 | E | 2.2 |
| Sept. | 9 | 1 | 5 | 6 | 151.6 | E | 1.8 |
| Oct. | 7 | <1 | 5 | 5 | 132.6 | E | 1.5 |
| Nov. | 5 | <1 | 7 | 5 | 121.3 | NE | 1.2 |
| Dec. | 5 | <1 | 7 | 5 | 110.9 | NE | 1.1 |
| Annual | 66 | 6 | 63 | 5 | 1,822.8 | E | 1.8 |

TABLE XV

CLIMATIC TABLE FOR GUAYMAS, SONORA, MEXICO
Latitude 27°55′N, longitude 110°53′W, elevation 4 m

| Month | Mean sta. press. (mbar) | Temperature (°C) | | | | Mean vap. press. (mbar) | Precipitation (mm) | |
|---|---|---|---|---|---|---|---|---|
| | | daily mean | mean daily range | extremes | | | mean | max. in 24 h |
| | | | | max. | min. | | | |
| Jan. | 1016.4 | 17.9 | 9.7 | 33.3 | 3.2 | 10.4 | 11.6 | 40.6 |
| Feb. | 1015.7 | 19.0 | 10.1 | 34.4 | 5.2 | 10.5 | 7.0 | 47.6 |
| Mar. | 1014.0 | 21.0 | 10.7 | 35.2 | 9.5 | 10.7 | 3.4 | 14.6 |
| Apr. | 1012.5 | 23.5 | 10.9 | 39.6 | 12.0 | 12.6 | 1.3 | 15.0 |
| May | 1010.9 | 26.4 | 11.0 | 41.3 | 13.5 | 15.4 | 0.0 | 0.4 |
| June | 1009.8 | 29.8 | 9.3 | 42.0 | 18.0 | 22.3 | 2.6 | 14.8 |
| July | 1010.8 | 31.2 | 8.1 | 42.0 | 20.6 | 27.7 | 38.8 | 56.8 |
| Aug. | 1010.8 | 31.1 | 8.1 | 42.2 | 20.5 | 28.6 | 60.2 | 74.3 |
| Sept. | 1009.4 | 30.7 | 8.3 | 42.0 | 17.7 | 27.6 | 50.7 | 152.3 |
| Oct. | 1011.6 | 27.7 | 9.5 | 40.7 | 4.7 | 19.8 | 21.7 | 309.3 |
| Nov. | 1014.5 | 22.7 | 10.0 | 37.5 | 3.8 | 13.1 | 6.7 | 43.0 |
| Dec. | 1016.1 | 19.4 | 9.8 | 32.5 | 6.0 | 11.3 | 16.7 | 74.8 |
| Annual | 1012.7 | 25.0 | 9.6 | 42.2 | 3.2 | 17.5 | 220.7 | 309.3 |

| Month | Number of days with | | | Mean cloud-iness (tenths) | Mean sun-shine (h) | Wind | |
|---|---|---|---|---|---|---|---|
| | precip. (>0.25 mm) | thunder-storm | fog | | | most freq. direct. | mean speed (m/sec) |
| Jan. | 3 | <1 | 1 | 3 | 202.2 | WNW | 2.1 |
| Feb. | 1 | <1 | <1 | 3 | 209.9 | NW | 2.5 |
| Mar. | 1 | 0 | <1 | 3 | 237.6 | W | 2.6 |
| Apr. | <1 | 0 | 1 | 2 | 270.9 | W | 2.4 |
| May | <1 | <1 | 2 | 2 | 305.9 | W | 2.1 |
| June | 1 | <1 | 1 | 2 | 301.9 | SW | 2.0 |
| July | 6 | 1 | 0 | 5 | 260.3 | SW | 1.9 |
| Aug. | 7 | 1 | <1 | 5 | 256.5 | SW | 2.1 |
| Sept. | 4 | <1 | 0 | 3 | 241.6 | SW | 1.9 |
| Oct. | 1 | <1 | <1 | 2 | 258.5 | W | 2.0 |
| Nov. | 1 | 0 | 1 | 3 | 224.9 | NW | 2.1 |
| Dec. | 2 | <1 | 1 | 3 | 197.6 | NW | 2.0 |
| Annual | 27 | 2 | 9 | 3 | 2,696.7 | SW | 2.1 |

# Reference Index

IRBE, J. C., *see* RICHARDS, T. L. and IRBE, J. C.
ISERI, K. T., *see* WILSON, A. and ISERI, K. T.
IVES, J. D., *see* BRYSON, R. A. et al.


JAMES, J. W., 246, 263
JETTON, E. V., OEY, H. S. and WOODS, C. E., 213, 263
JOHNSON, C. B., 63, 190
JONES, D. M. A., *see* CHAGNON JR., S. A. and JONES, D. M. A.
JORGENSEN, D. L., 207, 263
JULIAN, L. T. and JULIAN, P. R., 231, 263
JULIAN, P. R., *see* JULIAN, L. T. and JULIAN, P. R.


KANGIESER, P. C., 254, 263
KENDALL, G. R. and ANDERSON, S. R., 74, 190
KENDALL, G. R. and PETRIE, A. G., 97, 98
KENDREW, W. G. and CURRIE, B. W., 97, 98, 190
KERR, D. P., 90, 190
KESSELI, J. E., 246, 263
KISS, E., *see* MITCHELL JR., J. M. and KISS, E.
KLEIN, W. H., 31, 32, 47, 256, 257, 263, 360, 390
KOHLER, M. A., NORDENSON, T. J. and FOX, W. E., 233, 234, 263
KOPEC, R. J., 249, 263
KORVEN, H. C., *see* PELTON, W. L. and KORVEN, H. C.
KRAMER, F. L., 249, 263
KRAUSS, R. K., *see* BROWN, M. J. et al.
KREBS, S. and BARRY, R. G., 60, 61, 190
KUHN, P. M., *see* BRYSON, R. A. and KUHN, P. M.
KUNKEL, B., *see* REED, R. J. and KUNKEL, B.
KURFIS, K. R., *see* FLOWERS, F. C. et al.
KUTZBACH, J. E., 19, 47


LAHEY, J. F., BRYSON, R. A., CORZINE, H. A. and HUTCHINS, C. W., 24, 47
LALLY, V. E. and WATSON, B. F., 250, 263
LAMB, H. H., 38, 43, 47, 193, 263
LANDSBERG, H. E., 213, 225, 244, 245, 246, 263, 264
LAWSON, T. L., *see* WAHL, E. W. and LAWSON, T. L.
LAYCOCK, A. H., 118, 190
LEATHERWOOD, R. K., 258, 264
LESLIE, D., 123, 190
LETHBRIDGE, M. D., 255, 264
LETTAU, H., 102, 103, 190
LONG, M. J., 227, 264
LONGLEY, R. W., 55, 64, 191
LOPÉZ, E., 357, 368, 370, 390
LOWE, A. B., *see* MCKAY, G. A. and LOWE, A. B.
LOWRY, W. P., 255, 264
LOWRY, W. P., *see* BRYSON, R. A. and LOWRY, W. P.
LUDLUM, D., 244, 264
LUND, I. A., 255, 264
LYNOTT, R. E., *see* CRAMER, O. P. and LYNOTT, R. E.


MACDONALD, J. E., 258, 264
MACDONALD, N. J., *see* BRIER, G. W. et al.

MANNING, F. D., 93, 124, 191
MARCUS, M., 93, 191
MARKHAM, C. C., 212, 218, 264
MARLATT, W. and RIEHL, H., 213, 264
MATEER, C. L., 103, 191
MAYBANK, J., *see* MCKAY, G. A. et al.
MCCAUGHEY, J. H., *see* DAVIES, J. A. and MCCAUGHEY, J. H.
MCCLELLAN, D. E., *see* ANDERSON R. et al.
MCCOLLUM, R. D., *see* GRUBB, B. E. and MCCOLLUM, R. D.
MCCORMICK, R. A., *see* FLOWERS, E. C. et al.
MCFADDEN, J. D., 78, 80, 81, 82, 191
MCFADDEN, J. D., *see* RAGOTZKIE, R. A. and MCFADDEN, J. D.
MCINTYRE, D. P., 55, 191
MCKAY, G. A. and LOWE, A. B., 131, 132, 191
MCKAY, G. A. and THOMPSON, H. A., 93, 95, 130, 131, 191
MCKAY, G. A., MAYBANK, J., MOONEY, O. R. and PELTON, W. L., 191
MCKAY, G. A., *see* BROWN, D. M. et al.
MEHRINGER JR., P. J., 40, 47
MEIGS, P. and DE PERCIN, F., 250, 251, 264
MILLER, J. F., *see* PAULHUS, J. L. H. and MILLER, J. F.
MILLER, M. E., CANFIELD, N. L., RITTER, T. A. and WEAVER, C. R., 203, 264
MIRANDA, F., *see* GARCÍA, E. et al.
MITCHELL, V. L., 19, 21, 47
MITCHELL JR., J. M., 2, 43, 46, 47
MITCHELL JR., J. M. and KISS, E., 38, 47
MOKOSCH, E., 59, 191
MONTGOMERY, M. R., *see* HARE, F. K. and MONTGOMERY, M. R.
MOONEY, O. R., *see* MCKAY, G. A. et al.
MORAN, J. M., 40, 47
MORTON, F. I., 233, 264
MOSIÑO ALEMÁN, P. A., 356, 373, 374, 377, 379, 390
MOSIÑO ALEMÁN, P. A. and GARCÍA, E., 377, 390
MULLER, R. A., 224, 264
MURRAY, C. R., 232, 264
MURRAY, W. A., 99, 100, 191


NAGLER, K. M., *see* SOULES, S. D. and NAGLER, K. M.
NAMIAS, J., 236, 255, 264
NAVAIR, 6, 7, 8, 9, 10, 11, 12, 13, 14, 15, 16, 17, 19, 24, 26, 47
NEWMAN, C. J., *see* HOPE, J. R. and NEWMAN, C. J.
NEWMAN, E., 254, 264
NORDENSON, T. J., *see* KOHLER, M. A. et al.


OEY, H. S., *see* JETTON, J. V. et al.
OGDEN, J. G., III, 38, 47
OGDEN, R. M., *see* SERGIUS, L. A. et al.
OHTAKE, T., *see* BOWLING, S. A. et al.
OKIMOTO, G., *see* CHANG, JEN-HU and OKIMOTO, G.
OLTMAN, R. E. and TRACY, H. J., 256, 264
O'NEILL, A. D. J., *see* FERGUSON, H. L. et al.

ORVIG, S., *see* HARE, F. K. and ORVIG, S.
OUELLET, C. E., *see* BAIER, W. and OUELLET, C. E.
OWENS, G. V., *see* BERRY, F. A. et al.


PAGE, J. L., 368, 370, 389, 390
PALMER, W. C., 236, 264
PATRIC, J. H. and BLACK, P. E., 117, 191
PATTEN, H. L., *see* WRIGHT JR., H. E. et al.
PATTON, C. P., 246, 249
PAULHUS, J. L. H. and MILLER, J. F., 207, 216–217
PAUTZ, M. E., 230, 231, 264
PEACE JR., R. L., 203, 264
PELTON, W. L. and KORVEN, H. C., 118, 191
PELTON, W. L., *see* MCKAY, G. A. et al.
PENNER, C. M., 55, 191
PETERSON, J. T., 81, 191
PETRIE, A. G., *see* KENDALL, G. R. and PETRIE, A. G.
PÉWÉ, T. L., 52, 191
PHILLIPS, D. W., *see* SANDERSON, M. E. and PHILLIPS,
     D. W.
POTTER, J. G., 93, 94, 96, 191
POWE, N. N., 65, 191
PYKE, C. B., 213, 264


RADKE, L. F., *see* HOBBS, P. V. et al.
RAGOTZKIE, R. A. and MCFADDEN, J. D., 80, 191
RAHN, J. J., 253, 264
RAMASWAMY, C., 374, 390
RAMSAY, F. L., *see* ELLIOTT, W. P. and RAMSAY, F. L.
RANTZ, S. E., *see* HOFMANN, W. and RANTZ, S. E.
RASMUSSEN, J. L., 264
RASMUSSON, E. M., 28, 47, 84, 86, 191, 203, 213, 226,
     227, 232, 264, 265
REBMAN, E. J., 254, 265
REED, R. J. and KUNKEL, B., 60, 61
REITAN, C. H., 206, 265
RICHARDS, T. L., 79, 80, 82, 191
RICHARDS, T. L. and IRBE, J. C., 233, 265
RIEHL, H., 357, 375, 390
RIEHL, H., *see* MARLATT, W., and RIEHL, H.
RITTER, T. A., *see* MILLER, M. E. et al.
ROBERTSON, G. W., 178, 179, 191
ROBERTSON, G. W., *see* BAIER, W. and ROBERTSON,
     G. W.
ROBINSON, E., THUMAN, E. C. and WIGGINS, E. J., 126,
     191
ROBINSON, E., *see* THUMAN, W. C. and ROBINSON, E.
ROBINSON, T. W. and HUNT, C. B., 233, 265
ROONEY JR., J. F., 224, 265
ROUSE, W. R., 109, 192
ROUSE, W. R. and WILSON, R. G., 109, 192
RUDD, R. D., 213, 265
RUSSELO, D., *see* BAIER, W. et al.


SABBAGH, M. E. and BRYSON, R. A., 90, 91, 192
SANDERSON, M. E., 117, 192
SANDERSON, M. E. and PHILLIPS, D. W., 102, 117, 192

SANDS, R. D., 379, 383, 390
SATER, J. E., 52, 192
SATTERTHWAITE, J., *see* EMSLIE, J. H. and SATTERTH-
     WAITE, J.
SEARBY, H. W., 192
SERGIUS, L. A., ELLIS, G. R. and OGDEN, R. M., 200,
     265
SERVICIO METEOROLÓGICO MEXICANO, 390
SHAPIRO, R., *see* BRIER, G. W. et al.
SHARP, W., *see* BAIER, W. et al.
SHUMWAY, S. E., *see* HOBBS, P. V. et al.
SLUSSER, W. F., 197, 198–199, 200, 201, 207, 212
SLY, W. K., *see* COLIGADO, M. C. et al.
SMITH, R. M., *see* BROWN, M. J. et al.
SOTO, C., *see* GARCÍA, E. et al.
SOULES, S. D. and NAGLER, K. M., 36, 47
SPAR, J., 236, 265
SPAR, J., *see* GRILLO, J. N. and SPAR, J.
STIDD, C. K., 256
STOUT, G. E. and CHANGNON JR., S. A., 229, 265
STOUT, G. E., *see* CHANGNON JR., S. A. and STOUT, G.
     E.
SUGG, A. L., 227, 265


TATTELMAN, P. I., 246, 248, 265
TERJUNG, W. H., 250, 265
THOM, E. C., 250, 251, 265
THOM, H. C. S., 74, 192, 253, 265
THOM, H. C. S., *see* HOLZMAN, B. G. and THOM, H.
     C. S.
THOMAS, H. E., 236, 265
THOMAS, M. K., 93, 192
THOMAS, M. K., *see* CHAPMAN, L. J. and THOMAS, M. K.
THOMPSON, H. A., 79, 192
THOMPSON, H. A., *see* MCKAY, G. A. and THOMPSON,
     H. A.
THORNTHWAITE, C. W., 246, 265, 347, 390
THUMAN, W. C. and ROBINSON, E., 126, 192
THUMAN, W. C., *see* ROBINSON, E. et al.
TITUS, R. L. and TRUHLAR, E. J., 57, 109, 111, 192
TRACY, H. J., *see* OLTMAN, R. E. and TRACY, H. J.
TRENHOLM, C. H., *see* WRIGHT, J. B. and TRENHOLM,
     C. H.
TREWARTHA, G. T., 219, 249, 265
TRUHLAR, E. J., *see* TITUS, R. L. and TRUHLAR, E. J.
TULLER, S. E., 206, 265


U.S. WEATHER BUREAU, 77, 83, 88, 93, 95, 103, 106,
     109, 123, 192


VESTAL, C. K., 207
VIERECK, L., 53, 192
VILLMOW, J. R., 246, 265
VISHER, S. S., 260, 266
VIVÓ, J. A., 377, 390
VIVÓ, J. A. and GÓMEZ, J. C., 347, 390
VOSKRESENSKY, K. P., *see* BOCHKOF, A. P. et al.

# Geographical Index

# Subject Index